Problem Books in Mathematics

Series Editor:

Peter Winkler
Department of Mathematics
Dartmouth College
Hanover, NH
USA

More information about this series at http://www.springer.com/series/714

Alina Sîntămărian • Ovidiu Furdui

Sharpening Mathematical Analysis Skills

 Springer

Alina Sîntămărian
Department of Mathematics
Technical University of Cluj-Napoca
Cluj-Napoca, Romania

Ovidiu Furdui
Department of Mathematics
Technical University of Cluj-Napoca
Cluj-Napoca, Romania

ISSN 0941-3502   ISSN 2197-8506   (electronic)
Problem Books in Mathematics
ISBN 978-3-030-77141-6   ISBN 978-3-030-77139-3   (eBook)
https://doi.org/10.1007/978-3-030-77139-3

Mathematics Subject Classification: 26-XX, 26Axx, 26Bxx, 26Dxx, 26D15, 40-XX, 40Axx, 40Gxx, 41A58, 11M06, 11M35, 34A05, 45A05

This Springer imprint is published by the registered company Springer Nature Switzerland AG
The registered company address is: Gewerbestrasse 11, 6330 Cham, Switzerland

Gold, when multiplied, conspire against his master; but books, when multiplied, make great use of those who have them.
—Saint John Chrysostom (347–407)

In the memory of my father, Ioan
and
To my mother, Viorica

                      Alina

Preface

> The library is the friend that all those who
> want "to bring something new" must make.
> However, it cannot be revealed
> —an epsilon no matter how small—
> if you do not see and accumulate
> what others have done.
>
> *Viorel Gh. Vodă*

This book is the fruit of our work in the last decade teaching, researching, and solving and creating problems. Our material offers an unusual collection of problems of mathematical analysis that someone would study in a standard calculus course: limits, numerical and power series, derivatives, partial derivatives and their applications, and implicit functions. Working in a specific field of mathematics, new ideas come, but in order not to make the book too voluminous, you say "Now, I really do not add anything more!" many times, and you still add though..., because the ideas, the problems are too beautiful to keep them out.

The book is divided into two parts. Part I contains the problems, arranged by topics in the first seven chapters of the book, and Chap. 8, which collects two new proofs of the Sandham–Yeung series. Part II brings solutions to the proposed exercises in the first seven chapters of the book.

The first chapter of the book collects non-standard problems on limits of special sequences and integrals. Why limits? First, because in analysis most things reduce to the calculation of a limit, and second, because limits are the fundamental problems of analysis. Many problems are original, based on the ideas aroused during the research work, and not rarely quite unexpectedly. We mention that a few problems from this chapter, some elementary limits of sequences, have been taken from Romanian mathematical literature, especially from the famous *Gazeta Matematică*, which is the oldest Romanian journal of mathematics that appears without interruption since 1895.

The second chapter of the book contains series of real numbers. The problems in this chapter deal with two major topics on series of real numbers: the convergence of a numerical series and its exact calculation. Regarding the convergence, we have included in this chapter original problems that can be solved using the classical way, i.e., the well-known convergence tests for numerical series, or less-known ways . . . The problems about the exact calculation of the value of a series are, most of them, new in the literature and vary in diversity as well as in difficulty. Their topics range from the calculation of series involving factorial terms to exotic series involving the tail of Riemann zeta function values.

The third chapter of the book deals with power series. This chapter includes the standard formulae on the power series, Taylor series, and Maclaurin series of the classical elementary functions, and exercises dealing with finding the set of convergence of a power series and, whenever possible, determining the sum of the series. The problems in this chapter are as exotic and fascinating as possible. They cover topics such as the generating functions of special sequences, the calculation of power series with factorials, series involving polylogarithm functions, as well as nonlinear logarithmic series. The chapter also includes, as applications, the calculation of single and multiple series involving Riemann zeta function values. The last section of this chapter collects several nonlinear Euler series which are new in the literature (but, clearly, the original problems can be found in other sections, too).

Chapter 4 is about derivatives and their applications. This chapter contains problems on the classical topics of mathematical analysis: the calculation of the nth derivative of a function, Leibniz's formula, Taylor's formula, the extremum points of a function of a single variable, as well as non-standard topics of analysis: the generalized Leibniz formula, special differential equations, and the computation of exotic series involving the Maclaurin remainder of special functions.

In Chap. 5, we collect problems on partial derivatives and their applications. The problems included here deal with the chain rule, Euler's identity regarding homogeneous functions, Taylor's formula, and extrema of functions of several real variables.

Chapter 6 contains problems on implicit functions. Here is the place where a lot of curves appear, together with their interesting properties which are pointed out with the help of the implicit functions theory.

Chapter 7 collects *challenging problems* on several topics discussed in the previous chapters as well as *gems and mathematical beauties*. Special attention is given to problems concerning evaluation of linear and nonlinear series involving tails of Riemann zeta function values.

In Chap. 8, we give two new proofs of the famous Sandham–Yeung series.

This volume contains both theory and problems of analysis, many of the problems being new and original. We do not claim originality of all the problems included in the book and we are aware that some may be either known or very old.

We have not attempted to document the source of every problem. This would be a very difficult task: on the one hand, many of this volume's problems have been discovered by us over the last decade, some of them have been published in various journals with a problem column, others see the light of publication for the first time in this book. Also, there are problems whose history is either lost, with the passing of time, or the authors have not been aware of it. We have tried to avoid collecting too many problems that are well-known or published elsewhere, in order to keep *a high level of originality*. On the other hand, other problems of this book arose in a natural way: either as generalizations or motivated by known results that have long been forgotten (see, for example, in Problem 2.52, the alternating series due to Hardy). For such problems, when known, the source of the problem is mentioned, either in the solution or in the statement of the problem.

We have included throughout the book results or quick statements of mathematical analysis whose proofs have been given in detail. These were called *gems of mathematical analysis*, either because they have a very beautiful solution, which the authors discovered or revived and brought into the light, or because they are classical results in analysis that the reader should know. The intention of adding them in the book was to make the material more appealing, and also to inspire the reader's creativity for solving the problems.

The book contains theoretical results, added before a problem or a group of problems, which suggest to the reader the tools for solving the exercises.

The problems have been arranged on sections, each section dealing with a particular type of topic. Some problems, which are difficult or very challenging, have been divided into parts, (a), (b), (c), . . . to help the reader find a solution. Also, there are a couple of *open problems, as well as challenge problems*, that have been included in the book. These problems, motivated by knows results, may be considered as research problems or projects for students with a strong background in analysis and for the reader who enjoys mathematical research and discovery in mathematics. The intention of having some difficult problems and open problems, gathered in the book, is to stimulate creativity and discovery of original methods for proving known results and establishing new ones.

Many of the problems have been put in gray boxes and some of them have been colored in blue, either because these are very beautiful results, although the authors consider all of them to be very beautiful, or they are gems of mathematical analysis, or have unexpected elegant solutions, or, simply, these are results that should be studied by the reader who enjoys working on nonstandard problems of analysis.

This book is addressed to undergraduate students with a strong background in analysis who prepare for competitions like Putnam and SEEMOUS or other higher-level mathematical contests. Mathematicians and students interested in problem solving will find this collection of topics tempting. This volume is also useful for instructors who are involved in math contests, as well as for individuals who wish to enrich and test their knowledge by solving problems in analysis. It can be used by anyone for independent study courses, by students of mathematics, physics, and engineering, and by anyone who wants to explore the standard topics of mathematical analysis.

We also address this work to the first- and second-year graduate students who want to learn more about applications of a certain technique, to do calculations which happen to have interesting results. Pure and applied mathematicians, who confront certain difficult computations in their research, might find this book attractive.

We wish you success in going over this collection of analysis problems, as we hope that many of them *are challenging, worth studying, and splendid.* We also hope you enjoy reading the theory and solving the problems, even if some of them are quite tough. For questions, generalizations, remarks, or observations regarding the improvement of this material, please do not hesitate to contact us by regular or electronic mail.

Cluj-Napoca, Romania   Alina Sîntămărian
March 18, 2021   Ovidiu Furdui

Success in problem solving!!!

Contents

Notations

$\mathbb{N}$	The set of natural numbers ($\mathbb{N} = \{1, 2, 3, \ldots\}$)
$\mathbb{R}$	The set of real numbers
$\mathbb{R}^*$	The set of nonzero real numbers ($\mathbb{R}^* = \mathbb{R} \setminus \{0\}$)
$\overline{\mathbb{R}}$	$\mathbb{R} \cup \{\pm\infty\}$
$\mathbb{C}$	The set of complex numbers
$\Re(z)$	The real part of the complex number z
$\lfloor a \rfloor$	The floor of the real number a
$\{a\}$	The fractional part of the real number a, $\{a\} = a - \lfloor a \rfloor$
$n!$	$0! = 1$, $n! = 1 \cdot 2 \cdot 3 \cdots n$, $n \in \mathbb{N}$
$(2n)!!$	$0!! = 1$, $(2n)!! = 2 \cdot 4 \cdot 6 \cdots (2n)$, $n \in \mathbb{N}$
$(2n - 1)!!$	$(2n - 1)!! = 1 \cdot 3 \cdot 5 \cdots (2n - 1)$, $n \in \mathbb{N}$
$\sim$	$a_n \sim b_n$, as $n \to \infty$, if $\lim\limits_{n\to\infty} a_n/b_n$ exists and is finite
C_n^k	The binomial coefficient of n taken by k, $C_n^k = \frac{n!}{k!(n-k)!}$
e	$\lim_{n\to\infty}(1 + 1/n)^n = 2.71828\ldots$
γ	$\lim_{n\to\infty}(1 + 1/2 + \cdots + 1/n - \ln n) = 0.57721\ldots$
G	Catalan's constant $G = \sum_{n=0}^{\infty}(-1)^n/(2n + 1)^2$
H_n	The nth harmonic number $H_n = 1 + 1/2 + \cdots + 1/n$, $n \in \mathbb{N}$
ψ	The Digamma function (the Psi function) $\psi(z) = \Gamma'(z)/\Gamma(z)$, for $\Re(z) > 0$
ζ	The Riemann zeta function $\zeta(z) = \sum_{n=1}^{\infty} 1/n^z$, for $\Re(z) > 1$
$\zeta(2, a)$	The Hurwitz zeta function $\zeta(2, a) = \sum_{n=0}^{\infty} 1/(n + a)^2$, $a \neq 0, -1, -2, \ldots$
β	The Dirichlet beta function $\beta(s) = \sum_{n=0}^{\infty}(-1)^n/(2n + 1)^s$, $\Re(s) > 0$
Li_n	The Polylogarithm function $\text{Li}_n(z) = \sum_{k=1}^{\infty} z^k/k^n$, $|z| \leq 1$, $n \in \mathbb{N} \setminus \{1\}$, or $\text{Li}_n(z) = \int_0^z \text{Li}_{n-1}(t)/t \, dt$, $n \in \mathbb{N} \setminus \{1, 2\}$
$\mathscr{V}(x)$	The set of all neighborhoods of x, where $x \in \mathbb{R}^n$
$\text{int}(A)$	The interior of set $A \subseteq \mathbb{R}^n$
$f^{(n)}$	The nth derivative of function f

$C^k(\Omega, \mathbb{R})$ The set of all functions $f : \Omega \subseteq \mathbb{R}^n \to \mathbb{R}$ which have partial derivatives of order k continuous on Ω, called the set of functions of class C^k;

when $\Omega \subseteq \mathbb{R}$, the set of functions $f : \Omega \to \mathbb{R}$ which have derivative of order k continuous on Ω

$C^\infty(I, \mathbb{R})$ The set of all functions $f : I \to \mathbb{R}$, where $I \subseteq \mathbb{R}$ is an open interval, which are indefinite differentiable on I

Part I
Theory and Problems

Sequences of Real Numbers

1

If we prove that γ is irrational do you give us
a bonus point at the final exam?

A question of a UTCN student for Alina
during an analysis course

This chapter collects problems on sequences of real numbers. The problems included are challenging and the topics covered vary in diversity: from limits of sequences involving special numerical terms, applications of Stolz–Cesàro Theorem, both ∞/∞ and $0/0$ cases, Wolstenholme sequences, to the calculation of spectacular limits of integrals.

1.1 Limits of Sequences

The Constant e The sequence of rational numbers $(e_n)_{n\geq 1}$ defined by

$$e_n = \left(1 + \frac{1}{n}\right)^n, \quad n \geq 1,$$

is strictly increasing and bounded, hence convergent. Its limit is denoted by e, in honor of the Swiss mathematician Leonhard Euler (1707–1783). It is known that $e = 2.71828\ldots$.

Another sequence of rational numbers convergent to e is the sequence $(E_n)_{n\geq 0}$ defined by

$$E_n = 1 + \frac{1}{1!} + \frac{1}{2!} + \cdots + \frac{1}{n!}, \quad n \geq 0.$$

(continued)

© The Author(s), under exclusive license to Springer Nature Switzerland AG 2021
A. Sîntămărian, O. Furdui, *Sharpening Mathematical Analysis Skills*, Problem Books
in Mathematics, https://doi.org/10.1007/978-3-030-77139-3_1

Therefore, we have

$$e = \lim_{n \to \infty} \left(1 + \frac{1}{n}\right)^n;$$

$$e = \lim_{n \to \infty} \left(1 + \frac{1}{1!} + \frac{1}{2!} + \cdots + \frac{1}{n!}\right).$$

1.1. Prove that the sequence $\left(\dfrac{n! e^n}{n^{n+\frac{1}{2}}}\right)_{n \geq 1}$ is strictly decreasing.

1.2. Calculate $\displaystyle\lim_{n \to \infty} \sum_{k=1}^{n} \frac{9k^2 + 12k + 5}{(3k+2)!}$.

1.3. Calculate $\displaystyle\lim_{n \to \infty} \sum_{k=1}^{n} \frac{1}{k(k+1)(k+1)!}$.

1.4. Let $(x_n)_{n \geq 1}$ be the sequence defined by $x_n = \frac{1}{1 \cdot 3} + \frac{1}{2 \cdot 4} + \frac{1}{3 \cdot 5} + \cdots + \frac{1}{n(n+2)}$, $n \geq 1$. Calculate $\displaystyle\lim_{n \to \infty} \left(2x_n - \frac{1}{2}\right)^n$.

Some Algebraic Identities If $a, b \in \mathbb{R}$ and $n \in \mathbb{N}, n \geq 2$, then the following identities hold:

- $a^n - b^n = (a - b)(a^{n-1} + a^{n-2}b + \cdots + ab^{n-2} + b^{n-1});$
- $a^{2n-1} + b^{2n-1} = (a + b)(a^{2n-2} - a^{2n-3}b + \cdots - ab^{2n-3} + b^{2n-2}).$

Particular cases:

- $a^2 - b^2 = (a - b)(a + b);$
- $a^3 \pm b^3 = (a \pm b)(a^2 \mp ab + b^2).$

1.5. Let $k, m \in \mathbb{N}, k, m \geq 2$. Calculate

$$\lim_{n \to \infty} \frac{\sqrt[k]{n^k + 3} - \sqrt[k]{n^k + 1}}{\sqrt[m]{n^m + 2} - \sqrt[m]{n^m - 5}}.$$

1.6. Determine $a, b, c \in \mathbb{R}$ such that $\displaystyle\lim_{n \to \infty} n(an - \sqrt{bn^2 + cn - 3}) = 1$.

Three Sums of Natural Numbers If $n \geq 1$ is an integer, then:

- $1 + 2 + 3 + \cdots + n = \dfrac{n(n+1)}{2}$;

- $1^2 + 2^2 + 3^2 + \cdots + n^2 = \dfrac{n(n+1)(2n+1)}{6}$;

- $1^3 + 2^3 + 3^3 + \cdots + n^3 = \left[\dfrac{n(n+1)}{2}\right]^2$.

1.7. Calculate:

(a) $\displaystyle \lim_{n\to\infty} \left(\frac{1 + 3 + 5 + \cdots + (2n-1)}{n+1} - \frac{2n+1}{2} \right)$;

(b) $\displaystyle \lim_{n\to\infty} \frac{\displaystyle\sum_{k=1}^{2n} (-1)^k (1 + 2 + \cdots + k)}{\displaystyle\sum_{k=1}^{2n+1} (-1)^{k-1}(1 + 2 + \cdots + k)}$;

(c) $\displaystyle \lim_{n\to\infty} \frac{1}{n^2} \sum_{k=1}^{2n-1} (-1)^{k-1} k^2$;

(d) $\displaystyle \lim_{n\to\infty} \frac{1}{n^4} \sum_{k=1}^{2n} (-1)^k (1^3 + 2^3 + 3^3 + \cdots + k^3)$.

1.8. Let $\{x\}$ be the fractional part of x. Prove that:

(a) $\displaystyle \lim_{n\to\infty} \left\{ (2 + \sqrt{3})^n \right\} = 1$; (b) $\displaystyle \lim_{n\to\infty} \left\{ (3 + \sqrt{7})^n \right\} = 1$;

(c) $\displaystyle \lim_{n\to\infty} \left\{ (1 + \sqrt{2})^{2n} \right\} = 1$; (d) $\displaystyle \lim_{n\to\infty} \left\{ (1 + \sqrt{2})^{2n-1} \right\} = 0$.

The Squeeze Theorem Let $(a_n)_{n\geq 1}$, $(b_n)_{n\geq 1}$, $(c_n)_{n\geq 1}$ be sequences of real numbers for which there exists $n_0 \in \mathbb{N}$ such that $a_n \leq b_n \leq c_n$, for all $n \geq n_0$. If $\displaystyle\lim_{n\to\infty} a_n = \lim_{n\to\infty} c_n = l \in \mathbb{R}$, then $\displaystyle\lim_{n\to\infty} b_n = l$.

1.9. Calculate $\displaystyle \lim_{n\to\infty} n^{\frac{\sin n}{n}}$.

1.10. Calculate:

(a) $\displaystyle\lim_{n\to\infty}\left(\frac{1}{(n+1)^2}+\frac{1}{(n+2)^2}+\cdots+\frac{1}{(2n)^2}\right);$

(b) $\displaystyle\lim_{n\to\infty}\left(\frac{1}{\sqrt{n^2+1}}+\frac{1}{\sqrt{n^2+2}}+\cdots+\frac{1}{\sqrt{n^2+n}}\right);$

(c) $\displaystyle\lim_{n\to\infty}\sum_{k=1}^{n}\frac{5k^3+3k^2+2k+1}{n^4+k+7};$

(d) $\displaystyle\lim_{n\to\infty}\frac{(2n+1)(2n+3)\cdots(4n+1)}{(2n)(2n+2)\cdots(4n)};$

(e) $\displaystyle\lim_{n\to\infty}\frac{(n+2)(n+5)\cdots(4n-1)}{(n+1)(n+4)\cdots(4n-2)};$

(f) $\displaystyle\lim_{n\to\infty}\frac{C_{4n}^{2n}}{4^n C_{2n}^{n}};$

(g) $\displaystyle\lim_{n\to\infty}\frac{3^{3n}(C_{2n}^{n})^2}{C_{3n}^{n}C_{6n}^{3n}};$

(h) $\displaystyle\lim_{n\to\infty}\frac{1}{2^n}\sum_{k=0}^{n}\frac{C_n^k}{\sqrt{k+2}}.$

Remark 1.1. Regarding parts (d) and (e) of the above problem, more general limits are given in Problem 1.16 and part (i) of Problem 7.19.

1.11. Let $(x_n)_{n\in\mathbb{N}}$ be a bounded sequence of real numbers such that $\displaystyle\lim_{n\to\infty}(2x_n+x_{2n})=l\in\mathbb{R}$. Prove that $\displaystyle\lim_{n\to\infty}x_n=\frac{l}{3}$.

Remark 1.2. If $k\in\mathbb{N}$, $k\geq 2$, and $(x_n)_{n\in\mathbb{N}}$ is a bounded sequence of real numbers such that $\displaystyle\lim_{n\to\infty}(kx_n+x_{kn})=l\in\mathbb{R}$, then $\displaystyle\lim_{n\to\infty}x_n=\frac{l}{k+1}$.

1.12. [32] Prove that:

(a) $\displaystyle\lim_{n\to\infty}\int_0^1 \sqrt[n]{x^n+(1-x)^n}\,dx=\frac{3}{4};$

(b) $\displaystyle\lim_{n\to\infty}n^2\left(\int_0^1 \sqrt[n]{x^n+(1-x)^n}\,dx-\frac{3}{4}\right)=\frac{\pi^2}{48}.$

1.13. [73] Prove that:

(a) $\displaystyle \lim_{n\to\infty} \int_0^{\frac{\pi}{2}} \sqrt[n]{\sin^n x + \cos^n x}\, dx = \sqrt{2}$;

(b) $\displaystyle \lim_{n\to\infty} n^2 \left(\int_0^{\frac{\pi}{2}} \sqrt[n]{\sin^n x + \cos^n x}\, dx - \sqrt{2} \right) = \frac{\pi^2}{12\sqrt{2}}$.

1.14. [A. Sîntămărian, 2016] Calculate

$$\lim_{n\to\infty} \int_0^{\frac{\pi}{2}} \sqrt[n]{\sin^n x + \sin^n (2x) + \cos^n x}\, dx.$$

1.15.

(a) Let $(x_n)_{n\geq 1}$ be the sequence defined by

$$1,\ \underbrace{2, 2,}_{2\ \text{terms}}\ \underbrace{3, 3, 3,}_{3\ \text{terms}}\ \underbrace{4, 4, 4, 4,}_{4\ \text{terms}} \dots.$$

Calculate $\displaystyle \lim_{n\to\infty} \frac{x_n}{\sqrt{n}}$.

(b) *A problem of G. Boroica.* Let $(y_n)_{n\geq 1}$ be the sequence defined by

$$1, \underbrace{2, 2, 2, 2,}_{2^2\ \text{terms}}\ \underbrace{3, 3, \dots, 3,}_{3^2\ \text{terms}}\ \underbrace{4, 4, \dots, 4,}_{4^2\ \text{terms}} \dots.$$

Calculate $\displaystyle \lim_{n\to\infty} \frac{y_n}{\sqrt[3]{n}}$.

1.16. [128, pp. 82, 83], [130, Remark 4] **A Challenging Limit**
Let $p, q \in \mathbb{N}$, with $p \geq 2$. Calculate

$$\lim_{n\to\infty} \frac{(qn + 1)(qn + p + 1)\cdots(qn + np + 1)}{qn(qn + p)\cdots(qn + np)}.$$

Remark 1.3. For a generalization of the above limit, see part (i) of Problem 7.19.

1.17. Find the general term of the sequence $(a_n)_{n\geq 1}$, with $a_1 = a_2 = 1$, defined by the recurrence relation $a_{n+2} = a_n + \frac{1}{a_{n+1}}$, $n \geq 1$.

1.18. Let $(x_n)_{n\geq 1}$, with $x_1 = \sqrt{2}$, be the sequence defined by the recurrence relation $x_{n+1} = \sqrt{2 + x_n}$. Calculate $\lim\limits_{n\to\infty} 4^n (2 - x_n)$.

1.19. Let $(x_n)_{n\geq 0}$, with $x_0 > 0$, be the sequence defined by the recurrence relation $x_{n+1} = \frac{1+x_n}{2\sqrt{x_n}}$. Prove that the sequence converges and find its limit.

1.20.

(a) Study the convergence of the sequence $(x_n)_{n\geq 0}$, with $x_0 \geq 0$, defined by the recurrence relation

$$x_{n+1} = \sqrt{x_n} + \frac{1}{n+1}.$$

(b) If $k \in \mathbb{N}$, $k \geq 2$, study the convergence of the sequence $(x_n)_{n\geq 0}$, with $x_0 \geq 0$, defined by the recurrence relation

$$x_{n+1} = \sqrt[k]{x_n} + \frac{1}{n+1}.$$

Cauchy–d'Alembert's Criterion Let $(a_n)_{n\geq 1}$ be a sequence of positive real numbers such that $\lim\limits_{n\to\infty} \frac{a_{n+1}}{a_n} = l \in \overline{\mathbb{R}}$. Then $\lim\limits_{n\to\infty} \sqrt[n]{a_n} = l$.

1.21. Calculate:

(a) $\lim\limits_{n\to\infty} \dfrac{n}{\sqrt[n]{n!}}$;

(b) $\lim\limits_{n\to\infty} \sqrt[n]{\dfrac{(n!)^2}{(2n)!8^n}}$;

(c) $\lim\limits_{n\to\infty} \dfrac{\sqrt[n]{(n+1)(n+2)\cdots(2n)}}{n}$;

(d) $\lim\limits_{n\to\infty} \sqrt[n]{\dfrac{(2n)(2n+3)\cdots(8n-6)}{(n!)^2}}$;

(e) $\lim\limits_{n\to\infty} \dfrac{\sqrt[n]{(2n+1)(2n+4)\cdots(5n-2)}}{n}$.

Remark 1.4. For a generalization of parts (d) and (e) in the above problem see parts (ii) and (iii) of Problem 7.19.

1.22. Let $a \in (0, 1)$. Calculate $\lim\limits_{n\to\infty} \sqrt[n]{1 - \sqrt[n]{a}}$.

1.23. [103] *Traian Lalescu's Sequence* Calculate

$$\lim_{n\to\infty} \left(\sqrt[n+1]{(n+1)!} - \sqrt[n]{n!} \right).$$

The Fibonacci Sequence

One of the most important sequences defined by a recurrence relation with two terms is the sequence $(F_n)_{n\geq 0}$ defined by

$$F_0 = 0, \quad F_1 = 1, \quad F_{n+1} = F_n + F_{n-1}, \quad n \geq 1,$$

which bears the name of the Italian mathematician Leonardo Fibonacci (c. 1170–c. 1240). The Fibonacci number F_{n+1} is the sum of the other two previous Fibonacci numbers. It is well-known the following representation of the Fibonacci numbers:

$$F_n = \frac{1}{\sqrt{5}} \left[\left(\frac{1+\sqrt{5}}{2} \right)^n - \left(\frac{1-\sqrt{5}}{2} \right)^n \right], \quad n \geq 0,$$

the irrational number $\frac{1+\sqrt{5}}{2} = 1.61803\ldots$ being called the *golden number* or the *golden section*.

[117] An Equality with the Golden Section

$$\arctan \frac{2}{1+\sqrt{5}} = \frac{\arctan 2}{2}.$$

1.24. *Two special sums of binomial coefficients.* Prove that:

(a) $C_n^0 + C_{n+1}^2 + C_{n+2}^4 + \cdots + C_{2n}^{2n} = F_{2n+1}, \quad n \geq 0;$

(b) $C_n^1 + C_{n+1}^3 + C_{n+2}^5 + \cdots + C_{2n-1}^{2n-1} = F_{2n}, \quad n \geq 1.$

1.25. *Compositions of homographic functions.*

(a) Let $f : (0, \infty) \to \mathbb{R}$, $f(x) = \frac{3x+2}{2x+3}$. Calculate $\lim_{n\to\infty} \underbrace{f \circ f \circ \cdots \circ f}_{n \text{ times}}(x)$.

(b) Let $f : (0, \infty) \to \mathbb{R}$, $f(x) = \frac{2x+1}{x+3}$. Calculate $\lim_{n\to\infty} \underbrace{f \circ f \circ \cdots \circ f}_{n \text{ times}}(x)$.

1.2   Applications of Stolz–Cesàro Theorem, the ∞/∞ and the 0/0 Cases

Stolz–Cesàro Theorem, the ∞/∞ Case Let $(a_n)_{n\geq 1}$ and $(b_n)_{n\geq 1}$ be two sequences of real numbers such that:

(a) $0 < b_1 < b_2 < \ldots < b_n < \ldots$ and $\lim_{n\to\infty} b_n = \infty$;

(b) $\lim_{n\to\infty} \dfrac{a_{n+1} - a_n}{b_{n+1} - b_n} = l \in \overline{\mathbb{R}}$.

Then $\lim_{n\to\infty} \dfrac{a_n}{b_n}$ exists and it is equal to l.

This theorem, the discrete version of l'Hôpital's Theorem, is mentioned in various papers in the literature, either as a theoretical result or as a problem. In [10, pp. 413, 414], [26, Appendix B], it appears as a theoretical result regarding the calculation of limits of expressions x_n/y_n of the form ∞/∞, and as a problem it is mentioned, in a slightly modified form, in the famous problem book of Pólya and Szegö [121, problem 70, pp. 16, 17].

1.26. Calculate:

(a) $\lim_{n\to\infty} \dfrac{1 + \frac{1}{2} + \cdots + \frac{1}{n}}{\ln n}$;

(b) $\lim_{n\to\infty} \dfrac{1^p + 2^p + \cdots + n^p}{n^{p+1}}$, where $p \in \mathbb{N}$;

(c) $\lim_{n\to\infty} \dfrac{1 + 2\sqrt{2} + 3\sqrt{3} + \cdots + n\sqrt{n}}{n^2 \sqrt{n}}$;

(d) $\lim_{n\to\infty} \dfrac{1 + \sqrt[3]{2} + \sqrt[3]{3} + \cdots + \sqrt[3]{n}}{n\sqrt[3]{n}}$;

(e) $\lim_{n\to\infty} \dfrac{1 + 2^2\sqrt{2} + 3^2\sqrt[3]{3} + \cdots + n^2\sqrt[n]{n}}{n(n+1)(n+2)}$;

(f) $\lim_{n\to\infty} \dfrac{1 + \sqrt[4]{2!} + \sqrt[9]{3!} + \cdots + \sqrt[n^2]{n!}}{n}$;

(g) $\lim_{n\to\infty} \dfrac{n^n}{1 + 2^2 + 3^3 + \cdots + n^n}$;

(h) $\lim_{n\to\infty} \dfrac{\lfloor a \rfloor + \lfloor 2a \rfloor + \cdots + \lfloor na \rfloor}{n^2}$, where $a \in \mathbb{R}$ and $\lfloor x \rfloor$ is the floor of $x \in \mathbb{R}$;

(i) $\lim_{n\to\infty} \dfrac{\lfloor a \rfloor 1! + \lfloor 2a \rfloor 2! + \cdots + \lfloor na \rfloor n!}{(n+1)!}$, where $a \in \mathbb{R}$ and $\lfloor x \rfloor$ is the floor of $x \in \mathbb{R}$.

1.27. Let $(x_n)_{n\geq 1}$, with $x_1 > 0$, be the sequence defined by the recurrence relation $x_{n+1} = \ln(1 + x_n), n \geq 1$. Calculate $\lim\limits_{n\to\infty} nx_n$.

1.28. Let $(x_n)_{n\geq 1}$, with $x_1 > 0$, be the sequence defined by the recurrence relation $x_{n+1} = \arctan x_n, n \geq 1$. Calculate $\lim\limits_{n\to\infty} \sqrt{n}x_n$.

1.29. Let $(x_n)_{n\geq 1}$, with $x_1 \in \mathbb{R}$, be the sequence defined by the recurrence relation $x_{n+1} = x_n + e^{-x_n}, n \geq 1$. Calculate $\lim\limits_{n\to\infty} \frac{x_n}{\ln n}$.

1.30. Open Problem Let $\beta > 0$ and let $(x_n)_{n\geq 1}$ be the sequence defined by the recurrence relation

$$x_1 = a > 0, \quad x_{n+1} = x_n + \frac{n^{2\beta}}{x_1 + x_2 + \cdots + x_n}.$$

Prove that $\lim\limits_{n\to\infty} \frac{x_n}{n^{\beta}} \stackrel{?}{=} \sqrt{\frac{\beta + 1}{\beta}}$.

Stolz–Cesàro Theorem, the $0/0$ Case Let $(a_n)_{n\geq 1}$ and $(b_n)_{n\geq 1}$ be two sequences of real numbers such that:

(a) $\lim\limits_{n\to\infty} a_n = \lim\limits_{n\to\infty} b_n = 0$;
(b) $(b_n)_{n\geq 1}$ is strictly decreasing;
(c) $\lim\limits_{n\to\infty} \dfrac{a_{n+1} - a_n}{b_{n+1} - b_n} = l \in \overline{\mathbb{R}}$.

Then $\lim\limits_{n\to\infty} \dfrac{a_n}{b_n}$ exists and it is equal to l.

This version of Stolz–Cesàro Theorem is discussed in [94, p. 56], [26, pp. 265, 266], [124, pp. 281–284].

1.31. Euler's Constant Let $(x_n)_{n\geq 1}$ be the sequence defined by

$$x_n = 1 + \frac{1}{2} + \frac{1}{3} + \cdots + \frac{1}{n} - \ln n, \quad n \geq 1.$$

(continued)

Prove that $(x_n)_{n \geq 1}$ converges and, denoting by γ its limit, calculate $\lim_{n \to \infty} n(x_n - \gamma)$.

The constant $\gamma = 0.5772156649\ldots$ is called *Euler's constant*, being introduced in the literature by Leonhard Euler. It is not known whether γ is a rational or an irrational number, this being an open problem of great interest for mathematicians.

1.32.

(a) Calculate

$$\lim_{n \to \infty} n(2H_n - H_{n^2} - \gamma).$$

(b) *A generalization.* Let $k \geq 2$ be an integer. Calculate

$$\lim_{n \to \infty} n \left[k H_n - H_{n^k} - (k-1)\gamma \right].$$

1.33. Let H_n be the nth harmonic number and let $O_n = 1 + \frac{1}{3} + \cdots + \frac{1}{2n-1}, n \geq 1$. Calculate

$$\lim_{n \to \infty} \frac{1}{n} \left(1 + \frac{H_n}{n} \right)^n, \qquad \lim_{n \to \infty} \frac{1}{\sqrt{n}} \left(1 + \frac{O_n}{n} \right)^n \quad \text{and} \quad \lim_{n \to \infty} \frac{1}{n} \left(1 + \frac{2O_n}{n} \right)^n.$$

1.34. Calculate

$$\lim_{n \to \infty} \sqrt{n} \left(2e^{\frac{1}{2} + \frac{1}{4} + \cdots + \frac{1}{2n}} - e^{1 + \frac{1}{3} + \cdots + \frac{1}{2n-1}} \right).$$

1.35. Ioachimescu's Constant Let $(x_n)_{n \geq 1}$ be the sequence defined by

$$x_n = 1 + \frac{1}{\sqrt{2}} + \frac{1}{\sqrt{3}} + \cdots + \frac{1}{\sqrt{n}} - 2(\sqrt{n} - 1), \quad n \geq 1.$$

Prove that $(x_n)_{n \geq 1}$ converges and, denoting by $\mathscr{I}$ its limit, calculate $\lim_{n \to \infty} \sqrt{n}(x_n - \mathscr{I})$.

The constant $\mathscr{I} = 0.539645491\ldots$ is called *Ioachimescu's constant*. Romanian mathematician Andrei G. Ioachimescu (1868–1943), one of the

(continued)

founders of the famous Romanian magazine Gazeta Matematică, has proposed a problem [99] in which he was asking to prove that the sequence
$(x_n - 2)_{n \geq 1}$ converges to a finite limit between -2 and -1.

1.36. Calculate $\lim\limits_{n \to \infty} n \left(\frac{\pi^2}{6} - \sum\limits_{k=1}^{n} \frac{1}{k^2} \right)$.

1.37. Calculate $\lim\limits_{n \to \infty} n^3 \left(\frac{\pi^4}{90} - \sum\limits_{k=1}^{n} \frac{1}{k^4} \right)$.

1.38. [D. M. Bătineţu-Giurgiu, 1991]

Let $x_n = \sum\limits_{k=1}^{n} \frac{1}{(2k-1)^4}$, $n \geq 1$. Calculate $\lim\limits_{n \to \infty} \left(\frac{\pi^4}{96 x_n} \right)^{n^3}$.

1.39. Let $x \in \mathbb{R}$, $x > 1$.

(a) Prove that

$$\lim\limits_{n \to \infty} n^{x-1} \left(\zeta(x) - \sum\limits_{k=1}^{n} \frac{1}{k^x} \right) = \frac{1}{x-1}.$$

(b) Calculate

$$\lim\limits_{n \to \infty} n \left[n^{x-1} \left(\zeta(x) - \sum\limits_{k=1}^{n} \frac{1}{k^x} \right) - \frac{1}{x-1} \right].$$

The function ζ defined by $\zeta(z) = \sum_{n=1}^{\infty} 1/n^z$, for $\Re(z) > 1$, is called *the Riemann zeta function*. The most important open problem in mathematics, *the Riemann hypothesis*, is about proving that the nontrivial zeros of the zeta function are located on the vertical line that passes through $1/2$.

1.40. Let $k \in \mathbb{R}$, $k > 1$. Calculate

$$\lim\limits_{n \to \infty} \left(1 + \frac{1}{2} + \cdots + \frac{1}{n} \right) \left(\zeta(k) - 1 - \frac{1}{2^k} - \cdots - \frac{1}{n^k} \right).$$

1.41. *Limits with the zeta function.* Prove that:

(a) $\lim\limits_{x \to \infty} \zeta(x) = 1$;

(b) $\displaystyle \lim_{x \to \infty} \frac{\zeta(2x) - 1}{\zeta(x) - 1} = 0$;

(c) $\displaystyle \lim_{x \to \infty} \zeta(x)^{2^x} = e$;

(d) $\displaystyle \lim_{x \to \infty} \left(\frac{3}{2}\right)^x \left(\zeta(x)^{2^x} - e\right) = e$.

1.42. *Two limits involving the constant* e. Calculate

$$\lim_{n \to \infty} (n+1)! \left(e - \sum_{k=0}^{n} \frac{1}{k!}\right) \quad \text{and} \quad \lim_{n \to \infty} (-1)^{n-1} (n+1)! \left(\frac{1}{e} - \sum_{k=0}^{n} (-1)^k \frac{1}{k!}\right).$$

1.3   Wolstenholme Sequences

1.43. A Wolstenholme Limit

(a) Let $s > 0$. Prove that

$$\lim_{n \to \infty} \sum_{k=1}^{n} \left(\frac{k}{n}\right)^{sn} = \frac{e^s}{e^s - 1}.$$

(b) [120] **A Limit of Dumitru Popa**

Let $f : (0, 1] \to (0, \infty)$ be a differentiable function on $(0, 1]$, with $f'(1) > 0$ and $\ln f$ having decreasing derivative, and let $(x_n)_{n \geq 1}$ be the sequence defined by $x_n = \sum_{k=1}^{n} f^n\left(\frac{k}{n}\right)$.

Prove that

$$\lim_{n \to \infty} x_n = \begin{cases} 0 & \text{if } f(1) < 1 \\ \frac{1}{1 - e^{-f'(1)}} & \text{if } f(1) = 1 \\ \infty & \text{if } f(1) > 1. \end{cases}$$

Reference material regarding this problem can be found in [26, pp. 32–33]. More generally, it can be proved that if f is like in part (b) and $(y_n)_{n \geq 1}$ is the sequence defined by $y_n = \sum_{k=1}^{n} f^{\alpha n + \beta}\left(\frac{k}{n}\right), \alpha > 0, \beta \geq 0$, then

(continued)

$$\lim_{n\to\infty} y_n = \begin{cases} 0 & \text{if } f(1) < 1 \\ \dfrac{1}{1-e^{-\alpha f'(1)}} & \text{if } f(1) = 1 \\ \infty & \text{if } f(1) > 1. \end{cases}$$

1.44. Let $a \geq 1$. Prove that

$$\lim_{n\to\infty} \sum_{k=1}^{n} a^{k-n} \left(\frac{k}{n}\right)^n = \lim_{n\to\infty} \sum_{k=1}^{n} a^{k-n} \left(\frac{k}{n}\right)^k = \frac{ae}{ae-1}.$$

1.45.

(a) Let $s > 0$. Prove that

$$\lim_{n\to\infty} \sum_{k=1}^{n} \left(\frac{k}{n}\right)^{sk} = \frac{e^s}{e^s - 1}.$$

(b) Let $a, b \geq 0$, with $a + b > 0$. Prove that

$$\lim_{n\to\infty} \sum_{k=1}^{n} \left(\frac{k}{n}\right)^{ak+bn} = \frac{e^{a+b}}{e^{a+b} - 1}.$$

1.46. Let $f : (0, 1] \to (0, \infty)$ be a differentiable function on $(0, 1]$, with $f'(1) > 0$ and $\ln f$ having decreasing derivative, and let $(x_n)_{n\geq 1}$ be the sequence defined by

$$x_n = \sum_{k=1}^{n} a^{k-n} f^{\alpha n+\beta} \left(\frac{k}{n}\right),$$

where $a \geq 1$, $\alpha > 0$, and $\beta \geq 0$.
   Prove that

(continued)

$$\lim_{n\to\infty} x_n = \begin{cases} 0 & \text{if } f(1) < 1 \\ \dfrac{ae^{\alpha f'(1)}}{ae^{\alpha f'(1)} - 1} & \text{if } f(1) = 1 \\ \infty & \text{if } f(1) > 1. \end{cases}$$

1.4   Limits of Integrals

1.47. Spectacular Limits of Integrals

(a) Prove that:

- $$\lim_{n\to\infty} \int_0^1 \left(\frac{\sqrt[n]{x} + 1}{2} \right)^n dx = \frac{2}{3};$$
- $$\lim_{n\to\infty} n \left[\int_0^1 \left(\frac{\sqrt[n]{x} + 1}{2} \right)^n dx - \frac{2}{3} \right] = \frac{2}{27}.$$

Remark 1.5. If $k > 0$, then

- $$\lim_{n\to\infty} \int_0^1 \left(\frac{\sqrt[n]{x} + k - 1}{k} \right)^n dx = \frac{k}{k+1};$$
- $$\lim_{n\to\infty} n \left[\int_0^1 \left(\frac{\sqrt[n]{x} + k - 1}{k} \right)^n dx - \frac{k}{k+1} \right] = \frac{(k-1)k}{(k+1)^3}.$$

(b) Prove that:

- $$\lim_{n\to\infty} \int_0^1 \left(\frac{2}{\sqrt[n]{x} + 1} \right)^n dx = 2;$$
- $$\lim_{n\to\infty} n \left[2 - \int_0^1 \left(\frac{2}{\sqrt[n]{x} + 1} \right)^n dx \right] = 2.$$

Remark 1.6. [O. Furdui, A. Sîntămărian, Problem 2, SEEMOUS 2020]
   If $k > 1$, then

- $$\lim_{n\to\infty} \int_0^1 \left(\frac{k}{\sqrt[n]{x} + k - 1} \right)^n dx = \frac{k}{k-1};$$

(continued)

$$\bullet \quad \lim_{n \to \infty} n \left[\frac{k}{k-1} - \int_0^1 \left(\frac{k}{\sqrt[n]{x} + k - 1} \right)^n dx \right] = \frac{k}{(k-1)^2}.$$

1.48.

(a) Prove that

$$\lim_{n \to \infty} \frac{4^n}{\sqrt{n}} \int_0^1 \left(1 - \sqrt[n]{x} \right)^n dx = \sqrt{\pi}.$$

(b) If $f : [0, 1] \to \mathbb{R}$ is a Riemann integrable function which is continuous at 0, then

$$\lim_{n \to \infty} \frac{4^n}{\sqrt{n}} \int_0^1 \left(1 - \sqrt[n]{x} \right)^n f(x) dx = \sqrt{\pi} f(0).$$

1.49. Let $(a_n)_{n \geq 1}$ be a sequence of nonnegative real numbers such that $\lim_{n \to \infty} a_n = a$.

(a) Calculate

$$\lim_{n \to \infty} \sqrt[n]{\int_0^1 (1 + a_n x^n)^n dx}.$$

Application. Let $\alpha > 0$ be a real number. Calculate

$$\lim_{n \to \infty} \sqrt[n]{\int_0^1 (1 + \alpha x^n)^n dx}.$$

(b) Calculate

$$\lim_{n \to \infty} \sqrt[n]{\int_0^1 \left(1 + \frac{x + x^3 + \cdots + x^{2n-1}}{n} \right)^n dx}.$$

(c) Calculate

$$\lim_{n \to \infty} \sqrt[n]{\int_0^1 \left(1 + \frac{a_1 x + a_2 x^2 + \cdots + a_n x^n}{n} \right)^n dx}.$$

1.50. Let $f : [-1, 1] \to \mathbb{R}$ be a Riemann integrable function which is continuous at 0. Prove that

$$\lim_{h \to 0^+} \int_{-1}^{1} \frac{hf(x)}{x^2 + h^2} dx = \pi f(0).$$

1.51. *Two identities with the fractional part function.*

(a) Let $f : [0, 1] \to \mathbb{R}$ be a Riemann integrable function. Prove that

$$\int_{0}^{1} f(\{nx\}) dx = \int_{0}^{1} f(x) dx.$$

(b) Let $f : [0, 1] \to \mathbb{R}$ be a Riemann integrable function. Prove that

$$\int_{0}^{1} \lfloor nx \rfloor f(\{nx\}) dx = \frac{n-1}{2} \int_{0}^{1} f(x) dx.$$

1.52. Let $f : \mathbb{R} \to \mathbb{R}$ be a function and let $g : [0, 1] \to \mathbb{R}$ be a Riemann integrable function.

(a) Prove that

$$\int_{0}^{1} f(\lfloor nx \rfloor) g(\{nx\}) dx = \frac{1}{n} \sum_{k=0}^{n-1} f(k) \int_{0}^{1} g(x) dx.$$

(b) If f satisfies $\lim_{x \to \infty} \frac{f(x)}{x^\alpha} = L$, where $\alpha \geq 0$, then

$$\lim_{n \to \infty} \frac{1}{n^\alpha} \int_{0}^{1} f(\lfloor nx \rfloor) g(\{nx\}) dx = \frac{L}{1+\alpha} \int_{0}^{1} g(x) dx.$$

1.53. Let $f : [0, 1] \to \mathbb{R}$ be a Riemann integrable function. Calculate

$$\lim_{n \to \infty} \int_{0}^{1} f\left(\left\{\frac{n}{x}\right\}\right) dx,$$

where $\{x\}$ is the fractional part of x.

1.54.

(a) Calculate

$$\lim_{n\to\infty} \sqrt[n]{\int_0^1 \left\{\frac{1}{x}\right\}^n dx},$$

where $\{x\}$ is the fractional part of x.

(b) Let $f : [0, 1] \to (0, \infty)$ be a continuous function. Calculate

$$\lim_{n\to\infty} \sqrt[n]{\int_0^1 \left\{\frac{1}{x}\right\}^n f(x)dx}.$$

1.55.

(a) Prove that

$$\lim_{n\to\infty} \int_0^1 \frac{1}{\left\lfloor \frac{1}{x} \right\rfloor^n} dx = \frac{1}{2},$$

where $\lfloor a \rfloor$ is the floor of a.

(b) Let $f : [0, 1] \to \mathbb{R}$ be a Riemann integrable function which is continuous at 0. Prove that

$$\lim_{n\to\infty} \int_0^1 f\left(\frac{1}{\left\lfloor \frac{1}{x} \right\rfloor^n}\right) dx = \frac{f(0) + f(1)}{2}.$$

1.56.

(a) Let $k \in \mathbb{N}$. Prove that

$$\lim_{n\to\infty} \int_0^1 \frac{x^k}{\left\lfloor \frac{1}{x} \right\rfloor^n} dx = \frac{1}{k+1}\left(1 - \frac{1}{2^{k+1}}\right),$$

where $\lfloor a \rfloor$ is the floor of a.

(b) Let $f, g : [0, 1] \to \mathbb{R}$ be Riemann integrable functions, with f continuous at 0. Prove that

$$\lim_{n\to\infty} \int_0^1 f\left(\frac{1}{\left\lfloor \frac{1}{x} \right\rfloor^n}\right) g(x)dx = f(0)\int_0^{\frac{1}{2}} g(x)dx + f(1)\int_{\frac{1}{2}}^1 g(x)dx.$$

1.57. Let $f, g : [0, 1] \to \mathbb{R}$ be Riemann integrable functions, with f continuous at 0. Prove that

$$\lim_{n \to \infty} \int_0^1 f\left(\frac{1}{\left\lfloor \frac{n}{x} \right\rfloor}\right) g(x) dx = f(0) \int_0^1 g(x) dx.$$

1.58. Let $\alpha > 0$ and let $f : [-1, 0] \to \mathbb{R}$ be a continuous function. Prove that

$$\lim_{n \to \infty} n \int_{-1}^0 \left(x + e^{\alpha x}\right)^n f(x) dx = \frac{f(0)}{1 + \alpha}.$$

1.59. Let $\alpha > 0$. Calculate

$$\lim_{n \to \infty} n \int_{-1}^0 \left(x + \frac{x^2}{2} + e^{\alpha x}\right)^n dx.$$

1.60. Let $\alpha > 0$. Calculate

$$\lim_{n \to \infty} \left(\int_0^1 \sqrt[n]{x^n + \alpha}\, dx\right)^n.$$

1.61. Calculate

$$\lim_{n \to \infty} \int_0^n \frac{dx}{1 + n^2 \cos^2 x}.$$

1.62. Calculate

$$\lim_{n \to \infty} \int_0^n \frac{dx}{1 + n^2 \sin^2(x + na)}, \qquad a > 0.$$

1.63. [57] Calculate

$$\lim_{n \to \infty} \frac{1}{n} \int_0^n \frac{x}{1 + n^2 \cos^2 x} dx.$$

More generally (see [55]), it can be proved that if $f : [0, 1] \to \mathbb{R}$ is a continuous function, then

(continued)

$$\lim_{n\to\infty} \int_0^n \frac{f\left(\frac{x}{n}\right)}{1+n^2\cos^2 x}\,dx = \int_0^1 f(x)\,dx.$$

1.64. Calculate

$$\lim_{n\to\infty} \frac{1}{n\sqrt{n}} \int_0^n \frac{x}{1+n\cos^2 x}\,dx.$$

More generally, it can be proved that if $f : [0,1] \to \mathbb{R}$ is a continuous function, then

$$\lim_{n\to\infty} \frac{1}{\sqrt{n}} \int_0^n \frac{f\left(\frac{x}{n}\right)}{1+n\cos^2 x}\,dx = \int_0^1 f(x)\,dx.$$

1.65.

(a) Calculate

$$\lim_{n\to\infty} \frac{1}{n} \int_0^1 \ln\left(1+e^{nx}\right) dx.$$

(b) Let $k \geq 1$ be an integer. Calculate

$$\lim_{n\to\infty} \frac{1}{n^k} \int_0^1 \ln^k\left(1+e^{nx}\right) dx.$$

1.66. [46] Let $a, b \in \mathbb{R}$, with $0 < a < b$. Calculate

$$\lim_{n\to\infty} \frac{1}{n} \sqrt[n]{\int_a^b \ln^n\left(1+e^{nx}\right) dx}.$$

1.67.

(a) [45] Calculate

$$\lim_{n\to\infty} \sqrt[n]{\int_0^1 (1+x^n)^n dx}.$$

(b) [35] Let $f : [0, 1] \to (0, \infty)$ be a continuous function. Calculate

$$\lim_{n \to \infty} \sqrt[n]{\int_0^1 (1 + x^n)^n f(x) dx}.$$

1.68.

(a) Calculate

$$\lim_{n \to \infty} \frac{1}{n} \int_1^n n^{\frac{1}{x}} dx.$$

(b) Let $f : [1, \infty) \to \mathbb{R}$ be a continuous function such that $\lim_{x \to \infty} f(x) = L$. Calculate

$$\lim_{n \to \infty} \frac{1}{n} \int_1^n n^{\frac{1}{x}} f(x) dx.$$

1.69. Let $f : \left[0, \frac{\pi}{2}\right] \to \mathbb{R}$ be a continuous function. Prove that

$$\lim_{n \to \infty} n \int_0^{\frac{\pi}{2}} (\cos x - \sin x)^{2n} f(x) dx = \frac{f(0) + f\left(\frac{\pi}{2}\right)}{2}.$$

1.70. [37] Let $f : \left[0, \frac{\pi}{2}\right] \to \mathbb{R}$ be a continuous function. Prove that

$$\lim_{n \to \infty} n \int_0^{\frac{\pi}{2}} \left(\frac{\cos x - \sin x}{\cos x + \sin x}\right)^{2n} f(x) dx = \frac{f(0) + f\left(\frac{\pi}{2}\right)}{4}.$$

1.71. [26, Remark, p. 67] Let $f : [a, b] \to [0, \infty)$ be a continuous function. Calculate

$$\lim_{n \to \infty} n \left(\sqrt[n+1]{\int_a^b f^{n+1}(x) dx} - \sqrt[n]{\int_a^b f^n(x) dx} \right).$$

1.72. Let $f : [0, \infty) \to [0, \infty)$ be a bounded continuous function. Calculate

$$\lim_{n \to \infty} n \left(\sqrt[n]{\int_0^\infty f^{n+1}(x) e^{-x} dx} - \sqrt[n]{\int_0^\infty f^n(x) e^{-x} dx} \right).$$

1.73.

(a) Let $a, b \in \mathbb{R}$, $a < b$, and let $f : [a, b] \to \mathbb{R}$ be a Riemann integrable function. Calculate

$$L = \lim_{n \to \infty} \int_a^b \frac{f(x)}{1 + \sin x \sin(x + 1) \cdots \sin(x + n)} dx.$$

(b) Let $q \in [1, \sqrt{2})$. Calculate

$$\lim_{n \to \infty} q^n \left(\int_a^b \frac{f(x)}{1 + \sin x \sin(x + 1) \cdots \sin(x + n)} dx - L \right).$$

1.74.

(a) Let $a, b \in \mathbb{R}$. Calculate

$$\lim_{n \to \infty} \int_a^b |\sin^n x| \, dx.$$

(b) Let $f : [a, b] \to \mathbb{R}$ be a Riemann integrable function. Calculate

$$\lim_{n \to \infty} \int_a^b f(x) \sin^n x \, dx.$$

1.75. Limits with Exotic Integrals

(a) Let $f : [0, 1] \to \mathbb{R}$ be a continuous function. Prove that:

- $$\lim_{n \to \infty} \int_0^1 \frac{x^n}{x^n + (1 - x)^n} f(x) dx = \int_{\frac{1}{2}}^1 f(x) dx;$$

- $$\lim_{n \to \infty} n \left(\int_0^1 \frac{x^n}{x^n + (1 - x)^n} f(x) dx - \int_{\frac{1}{2}}^1 f(x) dx \right) = 0.$$

(b) If $f \in C^1([0, 1], \mathbb{R})$, then

$$\lim_{n \to \infty} n^2 \left(\int_0^1 \frac{x^n}{x^n + (1 - x)^n} f(x) dx - \int_{\frac{1}{2}}^1 f(x) dx \right) = -\frac{\pi^2}{96} f'\left(\frac{1}{2}\right).$$

(continued)

A *challenge.* Solve the same problem with $\frac{x^n}{x^n+(1-x)^n}$ replaced by $\frac{x^n-(1-x)^n}{x^n+(1-x)^n}$.

1.76.

(a) Let $k \geq 2$ be an integer. Prove that

$$\lim_{n\to\infty} \int_{\frac{1}{k}}^{1} \{nx\}\, dx = \frac{k-1}{2k},$$

where $\{x\}$ denotes the fractional part of x.

(b) Prove that

$$\lim_{n\to\infty} \int_{0}^{1} \frac{x^n}{x^n + (1-x)^n} \{nx\}\, dx = \frac{1}{4},$$

where $\{x\}$ denotes the fractional part of x.

1.77. Let $f : \mathbb{R} \to \mathbb{R}$ be a continuous function such that $\lim_{x\to-\infty} f(x) = f(-\infty)$ and $\lim_{x\to\infty} f(x) = f(\infty)$ exist and are finite. Prove that

$$\lim_{x\to\infty} \int_{-x}^{x} (f(t+1) - f(t))\, dt = f(\infty) - f(-\infty).$$

1.78. [66] Let $f : [0, 1] \to \mathbb{R}$ be a continuous function. Calculate

$$\lim_{n\to\infty} \frac{n}{\ln n} \int_{0}^{1} x^n f(x^n) \ln(1 - x)dx.$$

1.79. *A problem of Radu Gologan.* Let $k \geq 1$ be an integer and $f : [0, 1] \to [0, 1]$ be a Riemann integrable function. Prove that

$$\lim_{n\to\infty} n \int_{0}^{1} f^n(x)(1 - f^k(x))dx = 0.$$

1.80. An Open Problem Solved

(a) [95] Prove that

$$\lim_{n\to\infty} \sqrt[n]{\int_0^1 x(1-x)x^2(1-x^2)\cdots x^n(1-x^n)\mathrm{d}x} = b(1-b) = 0.185155\ldots,$$

where b is the unique solution of the equation

$$\int_0^1 \frac{x}{1-b^x}\mathrm{d}x = 1.$$

(b) Prove that

$$\lim_{n\to\infty} \sqrt[n]{\int_0^1 x^{1+\frac{1}{2}+\cdots+\frac{1}{n}}(1-x)(1-\sqrt{x})\cdots(1-\sqrt[n]{x})\mathrm{d}x} = 0.$$

Remark 1.7. Details regarding this problem due to O. Furdui and the motivation of studying these limits are given in [26, problem 1.79, p. 15].

Series of Real Numbers

2

Read to get wise and teach others when it will be needed.

Saint Basil the Great (329–379)

The problems in this chapter deal with two major topics on series of real numbers: the convergence of a numerical series and its exact calculation. Regarding the convergence, the problems can be solved by using the classical way, i.e. the well-known convergence tests for numerical series, or less known ways. Most of the problems about the exact calculation of the value of a series are new in the literature and vary in diversity as well as in difficulty. Their topics range from the calculation of series involving harmonic numbers and factorials to exotic series involving tails of Riemann zeta function values.

2.1   Miscellaneous Series

The Geometric Series Let $k \geq 0$ be an integer and let $q \in \mathbb{R}$. The geometric series $\sum_{n=k}^{\infty} q^n$ converges if and only if $|q| < 1$, and we have

$$\sum_{n=k}^{\infty} q^n = \frac{q^k}{1-q}.$$

(continued)

© The Author(s), under exclusive license to Springer Nature Switzerland AG 2021
A. Sîntămărian, O. Furdui, *Sharpening Mathematical Analysis Skills*, Problem Books in Mathematics, https://doi.org/10.1007/978-3-030-77139-3_2

Remark 2.1. This formula has an equivalent formulation

$$\text{the sum of the geometric series} = \frac{\text{the first term of the series}}{1 - \text{the ratio}}.$$

2.1. Calculate:

(a) $\displaystyle\sum_{n=1}^{\infty} (-1)^{n-1}\frac{1}{7^n};$

(b) $\displaystyle\sum_{n=1}^{\infty}\left(\frac{1+3+\cdots+(2n-1)}{(2n+1)+(2n+3)+\cdots+(4n-1)}\right)^n;$

(c) $\displaystyle\sum_{n=1}^{\infty}\frac{1}{n(n+3)};$

(d) $\displaystyle\sum_{n=1}^{\infty}\frac{1}{(2n-1)(2n+1)(2n+3)};$

(e) $\displaystyle\sum_{n=2}^{\infty}\ln\left(1-\frac{1}{n^2}\right);$

(f) $\displaystyle\sum_{n=1}^{\infty}\frac{n}{a^n},\quad |a|>1;$

(g) $\displaystyle\sum_{n=1}^{\infty}\frac{n}{n^4+n^2+1};$

(h) $\displaystyle\sum_{n=2}^{\infty}\frac{n}{(n-2)!+(n-1)!+n!};$

(i) $\displaystyle\sum_{n=1}^{\infty}\frac{n(n+1)(n+2)(n+3)+1}{(n-1)!(n^2+3n+1)^2};$

(j) $\displaystyle\sum_{n=0}^{\infty}\frac{n(n+3)(n+6)(n+9)+81}{n!(n^2+9n+9)}.$

2.2. Series with Fractional Part Function

Let $\{x\}$ be the fractional part of x. Prove that:

(a) $\displaystyle\sum_{n=1}^{\infty}\left\{(1+\sqrt{2})^{2n-1}\right\}=\frac{1}{2};$

(b) $\displaystyle\sum_{n=1}^{\infty}\left(1-\left\{(2+\sqrt{3})^n\right\}\right)=\frac{\sqrt{3}-1}{2};$

(c) $\displaystyle\sum_{n=1}^{\infty}\left(1-\left\{(3+\sqrt{7})^n\right\}\right)=\frac{\sqrt{7}-1}{3}.$

We record a theorem, *a gem of mathematical analysis*, which uses in its proof the geometric series.

Theorem 2.1. *The set of prime numbers is infinite.*

Proof. To prove that the set of prime numbers is infinite we use *Euler's series* $\sum_{n=1}^{\infty} \frac{1}{n^2} = \frac{\pi^2}{6}$. We assume, by way of contradiction, that the set of prime numbers P is finite. Let $P = \{p_1, p_2, \ldots, p_k\}$. Then, for all $n \in \mathbb{N}$ we have $n = p_1^{\alpha_1} p_2^{\alpha_2} \cdots p_k^{\alpha_k}$, where $\alpha_i \in \mathbb{N} \cup \{0\}$, $i = 1, 2, \ldots, k$. It follows that

$$\sum_{n=1}^{\infty} \frac{1}{n^2} = \sum_{\alpha_1, \alpha_2, \ldots, \alpha_k \geq 0} \frac{1}{p_1^{2\alpha_1} p_2^{2\alpha_2} \cdots p_k^{2\alpha_k}}$$

$$= \sum_{\alpha_1=0}^{\infty} \frac{1}{p_1^{2\alpha_1}} \sum_{\alpha_2=0}^{\infty} \frac{1}{p_2^{2\alpha_2}} \cdots \sum_{\alpha_k=0}^{\infty} \frac{1}{p_k^{2\alpha_k}}$$

$$= \frac{p_1^2}{p_1^2 - 1} \cdot \frac{p_2^2}{p_2^2 - 1} \cdots \frac{p_k^2}{p_k^2 - 1}.$$

Therefore,

$$\frac{p_1^2}{p_1^2 - 1} \cdot \frac{p_2^2}{p_2^2 - 1} \cdots \frac{p_k^2}{p_k^2 - 1} = \frac{\pi^2}{6},$$

which contradicts the irrationality of π^2. ∎

Remark 2.2. Using the formula $\sum_{n=1}^{\infty} \frac{1}{n^3} = \zeta(3)$ and *the irrationality* of $\zeta(3)$, one can prove similarly that the set of prime numbers is infinite.

2.3. Prove that

$$\sum_{n=1}^{\infty} \frac{1}{(2n+1)^2 - 1} = \frac{1}{4}.$$

2.4. Calculate:

(a) $\sum_{n=1}^{\infty} \frac{1}{n^2(n+1)^2}$;

(b) $\sum_{n=1}^{\infty} \frac{1}{n^3(n+1)^3}$.

2.5. Calculate:

(a) $\displaystyle\sum_{n=0}^{\infty} \frac{1}{(2n+1)(2n+2)}$;

(b) $\displaystyle\sum_{n=0}^{\infty} \frac{1}{(3n+1)(3n+2)(3n+3)}$;

(c)[1] $\displaystyle\sum_{n=0}^{\infty} \frac{1}{(4n+1)(4n+2)(4n+3)(4n+4)}$.

2.6. Let A be a finite set of natural numbers of the following form $2^m 3^n 5^p$. Prove that $\displaystyle\sum_{x\in A} \frac{1}{x} < 4$.

2.7. Prove that:

(a) $1 + \dfrac{1}{2} - \dfrac{1}{3} - \dfrac{1}{4} + \dfrac{1}{5} + \dfrac{1}{6} - \dfrac{1}{7} - \dfrac{1}{8} + \dfrac{1}{9} + \dfrac{1}{10} - \cdots = \dfrac{\pi}{4} + \dfrac{\ln 2}{2}$;

(b) $1 + \dfrac{1}{2} + \dfrac{1}{3} - \dfrac{1}{4} - \dfrac{1}{5} - \dfrac{1}{6} + \dfrac{1}{7} + \dfrac{1}{8} + \dfrac{1}{9} - \dfrac{1}{10} - \dfrac{1}{11} - \dfrac{1}{12} + \cdots = \dfrac{2\sqrt{3}}{9}\pi + \dfrac{\ln 2}{3}$.

2.8. [126] Prove that

$$\sum_{n=0}^{\infty} \frac{1}{n!(n^4 + n^2 + 1)} = \frac{e}{2}.$$

2.9.

(a) [97, problem 15, p. 230] Prove that $\sqrt{\underbrace{1\ldots1}_{2n \text{ digits}} - \underbrace{2\ldots2}_{n \text{ digits}}} = \underbrace{3\ldots3}_{n \text{ digits}}.$

(b) Determine the digits a, b, and c such that $\sqrt{\underbrace{a\ldots a}_{2n \text{ digits}} - \underbrace{b\ldots b}_{n \text{ digits}}} = \underbrace{c\ldots c}_{n \text{ digits}}.$

2.10. Prove that:

(a) $0.9999999\ldots = 1$;

(b) $0.90909090\ldots = \frac{10}{11}$;

(c) $0.0909090909\ldots = \frac{1}{11}$;

(d) $0.abababababab\ldots = \frac{10a+b}{99}$, where $a, b \in \{0, 1, 2, \ldots, 9\}$.

---

[1]This part of the problem was given in day 1 of the student contest IMC 2010, Blagoevgrad, Bulgaria.

2.11. Calculate

$$\sum_{n=1}^{\infty} \frac{1}{(3+(-1)^n)^n} \quad \text{and} \quad \sum_{n=1}^{\infty} \frac{(-1)^n}{(3+(-1)^n)^n}.$$

If $x, y \geq 0$ are real numbers, then

$$\arctan x - \arctan y = \arctan \frac{x-y}{1+xy}.$$

2.12. Calculate

$$\sum_{n=1}^{\infty} \arctan \frac{1}{n^2+n+1} \quad \text{and} \quad \sum_{n=1}^{\infty} \arctan \frac{1}{2n^2}.$$

2.13. Let $(F_n)_{n>0}$ be the Fibonacci sequence.

(a) Prove that $F_{2n} = F_{n+1}^2 - F_{n-1}^2$, for all $n \geq 1$.
(b) Calculate

$$\sum_{n=2}^{\infty} \frac{F_{2n}}{F_{n-1}^2 F_{n+1}^2}.$$

2.14. Calculate the sum of the series

$$\sum_{n=1}^{\infty} \left(\frac{2}{\left\lfloor \sqrt{n^2-n+1} + \sqrt{n^2+n+1} \right\rfloor} - \ln\left(1+\frac{1}{n}\right) \right),$$

where $\lfloor x \rfloor$ denotes the floor of x.

Mathematical Amusements: Dividing Integers!

- $\cancel{6}4 : 1\cancel{6} = 4$;
- $\cancel{9}5 : 1\cancel{9} = 5$;
- $\cancel{9}8 : 4\cancel{9} = 2$.

2.15. Calculate $\displaystyle\sum_{n=2}^{\infty} \ln\left(1 + \frac{(-1)^n}{n}\right)$.

2.16. Resembling a Series of Ramanujan

(a) Check that

$$\frac{1}{3n-1} + \frac{1}{3n} + \frac{1}{3n+1} = \frac{1}{n} + \frac{2}{(3n)^3 - 3n}, \quad n \geq 1.$$

(b) Prove that

$$\sum_{n=1}^{\infty} \frac{2}{(3n)^3 - 3n} = \ln 3 - 1.$$

2.17. Two Telescoping Series

(a) Find the polynomial P of degree 2 such that

$$\frac{n^2}{2^n} = \frac{P(n+1)}{2^{n+1}} - \frac{P(n)}{2^n}, \quad \forall n \in \mathbb{N}.$$

(b) Prove that $\sum_{n=1}^{\infty} \frac{n^2}{2^n} = 6$.

(c) Find the polynomial Q of degree 3 such that

$$\frac{n^3}{2^n} = \frac{Q(n+1)}{2^{n+1}} - \frac{Q(n)}{2^n}, \quad \forall n \in \mathbb{N}.$$

(d) Prove that $\sum_{n=1}^{\infty} \frac{n^3}{2^n} = 26$.

2.18. Let $k \in \mathbb{N}$. Prove that

$$\sum_{n=1}^{\infty} \frac{1}{n(n+k)(n+2k)} = \frac{1}{2k^2}(2H_k - H_{2k}).$$

A gem with a numerical series.

We prove that $\displaystyle\sum_{n=1}^{\infty} \frac{1}{n(n+1)(n+2)} = \frac{1}{4}$.

Since $\dfrac{1}{n(n+1)(n+2)} = \dfrac{1}{2}\left[\dfrac{1}{n(n+1)} - \dfrac{1}{(n+1)(n+2)}\right]$ we have that

---

(continued)

$$\sum_{n=1}^{\infty} \frac{1}{n(n+1)(n+2)} = \frac{1}{2}\left[\sum_{n=1}^{\infty}\frac{1}{n(n+1)} - \sum_{n=1}^{\infty}\frac{1}{(n+1)(n+2)}\right]$$

$$= \frac{1}{2}\left[\sum_{n=1}^{\infty}\frac{1}{n(n+1)} - \left(\sum_{i=1}^{\infty}\frac{1}{i(i+1)} - \frac{1}{2}\right)\right]$$

$$= \frac{1}{4}.$$

This is a particular case of Problem 2.18 and also, of part (a) of Problem 2.19.

2.19. Let $n \in \mathbb{N}$. Prove that:

(a) [2] $\displaystyle\sum_{k=1}^{\infty} \frac{1}{k(k+1)(k+2)\cdots(k+n)} = \frac{1}{n\cdot n!}$;

(b) $\displaystyle\sum_{k=1}^{\infty} \frac{1}{k^2(k+1)(k+2)\cdots(k+n)} = \frac{1}{n!}\left(\zeta(2) - 1 - \frac{1}{2^2} - \cdots - \frac{1}{n^2}\right).$

2.20. [17]

(a) Let $n \in \mathbb{N}$, $n \geq 2$. Check that

$$\frac{1}{C_{n+k}^n} = \frac{n!}{n-1}\left(\frac{k!}{(n+k-1)!} - \frac{(k+1)!}{(n+k)!}\right).$$

(b) Calculate $\displaystyle\sum_{k=0}^{\infty} \frac{1}{C_{n+k}^n}$.

2.21. Prove that

$$\int_0^1 x\left\{\frac{1}{x}\right\}\,dx = 1 - \frac{\zeta(2)}{2},$$

where $\{a\}$ is the fractional part of a.

2.22. Let $k \in \mathbb{N}$. Prove that

$$\int_0^1 \frac{dx}{\left\lfloor \frac{k}{x} \right\rfloor} = \begin{cases} \zeta(2) - 1 & \text{if } k = 1 \\ k \left(\zeta(2) - 1 - \frac{1}{2^2} - \cdots - \frac{1}{(k-1)^2} \right) - 1 & \text{if } k \geq 2, \end{cases}$$

where $\lfloor a \rfloor$ is the floor of a.

2.23. Let $n \in \mathbb{N}$. Prove that

$$\int_0^1 \frac{dx}{\left\lfloor \frac{1}{x} \right\rfloor^n} = \zeta(n+1) + \sum_{j=2}^{n} (-1)^{j+n+1} \zeta(j) + (-1)^n,$$

where $\lfloor a \rfloor$ is the floor of a.

2.24. [102] Prove that

$$\sum_{n=1}^{\infty} \frac{1}{\sqrt{n(n+1)}} < 2.$$

For the general case, see Problem 7.51.

2.25. [53] [O. Furdui, 2016] Let $(a_n)_{n \geq 1}$ be a strictly increasing sequence of natural numbers. Prove that the series

$$\sum_{n=1}^{\infty} \frac{\sqrt{a_n}}{[a_n, a_{n+1}]}$$

converges. Here $[a, b]$ denotes the least common multiple of a and b.

2.26. Prove that

$$\sum_{n=1}^{\infty} \frac{1}{n(n+1)} \left(1 - \frac{1}{2} + \frac{1}{3} - \cdots + \frac{(-1)^{n-1}}{n} \right) = \frac{\pi^2}{12}.$$

2.27. A Criterion of Mircea Ivan Let $(x_n)_{n \geq 1}$ be a sequence of positive real numbers such that $\lim_{n \to \infty} \left(\frac{x_{n+1}}{x_n} \right)^n < 1$. Prove that $\lim_{n \to \infty} x_n = 0$.

2.28. *A series that equals* 1. Prove that

$$\sum_{n=1}^{\infty} \frac{1 \cdot 3 \cdots (2n-1)}{2 \cdot 4 \cdots (2n)} \cdot \frac{1}{n+1} = 1.$$

2.29. Let $p, q \in \mathbb{R}$, with $0 < q < p$.

(a) Prove that

$$\lim_{n \to \infty} \frac{q(q+p)(q+2p) \cdots (q+np)}{p(2p) \cdots ((n+1)p)} = 0.$$

(b) [130, p. 896] Prove that

$$\sum_{n=1}^{\infty} \frac{q(q+p)(q+2p) \cdots (q+np-p)}{p(2p) \cdots (np)} \cdot \frac{1}{n+1} = \frac{q}{p-q}.$$

For the above problem, see also Problem 2.64

2.30. *A product with radicals.*

(a) Check that $n^4 + 4 = \left[(n+1)^2 + 1\right]\left[(n-1)^2 + 1\right]$, $\forall n \geq 1$.

(b) Prove that $\displaystyle\prod_{n=1}^{\infty} \frac{n^2+1}{\sqrt{n^4+4}} = \sqrt{2}$.

2.31. *A zeta series that equals* 1. Prove that $\displaystyle\sum_{n=2}^{\infty} (\zeta(n) - 1) = 1$.

2.32. Prove that

$$\sum_{n=2}^{\infty} (-1)^n (\zeta(n) - 1) = \frac{1}{2} \quad \text{and} \quad \sum_{n=1}^{\infty} (\zeta(2n) - 1) = \frac{3}{4}.$$

2.33. Prove that $\displaystyle\sum_{n=2}^{\infty} (-1)^n (\zeta(n+1) - \zeta(n)) = \zeta(2) - 2$.

2.34.

(a) Let $p \geq 3$ be a prime number. Prove that

$$\sum_{n=1}^{\infty}(-1)^{n-1}\frac{(n, p)}{n} = \frac{2p-1}{p}\ln 2.$$

(b) Prove that

$$\sum_{n=1}^{\infty}\frac{(n, 4)}{n^2} = \frac{11}{8}\zeta(2) \quad \text{and} \quad \sum_{n=1}^{\infty}\frac{(n, p)}{n^2} = \left(1 + \frac{1}{p} - \frac{1}{p^2}\right)\zeta(2),$$

where $p \geq 2$ is a prime number and (a, b) is the greatest common divisor of a and b.

Stirling's Formula for Approximating the Factorial

If $n \in \mathbb{N}$, then for large values of n we have

$$n! \sim \sqrt{2\pi n}\left(\frac{n}{e}\right)^n.$$

We mention that $a_n \sim b_n$, as $n \to \infty$, if $\lim_{n\to\infty}\frac{a_n}{b_n}$ exists and is finite. For reference materials about Stirling's formula the reader is referred to [26, p. 259].

2.35. Prove that

$$\sum_{n=1}^{\infty}\left(n\ln\left(1 + \frac{1}{n}\right) - 1 + \frac{1}{2n}\right) = 1 + \frac{\gamma}{2} - \ln\sqrt{2\pi}$$

and

$$\sum_{n=2}^{\infty}\left(n\ln\left(1 - \frac{1}{n}\right) + 1 + \frac{1}{2n}\right) = -\frac{3}{2} + \frac{\gamma}{2} + \ln\sqrt{2\pi}.$$

2.36. Prove that

$$\sum_{n=2}^{\infty}\left[n\ln\left(1 - \frac{1}{n^2}\right) + \frac{1}{n}\right] = \gamma - \ln 2$$

and

$$\sum_{n=2}^{\infty}\left(n\ln\frac{n+1}{n-1} - 2\right) = 3 - \ln(4\pi).$$

2.37. An Alternating Logarithmic Series Prove that

$$\sum_{n=2}^{\infty}(-1)^n \ln\left(1-\frac{1}{n^2}\right) = \ln\frac{8}{\pi^2}.$$

2.38. Prove that

$$\sum_{n=1}^{\infty} \ln\left(1+\frac{1}{2n}\right) \ln\left(1+\frac{1}{2n+1}\right) = \frac{\ln^2 2}{2}.$$

2.39. [151] Prove that

$$\sum_{n=1}^{\infty} \ln\left(1+\frac{1}{n}\right) \ln\left(1+\frac{1}{2n}\right) \ln\left(1+\frac{1}{2n+1}\right) = \frac{\ln^3 2}{3}.$$

2.40. *Remarkable harmonic series.*

Prove that:

(a) $\displaystyle\sum_{n=1}^{\infty} \frac{H_n}{n(n+1)} = \zeta(2);$

(b) $\displaystyle\sum_{n=1}^{\infty} \frac{H_{n+1}}{n(n+1)} = 2;$

(c) $\displaystyle\sum_{n=1}^{\infty} \frac{H_n}{(n+1)(n+2)} = 1;$

(d) $\displaystyle\sum_{n=1}^{\infty} \frac{H_n^2}{(n+1)(n+2)} = \zeta(2)+1.$

2.41. *Two series with Catalan's number.*

Let $C_n = \frac{1}{n+1}C_{2n}^n$ be the *n*th Catalan number. Prove that:

(a) $\displaystyle\int_0^1 x^n(1-x)^{n-1}dx = \frac{n!(n-1)!}{(2n)!}, \quad n \ge 1;$

(b) $\displaystyle\sum_{n=0}^{\infty} \frac{1}{C_n} = 2 + \frac{4\sqrt{3}}{27}\pi;$

(c) $\displaystyle\sum_{n=0}^{\infty} \frac{2^n}{C_n} = 5 + \frac{3}{2}\pi.$

2.42. Let $k, p \ge 1$ be integers such that $k+1 < p$. Prove that

$$\sum_{n=1}^{\infty} \frac{n^k}{(n+1)^p} = \sum_{i=0}^{k}(-1)^{k-i}C_k^i\zeta(p-i).$$

2.2   Applications of Abel's Summation Formula

Abel's Summation Formula [26, p. 258]

Let $(a_n)_{n \geq 1}$ and $(b_n)_{n \geq 1}$ be two sequences of real or complex numbers and let $A_n = \sum_{k=1}^{n} a_k$. Then

(a) *the finite version*

$$\sum_{k=1}^{n} a_k b_k = A_n b_{n+1} + \sum_{k=1}^{n} A_k (b_k - b_{k+1});$$

(b) *the infinite version*

$$\sum_{k=1}^{\infty} a_k b_k = \lim_{n \to \infty} A_n b_{n+1} + \sum_{k=1}^{\infty} A_k (b_k - b_{k+1}).$$

2.43.   Prove that:

(a) $\displaystyle \lim_{n \to \infty} \left(\frac{\cos 1}{n+1} + \frac{\cos 2}{n+2} + \cdots + \frac{\cos n}{2n} \right) = 0;$

(b) $\displaystyle \lim_{n \to \infty} \left(\frac{\cos^2 1}{n+1} + \frac{\cos^2 2}{n+2} + \cdots + \frac{\cos^2 n}{2n} \right) = \ln \sqrt{2}.$

Remark 2.3. More generally, it can be proved that if $k \geq 1$ is an integer, then

• $\displaystyle \lim_{n \to \infty} \left(\frac{\cos^{2k-1} 1}{n+1} + \frac{\cos^{2k-1} 2}{n+2} + \cdots + \frac{\cos^{2k-1} n}{2n} \right) = 0;$

• $\displaystyle \lim_{n \to \infty} \left(\frac{\cos^{2k} 1}{n+1} + \frac{\cos^{2k} 2}{n+2} + \cdots + \frac{\cos^{2k} n}{2n} \right) = \frac{C_{2k}^{k}}{2^{2k}} \ln 2.$

2.44.   [O. Furdui, Problem 4, SEEMOUS 2017]

Let $k \in \mathbb{N}$ and let $(x_n)_{n \geq k}$ be the sequence defined by

(continued)

$$x_n = \sum_{i=k}^{n} C_i^k \left(e - 1 - \frac{1}{1!} - \frac{1}{2!} - \cdots - \frac{1}{i!} \right).$$

Prove that:

(a) $(x_n)_{n \geq k}$ is strictly increasing;
(b) $(x_n)_{n \geq k}$ is bounded;
(c) $\displaystyle \lim_{n \to \infty} x_n = \frac{e}{(k+1)!}$.

2.45. A Bouquet of Exponential Series
Let $x \in \mathbb{R}$. Prove that:

(a) $\displaystyle \sum_{n=0}^{\infty} \left(e^x - 1 - \frac{x}{1!} - \frac{x^2}{2!} - \cdots - \frac{x^n}{n!} \right) = xe^x$;

(b) $\displaystyle \sum_{n=1}^{\infty} n \left(e^x - 1 - \frac{x}{1!} - \frac{x^2}{2!} - \cdots - \frac{x^n}{n!} \right) = \frac{x^2}{2} e^x$;

(c) $\displaystyle \sum_{n=1}^{\infty} n^2 \left(e^x - 1 - \frac{x}{1!} - \frac{x^2}{2!} - \cdots - \frac{x^n}{n!} \right) = \left(\frac{x^3}{3} + \frac{x^2}{2} \right) e^x$.

2.46. Prove that

$$\sum_{n=1}^{\infty} (-1)^{n-1} \left(\frac{1}{n^2} + \frac{1}{(n+1)^2} + \cdots \right) = \frac{\pi^2}{8}.$$

2.47. [20 October 2018]
Let $k \geq 0$ be an integer. Prove that:

(a) $\displaystyle \sum_{n=1}^{\infty} \left(\zeta(2) - 1 - \frac{1}{2^2} - \cdots - \frac{1}{n^2} - \frac{1}{n+k} \right) = H_{k+1} + \frac{k}{k+1} - \zeta(2)$;

(continued)

In particular, when $k = 0$ we have that

$$\sum_{n=1}^{\infty} \left(\zeta(2) - 1 - \frac{1}{2^2} - \cdots - \frac{1}{n^2} - \frac{1}{n} \right) = 1 - \zeta(2);$$

(b) [82] $\displaystyle\sum_{n=1}^{\infty} \left[\left(\frac{1}{n^2} + \frac{1}{(n+1)^2} + \cdots \right) - \frac{1}{n+k} \right] = H_{k+1} + \frac{k}{k+1};$

In particular, when $k = 0$ we have that

$$\sum_{n=1}^{\infty} \left[\left(\frac{1}{n^2} + \frac{1}{(n+1)^2} + \cdots \right) - \frac{1}{n} \right] = 1.$$

2.48.

(a) Let $n \in \mathbb{N}$ and $k > 0$ be a real number. Prove that

$$\frac{1}{n^2} + \frac{1}{(n+k)^2} + \frac{1}{(n+2k)^2} + \cdots = -\int_0^1 \frac{x^{n-1}}{1-x^k} \ln x \, dx.$$

(b) Let $k > 0$. Prove that

$$\lim_{n \to \infty} n \left(\frac{1}{n^2} + \frac{1}{(n+k)^2} + \frac{1}{(n+2k)^2} + \cdots \right) = \frac{1}{k}.$$

(c) [67] *A series with gaps.*
   Prove that

$$\sum_{n=1}^{\infty} \left[\left(\frac{1}{n^2} + \frac{1}{(n+2)^2} + \frac{1}{(n+4)^2} + \cdots \right) - \frac{1}{2n} \right] = \frac{\pi^2}{16} + \frac{1}{2}.$$

A challenge. Let $k \in \mathbb{N}$. Calculate

$$S_k = \sum_{n=1}^{\infty} \left[\left(\frac{1}{n^2} + \frac{1}{(n+k)^2} + \frac{1}{(n+2k)^2} + \cdots \right) - \frac{1}{kn} \right].$$

We mention that $S_1 = 1$, $S_2 = \frac{\pi^2}{16} + \frac{1}{2}$, and $S_4 = \frac{5\pi^2}{64} + \frac{1}{4} + \frac{G}{4}$.

2.49. *An alternating series with gaps.*
Prove that

$$\sum_{n=1}^{\infty}(-1)^{n-1}\left(\frac{1}{n^2}+\frac{1}{(n+2)^2}+\frac{1}{(n+4)^2}+\cdots\right)=\frac{\pi^2}{16}+\frac{\ln 2}{2}.$$

2.50.

(a) Let $n\in\mathbb{N}$ and $k>0$ be a real number. Prove that

$$\frac{1}{n^3}+\frac{1}{(n+k)^3}+\frac{1}{(n+2k)^3}+\cdots=\frac{1}{2}\int_0^1\frac{x^{n-1}}{1-x^k}\ln^2 x\,dx.$$

(b) [90] Let $k>0$. Prove that

$$\lim_{n\to\infty}n^2\left(\frac{1}{n^3}+\frac{1}{(n+k)^3}+\frac{1}{(n+2k)^3}+\cdots\right)=\frac{1}{2k}.$$

(c) *A series with gaps, $\zeta(2)$ and $\zeta(3)$.*

Prove that

$$\sum_{n=1}^{\infty}\left(\frac{1}{n^3}+\frac{1}{(n+2)^3}+\frac{1}{(n+4)^3}+\cdots\right)=\frac{\pi^2}{12}+\frac{7}{16}\zeta(3).$$

Open problem. Let $k>0$. Calculate

$$\sum_{n=1}^{\infty}\left(\frac{1}{n^3}+\frac{1}{(n+k)^3}+\frac{1}{(n+2k)^3}+\cdots\right).$$

Wallis Formula

$$\lim_{n\to\infty}\left(\frac{2}{1}\cdot\frac{4}{3}\cdots\frac{2n}{2n-1}\right)^2\cdot\frac{1}{2n+1}=\frac{\pi}{2}.$$

(continued)

Remark 2.4. The above formula implies that

$$\sum_{n=1}^{\infty}(-1)^{n-1}\ln\frac{n+1}{n}=\ln\frac{\pi}{2}.$$

2.51. A Wallis Product

Let $(x_n)_{n\geq2}$ be the sequence defined by

$$x_n=\sum_{k=2}^{n}(-1)^k\ln\left(1+\frac{(-1)^k}{k}\right)-\ln n.$$

Prove that:

(a) $x_{2n}=x_{2n+1},\ \forall n\geq1$;

(b) $\lim_{n\to\infty}x_n=\ln\frac{2}{\pi}$;

(c) $\sum_{k=2}^{\infty}\left((-1)^k\ln\left(1+\frac{(-1)^k}{k}\right)+\ln\left(1-\frac{1}{k}\right)\right)=\ln\frac{2}{\pi}.$

Remark 2.5. Part (c) of the problem has the following equivalent formulation:

$$\prod_{k=2}^{\infty}\left(\frac{k}{k+(-1)^k}\right)^{(-1)^k}\frac{k}{k-1}=\frac{\pi}{2}.$$

2.52. [10, p. 277] *An alternating series of Hardy*

Prove that

$$\sum_{n=1}^{\infty}(-1)^n\left(1+\frac{1}{2}+\frac{1}{3}+\cdots+\frac{1}{n}-\ln n-\gamma\right)=\frac{\gamma-\ln\pi}{2}.$$

2.53. Prove that:

(a) $\sum_{n=1}^{\infty}(-1)^n\left(1+\frac{1}{2}+\frac{1}{3}+\cdots+\frac{1}{n}-\ln\sqrt{n(n+1)}-\gamma\right)=\frac{\gamma-\ln2}{2}$;

(continued)

(b) $\displaystyle\sum_{n=1}^{\infty}(-1)^n \left(1 + \frac{1}{2} + \frac{1}{3} + \cdots + \frac{1}{n} - \ln \sqrt[3]{n(n+1)(n+2)} - \gamma\right)$

$= \dfrac{\gamma}{2} - \dfrac{\ln \pi}{6};$

(c) [43] $\displaystyle\sum_{n=1}^{\infty}\left(1 + \frac{1}{2} + \frac{1}{3} + \cdots + \frac{1}{n} - \ln \sqrt{n(n+1)} - \gamma\right)$

$= \dfrac{1}{2} - \ln \sqrt{2\pi} + \gamma.$

2.54. [50]

(a) Prove that

$$\sum_{n=1}^{\infty}\left(1 + \frac{1}{2} + \cdots + \frac{1}{n} - \ln\left(n + \frac{1}{2}\right) - \gamma\right) = \gamma + \frac{1}{2} - \frac{3}{2}\ln 2.$$

A challenge. Prove that

$$\sum_{n=1}^{\infty}(-1)^{n-1}\left(1 + \frac{1}{2} + \cdots + \frac{1}{n} - \ln\left(n + \frac{1}{2}\right) - \gamma\right)$$

$$= -\frac{\gamma}{2} + \frac{5}{2}\ln 2 + \ln \pi - 2\ln\Gamma\left(\frac{1}{4}\right).$$

(b) Let $p \in \mathbb{N}$. Prove that

$$\sum_{k=1}^{\infty}\left(1 + \frac{1}{2} + \cdots + \frac{1}{k} - \ln\left(k + \frac{1}{p}\right) - \gamma + \frac{2-p}{2pk}\right)$$

$$= \gamma\left(\frac{1}{2} + \frac{1}{p}\right) + \frac{1}{2} - \ln \sqrt{2\pi} + \ln\Gamma\left(1 + \frac{1}{p}\right),$$

where Γ denotes the Gamma function.

2.55. Let $O_n = 1 + \frac{1}{3} + \cdots + \frac{1}{2n-1}$, $n \in \mathbb{N}$. Prove that:

(a) $\displaystyle\sum_{n=1}^{\infty} \left(O_n - \frac{\gamma}{2} - \ln 2 - \frac{\ln n}{2} \right) = \frac{1}{4} \left(\gamma + \ln \frac{2}{\pi} \right)$;

(b) $\displaystyle\sum_{n=1}^{\infty} (-1)^n \left(O_n - \frac{\gamma}{2} - \ln 2 - \frac{\ln n}{2} \right) = \frac{1}{4} \left(\gamma + \ln \frac{8}{\pi} - \frac{\pi}{2} \right)$.

2.56. [31] **A Harmonic Series with the Tail of $\zeta(3)$**

(a) Prove that for any $n \in \mathbb{N}$ we have

$$H_1 + H_2 + \cdots + H_n = (n+1)H_n - n.$$

(b) Prove that

$$\sum_{n=1}^{\infty} H_n \left(\zeta(3) - 1 - \frac{1}{2^3} - \cdots - \frac{1}{n^3} \right) = 2\zeta(3) - \zeta(2),$$

where $H_n = 1 + \frac{1}{2} + \cdots + \frac{1}{n}$ denotes the nth harmonic number.

Remark 2.6. Other spectacular series involving the nth harmonic number H_n and tails of Riemann zeta function values, of which some are listed below

$$\sum_{n=1}^{\infty} H_n \left(\zeta(4) - 1 - \frac{1}{2^4} - \cdots - \frac{1}{n^4} \right) = \frac{5}{4}\zeta(4) - \zeta(3),$$

$$\sum_{n=1}^{\infty} n H_n \left(\zeta(4) - 1 - \frac{1}{2^4} - \cdots - \frac{1}{n^4} \right) = -\frac{1}{4}\zeta(2) + \frac{5}{4}\zeta(3) - \frac{5}{8}\zeta(4),$$

$$\sum_{n=1}^{\infty} H_n^2 \left(\zeta(3) - 1 - \frac{1}{2^3} - \cdots - \frac{1}{n^3} \right) = 3\zeta(4) - 4\zeta(3) + 2\zeta(2),$$

can be found in [30].

2.3 Series with Positive Terms

Comparison Criteria for Series with Positive Terms

A. Let $\sum_{n=1}^{\infty} a_n$ and $\sum_{n=1}^{\infty} b_n$ be two series with positive terms with the property that there exist $\alpha > 0$ and $n_0 \in \mathbb{N}$ such that $a_n \leq \alpha b_n$, for all $n \in \mathbb{N}$, $n \geq n_0$.

  (a) If the series $\sum_{n=1}^{\infty} b_n$ converges, then the series $\sum_{n=1}^{\infty} a_n$ converges.

  (b) If the series $\sum_{n=1}^{\infty} a_n$ diverges, then the series $\sum_{n=1}^{\infty} b_n$ diverges.

B. Let $\sum_{n=1}^{\infty} a_n$ and $\sum_{n=1}^{\infty} b_n$ be two series with positive terms such that the limit exists $\lim_{n \to \infty} \frac{a_n}{b_n} = l \in [0, \infty]$.

  (a) If $l \in (0, \infty)$, then the series $\sum_{n=1}^{\infty} a_n$ and $\sum_{n=1}^{\infty} b_n$ have the same nature.

  (b) If $l = 0$, then the convergence of the series $\sum_{n=1}^{\infty} b_n$ implies the convergence of the series $\sum_{n=1}^{\infty} a_n$.

  (c) If $l = \infty$, then the convergence of the series $\sum_{n=1}^{\infty} a_n$ implies the convergence of the series $\sum_{n=1}^{\infty} b_n$.

2.57. Study the nature of the series:

(a) $\sum_{n=1}^{\infty} \frac{1}{\sqrt[3]{n^5 + 1}}$;

(b) $\sum_{n=1}^{\infty} \frac{1}{\sqrt{n(n + 2)}}$;

(c) $\sum_{n=1}^{\infty} \frac{1}{\ln(n + 1)}$;

(d) $\sum_{n=1}^{\infty} \ln\left(1 + \frac{1}{n^3}\right)$;

(e) $\sum_{n=1}^{\infty} \arctan \frac{1}{n}$;

(f) $\sum_{n=1}^{\infty} \frac{1}{n} \sin \frac{\pi}{2n}$;

(g) $\sum_{n=1}^{\infty} \frac{1}{n\sqrt[n]{n + 1}}$;

(h) $\sum_{n=1}^{\infty} \frac{n^2}{n(n + 1)5^n + 3}$;

(i) $\sum_{n=1}^{\infty} \left(1 - \cos \frac{\pi}{n}\right)$.

2.58. Convergence of Series with Norms of Vectors in $\mathbb{R}^n$

Let $v = (v_1, v_2, \ldots, v_n) \in \mathbb{R}^n$, let $||v||_p = \sqrt[p]{|v_1|^p + |v_2|^p + \cdots + |v_p|^p}$ and $||v||_\infty = \max_{1 \le i \le n} |v_i|$. These are the usual p-norm and ∞-norm on $\mathbb{R}^n$.

(a) Prove that

$$\lim_{p \to \infty} p \left(||v||_p - ||v||_\infty \right) = ||v||_\infty \ln k,$$

where k is the number of entries of v which are equal to either $-||v||_\infty$ or $||v||_\infty$.

(b) [19] If $k = 1$, then prove that the series

$$\sum_{p=1}^{\infty} p^\alpha \left(||v||_p - ||v||_\infty \right)$$

converges for all $\alpha \ge 0$.

   Application. Prove that the series

$$\sum_{n=1}^{\infty} n^\alpha \left(\sqrt[n]{1 + 2^n + 3^n + \cdots + 2021^n} - 2021 \right)$$

converges for all $\alpha \ge 0$.

2.59. [25 April 2020] Let $k \in \mathbb{N}$ and $a_1, a_2, \ldots, a_k > 0$. Prove that

$$\lim_{n \to \infty} n \left[\left(\frac{\sqrt[n]{a_1} + \sqrt[n]{a_2} + \cdots + \sqrt[n]{a_k}}{k} \right)^n - \sqrt[k]{a_1 a_2 \cdots a_k} \right]$$

$$= \frac{\sqrt[k]{a_1 a_2 \cdots a_k}}{2} \left(\ln^2 a_1 + \ln^2 a_2 + \cdots + \ln^2 a_k - \ln^2 \sqrt[k]{a_1 a_2 \cdots a_k} \right).$$

Application. Study the nature of the series

$$\sum_{n=1}^{\infty} \left[\left(\frac{\sqrt[n]{a_1} + \sqrt[n]{a_2} + \cdots + \sqrt[n]{a_k}}{k} \right)^n - \sqrt[k]{a_1 a_2 \cdots a_k} \right].$$

A Consequence of d'Alembert's Ratio Test for Series with Positive Terms

Let $\sum\limits_{n=1}^{\infty} a_n$ be a series with positive terms such that the limit exists $\lim\limits_{n\to\infty} \frac{a_{n+1}}{a_n} = l \in [0, \infty]$.

(a) If $l < 1$, then the series $\sum\limits_{n=1}^{\infty} a_n$ converges.

(b) If $l > 1$, then the series $\sum\limits_{n=1}^{\infty} a_n$ diverges.

If $l = 1$, then the test is inconclusive.

2.60. Study the nature of the series:

(a) $\sum\limits_{n=1}^{\infty} \dfrac{a^n}{n^p}, \quad a > 0, \ p \in \mathbb{R};$

(b) $\sum\limits_{n=1}^{\infty} \dfrac{a^n}{n(7^n + 5^n)}, \quad a > 0;$

(c) $\sum\limits_{n=1}^{\infty} \dfrac{3^{2n}(n!)^3}{(3n)!};$

(d) $\sum\limits_{n=1}^{\infty} \dfrac{1 \cdot 4 \cdot 7 \cdots (3n-2)}{2^{2n} \cdot n!}.$

[12] A Convergence Criterion of Chen

Let $(a_n)_{n\geq 1}$ be a strictly decreasing sequence of positive numbers such that the limit exists

$$\lim_{n\to\infty} \frac{a_{2n}}{a_n} = l \in [0, \infty].$$

(a) If $l < \frac{1}{2}$, then the series $\sum\limits_{n=1}^{\infty} a_n$ converges.

(b) If $l > \frac{1}{2}$, then the series $\sum\limits_{n=1}^{\infty} a_n$ diverges.

If $l = \frac{1}{2}$, then the test is inconclusive.

2.61. Study the nature of the series:

(a) $\displaystyle\sum_{n=1}^{\infty} \frac{1}{3\sqrt{n}}$;   (b) $\displaystyle\sum_{n=2}^{\infty} \frac{1}{(\ln n)^{\ln n}}$;   (c) $\displaystyle\sum_{n=2}^{\infty} \frac{1}{(\ln n)^{\ln(\ln n)}}$.

A Consequence of Cauchy's Root Test for Series with Positive Terms

Let $\displaystyle\sum_{n=1}^{\infty} a_n$ be a series with positive terms such that the following limit exists
$\displaystyle\lim_{n\to\infty} \sqrt[n]{a_n} = l \in [0, \infty]$.

(a) If $l < 1$, then the series $\displaystyle\sum_{n=1}^{\infty} a_n$ converges.

(b) If $l > 1$, then the series $\displaystyle\sum_{n=1}^{\infty} a_n$ diverges.

If $l = 1$, then the test is inconclusive.

2.62. Study the nature of the series:

(a) $\displaystyle\sum_{n=1}^{\infty} \left(\frac{an^2}{n^2 - n + 1}\right)^n$, $a > 0$;   (b) $\displaystyle\sum_{n=1}^{\infty} \frac{1}{3^n}\left(1 + \frac{n}{n^2 + 1}\right)^{n^2}$;

(c) $\displaystyle\sum_{n=1}^{\infty} \left(n \sin \frac{1}{2n}\right)^n$;   (d) $\displaystyle\sum_{n=1}^{\infty} n^{n+1} \ln^n \left(1 + \frac{1}{3n}\right)$.

A Consequence of Raabe–Duhamel's Test for Series with Positive Terms

Let $\displaystyle\sum_{n=1}^{\infty} a_n$ be a series with positive terms such that the following limit exists
$\displaystyle\lim_{n\to\infty} n\left(\frac{a_n}{a_{n+1}} - 1\right) = l \in [0, \infty]$.

(a) If $l > 1$, then the series $\displaystyle\sum_{n=1}^{\infty} a_n$ converges.

(b) If $l < 1$, then the series $\displaystyle\sum_{n=1}^{\infty} a_n$ diverges.

If $l = 1$, then the test is inconclusive.

Series for which d'Alembert and Raabe–Duhamel's tests fail to hold:

- convergent $\sum_{n=1}^{\infty} \frac{1}{n H_n^k}$, $\sum_{n=2}^{\infty} \frac{1}{n \ln^k n}$, $k \in \mathbb{R}$, $k > 1$;
- divergent $\sum_{n=2}^{\infty} \frac{1}{n \ln n}$, $\sum_{n=1}^{\infty} n^{x-2} \left(\zeta(x) - 1 - \frac{1}{2^x} - \cdots - \frac{1}{n^x} \right)$, $x \in \mathbb{R}$, $x \geq 2$.

2.63. Study the nature of the series[2]:

(a) $\displaystyle\sum_{n=1}^{\infty} \frac{(2n-1)!}{2^{2n-1}(n-1)!(n+1)!}$; (b) $\displaystyle\sum_{n=1}^{\infty} \frac{1 \cdot 3 \cdot 5 \cdots (2n-1)}{2 \cdot 4 \cdot 6 \cdots (2n)} \cdot \frac{1}{2n+1}$;

(c) $\displaystyle\sum_{n=1}^{\infty} a^{\ln n}$, $a > 0$; (d) $\displaystyle\sum_{n=1}^{\infty} \frac{1 \cdot 5 \cdot 9 \cdots (4n-3)}{4^n \cdot a(a+1) \cdots (a+n-1)}$, $a > 0$;

(e) $\displaystyle\sum_{n=1}^{\infty} x^{1+\frac{1}{2}+\cdots+\frac{1}{n}}$, $x > 0$; (f) $\displaystyle\sum_{n=1}^{\infty} x^{\sin 1 + \sin \frac{1}{2} + \cdots + \sin \frac{1}{n}}$, $x > 0$.

2.64. Let $a, b, c > 0$, with $b > c$.

(a) Prove that

$$\frac{c(a+c) \cdots (an+c-a)}{b(a+b) \cdots (an+b-a)} \leq \frac{e^{-\frac{b-c}{\max\{a,b\}} (H_n - \ln n)}}{n^{\frac{b-c}{\max\{a,b\}}}}, \quad \forall n \geq 1.$$

(b) Let

$$a_n = \frac{c(a+c) \cdots (an+c-a)}{b(a+b) \cdots (an+b-a)} \cdot \frac{1}{an+b}, \quad n \geq 1.$$

Prove that

$$a_n = \frac{1}{b-c} [a_{n-1}(an+c-a) - a_n(an+c)], \quad \forall n \geq 2.$$

(continued)

---

[2] The sum of the series in part (b) is equal to $\frac{\pi}{2} - 1$. See the power series expansion of arcsin. The sum of the series in part (d) is equal to $\frac{1}{4a-5}$, when $a > \frac{5}{4}$. See Problem 2.64 for the general case.

(c) **A Telescoping Series** Prove that the series $\sum_{n=1}^{\infty} a_n$ converges and

$$\sum_{n=1}^{\infty} \frac{c(a+c)\cdots(an+c-a)}{b(a+b)\cdots(an+b-a)} \cdot \frac{1}{an+b} = \frac{c}{b(b-c)}.$$

Applications. Prove that:

(1) $\displaystyle\sum_{n=1}^{\infty} \frac{1\cdot 4\cdots(3n-2)}{2\cdot 5\cdots(3n-1)} \cdot \frac{1}{3n+2} = \frac{1}{2}$;

(2) $\displaystyle\sum_{n=1}^{\infty} \frac{1\cdot 5\cdots(4n-3)}{2\cdot 6\cdots(4n-2)} \cdot \frac{1}{4n+2} = \frac{1}{2}$;

(3) $\displaystyle\sum_{n=1}^{\infty} \frac{2\cdot 6\cdots(4n-2)}{3\cdot 7\cdots(4n-1)} \cdot \frac{1}{4n+3} = \frac{2}{3}$;

(4) [81] $\displaystyle\sum_{n=1}^{\infty} \frac{3\cdot 6\cdots 3n}{7\cdot 10\cdots(3n+4)} \cdot \frac{1}{3n+7} = \frac{3}{28}$.

2.65.

(a) [7, problem 3.35, p. 24] Let a, b, c be positive numbers. Prove that the series

$$\sum_{n=1}^{\infty} \left(a^{\frac{1}{n}} - \frac{b^{\frac{1}{n}} + c^{\frac{1}{n}}}{2} \right)$$

converges if and only if $a^2 = bc$.

(b) [51] Let a, b, c be positive numbers. Prove that the series

$$\sum_{n=1}^{\infty} \left[n \cdot \left(a^{\frac{1}{n}} - \frac{b^{\frac{1}{n}} + c^{\frac{1}{n}}}{2} \right) - \ln \frac{a}{\sqrt{bc}} \right]$$

converges if and only if $2 \ln^2 a = \ln^2 b + \ln^2 c$.

Cauchy's Integral Criterion Let $f : [1, \infty) \to (0, \infty)$ be a monotone decreasing function. The series $\sum_{n=1}^{\infty} f(n)$ converges if and only if the sequence $(v_n)_{n\geq 1}$, defined by $v_n = \int_1^n f(x)\,dx$, converges.

2.66. The Generalized Harmonic Series of Riemann

Let $p \in \mathbb{R}$. Prove that the series $\sum_{n=1}^{\infty} \frac{1}{n^p}$ converges for $p > 1$ and diverges for $p \leq 1$.

We record a gem of mathematical analysis regarding the convergence of the generalized harmonic series of Riemann.

- *The generalized harmonic series of Riemann* $\sum_{n=1}^{\infty} \frac{1}{n^p}$, $p \in (1, \infty)$, *converges.*

$$1 + \frac{1}{2^p} + \frac{1}{3^p} + \cdots + \frac{1}{n^p} < \frac{p}{p-1}, \quad \forall n \geq 1 \Rightarrow 1 < \zeta(p) \leq \frac{p}{p-1}$$

- *The generalized harmonic series of Riemann* $\sum_{n=1}^{\infty} \frac{1}{n^p}$, $p \in (0, 1]$, *diverges.*

If the series $\sum_{n=1}^{\infty} \frac{1}{n^p}$ would converge, then

$$\sum_{n=1}^{\infty} \frac{1}{n^p} = \sum_{n=0}^{\infty} \left(\frac{1}{(2n+1)^p} + \frac{1}{(2n+2)^p} \right) > \sum_{n=0}^{\infty} \frac{2}{(2n+2)^p} = \frac{1}{2^{p-1}} \sum_{n=0}^{\infty} \frac{1}{(n+1)^p},$$

and it follows that $2^{p-1} > 1$, which is a contradiction.

2.67. The Harmonic Series Diverges

(a) Use the inequality $e^x \geq 1 + x$, $\forall x \geq 0$, to prove that

$$\left(1 + \frac{1}{1} \right) \left(1 + \frac{1}{2} \right) \cdots \left(1 + \frac{1}{n} \right) \leq e^{1 + \frac{1}{2} + \cdots + \frac{1}{n}}, \quad \forall n \geq 1.$$

(b) Prove, by using part (a) of the problem, that $\sum_{n=1}^{\infty} \frac{1}{n} = \infty$.

Inequalities and Divergent Series

(a) $\dfrac{H_1}{1} + \dfrac{H_2}{2} + \cdots + \dfrac{H_n}{n} > \dfrac{n}{\sqrt[n]{n!}} \sqrt[n]{H_1 H_2 \cdots H_n} \Rightarrow \sum_{n=1}^{\infty} \dfrac{H_n}{n}$ diverges.

(continued)

(b) $\dfrac{\ln 2}{2} + \dfrac{\ln 3}{3} + \cdots + \dfrac{\ln(n+1)}{n+1} > \dfrac{n}{\sqrt[n]{(n+1)!}} \sqrt[n]{\ln 2 \ln 3 \cdots \ln(n+1)}$

$\Rightarrow \displaystyle\sum_{n=2}^{\infty} \dfrac{\ln n}{n}$ diverges.

(c) $\dfrac{1}{\ln 2} + \dfrac{1}{\ln 3} + \cdots + \dfrac{1}{\ln(n+1)} > \dfrac{n^2}{\ln((n+1)!)} \Rightarrow \displaystyle\sum_{n=2}^{\infty} \dfrac{1}{\ln n}$ diverges.

(d) $p \in (0, 1)$

$\dfrac{1}{1^p} + \dfrac{1}{2^p} + \cdots + \dfrac{1}{n^p} > \left(\dfrac{n}{\sqrt[n]{n!}}\right)^p n^{1-p} \Rightarrow \displaystyle\sum_{n=1}^{\infty} \dfrac{1}{n^p}$ diverges.

2.68. Let p be a positive number. Study the nature of the series:

(a) $\displaystyle\sum_{n=2}^{\infty} \dfrac{1}{n \ln n}$;   (b) $\displaystyle\sum_{n=2}^{\infty} \dfrac{1}{n \ln^p n}$;   (c) $\displaystyle\sum_{n=2}^{\infty} \dfrac{\ln n}{n}$;   (d) $\displaystyle\sum_{n=2}^{\infty} \dfrac{\ln n}{n^p}$.

2.69. Study the nature of the series

$$\sum_{n=2}^{\infty} \frac{\ln n}{1 + \sqrt{2} + \sqrt[3]{3} + \cdots + \sqrt[n]{n}}.$$

2.4   Alternating Series

Leibniz's Criterion for Alternating Series

If $(a_n)_{n\geq 1}$ is a monotone decreasing sequence convergent to 0, then the alternating series $\sum_{n=1}^{\infty}(-1)^{n-1}a_n$ converges.

2.70. Determine which of the following series are semiconvergent or absolutely convergent[3]:

---

[3]A series $\sum_{n=1}^{\infty} a_n$ is absolutely convergent if the series $\sum_{n=1}^{\infty} |a_n|$ converges. A series is semiconvergent if the series converges, but it does not converge absolutely. The sum of the series in part (d) is $\gamma \ln 2 - \frac{\ln^2 2}{2}$.

(a) $\displaystyle\sum_{n=1}^{\infty} (-1)^{n-1} \frac{1}{n}$;

(b) $\displaystyle\sum_{n=1}^{\infty} (-1)^{n} \frac{1}{n^2}$;

(c) $\displaystyle\sum_{n=1}^{\infty} (-1)^{n-1} \frac{1}{n\sqrt[n]{n}}$;

(d) $\displaystyle\sum_{n=2}^{\infty} (-1)^{n} \frac{\ln n}{n}$;

(e) $\displaystyle\sum_{n=1}^{\infty} (-1)^{n-1} \frac{(n+1)^{n+1}}{n^{n+2}}$;

(f) $\displaystyle\sum_{n=1}^{\infty} (-1)^{n} \tan \frac{1}{n\sqrt{n}}$.

2.71. Study the nature of the following series:

(a) $\displaystyle\sum_{n=1}^{\infty} \frac{(-1)^n}{n + \sin n}$;

(b) $\displaystyle\sum_{n=1}^{\infty} \frac{(-1)^n}{n + \cos n}$;

(c) $\displaystyle\sum_{n=1}^{\infty} \frac{(-1)^n}{n + (-1)^n \sin n}$;

(d) $\displaystyle\sum_{n=1}^{\infty} \frac{(-1)^n}{n + (-1)^n \cos n}$.

Remark 2.7. We mention that if $(a_n)_{n\geq 1}$ is a sequence which converges to 0, but not monotonically, then the alternating series $\sum_{n=1}^{\infty}(-1)^n a_n$ may or may not be convergent, as the next problem shows.

2.72. Let $a_n = \begin{cases} \frac{1}{n} & \text{if } n \text{ is even} \\ \frac{2}{n} & \text{if } n \text{ is odd} \end{cases}$ and $b_n = \begin{cases} \frac{1}{n^2} & \text{if } n \text{ is even} \\ \frac{2}{n^2} & \text{if } n \text{ is odd} \end{cases}$.

Prove that $\displaystyle\sum_{n=1}^{\infty} (-1)^{n-1} a_n = \infty$ and $\displaystyle\sum_{n=1}^{\infty} (-1)^{n-1} b_n = \frac{5\pi^2}{24}$.

2.73. Let $a, b > 0$, $p \in (0, 1]$ and let $a_n = \begin{cases} \frac{a}{n^p} & \text{if } n \text{ is even} \\ \frac{b}{n^p} & \text{if } n \text{ is odd} \end{cases}$.

Prove that $\displaystyle\sum_{n=1}^{\infty} (-1)^n a_n$ converges if and only if $a = b$.

2.74.

(a) *Botez–Catalan identity.* Let $n \in \mathbb{N}$. Prove that

$$1 - \frac{1}{2} + \frac{1}{3} - \frac{1}{4} + \cdots + \frac{1}{2n-1} - \frac{1}{2n} = \frac{1}{n+1} + \frac{1}{n+2} + \cdots + \frac{1}{2n}.$$

(b) Calculate $\displaystyle\sum_{n=1}^{\infty} (-1)^{n-1} \frac{1}{n}$.

2.75. Prove that

$$\lim_{x \to \infty} x \sum_{n=1}^{\infty} \frac{(-1)^{n-1}}{n+x} = \frac{1}{2}.$$

2.76. Limits of Special Series
Prove that:

(a) $\displaystyle\lim_{h \to 0^+} \sum_{n=1}^{\infty} \frac{h}{1+(nh)^2} = \frac{\pi}{2}$;

(b) $\displaystyle\lim_{h \to 0^+} \sum_{n=1}^{\infty} \frac{h}{1+(2n-1)^2h^2} = \frac{\pi}{4}$;

(c) $\displaystyle\lim_{h \to 0^+} \sum_{n=1}^{\infty} \frac{h}{1+(nh)^k} = \frac{\pi}{k \sin \frac{\pi}{k}}$, where $k \in (1, \infty)$;

(d) $\displaystyle\lim_{h \to 0^+} \sum_{n=1}^{\infty} \frac{h}{1+(2n-1)^k h^k} = \frac{\pi}{2k \sin \frac{\pi}{k}}$, where $k \in (1, \infty)$.

Remark 2.8. M. Ivan (2016) proved that, if $f : [0, \infty) \to \mathbb{R}$ is a decreasing function such that $\int_0^\infty f(x)\mathrm{d}x$ converges, then

$$\lim_{h \to 0^+} \sum_{n=1}^{\infty} hf(nh) = \int_0^\infty f(x)\mathrm{d}x \quad \text{and} \quad \lim_{h \to 0^+} \sum_{n=1}^{\infty} hf((2n-1)h) = \frac{1}{2} \int_0^\infty f(x)\mathrm{d}x.$$

2.5   Series with Harmonic Numbers and Factorials

2.77. Remarkable Series with Harmonic Numbers and Factorials
(a) Check that

$$\frac{(j-1)!}{(i+j)!} H_{i+j} = \frac{1}{i} \left(\frac{(j-1)!}{(i+j-1)!} H_{i+j-1} - \frac{j!}{(i+j)!} H_{i+j} \right)$$

$$+ \frac{1}{i^2} \left(\frac{(j-1)!}{(i+j-1)!} - \frac{j!}{(i+j)!} \right).$$

(b) [148, O. Oloa, 30 December 2005, entry 25]
   Prove that

(continued)

$$\sum_{i=1}^{\infty}\sum_{j=1}^{\infty}\frac{(i-1)!(j-1)!}{(i+j)!}H_{i+j}=3\zeta(3).$$

(c) [O. Furdui, 13 September 2019]
Prove that

$$\sum_{i=1}^{\infty}\sum_{j=1}^{\infty}(-1)^{i-1}\frac{(i-1)!(j-1)!}{(i+j)!}H_{i+j}=\frac{11}{8}\zeta(3).$$

(d) [O. Furdui, 17 September 2019]
An alternating series. Prove that

$$\sum_{i=1}^{\infty}\sum_{j=1}^{\infty}\left((-1)^{i-1}+(-1)^{j-1}\right)\frac{(i-1)!(j-1)!}{(i+j)!}H_{i+j}=\frac{11}{4}\zeta(3).$$

Remark 2.9. One can prove, and this is a challenging problem, that the alternating version of Oloa's series

$$\sum_{i=1}^{\infty}\sum_{j=1}^{\infty}(-1)^{i+j}\frac{(i-1)!(j-1)!}{(i+j)!}H_{i+j}$$

is equal to $3\zeta(3)+2\mathrm{Li}_3\left(-\frac{1}{3}\right)-2\mathrm{Li}_3\left(\frac{1}{3}\right)+2\mathrm{Li}_3\left(-\frac{1}{2}\right)+\frac{2}{3}\ln^3 2-\ln^2 2\ln 3+2\ln 2\mathrm{Li}_2\left(-\frac{1}{2}\right)$.

A Double Series with Harmonic Numbers and Factorials

We record as a gem another proof of Oloa's double series formula

$$\sum_{i=1}^{\infty}\sum_{j=1}^{\infty}\frac{(i-1)!(j-1)!}{(i+j)!}H_{i+j}=3\zeta(3).$$

We have

(continued)

$$\sum_{i=1}^{\infty} \sum_{j=1}^{\infty} \frac{(i-1)!(j-1)!}{(i+j)!} H_{i+j} = \sum_{i=1}^{\infty} (i-1)! \sum_{j=1}^{\infty} \frac{H_{j+i}}{j(j+1)\cdots(j+i)}$$

$$= \sum_{i=1}^{\infty} \frac{(i-1)!}{i \cdot i!} \left(\frac{1}{i} + H_i \right)$$

$$= \sum_{i=1}^{\infty} \frac{1}{i^3} + \sum_{i=1}^{\infty} \frac{H_i}{i^2}$$

$$= 3\zeta(3).$$

We used in the preceding calculations the formula

$$\sum_{n=1}^{\infty} \frac{H_{n+k}}{n(n+1)\cdots(n+k)} = \frac{1}{k \cdot k!} \left(\frac{1}{k} + H_k \right), \quad k \in \mathbb{N},$$

whose solution is given in detail in [26, problem 3.59 (b), p. 149], and Euler's series $\sum_{n=1}^{\infty} \frac{H_n}{n^2} = 2\zeta(3)$. ∎

2.78. Prove that

$$\sum_{i=1}^{\infty} \sum_{j=1}^{\infty} \frac{(i-1)!(j-1)!}{(i+j+1)!} H_{i+j+1} = 6 - \zeta(2) - 3\zeta(3).$$

2.6   A Mosaic of Series

2.79. [63] Prove that

$$\sum_{n=1}^{\infty} (-1)^{n-1} \left[n \left(\zeta(2) - 1 - \frac{1}{2^2} - \cdots - \frac{1}{n^2} \right) - 1 \right] = \frac{\pi^2}{16} - \frac{\ln 2}{2} - \frac{1}{2}.$$

Open problem. Let $k \geq 4$ be an integer. Calculate

$$\sum_{n=1}^{\infty} (-1)^{n-1} \left[n^{k-1} \left(\zeta(k) - 1 - \frac{1}{2^k} - \cdots - \frac{1}{n^k} \right) - \frac{1}{k-1} \right].$$

(continued)

The case when $k = 3$ is part (d) of Problem 2.82.

2.80. [64] **A Nonlinear Harmonic Series**

(a) Check that, for $n \geq 2$, we have

$$\frac{H_n H_{n+1}}{n^3 - n} = \frac{1}{2}\left(\frac{H_{n-1}H_n}{(n-1)n} - \frac{H_n H_{n+1}}{n(n+1)}\right) + \frac{H_{n-1}}{n-1} - \frac{H_n}{n}$$
$$+ \frac{3}{4}\left(\frac{1}{n-1} - \frac{1}{n}\right) - \frac{H_n}{2n^2} - \frac{1}{4}\left(\frac{H_{n-1}}{n-1} - \frac{H_{n+1}}{n+1}\right)$$
$$- \frac{1}{4(n+1)^2}.$$

(b) Prove that

$$\sum_{n=2}^{\infty} \frac{H_n H_{n+1}}{n^3 - n} = \frac{5}{2} - \frac{1}{4}\zeta(2) - \zeta(3).$$

2.81. [27 July 2018]

(a) Prove that

$$\frac{1}{n+1} - \frac{1}{n+2} + \frac{1}{n+3} - \cdots = \int_0^1 \frac{x^n}{1+x}dx, \quad n \in \mathbb{N}.$$

(b) Prove that

$$\sum_{n=1}^{\infty} \frac{H_n}{n}\left(\frac{1}{n+1} - \frac{1}{n+2} + \frac{1}{n+3} - \cdots\right) = \frac{1}{4}\zeta(3) + \frac{1}{2}\zeta(2)\ln 2 + \frac{1}{6}\ln^3 2.$$

(c) Prove that

$$\sum_{n=1}^{\infty}(-1)^n\frac{H_n}{n}\left(\frac{1}{n+1} - \frac{1}{n+2} + \frac{1}{n+3} - \cdots\right) = \frac{1}{4}\zeta(3) - \frac{1}{2}\zeta(2)\ln 2 + \frac{1}{6}\ln^3 2.$$

(continued)

(d) Prove that

$$\sum_{n=1}^{\infty} H_n \left(\frac{1}{n+1} - \frac{1}{n+2} + \frac{1}{n+3} - \cdots \right)^2 = \frac{\pi^2}{12} + \frac{1}{2} \ln^2 2 - \ln 2.$$

A challenge. Calculate

$$\sum_{n=1}^{\infty} (-1)^n H_n \left(\frac{1}{n+1} - \frac{1}{n+2} + \frac{1}{n+3} - \cdots \right)^2.$$

A Gem with an Alternating Quadratic Series

We prove that

$$\sum_{n=1}^{\infty} (-1)^{n-1} \left(\frac{1}{n} - \frac{1}{n+1} + \frac{1}{n+2} - \cdots \right)^2 = \frac{\pi^2}{24}.$$

We have

$$\sum_{n=1}^{\infty} (-1)^{n-1} \left(\frac{1}{n} - \frac{1}{n+1} + \frac{1}{n+2} - \cdots \right)^2$$

$$= \ln^2 2 + \sum_{n=2}^{\infty} (-1)^{n-1} \left(\frac{1}{n} - \frac{1}{n+1} + \frac{1}{n+2} - \cdots \right)^2$$

$$\overset{n-1=m}{=} \ln^2 2 + \sum_{m=1}^{\infty} (-1)^{m} \left(\frac{1}{m+1} - \frac{1}{m+2} + \frac{1}{m+3} - \cdots \right)^2$$

$$= \ln^2 2 - \sum_{m=1}^{\infty} (-1)^{m-1} \left[\frac{1}{m} - \left(\frac{1}{m} - \frac{1}{m+1} + \frac{1}{m+2} - \cdots \right) \right]^2$$

$$= \ln^2 2 + \sum_{m=1}^{\infty} \frac{(-1)^m}{m^2} + 2 \sum_{m=1}^{\infty} \frac{(-1)^{m-1}}{m} \left(\frac{1}{m} - \frac{1}{m+1} + \frac{1}{m+2} - \cdots \right)$$

$$- \sum_{m=1}^{\infty} (-1)^{m-1} \left(\frac{1}{m} - \frac{1}{m+1} + \frac{1}{m+2} - \cdots \right)^2$$

(continued)

and it follows that

$$2\sum_{n=1}^{\infty}(-1)^{n-1}\left(\frac{1}{n}-\frac{1}{n+1}+\frac{1}{n+2}-\cdots\right)^2$$

$$=\ln^2 2+\sum_{n=1}^{\infty}\frac{(-1)^n}{n^2}+2\sum_{n=1}^{\infty}\frac{(-1)^{n-1}}{n}\left(\frac{1}{n}-\frac{1}{n+1}+\frac{1}{n+2}-\cdots\right).$$

Since

$$\frac{1}{n}-\frac{1}{n+1}+\frac{1}{n+2}-\cdots=\int_0^1\frac{x^{n-1}}{1+x}dx,$$

we obtain that

$$\sum_{n=1}^{\infty}\frac{(-1)^{n-1}}{n}\left(\frac{1}{n}-\frac{1}{n+1}+\frac{1}{n+2}-\cdots\right)=\sum_{n=1}^{\infty}\frac{(-1)^{n-1}}{n}\int_0^1\frac{x^{n-1}}{1+x}dx$$

$$=-\int_0^1\frac{1}{x(1+x)}\sum_{n=1}^{\infty}\frac{(-x)^n}{n}dx$$

$$=\int_0^1\frac{\ln(1+x)}{x(1+x)}dx$$

$$=\int_0^1\frac{\ln(1+x)}{x}dx-\int_0^1\frac{\ln(1+x)}{1+x}dx$$

$$=\frac{\pi^2}{12}-\frac{\ln^2 2}{2}.$$

Therefore,

$$2\sum_{n=1}^{\infty}(-1)^{n-1}\left(\frac{1}{n}-\frac{1}{n+1}+\frac{1}{n+2}-\cdots\right)^2=\frac{\pi^2}{12}.$$

We mention that this quadratic series was calculated by a different method in [26, problem 3.45].

2.82. [28 January 2019]

(a) Prove that

$$(-1) \cdot 1 + (-1)^2 \cdot 2 + \cdots + (-1)^n \cdot n = \frac{-1 + (2n+1)(-1)^n}{4}, \quad n \geq 1.$$

(b) Prove that

$$\sum_{n=1}^{\infty} (-1)^n n \left(\zeta(2) - 1 - \frac{1}{2^2} - \cdots - \frac{1}{n^2} - \frac{1}{n} \right) = \frac{1}{4} \left(2 + 2\ln 2 - \frac{3}{2}\zeta(2) \right).$$

(c) Prove that

$$(-1) \cdot 1^2 + (-1)^2 \cdot 2^2 + \cdots + (-1)^n \cdot n^2 = (-1)^n \frac{n(n+1)}{2}, \quad n \geq 1.$$

(d) Prove that

$$\sum_{n=1}^{\infty} (-1)^n n^2 \left(\zeta(3) - 1 - \frac{1}{2^3} - \cdots - \frac{1}{n^3} - \frac{1}{2n^2} \right) = \frac{1}{2} \left(\ln 2 + \frac{1}{2} - \frac{\zeta(2)}{2} \right).$$

2.83.

(a) Prove that, if $k \in \mathbb{N}$, $k \geq 2$, then

$$\sum_{n=1}^{\infty} (-1)^{n-1} \left(\zeta(k) - 1 - \frac{1}{2^k} - \cdots - \frac{1}{n^k} \right) = \frac{\zeta(k)}{2^k}.$$

(b) Prove that, if $k \in \mathbb{N}$, $k \geq 3$, then

$$\sum_{n=1}^{\infty} (-1)^{n-1} n \left(\zeta(k) - 1 - \frac{1}{2^k} - \cdots - \frac{1}{n^k} \right)$$

$$= \zeta(k) \left(\frac{1}{2} - \frac{1}{2^{k+1}} \right) - \zeta(k-1) \left(\frac{1}{2} - \frac{1}{2^{k-1}} \right).$$

2.84. A Bouquet of Factorial Series

Prove that:

(a) $\displaystyle\sum_{n=1}^{\infty}\left(\frac{1}{n!}+\frac{2}{(n+1)!}+\frac{3}{(n+2)!}+\cdots\right)=\frac{3}{2}e;$

(b) $\displaystyle\sum_{n=1}^{\infty}n\left(\frac{1}{n!}+\frac{2}{(n+1)!}+\frac{3}{(n+2)!}+\cdots\right)=\frac{13}{6}e;$

(c) $\displaystyle\sum_{n=1}^{\infty}n^2\left(\frac{1}{n!}+\frac{2}{(n+1)!}+\frac{3}{(n+2)!}+\cdots\right)=\frac{47}{12}e.$

2.85. [16 August 2019]

(a) Let $k > 3$. Prove that

$$\sum_{n=1}^{\infty}\left(\frac{1}{n^k}+\frac{2}{(n+1)^k}+\frac{3}{(n+2)^k}+\cdots\right)=\frac{1}{2}(\zeta(k-2)+\zeta(k-1)).$$

(b) Let $k > 4$. Prove that

$$\sum_{n=1}^{\infty}n\left(\frac{1}{n^k}+\frac{2}{(n+1)^k}+\frac{3}{(n+2)^k}+\cdots\right)=\frac{1}{6}\left(\zeta(k-3)+3\zeta(k-2)+2\zeta(k-1)\right).$$

2.86. [28 January 2019]

(a) **A Quadratic Series with the Tail of $\zeta(2)$**

Prove that

$$\sum_{n=1}^{\infty}n\left(\zeta(2)-1-\frac{1}{2^2}-\cdots-\frac{1}{n^2}-\frac{1}{n}\right)^2=\frac{3}{2}\zeta(2)-\frac{3}{2}\zeta(3)-\frac{1}{2}.$$

(b) **A Quadratic Series with the Tail of $\zeta(3)$**

Prove that

$$\sum_{n=1}^{\infty}n^3\left(\zeta(3)-1-\frac{1}{2^3}-\cdots-\frac{1}{n^3}-\frac{1}{2n^2}\right)^2=-\frac{1}{2}\zeta(2)+\frac{1}{2}\zeta(3)+\frac{3}{8}\zeta(4)-\frac{1}{16}.$$

2.87. An Amazing Series with Double Factorials and the Tail of $\zeta(2)$
  Let

$$I_n = \int_0^{\frac{\pi}{2}} x^2 \cos^{2n} x \, dx, \quad n \geq 0.$$

(a) Prove that

$$I_n = \frac{2n-1}{2n} I_{n-1} - \frac{1}{n^2} \cdot \frac{1 \cdot 3 \cdots (2n-1)}{2 \cdot 4 \cdots (2n)} \cdot \frac{\pi}{4}, \quad n \geq 1.$$

(b) Prove that

$$I_n = \frac{1 \cdot 3 \cdots (2n-1)}{2 \cdot 4 \cdots (2n)} \cdot \left(\zeta(2) - 1 - \frac{1}{2^2} - \cdots - \frac{1}{n^2} \right) \cdot \frac{\pi}{4}, \quad n \geq 1.$$

(c) Prove that

$$\sum_{n=1}^{\infty} \frac{1 \cdot 3 \cdots (2n-1)}{2 \cdot 4 \cdots (2n)} \cdot \left(\zeta(2) - 1 - \frac{1}{2^2} - \cdots - \frac{1}{n^2} \right) = 4 \ln 2 - \zeta(2).$$

Power Series

3

Learned is a man who is happy to keep learning.

Nicolae Iorga (1871–1940)

This chapter is concentrated on power series, as the title says. It includes the standard formulae on the power series, Taylor and Maclaurin series of the classical elementary functions, and exercises dealing with finding the set of convergence of a power series and, whenever possible, determining the sum of the series. The problems in this chapter are as exotic and fascinating as possible. They cover topics such as the generating functions of special sequences, the calculation of power series with factorials, series involving polylogarithm functions, and nonlinear logarithmic series. The chapter also includes, as applications, the calculation of single and multiple series involving Riemann zeta function values. The last section of this chapter collects several nonlinear Euler series which are new in the literature, but the original problems can be found in other sections, too.

3.1   Convergence and Sum of Power Series

The Radius and the Convergence set of a Power Series

Let $\sum_{n=0}^{\infty} a_n x^n$, $a_n \in \mathbb{R}$, be a power series and let R be its radius of convergence. Then

$$R = \frac{1}{\lim\limits_{n \to \infty} \sqrt[n]{|a_n|}} \in [0, +\infty].$$

(continued)

A. Sîntămărian, O. Furdui, *Sharpening Mathematical Analysis Skills*, Problem Books in Mathematics, https://doi.org/10.1007/978-3-030-77139-3_3

If $R = 0$, then the convergence set of the power series is $\{0\}$.

If $R \in (0, +\infty)$, then $(-R, R)$ is the interval of convergence of the power series. The nature of the power series has to be studied for $x = -R$ and $x = R$, and the convergence set can be $(-R, R)$, $[-R, R)$, $(-R, R]$ or $[-R, R]$.

If $R = +\infty$, then the convergence set of the power series is $\mathbb{R}$.

If $\lim_{n\to\infty} \left|\frac{a_n}{a_{n+1}}\right|$ exists, then the radius of convergence of the power series can be calculated by the formula

$$R = \lim_{n\to\infty} \left|\frac{a_n}{a_{n+1}}\right|.$$

3.1. Find the convergence set of the following power series:

(a) $\displaystyle\sum_{n=0}^{\infty} \frac{x^n}{2^n + 5^n}$;

(b) $\displaystyle\sum_{n=0}^{\infty} \frac{x^n}{(n+1)3^n}$;

(c) $\displaystyle\sum_{n=1}^{\infty} (-1)^n \frac{(x+1)^n}{n+7}$;

(d) $\displaystyle\sum_{n=1}^{\infty} \frac{(x-2)^n}{(3n-1)5^n}$;

(e) $\displaystyle\sum_{n=1}^{\infty} \frac{(x-1)^{2n}}{n4^n}$;

(f) $\displaystyle\sum_{n=1}^{\infty} \frac{x^n}{n^{n+1}}$;

(g) $\displaystyle\sum_{n=1}^{\infty} \left(1 - \frac{1}{n}\right)^{n^2 - n} x^n$;

(h) $\displaystyle\sum_{n=0}^{\infty} \frac{n^{2n}}{(n!)^3} x^n$;

(i) $\displaystyle\sum_{n=1}^{\infty} H_n x^n$;

(j) $\displaystyle\sum_{n=1}^{\infty} n^n x^n$;

(k) $\displaystyle\sum_{n=1}^{\infty} \frac{x^{n^2}}{n^{n^2}}$;

(l) $\displaystyle\sum_{n=1}^{\infty} \frac{x^{n^5}}{3^{2n} n^{n^3}}$.

3.2. Find the convergence set of the power series

$$\sum_{n=0}^{\infty} x^{n!} = 1 + x + x^2 + x^6 + x^{24} + x^{120} + \cdots.$$

3.3. Find the convergence set and the sum of the following power series:

(a) $\displaystyle\sum_{n=0}^{\infty} (n+1)x^{n+1}$;

(b) $\displaystyle\sum_{n=0}^{\infty} (n+1)(n+2)x^n$;

(c) $\displaystyle\sum_{n=1}^{\infty} (-1)^{n-1} \frac{x^{3n-1}}{3n-1}$;

(d) $\displaystyle\sum_{n=1}^{\infty} (-1)^{n-1} (5n-1) x^{5n-2}$;

(e) $\displaystyle\sum_{n=0}^{\infty} (-1)^n (n+1)^3 x^n$;

(f) $\displaystyle\sum_{n=1}^{\infty} \frac{x^{4n-1}}{4n-3}$.

3.4.

A. Find the convergence set of the following power series:

(a) $\displaystyle\sum_{n=1}^{\infty}\left(1+\frac{1}{2}+\cdots+\frac{1}{n}-\ln n-\gamma\right)x^{n}$;

(b) $\displaystyle\sum_{n=1}^{\infty}\left(e-1-\frac{1}{1!}-\frac{1}{2!}-\cdots-\frac{1}{n!}\right)x^{n}$;

(c) $\displaystyle\sum_{n=1}^{\infty}\left(\zeta(k)-\frac{1}{1^{k}}-\frac{1}{2^{k}}-\cdots-\frac{1}{n^{k}}\right)x^{n}$, $k\geq 2$.

B. Prove that

$$\sum_{n=1}^{\infty}\left(e-1-\frac{1}{1!}-\frac{1}{2!}-\cdots-\frac{1}{n!}\right)x^{n}=\begin{cases}\dfrac{e^{x}-e\,x}{x-1}+1 & \text{if}\quad x\neq 1\\[2mm] 1 & \text{if}\quad x=1.\end{cases}$$

The polylogarithm function Li_{k}, $k>2$, is the special function defined by

$$\mathrm{Li}_{k}(x)=\sum_{n=1}^{\infty}\frac{x^{n}}{n^{k}},\quad x\in[-1,1].$$

C. Prove that, for $k>2$, we have

$$\sum_{n=1}^{\infty}\left(\zeta(k)-\frac{1}{1^{k}}-\frac{1}{2^{k}}-\cdots-\frac{1}{n^{k}}\right)x^{n}=\begin{cases}\dfrac{x\zeta(k)-\mathrm{Li}_{k}(x)}{1-x} & \text{if}\quad x\in[-1,1)\\[2mm] \zeta(k-1)-\zeta(k) & \text{if}\quad x=1\end{cases}$$

and

$$\sum_{n=1}^{\infty}\left(\zeta(2)-\frac{1}{1^{2}}-\frac{1}{2^{2}}-\cdots-\frac{1}{n^{2}}\right)x^{n}=\frac{x\zeta(2)-\mathrm{Li}_{2}(x)}{1-x},\quad x\in[-1,1).$$

3.2 Maclaurin Series of Elementary Functions

The Maclaurin Series of Some Elementary Functions

$$\sin x = \sum_{n=0}^{\infty} (-1)^n \frac{x^{2n+1}}{(2n+1)!}, \quad x \in \mathbb{R};$$

$$\cos x = \sum_{n=0}^{\infty} (-1)^n \frac{x^{2n}}{(2n)!}, \quad x \in \mathbb{R};$$

$$\arcsin x = x + \sum_{n=1}^{\infty} \frac{(2n-1)!!}{(2n)!!} \cdot \frac{x^{2n+1}}{2n+1}, \quad x \in [-1, 1];$$

$$\arctan x = \sum_{n=0}^{\infty} (-1)^n \frac{x^{2n+1}}{2n+1}, \quad x \in [-1, 1];$$

the binomial series

$$(1+x)^\alpha = 1 + \frac{\alpha}{1!}x + \frac{\alpha(\alpha-1)}{2!}x^2 + \cdots + \frac{\alpha(\alpha-1)\cdots(\alpha-n+1)}{n!}x^n + \cdots,$$

$$x \in (-1, 1), \ \alpha \in \mathbb{R};$$

$$\frac{1}{1+x} = \sum_{n=0}^{\infty} (-1)^n x^n, \quad x \in (-1, 1); \quad \frac{1}{1-x} = \sum_{n=0}^{\infty} x^n, \quad x \in (-1, 1);$$

$$\frac{1}{(1-x)^\alpha} = \sum_{n=0}^{\infty} \frac{\Gamma(\alpha+n)}{\Gamma(\alpha)n!}x^n, \ \alpha > 0, \ x \in (-1, 1),$$

where Γ is the Gamma function of Euler;
the logarithmic series

$$\ln(1+x) = \sum_{n=1}^{\infty} (-1)^{n-1} \frac{x^n}{n}, \quad x \in (-1, 1];$$

$$\ln(1-x) = -\sum_{n=1}^{\infty} \frac{x^n}{n}, \quad x \in [-1, 1);$$

the exponential series

$$e^x = \sum_{n=0}^{\infty} \frac{x^n}{n!}, \quad x \in \mathbb{R}; \qquad e^{-x} = \sum_{n=0}^{\infty} (-1)^n \frac{x^n}{n!}, \quad x \in \mathbb{R}.$$

Infinite Nested Product Formulae

(a) *the geometric series I*

$$\frac{1}{1-x} = 1 + x\left(1 + x\left(1 + x\left(1 + x\left(1 + \cdots\right)\right)\right)\right), \quad x \in (-1, 1);$$

(b) *the geometric series II*

$$\frac{1}{1+x} = 1 - x\left(1 - x\left(1 - x\left(1 - x\left(1 - \cdots\right)\right)\right)\right), \quad x \in (-1, 1);$$

(c) *the logarithmic series I*

$$-\ln(1-x) = x\left(1 + \frac{x}{2}\left(1 + \frac{2x}{3}\left(1 + \frac{3x}{4}\left(1 + \cdots\right)\right)\right)\right), \quad x \in [-1, 1);$$

(d) *the logarithmic series II*

$$\ln(1+x) = x\left(1 - \frac{x}{2}\left(1 - \frac{2x}{3}\left(1 - \frac{3x}{4}\left(1 - \cdots\right)\right)\right)\right), \quad x \in (-1, 1];$$

(e) *the binomial series*

$$(1+x)^{\alpha} = 1 + \frac{\alpha x}{1}\left(1 + \frac{(\alpha-1)x}{2}\left(1 + \frac{(\alpha-2)x}{3}\left(1 + \cdots\right)\right)\right),$$

$$x \in (-1, 1), \quad \alpha \in \mathbb{R};$$

(f) *the constant* e

$$e = 1 + \frac{1}{1}\left(1 + \frac{1}{2}\left(1 + \frac{1}{3}\left(1 + \frac{1}{4}\left(1 + \frac{1}{5}(1 + \cdots)\right)\right)\right)\right);$$

(g) *the Maclaurin series of the function* e^x

$$e^x = 1 + \frac{x}{1}\left(1 + \frac{x}{2}\left(1 + \frac{x}{3}\left(1 + \frac{x}{4}\left(1 + \frac{x}{5}(1 + \cdots)\right)\right)\right)\right), \quad x \in \mathbb{R};$$

(h) *the Maclaurin series of the function* e^{-x}

$$e^{-x} = 1 - \frac{x}{1}\left(1 - \frac{x}{2}\left(1 - \frac{x}{3}\left(1 - \frac{x}{4}\left(1 - \frac{x}{5}(1 - \cdots)\right)\right)\right)\right), \quad x \in \mathbb{R}.$$

Beautiful Numerical Series

- $\displaystyle\sum_{n=1}^{\infty} n \left(\frac{3-\sqrt{5}}{2}\right)^n = 1;$

- $\displaystyle\sum_{n=1}^{\infty} n \left(\frac{2+k-\sqrt{k^2+4k}}{2}\right)^n = \frac{1}{k}, \ k > 0;$

- $\displaystyle\sum_{n=1}^{\infty} n \left(2-\sqrt{3}\right)^n = \frac{1}{2};$

- $\displaystyle\sum_{n=1}^{\infty} n \left(1+k-\sqrt{k^2+2k}\right)^n = \frac{1}{2k}, \ k > 0.$

3.5. The Power Series Expansion of f^k

If $f : (-r, r) \to \mathbb{R}$ has the power series expansion $f(x) = \displaystyle\sum_{n=0}^{\infty} a_n x^n$ and $k \in \mathbb{N}$,

then $f^k(x) = \displaystyle\sum_{n=0}^{\infty} c_n x^n$, where $c_n = \sum a_{i_1} a_{i_2} \cdots a_{i_k}$ and the sum is over all k-tuples

of integers $(i_1, i_2, \ldots, i_k)$, with $i_1, i_2, \ldots, i_k \geq 0$ and $i_1 + i_2 + \cdots + i_k = n$.

Application. If $k \in \mathbb{N}$, then

$$\frac{1}{(1-x)^k} = \sum_{n=0}^{\infty} C_{n+k-1}^n x^n, \quad |x| < 1.$$

3.6. Calculate:

(a) [134] $\displaystyle\sum_{n=0}^{\infty} \frac{16n^2 + 4n - 1}{(4n+2)!};$ 　　　　　　(b) $\displaystyle\sum_{n=0}^{\infty} \frac{16n^2 + 12n + 1}{(4n+3)!}.$

3.7. [A. Sîntămărian, 2014] Calculate

$$\sum_{n=1}^{\infty} \frac{1}{n(2n+1)(4n+1)}.$$

Next we record a proof, a gem of mathematical analysis, regarding the divergence of the harmonic series which uses Abel's theorem for power series.

The Harmonic Series Diverges

If the harmonic series $\sum\limits_{n=1}^{\infty} \frac{1}{n}$ would converge, then we have, based on Abel's Theorem for power series, that

$$\sum_{n=1}^{\infty} \frac{1}{n} = \lim_{x \to 1^-} \sum_{n=1}^{\infty} \frac{x^n}{n} = -\lim_{x \to 1^-} \ln(1-x) = \infty.$$

3.8. Find the Maclaurin series expansion for the following functions and specify the convergence set in each case:

(a) $f : \mathbb{R} \to \mathbb{R}, \quad f(x) = \sinh x;$ (b) $f : \mathbb{R} \to \mathbb{R}, \quad f(x) = \cos^2 x;$

(c) $f : \mathbb{R} \to \mathbb{R}, \quad f(x) = \sin^3 x;$ (d) $f : \mathbb{R} \to \mathbb{R}, \quad f(x) = x^2 e^{-3x};$

(e) $f : \mathbb{R} \setminus \{1\} \to \mathbb{R}, \quad f(x) = \frac{3x-1}{(x-1)^2};$

(f) $f : (-3, 3) \to \mathbb{R}, \quad f(x) = \frac{1}{\sqrt{9-x^2}};$

(g) $f : (-1, 1) \to \mathbb{R}, \quad f(x) = \frac{1}{2} \ln \frac{1+x}{1-x};$

(h) $f : \left(-\frac{1}{5}, 1\right) \to \mathbb{R}, \quad f(x) = \ln(1 + 4x - 5x^2);$

(i) $f : \mathbb{R} \to \mathbb{R}, \quad f(x) = \ln(x + \sqrt{1+x^2});$

(j) $f : \mathbb{R} \to \mathbb{R}, \quad f(x) = \cos(3x) + x \sin(3x);$

(k) $f : \mathbb{R} \setminus \{1\} \to \mathbb{R}, \quad f(x) = \frac{1+x^2}{1-x}.$

3.9. Prove that:

(a) $\dfrac{1}{\sqrt{1-x}} = 1 + \sum\limits_{n=1}^{\infty} \dfrac{1 \cdot 3 \cdots (2n-1)}{2 \cdot 4 \cdots (2n)} x^n, \ x \in [-1, 1);$

(b) $\sum\limits_{n=1}^{\infty} \dfrac{1 \cdot 3 \cdots (2n-1)}{2 \cdot 4 \cdots (2n)} \cdot \dfrac{1}{n} = \ln 4;$

(c) $\sum\limits_{n=1}^{\infty} (-1)^n \dfrac{1 \cdot 3 \cdots (2n-1)}{2 \cdot 4 \cdots (2n)} \cdot \dfrac{1}{n} = 2 \ln \dfrac{2}{1 + \sqrt{2}}.$

3.10. Prove that

$$\frac{1}{x^2 + x + 1} = \frac{2}{\sqrt{3}} \sum_{n=0}^{\infty} \sin \frac{2\pi(n+1)}{3} x^n, \quad |x| < 1.$$

3.11. Let f be a function which has the power series expansion $f(x) = \sum\limits_{n=0}^{\infty} a_n x^n$, $|x| < R$, $R \in (0, \infty]$. Prove that

$$\sum_{n=0}^{\infty} a_{3n} x^{3n} = \frac{f(x) + f(\epsilon x) + f(\epsilon^2 x)}{3}, \quad |x| < R,$$

where $\epsilon \neq 1$ is a root of order 3 of unity, i.e. $\epsilon^3 = 1$.

Application. Prove that

$$\sum_{n=0}^{\infty} \frac{x^{3n}}{(3n)!} = \frac{1}{3}\left(e^x + 2e^{-\frac{x}{2}}\cos\frac{x\sqrt{3}}{2}\right), \quad x \in \mathbb{R}.$$

3.12.

(a) Prove that

$$\sum_{n=2}^{\infty} \frac{(n^2 + n - 1)^3}{(n-2)! + (n+2)!} = 12e \quad \text{and} \quad \sum_{n=2}^{\infty}(-1)^n \frac{(n^2 + n - 1)^3}{(n-2)! + (n+2)!} = 0.$$

(b) Prove that

$$\sum_{n=2}^{\infty} \frac{(n^2 + n - 1)^3}{(n-2)! + (n+2)!} x^{n-2} = (x^2 + 6x + 5)e^x, \quad x \in \mathbb{R}.$$

3.13. *Series with the floor function.* Prove that:

(a) $\displaystyle\sum_{n=1}^{\infty} \frac{\lfloor \log_2 n \rfloor}{n(n+1)} = 1$;

(b) $\displaystyle\sum_{n=2}^{\infty} \frac{1}{n(n+1)\lfloor \log_2 n \rfloor} = \frac{\ln 2}{2}$;

(c) $\displaystyle\sum_{n=2}^{\infty} \frac{1}{n(n+1)\lfloor \log_2 n \rfloor^2} = \frac{\pi^2}{24} - \frac{\ln^2 2}{4}$;

(d) $\displaystyle\sum_{n=2}^{\infty} \frac{(-1)^n}{\lfloor \log_2 n \rfloor^k} = 0, \ k \in \mathbb{N}.$

Curiosities with Two Series

$$\sum_{n=1}^{\infty} \frac{\lfloor \log_2 n \rfloor}{n(n+1)} = 1 \quad \text{and} \quad \sum_{n=1}^{\infty} \frac{1}{n(n+1)} = 1.$$

For Problems 3.14–3.21 write the power series expansion in terms of $x - x_0$ and determine the convergence set for which the expansion holds.

3.14. $f : \mathbb{R} \setminus \{-5\} \to \mathbb{R}, \quad f(x) = \frac{1}{x+5}, \quad x_0 = 2.$

3.15. $f : (-1, \infty) \to \mathbb{R}, \quad f(x) = \ln(x+1), \quad x_0 = 3.$

3.16. $f : \mathbb{R} \setminus \{1\} \to \mathbb{R}, \quad f(x) = \frac{1}{(x-1)^2}, \quad x_0 = -1.$

3.17. $f : \mathbb{R} \setminus \{-3, -1\} \to \mathbb{R}, \quad f(x) = \frac{1}{x^2+4x+3}, \quad x_0 = -5.$

3.18. $f : \mathbb{R} \to \mathbb{R}, \quad f(x) = \frac{1}{x^2-4x+8}, \quad x_0 = 2.$

3.19. $f : \mathbb{R} \to \mathbb{R}, \quad f(x) = e^{2x-1}, \quad x_0 = -1.$

3.20. $f : \mathbb{R} \to \mathbb{R}, \quad f(x) = \sin(3x + \pi), \quad x_0 = \pi.$

3.21. $f : \mathbb{R} \to \mathbb{R}, \quad f(x) = \arctan x - \arctan \frac{1}{x^2-x+1}, \quad x_0 = 1.$

3.3   Gems with Numerical and Power Series

3.22.   [22 October 2018] **Power Series with Factorials**

(a) Prove that

$$\sum_{n=1}^{\infty} \frac{x^n}{(n-1)! + n!} = \frac{1 - e^x}{x} + e^x, \quad x \in \mathbb{R}.$$

(continued)

In particular, for $x = -1$ we obtain the series

$$\sum_{n=1}^{\infty} \frac{(-1)^n}{(n-1)! + n!} = 2e^{-1} - 1.$$

(b) Prove that

$$\sum_{n=1}^{\infty} \frac{x^{n-1} + x^n}{(n-1)! + n!} = (1+x)\frac{1 - e^x + xe^x}{x^2}, \quad x \in \mathbb{R}.$$

In particular, for $x = 1$ we obtain the series

$$\sum_{n=1}^{\infty} \frac{1}{(n-1)! + n!} = 1.$$

(c) Prove that

$$\sum_{n=2}^{\infty} \frac{n}{(n-2)! + (n-1)! + n!} x^n = (x-1)e^x + 1, \quad x \in \mathbb{R}.$$

In particular, for $x = 1$ we obtain the series

$$\sum_{n=2}^{\infty} \frac{n}{(n-2)! + (n-1)! + n!} = 1.$$

(d) Prove that

$$\sum_{n=2}^{\infty} \frac{x^n}{(n-2)! + (n-1)! + n!} = e^x - 1 + \int_0^x \frac{1 - e^t}{t} dt, \quad x \in \mathbb{R}.$$

In particular, for $x = 1$ we obtain the series

$$\sum_{n=2}^{\infty} \frac{1}{(n-2)! + (n-1)! + n!} = e - 1 + \int_0^1 \frac{1 - e^t}{t} dt.$$

(e) Prove that

$$\sum_{n=2}^{\infty} \frac{x^{n-2} + x^{n-1} + x^n}{(n-2)! + (n-1)! + n!}$$

(continued)

$$= \frac{1 + x + x^2}{x^2} \left[e^x - 1 + \int_0^x \frac{1 - e^t}{t} dt \right], \quad x \in \mathbb{R}.$$

3.23.

(a) Prove that

$$\sum_{n=0}^{\infty} (-1)^n \left(\frac{1}{n+1} - \frac{1}{n+2} + \frac{1}{n+3} - \cdots \right) = \frac{1}{2}.$$

(b) Prove that

$$\sum_{n=0}^{\infty} (-1)^{\lfloor \frac{n}{2} \rfloor} x^n = \frac{1 + x}{1 + x^2}, \quad x \in (-1, 1),$$

where $\lfloor x \rfloor$ is the floor of the real number x.

Proof. We give here the solution of this part of the problem. First, we observe that the power series converges for $x \in (-1, 1)$. We calculate the sum of the power series. We have

$$\sum_{n=0}^{\infty} (-1)^{\lfloor \frac{n}{2} \rfloor} x^n = 1 + x + \sum_{n=2}^{\infty} (-1)^{\lfloor \frac{n}{2} \rfloor} x^n$$

$$= 1 + x + \sum_{m=0}^{\infty} (-1)^{\lfloor \frac{m+2}{2} \rfloor} x^{m+2}$$

$$= 1 + x - x^2 \sum_{m=0}^{\infty} (-1)^{\lfloor \frac{m}{2} \rfloor} x^m,$$

and the result follows. An alternative solution, based on calculating the $2n$th partial sum of the series, is left to the reader.   ∎

(c) Prove that

$$\sum_{n=0}^{\infty} (-1)^{\lfloor \frac{n}{2} \rfloor} \left(\frac{1}{n+1} - \frac{1}{n+2} + \frac{1}{n+3} - \cdots \right) = \frac{\pi}{4}.$$

(d) Let $k \in \mathbb{N}$. Prove that

$$\sum_{n=0}^{\infty} (-1)^{\lfloor \frac{n}{k} \rfloor} x^n = \frac{1 + x + \cdots + x^{k-1}}{1 + x^k}, \quad x \in (-1, 1).$$

(e) Let $k \in \mathbb{N}$. Prove that

$$\sum_{n=0}^{\infty} (-1)^{\lfloor \frac{n}{k} \rfloor} \left(\frac{1}{n+1} - \frac{1}{n+2} + \frac{1}{n+3} - \cdots \right) = \int_0^1 \frac{1 + x + \cdots + x^{k-1}}{(1+x)(1+x^k)} \, dx.$$

(f) Prove that

$$\sum_{n=0}^{\infty} (-1)^{\lfloor \frac{n}{2} \rfloor} \frac{x^{n+1}}{n+1} = \arctan x + \frac{1}{2} \ln(1 + x^2), \quad x \in [-1, 1].$$

3.24. [19 December 2019] Prove that

$$\sum_{n=1}^{\infty} \left(\frac{1}{n+1} - \frac{1}{n+2} + \cdots \right) \left(\zeta(2) - 1 - \frac{1}{2^2} - \cdots - \frac{1}{n^2} \right) = \frac{21}{16} \zeta(3) - \zeta(2) \ln 2.$$

3.25. [10 January 2020] Prove that:

An alternating series

(a) $\displaystyle\sum_{n=1}^{\infty} (-1)^{n-1} \left(\frac{1}{n+1} + \frac{1}{n+2} - \frac{1}{n+3} - \frac{1}{n+4} + \cdots \right) = \frac{\ln 2}{2}$;

Quadratic series

(b) [91] $\displaystyle\sum_{n=1}^{\infty} (-1)^{n-1} H_n \left(\frac{1}{n+1} + \frac{1}{n+2} - \frac{1}{n+3} - \frac{1}{n+4} + \cdots \right) = \frac{\pi \ln 2}{8}$;

(c) $\displaystyle\sum_{n=1}^{\infty} \left(\frac{1}{n} + \frac{1}{n+1} - \frac{1}{n+2} - \frac{1}{n+3} + \cdots \right)^2 = \frac{\pi}{4} + G + \frac{\ln 2}{2}$,

where G denotes Catalan's constant;

(d) $\displaystyle\sum_{n=1}^{\infty} (-1)^{n-1} \left(\frac{1}{n} + \frac{1}{n+1} - \frac{1}{n+2} - \frac{1}{n+3} + \cdots \right)^2 = \frac{\pi^2}{16} + \frac{\pi}{4} - \frac{\ln 2}{2}.$

3.26. [12 January 2020]
Prove that

$$\sum_{n=1}^{\infty} \left(\frac{1}{n} + \frac{1}{n+1} - \frac{1}{n+2} - \frac{1}{n+3} + \cdots \right) \left(\frac{1}{n} - \frac{1}{n+1} + \frac{1}{n+2} - \cdots \right)$$

$$= \frac{\pi \ln 2}{8} + G,$$

where G denotes Catalan's constant.

A challenge. Prove that

$$\sum_{n=1}^{\infty} (-1)^{n-1} \left(\frac{1}{n} + \frac{1}{n+1} - \frac{1}{n+2} - \frac{1}{n+3} + \cdots \right)$$

$$\times \left(\frac{1}{n} - \frac{1}{n+1} + \frac{1}{n+2} - \cdots \right)$$

$$= \frac{7\pi^2}{96} - \frac{\ln^2 2}{8}.$$

3.4 Single Zeta Series

3.27. Prove that

$$\sum_{n=1}^{\infty} n \left(\zeta(n+1) - 1 \right) = \zeta(2) \quad \text{and} \quad \sum_{n=1}^{\infty} (-1)^{n-1} n \left(\zeta(n+1) - 1 \right) = \zeta(2) - \frac{5}{4}.$$

3.28. Prove that

$$\sum_{n=2}^{\infty} \frac{\zeta(n) - 1}{n} = 1 - \gamma \quad \text{and} \quad \sum_{n=1}^{\infty} \frac{\zeta(2n) - 1}{n} = \ln 2.$$

3.29. Series Mirabili Prove that:

(a) [72] $\displaystyle\sum_{n=2}^{\infty} \left[2^n \left(\zeta(n) - 1 \right) - 1 \right] = 3;$

(continued)

(b) $\displaystyle\sum_{n=2}^{\infty}\left[3^n\left(\zeta(n)-1-\frac{1}{2^n}\right)-1\right]=\frac{11}{2}$;

(c) [84] $\displaystyle\sum_{n=1}^{\infty}\left[2^{2n}\left(\zeta(2n)-1\right)-1\right]=\frac{25}{12}$;

(d) $\displaystyle\sum_{n=1}^{\infty}\left[3^{2n}\left(\zeta(2n)-1-\frac{1}{2^{2n}}\right)-1\right]=\frac{3}{2}H_6$;

(e) $\displaystyle\sum_{n=2}^{\infty}\left[(k+1)^n\left(\zeta(n)-1-\frac{1}{2^n}-\cdots-\frac{1}{k^n}\right)-1\right]=(k+1)H_{k+1},\ k\in\mathbb{N}$;

(f) $\displaystyle\sum_{n=1}^{\infty}\left[(k+1)^{2n}\left(\zeta(2n)-1-\frac{1}{2^{2n}}-\cdots-\frac{1}{k^{2n}}\right)-1\right]=\frac{k+1}{2}H_{2k+2}$,

$$k\in\mathbb{N}.$$

3.30. Prove that:

(a) $\displaystyle\sum_{n=2}^{\infty}\left[2^{2n-1}\left(\zeta(2n-1)-1\right)-1\right]=\frac{11}{12}$;

(b) $\displaystyle\sum_{n=2}^{\infty}n\left[2^{2n-1}\left(\zeta(2n-1)-1\right)-1\right]=\frac{337}{144}$.

3.31.

(a) Prove that

$$\sum_{n=2}^{\infty}(-1)^n\left[2^n\left(\zeta(n)-1\right)-1\right]=\frac{7}{6}.$$

(b) Let $k\in\mathbb{N}$. Prove that

$$\sum_{n=2}^{\infty}(-1)^n\left[(k+1)^n\left(\zeta(n)-1-\frac{1}{2^n}-\cdots-\frac{1}{k^n}\right)-1\right]$$

$$=(k+1)\left(H_{2k+2}-H_{k+1}\right).$$

3.32. [40]

(a) Prove that

$$\sum_{n=2}^{\infty} (n-1) \left[2^n \left(\zeta(n) - 1 \right) - 1 \right] = 4\zeta(2).$$

(b) Let $k \in \mathbb{N}$. Prove that

$$\sum_{n=2}^{\infty} (n-1) \left[(k+1)^n \left(\zeta(n) - 1 - \frac{1}{2^n} - \cdots - \frac{1}{k^n} \right) - 1 \right] = (k+1)^2 \zeta(2).$$

3.33. Prove that

$$\sum_{n=2}^{\infty} \frac{1}{n} \left[2^n \left(\zeta(n) - 1 \right) - 1 \right] = 3 - 2\gamma - \ln 2$$

and

$$\sum_{n=1}^{\infty} \frac{1}{n} \left[4^n \left(\zeta(2n) - 1 \right) - 1 \right] = \ln 6.$$

3.34. Prove that $\displaystyle\sum_{n=2}^{\infty} H_n (\zeta(n) - \zeta(n+1)) = \zeta(2) - \gamma.$

3.35. Prove that:

(a) [39] $\displaystyle\sum_{n=2}^{\infty} \left(H_n - \gamma - \frac{\zeta(2)}{2} - \frac{\zeta(3)}{3} - \cdots - \frac{\zeta(n)}{n} \right) = 2\gamma - 1;$

(b) $\displaystyle\sum_{n=2}^{\infty} n \left(H_n - \gamma - \frac{\zeta(2)}{2} - \frac{\zeta(3)}{3} - \cdots - \frac{\zeta(n)}{n} \right) = \frac{\zeta(2)}{2} - 1 + \gamma;$

(c) $\displaystyle\sum_{n=2}^{\infty} (2n-1) \left(H_n - \gamma - \frac{\zeta(2)}{2} - \frac{\zeta(3)}{3} - \cdots - \frac{\zeta(n)}{n} \right) = \zeta(2) - 1.$

3.36. Prove that

$$\sum_{n=1}^{\infty} \frac{\zeta(2n+1)-1}{n+1} = -\sum_{n=2}^{\infty} \left[n \ln \left(1 - \frac{1}{n^2} \right) + \frac{1}{n} \right] = \ln 2 - \gamma.$$

3.37. Prove that

$$\sum_{n=1}^{\infty} \frac{n \zeta(2n)}{4^{n-2}} = \pi^2.$$

3.38. Series Delights Prove that:

(a) $\displaystyle \sum_{k=1}^{\infty} \frac{1}{k(k+1)^n} = n - \zeta(2) - \zeta(3) - \cdots - \zeta(n), \quad n \geq 2;$

(b) $\displaystyle \sum_{n=1}^{\infty} (n + 1 - \zeta(2) - \zeta(3) - \cdots - \zeta(n+1)) = \zeta(2) - 1;$

(c) [42] $\displaystyle \sum_{n=1}^{\infty} n \left(n + 1 - \zeta(2) - \zeta(3) - \cdots - \zeta(n+1) \right) = \zeta(3);$

(d) $\displaystyle \sum_{n=1}^{\infty} (-1)^{n-1} \left(n + 1 - \zeta(2) - \zeta(3) - \cdots - \zeta(n+1) \right) = \frac{1}{4};$

(e) $\displaystyle \sum_{n=2}^{\infty} \left[2^n \left(n - \zeta(2) - \zeta(3) - \cdots - \zeta(n) \right) - 1 \right] = 1;$

(f) $\displaystyle \sum_{n=2}^{\infty} n \left[2^n \left(n - \zeta(2) - \zeta(3) - \cdots - \zeta(n) \right) - 1 \right] = 4\zeta(2) - 3.$

The Dirichlet beta function is defined by

$$\beta(x) = \sum_{n=0}^{\infty} \frac{(-1)^n}{(2n+1)^x}, \quad x > 0.$$

We mention the special values $\beta(1) = \frac{\pi}{4}$ and $\beta(2) = G$.

3.39. Series with the Dirichlet Beta Function
Prove that:

(continued)

(a) $\displaystyle\sum_{n=1}^{\infty} (\beta(n) - 1) = -\frac{\ln 2}{2}$;

(b) $\displaystyle\sum_{n=1}^{\infty} (-1)^n (\beta(n) - 1) = \frac{1 - \ln 2}{2}$;

(c) $\displaystyle\sum_{n=1}^{\infty} n (\beta(n) - 1) = -\frac{\ln 2}{2} - \frac{\pi^2}{48}$;

(d) $\displaystyle\sum_{n=1}^{\infty} (-1)^n n (\beta(n) - 1) = \frac{1}{4} - \frac{\ln 2}{2} + \frac{\pi^2}{48}$;

(e) $\displaystyle\sum_{n=1}^{\infty} \left[5^n \left(\beta(n) - 1 + \frac{1}{3^n} \right) - 1 \right] = -\frac{5}{2} \ln 2$;

(f) $\displaystyle\sum_{n=1}^{\infty} (-1)^n \left[5^n \left(\beta(n) - 1 + \frac{1}{3^n} \right) - 1 \right] = \frac{5}{2} \left(\frac{47}{60} - \ln 2 \right)$;

(g) $\displaystyle\sum_{n=1}^{\infty} (\beta(2n) - 1) = \frac{1}{4} - \frac{\ln 2}{2}$.

Remark 3.1. We obtain, based on parts (a) and (c), the remarkable series formula

$$-8 \sum_{n=2}^{\infty} (n - 1)(\beta(n) - 1) = \zeta(2).$$

3.40. Prove that

$$\sum_{n=2}^{\infty} (\zeta(n) - \beta(n)) = \frac{\pi}{4} + \frac{\ln 2}{2}.$$

3.5 Polylogarithm Series

3.41. A Bouquet of Polylogarithm Series

(a) Prove that, for $x \in [-1, 1]$, we have

$$\sum_{n=2}^{\infty} (\mathrm{Li}_n(x) - x) = (1 - x) \ln(1 - x) + x.$$

(continued)

In particular, for $x = 1$ we obtain the series of Problem 2.31

$$\sum_{n=2}^{\infty}(\zeta(n) - 1) = 1.$$

(b) Prove that, for $x \in [-1, 1]$, we have

$$\sum_{n=2}^{\infty}(-1)^n \left(\text{Li}_n(x) - x\right) = \frac{1-x}{x}\ln(1-x) - \frac{x}{2} + 1.$$

In particular, for $x = 1$ we obtain the first series of Problem 2.32

$$\sum_{n=2}^{\infty}(-1)^n(\zeta(n) - 1) = \frac{1}{2}.$$

(c) Prove that, for $x \in [-1, 1]$, we have

$$\sum_{n=1}^{\infty}(\text{Li}_{2n}(x) - x) = \frac{1-x^2}{2x}\ln(1-x) + \frac{x}{4} + \frac{1}{2}.$$

In particular, for $x = 1$ we obtain the second series of Problem 2.32

$$\sum_{n=1}^{\infty}(\zeta(2n) - 1) = \frac{3}{4}.$$

(d) Prove that, for $x \in [-1, 1]$, we have

$$\sum_{n=1}^{\infty}(\text{Li}_{2n+1}(x) - x) = -\frac{(1-x)^2}{2x}\ln(1-x) + \frac{3x}{4} - \frac{1}{2}.$$

In particular, for $x = 1$ we have

$$\sum_{n=1}^{\infty}(\zeta(2n+1) - 1) = \frac{1}{4}.$$

(e) Prove that, for $x \in [-1, 1]$, we have

$$\sum_{n=1}^{\infty}(\text{Li}_{2n}(x) - \text{Li}_{2n+1}(x)) = \frac{1-x}{x}\ln(1-x) - \frac{x}{2} + 1.$$

(continued)

In particular, for $x = 1$ we obtain the series

$$\sum_{n=1}^{\infty}(\zeta(2n) - \zeta(2n+1)) = \frac{1}{2}.$$

(f) Prove that, for $x \in [-1, 1]$, we have

$$\sum_{n=2}^{\infty}(-1)^n \left(\text{Li}_n(x) - \text{Li}_{n+1}(x)\right) = 2\frac{1-x}{x}\ln(1-x) - \text{Li}_2(x) + 2.$$

In particular, for $x = 1$ we obtain the series of Problem 2.33

$$\sum_{n=1}^{\infty}(-1)^n(\zeta(n) - \zeta(n+1)) = 2 - \zeta(2).$$

(g) Prove that, for $x \in [-1, 1]$, we have

$$\sum_{n=2}^{\infty} n \left(\text{Li}_n(x) - x\right) = (1-x)\ln(1-x) + x\,\text{Li}_2(x) + x.$$

In particular, for $x = 1$ we obtain the series

$$\sum_{n=2}^{\infty} n(\zeta(n) - 1) = \zeta(2) + 1.$$

(h) Prove that, for $x \in [-1, 1]$, we have

$$\sum_{n=2}^{\infty} n \left(\text{Li}_n(x) - \text{Li}_{n+1}(x)\right) = (1-x)\ln(1-x) + \text{Li}_2(x).$$

In particular, for $x = 1$ we obtain the series

$$\sum_{n=2}^{\infty} n(\zeta(n) - \zeta(n+1)) = \zeta(2).$$

(i) Prove that, for $x \in [-1, 1]$, we have

$$\sum_{n=2}^{\infty}\left[2^n \left(\text{Li}_n(x) - x\right) - x^2\right] = 2\left[(1-x^2)\ln(1-x) + x + \frac{x^2}{2}\right].$$

(continued)

In particular, for $x = 1$ we obtain part (a) of Problem 3.29

$$\sum_{n=2}^{\infty} \left[2^n (\zeta(n) - 1) - 1 \right] = 3.$$

(j) Prove that, for $x \in [-1, 1]$, we have

$$\sum_{n=1}^{\infty} \left[4^n \left(\text{Li}_{2n}(x) - x \right) - x^2 \right] = \frac{1 - x^4}{x^2} \ln(1-x) + \frac{1}{x} \left(1 + \frac{x}{2} + \frac{x^2}{3} + \frac{x^3}{4} \right).$$

In particular, for $x = 1$ we obtain part (c) of Problem 3.29

$$\sum_{n=1}^{\infty} \left[4^n (\zeta(2n) - 1) - 1 \right] = \frac{25}{12}.$$

(k) Prove that, for $x \in [-1, 1]$, we have

$$\sum_{n=2}^{\infty} \left[3^n \left(\text{Li}_n(x) - x - \frac{x^2}{2^n} \right) - x^3 \right]$$

$$= 3 \left[(1 - x^3) \ln(1 - x) + x + \frac{x^2}{2} + \frac{x^3}{3} \right].$$

In particular, for $x = 1$ we obtain part (b) of Problem 3.29

$$\sum_{n=2}^{\infty} \left[3^n \left(\zeta(n) - 1 - \frac{1}{2^n} \right) - 1 \right] = \frac{11}{2}.$$

(l) *A generalization of parts (i) and (k).*
Let $k \in \mathbb{N}$. Prove that, for $x \in [-1, 1]$, we have

$$\sum_{n=2}^{\infty} \left[(k+1)^n \left(\text{Li}_n(x) - x - \frac{x^2}{2^n} - \cdots - \frac{x^k}{k^n} \right) - x^{k+1} \right]$$

$$= (k+1) \left[\left(1 - x^{k+1} \right) \ln(1 - x) + x + \frac{x^2}{2} + \cdots + \frac{x^{k+1}}{k+1} \right].$$

In particular, for $x = 1$ we obtain part (e) of Problem 3.29

(continued)

$$\sum_{n=2}^{\infty} \left[(k+1)^n \left(\zeta(n) - 1 - \frac{1}{2^n} - \cdots - \frac{1}{k^n} \right) - 1 \right] = (k+1)H_{k+1}.$$

(m) Let $k \in \mathbb{N}$. Prove that, for $x \in [-1, 1]$, we have

$$\sum_{n=2}^{\infty} (n-1) \left[(k+1)^n \left(\mathrm{Li}_n(x) - x - \frac{x^2}{2^n} - \cdots - \frac{x^k}{k^n} \right) - x^{k+1} \right]$$

$$= (k+1)^2 x^{k+1} \mathrm{Li}_2(x).$$

In particular, for $x = 1$ we obtain part (b) of Problem 3.32

$$\sum_{n=2}^{\infty} (n-1) \left[(k+1)^n \left(\zeta(n) - 1 - \frac{1}{2^n} - \cdots - \frac{1}{k^n} \right) - 1 \right] = (k+1)^2 \zeta(2).$$

Remark 3.2. From parts (a) and (g) we have, for $x \in [-1, 1]$, that

$$\sum_{n=2}^{\infty} (n-1) \left(\mathrm{Li}_n(x) - x \right) = x \, \mathrm{Li}_2(x).$$

Multiple Polylogarithm Series

(n) Prove that, for $x \in [-1, 1]$, we have

$$\sum_{n=1}^{\infty} \sum_{m=1}^{\infty} \left(\mathrm{Li}_{n+m}(x) - x \right) = x \, \mathrm{Li}_2(x).$$

In particular, for $x = 1$ we obtain part (a) of Problem 3.96

$$\sum_{n=1}^{\infty} \sum_{m=1}^{\infty} \left(\zeta(n+m) - 1 \right) = \zeta(2).$$

(o) Prove that, for $x \in [-1, 1]$, we have

$$\sum_{n=1}^{\infty} \sum_{m=1}^{\infty} n \left(\mathrm{Li}_{n+m}(x) - x \right) = x \left(\mathrm{Li}_2(x) + \mathrm{Li}_3(x) \right).$$

(continued)

In particular, for $x = 1$ we obtain part (a) of Problem 3.99

$$\sum_{n=1}^{\infty} \sum_{m=1}^{\infty} n \left(\zeta(n+m) - 1 \right) = \zeta(2) + \zeta(3).$$

(p) Prove that, for $x \in [-1, 1]$, we have

$$\sum_{n=1}^{\infty} \sum_{m=1}^{\infty} nm \left(\mathrm{Li}_{n+m}(x) - x \right) = x \left(\mathrm{Li}_2(x) + 2\mathrm{Li}_3(x) + \mathrm{Li}_4(x) \right).$$

In particular, for $x = 1$ we obtain part (a) of Problem 3.100

$$\sum_{n=1}^{\infty} \sum_{m=1}^{\infty} nm \left(\zeta(n+m) - 1 \right) = \zeta(2) + 2\zeta(3) + \zeta(4).$$

(q) Let $k \in \mathbb{N}$, $k \geq 2$. Prove that, for $x \in [-1, 1]$, we have

$$\sum_{n_1=1}^{\infty} \cdots \sum_{n_k=1}^{\infty} \left(\mathrm{Li}_{n_1 + \cdots + n_k}(x) - x \right) = x \, \mathrm{Li}_k(x).$$

In particular, for $x = 1$ we obtain part (b) of Problem 3.96

$$\sum_{n_1=1}^{\infty} \cdots \sum_{n_k=1}^{\infty} \left(\zeta(n_1 + \cdots + n_k) - 1 \right) = \zeta(k).$$

(r) Prove that, for $x \in [-1, 1]$, we have

$$\sum_{n=1}^{\infty} \sum_{m=1}^{\infty} \left(\mathrm{Li}_{2n+m}(x) - \mathrm{Li}_{n+2m}(x) \right) = 0.$$

In particular, for $x = 1$ we obtain the series

$$\sum_{n=1}^{\infty} \sum_{m=1}^{\infty} \left(\zeta(2n+m) - \zeta(n+2m) \right) = 0.$$

(continued)

(s) Let $k \in \mathbb{N}$ and $1 \le i \le k$. Prove that, for $x \in [-1, 1]$, we have

$$\sum_{n_1=1}^{\infty} \cdots \sum_{n_k=1}^{\infty} n_1 \cdots n_i \left(\mathrm{Li}_{n_1+\cdots+n_k}(x) - x \right) = x \sum_{j=0}^{i} C_i^j \mathrm{Li}_{i+k-j}(x).$$

In particular, for $x = 1$ and $i = k$ we obtain part (b) of Problem 3.100

$$\sum_{n_1=1}^{\infty} \cdots \sum_{n_k=1}^{\infty} n_1 \cdots n_k \left(\zeta(n_1 + \cdots + n_k) - 1 \right) = \sum_{j=0}^{k} C_k^j \zeta(2k - j).$$

3.42. Polylogarithm Delights

Let $x \in [-1, 1]$. Prove that:

(a) *A remarkable power series*

$$\sum_{i=2}^{\infty} \frac{x^i}{i^n(i-1)}$$

$$= nx + (1-x)\ln(1-x) - \mathrm{Li}_2(x) - \mathrm{Li}_3(x) - \cdots - \mathrm{Li}_n(x), \quad n \ge 2;$$

(b) *polylogarithm series I*

$$\sum_{n=1}^{\infty} \left(\mathrm{Li}_2(x) + \mathrm{Li}_3(x) + \cdots + \mathrm{Li}_{n+1}(x) - (n+1)x - (1-x)\ln(1-x) \right)$$

$$= (1-x)\ln(1-x) - x\mathrm{Li}_2(x) + x;$$

(c) *polylogarithm series II*

$$\sum_{n=1}^{\infty} n \left(\mathrm{Li}_2(x) + \mathrm{Li}_3(x) + \cdots + \mathrm{Li}_{n+1}(x) - (n+1)x - (1-x)\ln(1-x) \right)$$

$$= -x\mathrm{Li}_3(x);$$

(continued)

(d) *polylogarithm series III*

$$\sum_{n=1}^{\infty}(-1)^{n-1}$$

$$\times \left(\mathrm{Li}_2(x) + \mathrm{Li}_3(x) + \cdots + \mathrm{Li}_{n+1}(x) - (n+1)x - (1-x)\ln(1-x)\right)$$

$$= \frac{(1-x)^2}{2x}\ln(1-x) + \frac{1}{2} - \frac{3x}{4};$$

(e) *polylogarithm series IV*

$$\sum_{n=2}^{\infty}\left[2^n\left(\mathrm{Li}_2(x) + \mathrm{Li}_3(x) + \cdots + \mathrm{Li}_n(x) - nx - (1-x)\ln(1-x)\right) + x^2\right]$$

$$= 2(1-x)^2\ln(1-x) - 3x^2 + 2x;$$

(f) *polylogarithm series V*

$$\sum_{n=2}^{\infty}n\left[2^n\left(\mathrm{Li}_2(x) + \mathrm{Li}_3(x) + \cdots + \mathrm{Li}_n(x) - nx - (1-x)\ln(1-x)\right) + x^2\right]$$

$$= 2(1-x^2)\ln(1-x) - 4x^2\mathrm{Li}_2(x) + x^2 + 2x;$$

(g) *polylogarithm series VI*
     Let $k \in \mathbb{N}$.

$$\sum_{n_1=1}^{\infty}\cdots\sum_{n_k=1}^{\infty}$$

$$\left(\mathrm{Li}_2(x) + \cdots + \mathrm{Li}_{n_1+\cdots+n_k}(x) - (n_1 + \cdots + n_k)x - (1-x)\ln(1-x)\right)$$

$$= -x\,\mathrm{Li}_{k+1}(x).$$

Remark 3.3. The first five parts generalize Problem 3.38 to the case of polylogarithm functions. When $x = 1$ we obtain Problem 3.38.

3.43. Alternating Series with Polylogarithm Functions

Let $x \in [-1, 1] \setminus \{0\}$ and $n \in \mathbb{N}$, $n \geq 2$. Prove that:

(a) *A remarkable power series*

$$\sum_{i=2}^{\infty} \frac{x^i}{i^n (i+1)}$$

$$= (-1)^n \left(\mathrm{Li}_2(x) - \mathrm{Li}_3(x) + \cdots + (-1)^n \mathrm{Li}_n(x) - \frac{1-x}{x} \ln(1-x) - 1 \right)$$

$$- \frac{x}{2};$$

(b) *A limit with polylogarithm functions*

$$\lim_{n \to \infty} \left[(-1)^n \left(\mathrm{Li}_2(x) + \cdots + (-1)^n \mathrm{Li}_n(x) - \frac{1-x}{x} \ln(1-x) - 1 \right) - \frac{x}{2} \right]$$

$$= 0;$$

In particular, for $x = 1$ we obtain the limit

$$\lim_{n \to \infty} \left[(-1)^n \left(\zeta(2) - \zeta(3) + \cdots + (-1)^n \zeta(n) - 1 \right) - \frac{1}{2} \right] = 0;$$

(c) *Alternating series I*

$$\sum_{n=2}^{\infty} \left[(-1)^n \left(\mathrm{Li}_2(x) + \cdots + (-1)^n \mathrm{Li}_n(x) - \frac{1-x}{x} \ln(1-x) - 1 \right) - \frac{x}{2} \right]$$

$$= -\frac{(1-x)^2}{2x} \ln(1-x) + \frac{3x}{4} - \frac{1}{2};$$

In particular, for $x = 1$ we obtain the series

$$\sum_{n=2}^{\infty} \left[(-1)^n \left(\zeta(2) - \zeta(3) + \cdots + (-1)^n \zeta(n) - 1 \right) - \frac{1}{2} \right] = \frac{1}{4};$$

(continued)

(d) *Alternating series II*

$$\sum_{n=2}^{\infty}(-1)^n\left[(-1)^n\left(\mathrm{Li}_2(x)+\cdots+(-1)^n\mathrm{Li}_n(x)-\frac{1-x}{x}\ln(1-x)-1\right)-\frac{x}{2}\right]$$

$$=\frac{1-x}{x}\ln(1-x)-\frac{\mathrm{Li}_2(x)}{x}-\frac{x}{4}+2;$$

In particular, for $x = 1$ we obtain the series

$$\sum_{n=2}^{\infty}(-1)^n\left[(-1)^n\left(\zeta(2)-\zeta(3)+\cdots+(-1)^n\zeta(n)-1\right)-\frac{1}{2}\right]=\frac{7}{4}-\zeta(2).$$

3.6   Inequalities and Integrals

3.44. [115] Prove that $x^x \le x^2 - x + 1$, for all $0 < x \le 1$.

3.45. Prove that $\frac{e^x+e^{-x}}{2} \le e^{\frac{x^2}{2}}$, for all $x \in \mathbb{R}$.

3.46. Young's Inequality ((a) and (c))

(a) Let $p, q > 1$, with $\frac{1}{p} + \frac{1}{q} = 1$, and $x, y \ge 0$. Prove that

$$\frac{x^p}{p} + \frac{y^q}{q} \ge xy.$$

(b) [111] Let $p, q > 1$, with $\frac{1}{p} + \frac{1}{q} = 1$, and $x, y \in (0, 1)$. Prove that

$$\frac{q}{1-x^p} + \frac{p}{1-y^q} \ge \frac{pq}{1-xy}.$$

(c) Let $p, q, r > 1$, with $\frac{1}{p} + \frac{1}{q} + \frac{1}{r} = 1$, and $x, y, t \ge 0$. Prove that

$$\frac{x^p}{p} + \frac{y^q}{q} + \frac{t^r}{r} \ge xyt.$$

(d) Let $p, q, r > 1$, with $\frac{1}{p} + \frac{1}{q} + \frac{1}{r} = 1$, and $x, y, t \in (0, 1)$. Prove that

$$\frac{qr}{1-x^p} + \frac{pr}{1-y^q} + \frac{pq}{1-t^r} \ge \frac{pqr}{1-xyt}.$$

3.47. Nesbitt's Inequality

(a) If a, b, c are positive real numbers, then

$$\frac{a}{b+c} + \frac{b}{a+c} + \frac{c}{a+b} \geq \frac{3}{2}.$$

(b) If $n \in \mathbb{N}$, $n \geq 2$, and $x_1, x_2, \ldots, x_n$ are positive real numbers, then

$$\frac{x_1}{x_2 + x_3 + \cdots + x_n} + \frac{x_2}{x_1 + x_3 + \cdots + x_n} + \cdots + \frac{x_n}{x_1 + x_2 + \cdots + x_{n-1}}$$

$$\geq \frac{n}{n-1}.$$

3.48. Circular Inequalities

(a) [111] Prove that, for $a, b, c \in (-1, 1)$, the following inequality holds:

$$\frac{1}{1-a^2} + \frac{1}{1-b^2} + \frac{1}{1-c^2} \geq \frac{1}{1-ab} + \frac{1}{1-bc} + \frac{1}{1-ac}.$$

(b) Prove that, for $a, b, c, d \in (0, 1)$, the following inequality holds:

$$\frac{1}{1-a^3} + \frac{1}{1-b^3} + \frac{1}{1-c^3} + \frac{1}{1-d^3} \geq \frac{1}{1-abc} + \frac{1}{1-bcd} + \frac{1}{1-cda} + \frac{1}{1-dab}.$$

3.49. Let $n \in \mathbb{N}$ and

$$I_n = \int_0^\infty \frac{1}{(1+x^2)^n} dx.$$

Prove that:

(a) $\displaystyle\sum_{n=1}^\infty \frac{I_n}{n} = \pi$;

(b) $\displaystyle\sum_{n=2}^\infty \frac{(2n-3)!!}{(2n)!!} = \frac{1}{2}.$

3.50. [O. Furdui, Problem 4, SEEMOUS 2016 (parts (a) and (b))]
Let $n \in \mathbb{N}$ and

$$I_n = \int_0^\infty \frac{\arctan x}{(1+x^2)^n} dx.$$

Prove that:

(a) $\displaystyle\sum_{n=1}^\infty \frac{I_n}{n} = \zeta(2);$

(b) $\displaystyle\int_0^\infty \arctan x \ln\left(1 + \frac{1}{x^2}\right) dx = \zeta(2);$

(c) $\displaystyle\int_0^\infty \operatorname{arccot} x \ln\left(1 + \frac{1}{x^2}\right) dx = 2\zeta(2).$

3.51. Let $a > 1$. Prove that:

(a) $\displaystyle\sum_{n=-\infty}^\infty \frac{(-1)^n}{an+1} = \frac{\pi}{a \sin \frac{\pi}{a}};$

(b) $\displaystyle\sum_{n=-\infty}^\infty \frac{1}{(an+1)^2} = -\int_0^1 \frac{1+x^{a-2}}{1-x^a} \ln x \, dx.$

When $a = 2$ one obtains sophomore's dream for series [9, problem 11, part a), p. 381]

$$\left(\sum_{n=-\infty}^\infty \frac{(-1)^n}{2n+1}\right)^2 = \sum_{n=-\infty}^\infty \frac{1}{(2n+1)^2}.$$

3.52. *Sophomore's dream.* Using the power series expansion of the exponential function, prove that

$$\sum_{n=1}^\infty \frac{1}{n^n} = \int_0^1 \frac{1}{x^x} dx \quad \text{and} \quad \sum_{n=1}^\infty (-1)^{n-1} \frac{1}{n^n} = \int_0^1 x^x \, dx.$$

Remark 3.4. Formula

$$\sum_{n=1}^{\infty} \frac{1}{n^n} = \int_0^1 \frac{1}{x^x}\,dx$$

is known in the mathematical literature as sophomore's dream because of the likeness of the terms in the two members of the formula [112, p. 189]. The equality is too good to be true, however, it is true.

The second integral formula

$$\sum_{n=1}^{\infty} (-1)^{n-1} \frac{1}{n^n} = \int_0^1 x^x\,dx$$

was discovered by the Swiss mathematician John Bernoulli (1667–1748) in 1697, being so fascinated by this result that he called it "*series mirabili*", which means *marvelous series* [112, p. xvi].

3.53.

(a) *An exponential integral with the floor function.*
   Let $a \in (0, 1)$. Calculate

$$\int_0^1 a^{\lfloor \frac{1}{x} \rfloor}\,dx,$$

   where $\lfloor x \rfloor$ is the floor of the real number x.
(b) *A particular case.* Prove that

$$\int_0^1 2^{-\lfloor \frac{1}{x} \rfloor}\,dx = 1 - \ln 2.$$

3.54. *Curiosities with integrals.* Prove that

$$\int_0^{\infty} 2^{-\lfloor x \rfloor}\,dx = 2 \quad \text{and} \quad \int_0^{\infty} x\, 2^{-\lfloor x \rfloor}\,dx = 3,$$

where $\lfloor x \rfloor$ is the floor of the real number x.

3.55. Prove that

$$\int_0^{\infty} \lfloor x \rfloor e^{-x}\,dx = \frac{1}{e-1},$$

where $\lfloor x \rfloor$ is the floor of the real number x.

3.56. Prove that

$$\int_0^1 \frac{\ln(1-x)}{x}dx = -\zeta(2) \quad \text{and} \quad \int_0^1 \left(\frac{\ln(1-x)}{x}\right)^2 dx = 2\zeta(2).$$

3.57. Prove that

$$\int_0^1 \frac{\ln(1+x)}{x}dx = \frac{\pi^2}{12} \quad \text{and} \quad \int_0^1 \left(\frac{\ln(1+x)}{x}\right)^2 dx = \frac{\pi^2}{6} - 2\ln^2 2.$$

3.58.

(a) [139] Let $n \geq 2$ be an integer. Prove that

$$\int_0^1 \frac{\ln(1 + x + x^2 + \cdots + x^{n-1})}{x}dx = \frac{\pi^2(n-1)}{6n}.$$

(b) Let $n \geq 2$ and $k \geq 0$ be integers. Prove that

$$\int_0^1 \frac{\ln^k x \ln(1 + x + x^2 + \cdots + x^{n-1})}{x}dx = (-1)^k k! \left(1 - \frac{1}{n^{k+1}}\right)\zeta(k+2).$$

3.59. Calculate

$$\int_0^1 \frac{\ln(1 - x + x^2)}{x}dx.$$

3.60.

(a) Let $n \geq 3$ be an odd integer. Prove that

$$\int_0^1 \frac{\ln(1 - x + x^2 - x^3 + \cdots + x^{n-1})}{x}dx = \left(\frac{1}{n} - 1\right)\frac{\pi^2}{12}.$$

(b) Let $n \geq 3$ be an odd integer and $k \geq 0$ be an integer. Prove that

$$\int_0^1 \frac{\ln^k x \ln(1 - x + x^2 - x^3 + \cdots + x^{n-1})}{x}dx$$

$$= (-1)^k k! \left(1 - \frac{1}{2^{k+1}}\right)\left(\frac{1}{n^{k+1}} - 1\right)\zeta(k+2).$$

3.61. *A surprising series with binomial coefficients.*

(a) [49] Prove that

$$\int_0^1 \frac{\ln(1 - x + x^2)}{x - x^2} \, dx = -\frac{\pi^2}{9}.$$

(b) Deduce the surprising series giving $\zeta(2)$

$$3 \sum_{n=1}^{\infty} \frac{1}{n^2 C_{2n}^n} = \zeta(2).$$

3.7 Generating Functions

The Cauchy Product of Two Power Series

The Cauchy product of the power series $\sum_{n=0}^{\infty} a_n x^n$ and $\sum_{n=0}^{\infty} b_n x^n$, with $a_n, b_n \in \mathbb{R}$, is the power series

$$\sum_{n=0}^{\infty} c_n x^n, \quad \text{where} \quad c_n = \sum_{k=0}^{n} a_k b_{n-k}.$$

3.62. Write the first three nonzero terms of the power series expansion in x of the following functions:

(a) $f : \mathbb{R} \to \mathbb{R}, \quad f(x) = e^{-x} \sin x;$ (b) $f : \mathbb{R} \to \mathbb{R}, \quad f(x) = e^{3x} \cos x;$
(c) $f : \left(-\frac{\pi}{2}, \frac{\pi}{2}\right) \to \mathbb{R}, \quad f(x) = \tan x;$ (d) $f : \mathbb{R} \to \mathbb{R}, \quad f(x) = \tanh(2x).$

3.63. *The generating function of the nth harmonic number H_n.*

(a) Prove that

$$-\frac{\ln(1 - x)}{1 - x} = \sum_{n=1}^{\infty} H_n x^n, \quad x \in (-1, 1).$$

(b) Prove, by differentiating the equality in part (a), that

$$\frac{1 - \ln(1 - x)}{(1 - x)^2} = \sum_{n=1}^{\infty} n H_n x^{n-1}, \quad x \in (-1, 1).$$

(c) Prove, by integrating the equality in part (a), that

$$\ln^2(1-x) = 2\sum_{n=1}^{\infty} \frac{H_n}{n+1} x^{n+1}, \quad x \in [-1, 1).$$

(d) Prove the identity

$$\sum_{k=2}^{n} \frac{1}{k-1} \cdot \frac{1}{n+1-k} = 2\frac{H_{n-1}}{n}, \quad n \geq 2.$$

3.64.

(a) *Computing* $-\mathrm{Li}_2\left(\frac{1}{2}\right)$. Prove that

$$\int_0^{\frac{1}{2}} \frac{\ln(1-x)}{x} dx = \frac{\ln^2 2}{2} - \frac{\pi^2}{12}.$$

(b) *Two series with the harmonic number.* Prove that

$$\sum_{n=1}^{\infty} \frac{H_n}{n2^{n-1}} = \zeta(2) \quad \text{and} \quad \sum_{n=1}^{\infty} \frac{H_n}{(n+1)2^n} = \ln^2 2.$$

3.65. *The generating function of the harmonic number* O_n.

(a) Prove that

$$\frac{1}{4}\ln^2 \frac{1+x}{1-x} = \sum_{n=1}^{\infty}\left(1 + \frac{1}{3} + \cdots + \frac{1}{2n-1}\right)\frac{x^{2n}}{n}, \quad x \in (-1, 1).$$

(b) Prove that

$$\frac{x}{2(1-x^2)}\ln\frac{1+x}{1-x} = \sum_{n=1}^{\infty}\left(1 + \frac{1}{3} + \cdots + \frac{1}{2n-1}\right)x^{2n}, \quad x \in (-1, 1).$$

(c) Prove that

$$\sum_{n=1}^{\infty} \frac{O_n}{n^2} = \frac{7}{4}\zeta(3).$$

3.66. *The generating function of the skew-harmonic number H_n^-.*

Let $n \in \mathbb{N}$. The skew-harmonic number H_n^- is defined by

$$H_n^- = 1 - \frac{1}{2} + \frac{1}{3} - \cdots + \frac{(-1)^{n-1}}{n}.$$

(a) Prove that

$$H_n^- = \int_0^1 \frac{1 - (-t)^n}{1 + t}\,dt, \quad n \geq 1.$$

(b) Prove that

$$\sum_{n=1}^{\infty} H_n^- x^n = \frac{\ln(1 + x)}{1 - x}, \quad x \in (-1, 1).$$

3.67. *The generating function of the sequence $(E_n)_{n \geq 0}$.*

Find the Maclaurin series expansion of the function $f : \mathbb{R} \setminus \{1\} \to \mathbb{R}$, $f(x) = \frac{e^x}{1-x}$, and mention the set on which the expansion holds.

3.68. *The generating function of Bernoulli numbers.*

The Bernoulli numbers, B_n, $n \geq 0$, are defined by the equality

$$\frac{x}{e^x - 1} = \sum_{n=0}^{\infty} \frac{B_n}{n!} x^n.$$

Determine the first five nonzero Bernoulli numbers.

Remark 3.5. The Bernoulli numbers satisfy the recurrence relation

$$C_{n+1}^1 B_n + C_{n+1}^2 B_{n-1} + \cdots + C_{n+1}^k B_{n-k+1} + \cdots + C_{n+1}^n B_1 + 1 = 0, \quad n \geq 1,$$

which can be written in the following form:

$$(B + 1)^{n+1} - B^{n+1} = 0, \quad n \geq 1,$$

where B^k is a notation for B_k.

All Bernoulli numbers of odd index, except for B_1, are equal to 0.

3.69. *The generating function of the Fibonacci sequence.*

Knowing that, for $x \in \left(\frac{1-\sqrt{5}}{2}, \frac{\sqrt{5}-1}{2} \right)$, one has the power series expansion

$$\frac{1}{1-x-x^2} = \sum_{n=1}^{\infty} a_n x^{n-1},$$

prove that $(a_n)_{n\geq 1}$ equals the Fibonacci sequence.

The Lucas sequence $(L_n)_{n\geq 0}$ is defined by the recurrence relation

$$L_0 = 2, \quad L_1 = 1, \quad L_{n+2} = L_{n+1} + L_n, \quad \forall n \geq 0.$$

3.70. *The generating function of the Lucas sequence.*

(a) Prove that, for $x \in \left(\frac{1-\sqrt{5}}{2}, \frac{\sqrt{5}-1}{2}\right)$, the following equality holds

$$\frac{x-2}{x^2+x-1} = \sum_{n=0}^{\infty} L_n x^n.$$

(b) Prove that, for $x \in \left(\frac{1-\sqrt{5}}{2}, \frac{\sqrt{5}-1}{2}\right)$, the following equality holds

$$\ln\frac{1}{1-x-x^2} = \sum_{n=1}^{\infty} \frac{L_n}{n} x^n.$$

3.71.

(a) *The generating function of the sequence* $(\sin n)_{n\geq 1}$.
    Prove that, for $x \in (-1, 1)$, the following equality holds

$$\sum_{n=1}^{\infty} (\sin n) x^n = \frac{x \sin 1}{1 - 2x \cos 1 + x^2}.$$

(b) *The generating function of the sequence* $\left(\frac{\sin n}{n}\right)_{n\geq 1}$.
    Prove that, for $x \in [-1, 1]$, the following equality holds

$$\sum_{n=1}^{\infty} \frac{\sin n}{n} x^n = \arctan\frac{x - \cos 1}{\sin 1} + \frac{\pi}{2} - 1.$$

(c) *Two special series with sine.*
    Prove that

$$\sum_{n=1}^{\infty} \frac{\sin n}{n} = \frac{\pi - 1}{2} \quad \text{and} \quad \sum_{n=1}^{\infty} (-1)^{n-1} \frac{\sin n}{n} = \frac{1}{2}.$$

3.72.

(a) *The generating function of the sequence* $(\cos n)_{n \geq 1}$.
   Prove that, for $x \in (-1, 1)$, the following equality holds

$$\sum_{n=1}^{\infty} (\cos n) x^n = \frac{x \cos 1 - x^2}{1 - 2x \cos 1 + x^2}.$$

(b) *The generating function of the sequence* $\left(\frac{\cos n}{n}\right)_{n \geq 1}$.
   Prove that, for $x \in [-1, 1]$, the following equality holds

$$\sum_{n=1}^{\infty} \frac{\cos n}{n} x^n = -\frac{1}{2} \ln(1 - 2x \cos 1 + x^2).$$

(c) *Two special series with cosine.*
   Prove that

$$\sum_{n=1}^{\infty} \frac{\cos n}{n} = -\ln\left(2 \sin \frac{1}{2}\right) \quad \text{and} \quad \sum_{n=1}^{\infty} (-1)^{n-1} \frac{\cos n}{n} = \ln\left(2 \cos \frac{1}{2}\right).$$

3.73. Let $(a_n)_{n \geq 0}$ be the sequence defined by

$$\frac{1}{x^2 - 2x + 1} = \sum_{n=0}^{\infty} a_n x^n, \quad x \in (-1, 1).$$

Prove that the numbers $m \in \mathbb{N}$, with the property that there exists $n \in \mathbb{N} \cup \{0\}$ such that

$$a_n a_{n+1} a_{n+2} a_{n+3} = a_{m-2},$$

are perfect squares.

3.74. Let $(a_n)_{n \geq 0}$ and $(b_n)_{n \geq 0}$ be the sequences defined by

$$\sinh x = \sum_{n=0}^{\infty} a_n x^n, \quad \cosh x = \sum_{n=0}^{\infty} b_n x^n, \quad x \in \mathbb{R}.$$

Eliminate a factor of the form $\frac{1}{a_{2k-1}}$ or $\frac{1}{b_{2k}}$ from the product $\prod_{k=1}^{2n} \frac{1}{a_{2k-1}b_{2k}}$ such that the remaining product be a perfect square.

3.8   Series with Harmonic and Skew-Harmonic Numbers

3.75. *Skew-harmonic numbers and tails of zeta function.*
Prove that:

(a) $\displaystyle\sum_{n=1}^{\infty} H_n^- \left(H_n^- - \ln 2 \right) = \frac{\ln 2}{2};$

(b) $\displaystyle\sum_{n=1}^{\infty} \left(H_n^- - \ln 2 \right) \left(\zeta(2) - 1 - \frac{1}{2^2} - \cdots - \frac{1}{n^2} \right) = -\frac{\pi^2}{8} + \frac{\pi^2 \ln 2}{6} + \frac{\ln^2 2}{2};$

(c) $\displaystyle\sum_{n=1}^{\infty} (-1)^n H_n^- \left(\zeta(2) - 1 - \frac{1}{2^2} - \cdots - \frac{1}{n^2} \right) = \frac{\pi^2 \ln 2}{8} - \frac{21}{16}\zeta(3);$

(d) $\displaystyle\sum_{n=1}^{\infty} H_n^- \left(\zeta(3) - 1 - \frac{1}{2^3} - \cdots - \frac{1}{n^3} \right) = \frac{\pi^2 \ln 2}{4} - \frac{9}{8}\zeta(3).$

3.76. *A mosaic of series.* Prove that:

(a) $\displaystyle\sum_{n=1}^{\infty} (-1)^n \frac{H_n^-}{n} = -\frac{\pi^2}{12} - \frac{\ln^2 2}{2};$

(b) $\displaystyle\sum_{n=1}^{\infty} \frac{H_n^-}{n^2} = \frac{\pi^2}{4} \ln 2 - \frac{\zeta(3)}{4};$

(c) $\displaystyle\sum_{n=1}^{\infty} \frac{H_n^-}{n^3} = \frac{7}{4}\zeta(3) \ln 2 - \frac{\pi^4}{288};$

(d) $\displaystyle\sum_{n=1}^{\infty} \frac{H_n^-}{n} \cdot \frac{H_{n+1}}{n+1} = \frac{5}{8}\zeta(3) + \frac{\pi^2}{12};$

(e) $\displaystyle\sum_{n=1}^{\infty} (-1)^n \frac{H_n^-}{n} \cdot \frac{H_{n+1}}{n+1} = \frac{3}{2}\zeta(3) - \frac{\pi^2}{3} \ln 2 - \ln^2 2 + \frac{2}{3} \ln^3 2.$

3.77.

(a) Prove that

$$\sum_{n=1}^{\infty}(-1)^{n-1}\frac{H_n^- H_{n+1}}{n+1} = 2\operatorname{Li}_3\left(\frac{1}{2}\right) = -\frac{\pi^2}{6}\ln 2 + \frac{\ln^3 2}{3} + \frac{7}{4}\zeta(3).$$

(b) Prove that

$$\sum_{n=1}^{\infty}(-1)^n\frac{H_n^- H_n}{n} = \frac{\ln^3 2}{3} - \frac{\pi^2}{6}\ln 2 - \frac{\zeta(3)}{4}.$$

3.78. Prove that:

(a) $\displaystyle\sum_{n=1}^{\infty}\frac{H_n}{n}\left(H_n^- - \ln 2\right) = \frac{\pi^2\ln 2}{12} - \frac{\zeta(3)}{4} - \frac{\ln^3 2}{6};$

(b) $\displaystyle\sum_{n=1}^{\infty}(-1)^n\frac{H_n}{n}\left(H_n^- - \ln 2\right) = -\frac{\pi^2\ln 2}{12} - \frac{\zeta(3)}{4} - \frac{\ln^3 2}{6};$

(c) $\displaystyle\sum_{n=1}^{\infty}\frac{H_{2n}}{n}\left(H_{2n}^- - \ln 2\right) = -\frac{\zeta(3)}{2} - \frac{\ln^3 2}{3};$

(d) $\displaystyle\sum_{n=1}^{\infty}\frac{H_{2n-1}}{2n-1}\left(H_{2n-1}^- - \ln 2\right) = \frac{\zeta(2)\ln 2}{2}.$

3.79. *Euler's related series.* Prove that

$$\sum_{n=1}^{\infty}\frac{H_{2n-1}}{(2n-1)^2} = \frac{21}{16}\zeta(3) \quad \text{and} \quad \sum_{n=1}^{\infty}\frac{H_{2n}}{n^2} = \frac{11}{4}\zeta(3).$$

3.80. *Gems with skew-harmonic numbers.*

(a) Let $n \in \mathbb{N}$. Prove that

(continued)

$$\int_0^1 x^{n-1} \ln(1+x)\,\mathrm{d}x = \frac{1-(-1)^n}{n}\ln 2 + \frac{(-1)^n}{n}H_n^-.$$

(b) Prove that

$$\sum_{n=1}^{\infty}(-1)^{n-1}\frac{H_{2n-1}^-}{2n-1} = \frac{3\pi \ln 2}{8}.$$

(c) Prove that

$$\sum_{n=1}^{\infty}\frac{H_n^-}{n(n+1)} = \frac{\pi^2}{12} \quad \text{and} \quad \sum_{n=1}^{\infty}(-1)^{n-1}\frac{H_n^-}{n(n+1)} = \ln^2 2.$$

(d) *A nonlinear skew-harmonic sum I.*
    Prove that

$$\sum_{n=1}^{\infty}\frac{H_n^-}{n}\cdot\frac{H_{n+1}^-}{n+1} = \frac{\pi^2 \ln 2}{4} - \ln^2 2 - \frac{5}{8}\zeta(3).$$

(e) *A nonlinear skew-harmonic sum II.*
    Prove that

$$\sum_{n=1}^{\infty}\frac{H_n H_n^-}{n^2} = \frac{43\pi^4}{1440} + \frac{\pi^2 \ln^2 2}{8} - \frac{\ln^4 2}{8} - 3\mathrm{Li}_4\left(\frac{1}{2}\right).$$

3.81.

(a) Prove that

$$\sum_{n=1}^{\infty}\frac{(-1)^n}{n}\left(\zeta(2) - 1 - \frac{1}{2^2} - \cdots - \frac{1}{n^2}\right) = \zeta(3) - \frac{\pi^2}{4}\ln 2.$$

(b) Prove that

$$\sum_{n=1}^{\infty}\frac{(-1)^n}{n}\left(\zeta(3) - 1 - \frac{1}{2^3} - \cdots - \frac{1}{n^3}\right) = \frac{19\pi^4}{1440} - \frac{7}{4}\zeta(3)\ln 2.$$

(continued)

(c) Prove that

$$\sum_{n=1}^{\infty} \frac{H_n^-}{n}\left(\zeta(2) - 1 - \frac{1}{2^2} - \cdots - \frac{1}{n^2}\right)$$

$$= \frac{53\pi^4}{1440} + \frac{\pi^2 \ln^2 2}{4} - \frac{\ln^4 2}{8} - \frac{21 \ln 2}{8}\zeta(3) - 3\operatorname{Li}_4\left(\frac{1}{2}\right).$$

Open problem. Calculate

$$\sum_{n=1}^{\infty} \frac{H_n^-}{n}\left(\zeta(3) - 1 - \frac{1}{2^3} - \cdots - \frac{1}{n^3}\right).$$

3.9   Remarkable Numerical and Function Series

3.82. Let

$$u(x) = 1 + \frac{x^3}{3!} + \frac{x^6}{6!} + \cdots,$$

$$v(x) = x + \frac{x^4}{4!} + \frac{x^7}{7!} + \cdots,$$

$$w(x) = \frac{x^2}{2!} + \frac{x^5}{5!} + \frac{x^8}{8!} + \cdots.$$

Prove that

$$u^2(x) + v^2(x) + w^2(x) - u(x)v(x) - v(x)w(x) - w(x)u(x) = e^{-x}$$

and

$$u^3(x) + v^3(x) + w^3(x) - 3u(x)v(x)w(x) = 1.$$

3.83.

(a) Prove that

$$\ln\frac{1}{1-x} - x - \frac{x^2}{2} - \cdots - \frac{x^n}{n} = \int_0^x \frac{t^n}{1-t}dt, \quad x \in [-1, 1).$$

(b) Prove that

$$\sum_{n=1}^{\infty} \frac{1}{n} \left(\ln \frac{1}{1-x} - x - \frac{x^2}{2} - \cdots - \frac{x^n}{n} \right) = \frac{\ln^2(1-x)}{2}, \quad x \in [-1, 1).$$

3.84. Calculate:

(a) $\displaystyle\sum_{n=1}^{\infty} \left(\ln \frac{1}{1-x} - x - \frac{x^2}{2} - \cdots - \frac{x^n}{n} \right), \quad x \in [-1, 1);$

(b) $\displaystyle\sum_{n=0}^{\infty} \left(\arctan x - x + \frac{x^3}{3} + \cdots + (-1)^{n+1} \frac{x^{2n+1}}{2n+1} \right), \quad x \in [-1, 1].$

3.85. Prove that:

(a) $H_1 + H_2 + \cdots + H_n = (n+1)H_n - n, \ \forall n \geq 1;$
(b) *harmonic numbers and the tail of* $-\ln(1-x)$.

$$\sum_{n=1}^{\infty} H_n \left(\ln \frac{1}{1-x} - x - \frac{x^2}{2} - \cdots - \frac{x^n}{n} \right)$$

$$= -\frac{\ln(1-x)}{1-x} - \frac{x}{1-x}, \quad x \in [-1, 1);$$

(c) *harmonic numbers and the tail of* $\text{Li}_2(x)$.

$$\sum_{n=1}^{\infty} H_n \left(\text{Li}_2(x) - x - \frac{x^2}{2^2} - \cdots - \frac{x^n}{n^2} \right)$$

$$= \text{Li}_2(x) + \frac{\ln^2(1-x)}{2} + \ln(1-x), \quad x \in [-1, 1);$$

(d) *harmonic numbers and the tail of* $\text{Li}_3(x)$.

$$\sum_{n=1}^{\infty} H_n \left(\text{Li}_3(x) - x - \frac{x^2}{2^3} - \cdots - \frac{x^n}{n^3} \right)$$

$$= \text{Li}_3(x) - \text{Li}_2(x) + \frac{1}{2} \int_0^x \frac{\ln^2(1-t)}{t} dt, \quad x \in [-1, 1];$$

In particular, when $x = 1$ we obtain the series of Problem 2.56

(continued)

$$\sum_{n=1}^{\infty} H_n \left(\zeta(3) - 1 - \frac{1}{2^3} - \cdots - \frac{1}{n^3} \right) = 2\zeta(3) - \zeta(2).$$

(e) *harmonic numbers and the tail of* $\mathrm{Li}_4(x)$.

$$\sum_{n=1}^{\infty} H_n \left(\mathrm{Li}_4(x) - x - \frac{x^2}{2^4} - \cdots - \frac{x^n}{n^4} \right)$$

$$= \mathrm{Li}_4(x) - \mathrm{Li}_3(x) + \frac{\ln x}{2} \int_0^x \frac{\ln^2(1-t)}{t} dt$$

$$- \frac{1}{2} \int_0^x \frac{\ln t \ln^2(1-t)}{t} dt, \quad x \in [-1, 1];$$

In particular, when $x = 1$ we obtain the first series in Remark 2.6

$$\sum_{n=1}^{\infty} H_n \left(\zeta(4) - 1 - \frac{1}{2^4} - \cdots - \frac{1}{n^4} \right) = \frac{5}{4}\zeta(4) - \zeta(3).$$

3.86.

(a) Prove that, for $x \in (-1, 1]$, we have

$$\sum_{n=1}^{\infty} \left(\frac{1}{n^2} - \frac{1}{(n+1)^2} + \frac{1}{(n+2)^2} - \cdots \right) x^n = \frac{x}{1+x} \left(\mathrm{Li}_2(x) - \mathrm{Li}_2(-1) \right).$$

For $x = -1$ we obtain [O. Furdui, A. Sîntămărian, Problem 4, part (b), SEEMOUS 2019]

$$\sum_{n=1}^{\infty} (-1)^{n-1} \left(\frac{1}{n^2} - \frac{1}{(n+1)^2} + \frac{1}{(n+2)^2} - \cdots \right) = \ln 2$$

and, for $x = 1$ we have the series

$$\sum_{n=1}^{\infty} \left(\frac{1}{n^2} - \frac{1}{(n+1)^2} + \frac{1}{(n+2)^2} - \cdots \right) = \frac{\pi^2}{8}.$$

(continued)

(b) Prove that

$$\sum_{n=1}^{\infty} \frac{1}{n} \left(\frac{1}{n^2} - \frac{1}{(n+1)^2} + \frac{1}{(n+2)^2} - \cdots \right) x^n$$

$$= \ln(1+x) \left(\mathrm{Li}_2(x) - \mathrm{Li}_2(-1) \right) + \int_0^x \frac{\ln(1-t)\ln(1+t)}{t} dt, \quad x \in [-1, 1].$$

In particular, for $x = -1$ we have

$$\sum_{n=1}^{\infty} \frac{(-1)^{n-1}}{n} \left(\frac{1}{n^2} - \frac{1}{(n+1)^2} + \frac{1}{(n+2)^2} - \cdots \right) = \frac{5}{8}\zeta(3)$$

and, for $x = 1$ we obtain the series

$$\sum_{n=1}^{\infty} \frac{1}{n} \left(\frac{1}{n^2} - \frac{1}{(n+1)^2} + \frac{1}{(n+2)^2} - \cdots \right) = \frac{\pi^2 \ln 2}{4} - \frac{5}{8}\zeta(3).$$

3.87. [5 April 2019] **Remarkable Series of Functions**

(a) Prove that

$$\sum_{n=1}^{\infty} \left(\ln \frac{1}{1-x} - x - \frac{x^2}{2} - \cdots - \frac{x^n}{n} \right) \left(\zeta(2) - 1 - \frac{1}{2^2} - \cdots - \frac{1}{n^2} \right)$$

$$= \frac{x}{1-x} \left(\zeta(2) - \mathrm{Li}_2(x) \right) + \zeta(2) \ln(1-x) + \frac{\ln^2(1-x)}{2}, \quad -1 \le x < 1.$$

In particular, for $x = -1$ we have

$$\sum_{n=1}^{\infty} \left(\ln \frac{1}{2} - (-1) - \frac{(-1)^2}{2} - \cdots - \frac{(-1)^n}{n} \right) \left(\zeta(2) - 1 - \frac{1}{2^2} - \cdots - \frac{1}{n^2} \right)$$

$$= -\frac{\pi^2}{8} + \frac{\pi^2}{6} \ln 2 + \frac{\ln^2 2}{2},$$

and, for $x = \frac{1}{2}$ we obtain the series

(continued)

$$\sum_{n=1}^{\infty} \left(\ln 2 - \frac{1}{2} - \frac{1}{2^2 2} - \cdots - \frac{1}{2^n n} \right) \left(\zeta(2) - 1 - \frac{1}{2^2} - \cdots - \frac{1}{n^2} \right)$$

$$= \frac{\pi^2}{12} + \ln^2 2 - \frac{\pi^2}{6} \ln 2.$$

(b) Prove that

$$\sum_{n=1}^{\infty} \left(\mathrm{Li}_2(x) - x - \frac{x^2}{2^2} - \cdots - \frac{x^n}{n^2} \right) \left(\zeta(2) - 1 - \frac{1}{2^2} - \cdots - \frac{1}{n^2} \right)$$

$$= -\ln(1-x)(\zeta(2) - \mathrm{Li}_2(x)) - \zeta(2)\mathrm{Li}_2(x)$$

$$+ \frac{3}{2} \int_0^x \frac{\ln^2(1-t)}{t} \, dt, \quad -1 \le x \le 1.$$

In particular, for $x = 1$ we obtain the remarkable quadratic series [26, problem 3.22, p. 142]

$$\sum_{n=1}^{\infty} \left(\zeta(2) - 1 - \frac{1}{2^2} - \cdots - \frac{1}{n^2} \right)^2 = 3\zeta(3) - \frac{5}{2}\zeta(4).$$

3.88. [71] A Logarithmic Power Series

(a) Prove that

$$\ln \frac{1}{2} - (-1) - \frac{(-1)^2}{2} - \cdots - \frac{(-1)^n}{n} = (-1)^{n-1} \int_0^1 \frac{x^n}{1+x} \, dx.$$

(b) Prove that the convergence set of the power series

$$\sum_{n=1}^{\infty} \left(\ln \frac{1}{2} - (-1) - \frac{(-1)^2}{2} - \cdots - \frac{(-1)^n}{n} \right) x^n$$

is $(-1, 1]$.

(continued)

(c) Prove that

$$\sum_{n=1}^{\infty}\left(\ln\frac{1}{2}-(-1)-\frac{(-1)^2}{2}-\cdots-\frac{(-1)^n}{n}\right)x^n$$

$$=\begin{cases}\ln 2-\frac{1}{2} & \text{if } x=1\\[2mm]\dfrac{\ln(1+x)-x\ln 2}{1-x} & \text{if } x\in(-1,1).\end{cases}$$

3.89. Let $f(x)=\sum\limits_{n=1}^{\infty}a_nx^n$, $|x|<1$. Prove that if both series $\sum\limits_{n=1}^{\infty}a_n$ and
$\sum\limits_{n=1}^{\infty}(f(1)-a_1-a_2-\cdots-a_n)$ converge, then

$$\sum_{n=1}^{\infty}(f(1)-a_1-a_2-\cdots-a_n)x^n=\begin{cases}\dfrac{f(1)x-f(x)}{1-x} & \text{if } x\neq 1\\[2mm]f'(1)-f(1) & \text{if } x=1.\end{cases}$$

Applications. Problems 3.4 and 3.88.

3.90. Marvelous Numerical and Power Series

(a) Prove that

$$\frac{1}{n}-\frac{2}{n+1}+\frac{2}{n+2}-\frac{2}{n+3}+\cdots=\int_0^1 x^{n-1}\frac{1-x}{1+x}\,dx,\quad n\geq 1.$$

(b) [88] Prove that

$$\sum_{n=1}^{\infty}x^n\left(\frac{1}{n}-\frac{2}{n+1}+\frac{2}{n+2}-\frac{2}{n+3}+\cdots\right)$$

$$=\begin{cases}\dfrac{2x\ln 2+(1-x)\ln(1-x)}{1+x} & \text{if } x\in(-1,1)\\[2mm]\ln 2-1 & \text{if } x=-1\\[1mm]\ln 2 & \text{if } x=1.\end{cases}$$

(continued)

We mention the interesting formula

$$\sum_{n=1}^{\infty}\left(\frac{1}{n}-\frac{2}{n+1}+\frac{2}{n+2}-\frac{2}{n+3}+\cdots\right)=\ln 2.$$

(c) Prove that

$$\sum_{n=1}^{\infty}\left(\frac{1}{n}-\frac{2}{n+1}+\frac{2}{n+2}-\frac{2}{n+3}+\cdots\right)^2=4\ln 2-2\ln^2 2-\zeta(2).$$

(d) Prove that

$$\sum_{n=1}^{\infty}(-1)^n\left(\frac{1}{n}-\frac{2}{n+1}+\frac{2}{n+2}-\frac{2}{n+3}+\cdots\right)^2=\frac{\pi^2}{12}-2\ln^2 2.$$

Open problem. Calculate:

- $$\sum_{n=1}^{\infty}\left(\frac{1}{n}-\frac{2}{n+1}+\frac{2}{n+2}-\frac{2}{n+3}+\cdots\right)^3;$$
- $$\sum_{n=1}^{\infty}(-1)^n\left(\frac{1}{n}-\frac{2}{n+1}+\frac{2}{n+2}-\frac{2}{n+3}+\cdots\right)^3.$$

A Jewel with a Quadratic Series

We prove that

$$\sum_{n=1}^{\infty}\left(\frac{1}{n}-\frac{1}{n+1}+\frac{1}{n+2}-\cdots\right)^2=\ln 2.$$

We apply in our proof Abel's summation formula with $a_n=1$ and $b_n=\left(\frac{1}{n}-\frac{1}{n+1}+\frac{1}{n+2}-\cdots\right)^2$. We have

$$b_n-b_{n+1}=\frac{1}{n}\left(\frac{1}{n}-\frac{2}{n+1}+\frac{2}{n+2}-\frac{2}{n+3}+\cdots\right).$$

Therefore,

(continued)

$$\sum_{n=1}^{\infty}\left(\frac{1}{n}-\frac{1}{n+1}+\frac{1}{n+2}-\cdots\right)^2$$

$$=\lim_{n\to\infty}n\left(\frac{1}{n+1}-\frac{1}{n+2}+\frac{1}{n+3}-\cdots\right)^2$$

$$+\sum_{n=1}^{\infty}\left(\frac{1}{n}-\frac{2}{n+1}+\frac{2}{n+2}-\frac{2}{n+3}+\cdots\right)$$

$$=\sum_{n=1}^{\infty}\left(\frac{1}{n}-\frac{2}{n+1}+\frac{2}{n+2}-\frac{2}{n+3}+\cdots\right)$$

$$\overset{3.90\,(b)}{=}\ln 2.$$

This series was calculated using a different method in [26, problem 3.29].   ■

Another Jewel with a Quadratic Series

We prove that

$$\sum_{n=1}^{\infty}\left(\frac{1}{n^2}-\frac{1}{(n+1)^2}+\frac{1}{(n+2)^2}-\cdots\right)^2=\frac{\pi^2\ln 2}{2}-\frac{9}{4}\zeta(3). \qquad (3.1)$$

We apply in our proof Abel's summation formula with $a_n = 1$ and $b_n = \left(\frac{1}{n^2}-\frac{1}{(n+1)^2}+\frac{1}{(n+2)^2}-\cdots\right)^2$. We have

$$b_n - b_{n+1} = \frac{1}{n^2}\left(\frac{1}{n^2}-\frac{2}{(n+1)^2}+\frac{2}{(n+2)^2}-\cdots\right).$$

Therefore,

(continued)

$$\sum_{n=1}^{\infty} \left(\frac{1}{n^2} - \frac{1}{(n+1)^2} + \frac{1}{(n+2)^2} - \cdots \right)^2$$

$$= \lim_{n \to \infty} n \left(\frac{1}{(n+1)^2} - \frac{1}{(n+2)^2} + \frac{1}{(n+3)^2} - \cdots \right)^2$$

$$+ \sum_{n=1}^{\infty} \frac{1}{n} \left(\frac{1}{n^2} - \frac{2}{(n+1)^2} + \frac{2}{(n+2)^2} - \cdots \right)$$

$$= \sum_{n=1}^{\infty} \frac{1}{n} \left(\frac{1}{n^2} - \frac{2}{(n+1)^2} + \frac{2}{(n+2)^2} - \cdots \right) \tag{3.2}$$

$$= \sum_{n=1}^{\infty} \frac{1}{n} \left[\left(\frac{2}{n^2} - \frac{2}{(n+1)^2} + \frac{2}{(n+2)^2} - \cdots \right) - \frac{1}{n^2} \right]$$

$$= 2 \sum_{n=1}^{\infty} \frac{1}{n} \left(\frac{1}{n^2} - \frac{1}{(n+1)^2} + \frac{1}{(n+2)^2} - \cdots \right) - \zeta(3).$$

Using that $\frac{1}{k^2} = - \int_0^1 x^{k-1} \ln x \, dx$, we obtain

$$\frac{1}{n^2} - \frac{1}{(n+1)^2} + \frac{1}{(n+2)^2} - \cdots = - \int_0^1 \frac{x^{n-1}}{1+x} \ln x \, dx, \quad n \geq 1.$$

It follows that

$$\sum_{n=1}^{\infty} \frac{1}{n} \left(\frac{1}{n^2} - \frac{1}{(n+1)^2} + \frac{1}{(n+2)^2} - \cdots \right) = - \sum_{n=1}^{\infty} \frac{1}{n} \int_0^1 \frac{x^{n-1}}{1+x} \ln x \, dx$$

$$= - \int_0^1 \frac{\ln x}{1+x} \sum_{n=1}^{\infty} \frac{x^{n-1}}{n} \, dx$$

$$= \int_0^1 \frac{\ln x \, \ln(1-x)}{x(1+x)} \, dx.$$

We calculate the preceding integral and we have

(continued)

$$\int_0^1 \frac{\ln x \ln(1-x)}{x(1+x)}dx = \int_0^1 \ln x \ln(1-x)\left(\frac{1}{x} - \frac{1}{1+x}\right)dx$$

$$= \int_0^1 \frac{\ln x \ln(1-x)}{x}dx - \int_0^1 \frac{\ln x \ln(1-x)}{1+x}dx$$

$$= \zeta(3) - \frac{1}{2}\int_0^1 \frac{\ln^2 x + \ln^2(1-x) - \ln^2\left(\frac{1-x}{x}\right)}{1+x}dx$$

$$= \frac{\pi^2 \ln 2}{4} - \frac{5}{8}\zeta(3),$$

since

- $$\int_0^1 \frac{\ln^2 x}{1+x}dx = \frac{3}{2}\zeta(3);$$
- $$\int_0^1 \frac{\ln^2(1-x)}{1+x}dx = \frac{7}{4}\zeta(3) - \frac{\pi^2 \ln 2}{6} + \frac{\ln^3 2}{3};$$
- $$\int_0^1 \frac{\ln^2\left(\frac{1-x}{x}\right)}{1+x}dx \overset{\frac{1-x}{x}=t}{=} \int_0^\infty \frac{\ln^2 t}{(1+t)(2+t)}dt = \frac{\pi^2 \ln 2}{3} + \frac{\ln^3 2}{3}.$$

It follows that

$$\sum_{n=1}^\infty \frac{1}{n}\left(\frac{1}{n^2} - \frac{1}{(n+1)^2} + \frac{1}{(n+2)^2} - \cdots\right) = \frac{\pi^2 \ln 2}{4} - \frac{5}{8}\zeta(3), \qquad (3.3)$$

and the quadratic series is calculated based on formula (3.2). ∎

A Series with Harmonic Numbers
We prove that

$$\sum_{n=1}^\infty H_n\left(\frac{1}{n^2} - \frac{1}{(n+1)^2} + \frac{1}{(n+2)^2} - \cdots\right) = \frac{11}{16}\zeta(3) + \frac{\pi^2 \ln 2}{8}.$$

(continued)

We have

$$S = \sum_{n=1}^{\infty} H_n \left(\frac{1}{n^2} - \frac{1}{(n+1)^2} + \frac{1}{(n+2)^2} - \cdots \right)$$

$$= \left(\frac{1}{1^2} - \frac{1}{2^2} + \frac{1}{3^2} + \cdots \right) + \sum_{n=2}^{\infty} H_n \left(\frac{1}{n^2} - \frac{1}{(n+1)^2} + \frac{1}{(n+2)^2} - \cdots \right)$$

$$\overset{n-1=m}{=} \frac{\pi^2}{12} + \sum_{m=1}^{\infty} H_{m+1} \left(\frac{1}{(m+1)^2} - \frac{1}{(m+2)^2} + \frac{1}{(m+3)^2} - \cdots \right)$$

$$= \frac{\pi^2}{12} + \sum_{m=1}^{\infty} \left(H_m + \frac{1}{m+1} \right) \left[\frac{1}{m^2} - \left(\frac{1}{m^2} - \frac{1}{(m+1)^2} + \frac{1}{(m+2)^2} - \cdots \right) \right]$$

$$= \frac{\pi^2}{12} + \sum_{m=1}^{\infty} \frac{H_m}{m^2} - S + \sum_{m=1}^{\infty} \frac{1}{m+1} \left(\frac{1}{(m+1)^2} - \frac{1}{(m+2)^2} + \frac{1}{(m+3)^2} - \cdots \right)$$

$$\overset{m+1=n}{=} \frac{\pi^2}{12} + 2\zeta(3) - S + \sum_{n=2}^{\infty} \frac{1}{n} \left(\frac{1}{n^2} - \frac{1}{(n+1)^2} + \frac{1}{(n+2)^2} - \cdots \right)$$

$$\overset{(3.3)}{=} 2\zeta(3) - S + \frac{\pi^2 \ln 2}{4} - \frac{5}{8}\zeta(3)$$

$$= \frac{11}{8}\zeta(3) + \frac{\pi^2 \ln 2}{4} - S,$$

and the result follows. ∎

A challenge. Prove that

$$\sum_{n=1}^{\infty} (-1)^n H_n \left(\frac{1}{n^2} - \frac{1}{(n+1)^2} + \frac{1}{(n+2)^2} - \cdots \right)$$

$$= -\frac{5}{8}\zeta(3) - \frac{\pi^2}{12} + \ln 2 + \frac{\ln^2 2}{2}.$$

Two Nonlinear Series with Gaps

We prove that

$$\sum_{n=1}^{\infty} \left(\frac{1}{n^2} + \frac{1}{(n+2)^2} + \frac{1}{(n+4)^2} + \cdots \right)$$

$$\times \left(\frac{1}{(n+1)^2} + \frac{1}{(n+3)^2} + \frac{1}{(n+5)^2} + \cdots \right)$$

$$= \frac{21}{16} \zeta(3) - \frac{\pi^2 \ln 2}{8}$$

and the quadratic series with gaps

$$\sum_{n=1}^{\infty} \left(\frac{1}{n^2} + \frac{1}{(n+2)^2} + \frac{1}{(n+4)^2} + \cdots \right)^2 = \frac{3}{16} \zeta(3) + \frac{\pi^4}{128} + \frac{\pi^2 \ln 2}{8}.$$

Let

$$a_n = \frac{1}{n^2} + \frac{1}{(n+2)^2} + \frac{1}{(n+4)^2} + \cdots$$

and let

$$S_1 = \sum_{n=1}^{\infty} a_n a_{n+1} \quad \text{and} \quad S_2 = \sum_{n=1}^{\infty} a_n^2.$$

On the one hand, using that $ab = \frac{(a+b)^2 - a^2 - b^2}{2}$, we have

$$S_1 = \frac{1}{2} \sum_{n=1}^{\infty} \left(\frac{1}{n^2} + \frac{1}{(n+1)^2} + \frac{1}{(n+2)^2} + \cdots \right)^2$$

$$- \frac{S_2}{2} - \frac{1}{2} \sum_{n=1}^{\infty} \left(\frac{1}{(n+1)^2} + \frac{1}{(n+3)^2} + \cdots \right)^2$$

$$\overset{(7.2)}{=} \frac{3}{2} \zeta(3) - \frac{S_2}{2} - \frac{1}{2} \sum_{m=2}^{\infty} \left(\frac{1}{m^2} + \frac{1}{(m+2)^2} + \cdots \right)^2$$

$$= \frac{3}{2} \zeta(3) - \frac{S_2}{2} - \frac{1}{2} \left[S_2 - \left(\frac{1}{1^2} + \frac{1}{3^2} + \frac{1}{5^2} + \cdots \right)^2 \right]$$

$$= \frac{3}{2} \zeta(3) - S_2 + \frac{\pi^4}{128},$$

(continued)

since $\frac{1}{1^2} + \frac{1}{3^2} + \frac{1}{5^2} + \cdots = \frac{3}{4}\zeta(2)$.

It follows that

$$S_1 + S_2 = \frac{3}{2}\zeta(3) + \frac{\pi^4}{128}.$$

On the other hand, using that $ab = \frac{a^2+b^2-(a-b)^2}{2}$, we have

$$S_1 = \frac{S_2}{2} + \frac{1}{2}\sum_{n=1}^{\infty}\left(\frac{1}{(n+1)^2} + \frac{1}{(n+3)^2} + \cdots\right)^2$$

$$- \frac{1}{2}\sum_{n=1}^{\infty}\left(\frac{1}{n^2} - \frac{1}{(n+1)^2} + \frac{1}{(n+2)^2} - \frac{1}{(n+3)^2} + \cdots\right)^2$$

$$\overset{(3.1)}{=} \frac{S_2}{2} + \frac{1}{2}\left(S_2 - \frac{9}{16}\zeta^2(2)\right) - \frac{1}{2}\left(\frac{\pi^2\ln 2}{2} - \frac{9}{4}\zeta(3)\right)$$

$$= S_2 - \frac{\pi^4}{128} - \frac{\pi^2\ln 2}{4} + \frac{9}{8}\zeta(3).$$

It follows that

$$S_1 - S_2 = -\frac{\pi^4}{128} - \frac{\pi^2\ln 2}{4} + \frac{9}{8}\zeta(3).$$

Solving the system of equations we obtain the values of series S_1 and S_2. ∎

3.91. Open Problem. Calculate

$$\sum_{n=1}^{\infty}(-1)^n\left(\frac{1}{n^2} + \frac{1}{(n+2)^2} + \cdots\right)\left(\frac{1}{(n+1)^2} + \frac{1}{(n+3)^2} + \cdots\right)$$

and

$$\sum_{n=1}^{\infty}(-1)^n\left(\frac{1}{n^2} + \frac{1}{(n+2)^2} + \frac{1}{(n+4)^2} + \cdots\right)^2.$$

3.92. Prove that

$$\sum_{n=1}^{\infty} \frac{1}{n^2} \left(\frac{1}{n^2} - \frac{1}{(n+1)^2} + \frac{1}{(n+2)^2} - \cdots \right) = \frac{13}{1440}\pi^4.$$

3.93. [15 April 2019]

(a) Prove that

$$\sum_{n=1}^{\infty} \left(\ln \frac{1}{1-x} - x - \frac{x^2}{2} - \cdots - \frac{x^n}{n} \right)$$

$$\times \left(\ln \frac{1}{2} - (-1) - \frac{(-1)^2}{2} - \cdots - \frac{(-1)^n}{n} \right)$$

$$= \begin{cases} \frac{\ln(1+x) - x\ln 2}{1-x} + \frac{1}{2}\ln\frac{1-x}{1+x} - \ln 2 \ln(1-x) & \text{if } x \in (-1, 1) \\ \ln 2 - \ln^2 2 & \text{if } x = -1. \end{cases}$$

In particular, for $x = -1$, we obtain the remarkable **quadratic series**

$$\sum_{n=1}^{\infty} \left(\ln \frac{1}{2} - (-1) - \frac{(-1)^2}{2} - \cdots - \frac{(-1)^n}{n} \right)^2 = \ln 2 - \ln^2 2.$$

(b) Prove that

$$\sum_{n=1}^{\infty} (-1)^n \left(\ln \frac{1}{1-x} - x - \frac{x^2}{2} - \cdots - \frac{x^n}{n} \right)$$

$$\times \left(\ln \frac{1}{2} - (-1) - \frac{(-1)^2}{2} - \cdots - \frac{(-1)^n}{n} \right)$$

$$= \begin{cases} -\frac{\ln^2(1-x)}{4} + \frac{1}{2}\mathrm{Li}_2\left(\frac{1}{2}\right) - \frac{1}{2}\mathrm{Li}_2\left(\frac{1+x}{2}\right) - \frac{\ln 2 \ln(1-x)}{2} & \text{if } x \in (-1, 1) \\ \frac{\pi^2}{24} - \ln^2 2 & \text{if } x = -1. \end{cases}$$

(continued)

In particular, for $x = -1$, we obtain the remarkable **alternating quadratic series**

$$\sum_{n=1}^{\infty}(-1)^n\left(\ln\frac{1}{2}-(-1)-\frac{(-1)^2}{2}-\cdots-\frac{(-1)^n}{n}\right)^2=\frac{\pi^2}{24}-\ln^2 2.$$

3.94. **[22 April 2019] A Nonlinear Logarithmic Series**
  Prove that

$$\sum_{n=1}^{\infty}\left(\ln\frac{1}{1-x}-x-\frac{x^2}{2}-\cdots-\frac{x^n}{n}\right)\left(\ln\frac{1}{1-y}-y-\frac{y^2}{2}-\cdots-\frac{y^n}{n}\right)$$

$$=-\ln(1-x)\ln(1-y)-\frac{x}{1-x}\ln(1-y)-\frac{y}{1-y}\ln(1-x)$$

$$+\frac{1-xy}{(1-x)(1-y)}\ln(1-xy),\quad\forall x,y\in[-1,1).$$

A particular case. When $x = y \in [-1, 1)$, we recover the quadratic series involving the tail of the logarithmic function [26, problem 3.79, p. 152]

$$\sum_{n=1}^{\infty}\left(\ln\frac{1}{1-x}-x-\frac{x^2}{2}-\cdots-\frac{x^n}{n}\right)^2$$

$$=-\ln^2(1-x)-\frac{2x}{1-x}\ln(1-x)+\frac{1+x}{1-x}\ln(1-x^2).$$

3.10   Multiple Series with the Riemann Zeta Function

3.95. *An alternating double series.*
  Prove that

$$\sum_{n=1}^{\infty}\sum_{m=1}^{\infty}\frac{(-1)^{n+m}}{nm(n+m)}=\frac{\zeta(3)}{4}.$$

3.96.

(a) Prove that

$$\sum_{i=1}^{\infty}\sum_{j=1}^{\infty}(\zeta(i+j)-1)=\zeta(2).$$

(b) Let $k \geq 2$ be an integer. Prove that

$$\sum_{i_1=1}^{\infty}\cdots\sum_{i_k=1}^{\infty}(\zeta(i_1+\cdots+i_k)-1)=\zeta(k).$$

3.97. [4 April 2018]

(a) Prove that

$$\sum_{i=1}^{\infty}\sum_{j=1}^{\infty}i\frac{\zeta(i+j)-1}{i+j}=\frac{\zeta(2)}{2}.$$

(b) Let $k \geq 2$ be an integer. Prove that

$$\sum_{i_1=1}^{\infty}\cdots\sum_{i_k=1}^{\infty}i_1\frac{\zeta(i_1+\cdots+i_k)-1}{i_1+\cdots+i_k}=\frac{\zeta(k)}{k}.$$

3.98. Prove that:

(a) $\displaystyle\sum_{i=1}^{\infty}\sum_{j=1}^{\infty}(\zeta(2i+j)-1)=\frac{\pi^2}{12}-\frac{3}{8};$

(b) $\displaystyle\sum_{i=1}^{\infty}\sum_{j=1}^{\infty}\left[2^{2i+j}\left(\zeta(2i+j)-1\right)-1\right]=\frac{\pi^2}{3}-\frac{25}{24}.$

3.99.

(a) Prove that

$$\sum_{i=1}^{\infty}\sum_{j=1}^{\infty}i\,(\zeta(i+j)-1)=\zeta(2)+\zeta(3).$$

(continued)

(b) Let $n, k \in \mathbb{N}$ be such that $n \geq 2$ and $1 \leq k \leq n$. Prove that

$$\sum_{i_1=1}^{\infty} \cdots \sum_{i_n=1}^{\infty} i_1 \cdots i_k \left(\zeta(i_1 + \cdots + i_n) - 1 \right) = \sum_{j=0}^{k} C_k^j \zeta(j+n).$$

3.100.

(a) Prove that

$$\sum_{i=1}^{\infty} \sum_{j=1}^{\infty} ij \left(\zeta(i+j) - 1 \right) = \zeta(2) + 2\zeta(3) + \zeta(4).$$

(b) Let $k \geq 2$ be an integer. Prove that

$$\sum_{i_1=1}^{\infty} \cdots \sum_{i_k=1}^{\infty} i_1 \cdots i_k \left(\zeta(i_1 + \cdots + i_k) - 1 \right) = \sum_{j=0}^{k} C_k^j \zeta(k+j).$$

3.101. [5 May 2018]

(a) Prove that

$$\sum_{i=1}^{\infty} \sum_{j=1}^{\infty} \left[2^{i+j} (\zeta(i+j) - 1) - 1 \right] = 4\zeta(2).$$

(b) Let $k \geq 2$ be an integer. Prove that

$$\sum_{i_1=1}^{\infty} \cdots \sum_{i_k=1}^{\infty} \left[2^{i_1 + \cdots + i_k} (\zeta(i_1 + \cdots + i_k) - 1) - 1 \right] = 2^k \zeta(k).$$

3.102. [6 September 2018]

(a) Prove that

$$\sum_{i=1}^{\infty}\sum_{j=1}^{\infty}\left[3^{i+j}\left(\zeta(i+j)-1-\frac{1}{2^{i+j}}\right)-1\right]=9\zeta(2).$$

(b) Prove that

$$\sum_{i=1}^{\infty}\sum_{j=1}^{\infty}ij\left[3^{i+j}\left(\zeta(i+j)-1-\frac{1}{2^{i+j}}\right)-1\right]=9\zeta(2)+54\zeta(3)+81\zeta(4).$$

A challenge. Let $n \geq 2$ be an integer. Calculate

$$\sum_{i_1=1}^{\infty}\cdots\sum_{i_n=1}^{\infty}\left[(n+1)^{i_1+\cdots+i_n}\left(\zeta(i_1+\cdots+i_n)-1-\cdots-\frac{1}{n^{i_1+\cdots+i_n}}\right)-1\right].$$

3.103. Prove that:

(a) $\displaystyle\sum_{i=1}^{\infty}\sum_{j=1}^{\infty}(-1)^{i-1}(\zeta(i+j)-1)=\frac{3}{4};$

(b) $\displaystyle\sum_{i=1}^{\infty}\sum_{j=1}^{\infty}(-1)^{i-1}\left[2^{i+j}(\zeta(i+j)-1)-1\right]=\frac{25}{12}.$

3.104.

(a) Prove that

$$\sum_{i=1}^{\infty}\sum_{j=1}^{\infty}(-1)^{i+j}(\zeta(i+j)-1)=\zeta(2)-\frac{5}{4}.$$

(continued)

(b) Let $k \geq 2$ be an integer. Prove that

$$\sum_{i_1=1}^{\infty} \cdots \sum_{i_k=1}^{\infty} (-1)^{i_1+\cdots+i_k} \left(\zeta(i_1 + \cdots + i_k) - 1 \right) = (-1)^k \left(\zeta(k) - 1 - \frac{1}{2^k} \right).$$

3.11   Series Involving Products of Harmonic Numbers

The next problems, which are new in the literature, are about the calculations of special nonlinear Euler sums. A series which contains a product of at least two harmonic numbers is called a nonlinear Euler sum. The investigation of nonlinear series has been motivated by the discovery of the famous quadratic series $\sum_{n=1}^{\infty} \frac{H_n^2}{n^2} = \frac{17}{4}\zeta(4)$. This series was introduced in the literature by Sandham and was rediscovered 45 years later by Au-Yeung. See Chap. 8 for two new proofs and for more information about the history of this formula.

3.105.   Series with Consecutive Harmonic Numbers

(a) [25] *A harmonic sum.*
     Prove that

$$\sum_{n=1}^{\infty} \frac{H_n}{n} \cdot \frac{H_{n+1}}{n+1} = \zeta(2) + 2\zeta(3).$$

(b) *An alternating harmonic sum.*
     Prove that

$$\sum_{n=1}^{\infty} (-1)^n \frac{H_n}{n} \cdot \frac{H_{n+1}}{n+1} = -\frac{7}{8}\zeta(3) - \frac{\pi^2}{12} - \frac{2}{3}\ln^3 2 + \ln^2 2 + \zeta(2)\ln 2.$$

Open problem. Calculate

$$\sum_{n=2}^{\infty} \frac{H_{n-1}}{n-1} \cdot \frac{H_n}{n} \cdot \frac{H_{n+1}}{n+1} \quad \text{and} \quad \sum_{n=2}^{\infty} (-1)^n \frac{H_{n-1}}{n-1} \cdot \frac{H_n}{n} \cdot \frac{H_{n+1}}{n+1}.$$

3.106.   Prove that

$$\sum_{n=1}^{\infty} \frac{H_n}{n^2} \cdot \frac{H_{n+1}}{n+1} = -2\zeta(2) + \frac{17}{4}\zeta(4).$$

3.107. Telescoping Series

(a) Prove that

$$\frac{H_n^2}{n} \cdot \frac{H_{n+1}}{n+1} = \frac{H_n^3}{n} - \frac{H_{n+1}^3}{n+1} + \frac{H_n^2}{n} - \frac{H_{n+1}^2}{n+1} + \frac{2H_{n+1}^2}{(n+1)^2} + \frac{2H_{n+1}}{(n+1)^2}$$
$$- \frac{H_{n+1}}{(n+1)^3} - \frac{1}{(n+1)^3}, \quad n \geq 1.$$

(b) Prove that

$$\sum_{n=1}^{\infty} \frac{H_n^2}{n} \cdot \frac{H_{n+1}}{n+1} = 3\zeta(3) + \frac{29}{4}\zeta(4).$$

(c) Prove that

$$\frac{H_n}{n} \cdot \frac{H_{n+1}^2}{n+1} = \frac{H_n^3}{n} - \frac{H_{n+1}^3}{n+1} + 2\left(\frac{H_n^2}{n} - \frac{H_{n+1}^2}{n+1}\right) + 3\frac{H_{n+1}}{(n+1)^2}$$
$$- \frac{1}{(n+1)^3} + \frac{H_n}{n} - \frac{H_{n+1}}{n+1} + \frac{1}{(n+1)^2} + \frac{H_{n+1}^2}{(n+1)^2}, \quad n \geq 1.$$

(d) Prove that

$$\sum_{n=1}^{\infty} \frac{H_n}{n} \cdot \frac{H_{n+1}^2}{n+1} = \zeta(2) + 5\zeta(3) + \frac{17}{4}\zeta(4).$$

A challenge. Prove that

$$\sum_{n=1}^{\infty} \frac{H_n^2}{n} \cdot \frac{H_{n+1}^2}{n+1} = 3\zeta(3) + \frac{69}{4}\zeta(4) + \frac{33}{2}\zeta(5) + 3\zeta(2)\zeta(3).$$

3.108. Prove that

$$\sum_{n=1}^{\infty} \frac{H_n}{n} \cdot \frac{H_{n+1}}{(n+1)^2} = \zeta(2) + 2\zeta(3) - 3\zeta(4).$$

Derivatives and Applications

4

> *The teacher you will be studying with,*
> *such knowledge you will be learning.*
>
> A Greek Proverb

Chapter 4 is about derivatives and their applications. This chapter contains problems on the classical topics of mathematical analysis: the calculation of the nth derivative of a function, Leibniz's formula, Taylor's formula, the extremum points of a function of a single variable, and non-standard topics of analysis such as the generalized Leibniz formula, special differential equations, and the computation of exotic series involving the Maclaurin remainder of special functions.

4.1 Apéritif

4.1. Let $h : \mathbb{R} \to \mathbb{R}$, $h(x) = x^3 - 3x$. Determine $J = h(I)$, where $I = (1, \infty)$. Prove that the function $f : I \to J$, $f(x) = h(x)$ is bijective. Let g be the inverse of f. Calculate $g'(2)$ and $g''(2)$.

4.2. Let $a \in \mathbb{R}$. Find all differentiable functions $f : \mathbb{R} \to \mathbb{R}$ such that $f(x + y) = f(x) + ay$, for all $x \in \mathbb{R}$ and $y \in \mathbb{R}^*$.

4.3. [113] Find all functions $f : \mathbb{R} \to \mathbb{R}$, derivable at 0, with $f'(0) = 1$, such that

$$f(2x) = 2f^2(x) - 4f(x) + 3, \quad \forall x \in \mathbb{R}.$$

4.4. Let $f(x) = x(x + 1) \cdots (x + 2021)$ and let $g = f \circ f \circ f$. Calculate $g'(0)$.

© The Author(s), under exclusive license to Springer Nature Switzerland AG 2021   121
A. Sîntămărian, O. Furdui, *Sharpening Mathematical Analysis Skills*, Problem Books
in Mathematics, https://doi.org/10.1007/978-3-030-77139-3_4

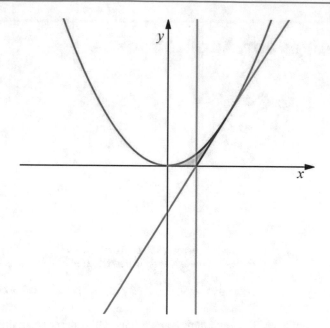

Fig. 4.1 Equal areas

4.5. *The areas of two regions determined by a parabola and two tangents.*

Let (x_0, y_0), $x_0 \neq 0$, be a point on the graph of the parabola $y = ax^2$, $a > 0$. The tangent to the parabola at (x_0, y_0) intersects the x axis at $(x_1, 0)$. Prove that the area of the region bounded by the tangent to the parabola at (x_0, y_0), parabola, and the vertical line $x = x_1$ is equal to the area of the region bounded by the parabola, the x axis, and the vertical line $x = x_1$. (See Fig. 4.1.)

4.6. Prove that $(1 + x)^{1-x} \leq 1 + x - x^2$, $\forall x \in [0, 1]$.

4.7. Let $n \in \mathbb{N}$. Calculate:

(a) $1 + x + 2x^2 + 3x^3 + \cdots + nx^n$, $\quad x \in \mathbb{R}$;
(b) $1 + 2^2 x + 3^2 x^2 + \cdots + n^2 x^{n-1}$, $\quad x \in \mathbb{R}$;
(c) $2 + 3 \cdot 2x + 4 \cdot 3x^2 + \cdots + n(n-1)x^{n-2}$, $\quad x \in \mathbb{R}$.

4.8. *Two inequalities and a limit with* e.

(a) Prove that

$$\frac{e}{2n + 2} < e - \left(1 + \frac{1}{n}\right)^n < \frac{e}{2n + 1}, \quad \forall n \geq 1.$$

(continued)

(b) Calculate

$$\lim_{n \to \infty} n \left(e - \left(1 + \frac{1}{n} \right)^n \right).$$

4.9. Let $(x_n)_{n \geq 1}$ be the sequence defined by $x_n = 1 + \frac{1}{2} + \cdots + \frac{1}{n} - \ln n$ and let $\gamma = \lim_{n \to \infty} x_n$. Prove that

$$\frac{1}{2n+1} < x_n - \gamma < \frac{1}{2n}, \quad \forall n \geq 1.$$

4.10. [18] *Identities with derivatives.* Check that:

(a) $\dfrac{(fg)''}{fg} = \dfrac{f''}{f} + \dfrac{g''}{g} + 2\dfrac{f'}{f} \cdot \dfrac{g'}{g}$;

(b) $\dfrac{(f/g)''}{f/g} = \dfrac{f''}{f} - \dfrac{g''}{g} - 2\dfrac{(f/g)'}{f/g} \cdot \dfrac{g'}{g}$.

We record, as gems, the proofs of two amazing formulae of mathematical analysis.

(a) **Euler's Formula** $e^{ix} = \cos x + i \sin x$, $\forall x \in \mathbb{R}$.

   Let $f : \mathbb{R} \to \mathbb{C}$, $f(x) = e^{ix} - \cos x - i \sin x$. We have $f'(x) = if(x) \Rightarrow \left(e^{-ix} f(x) \right)' = 0 \Rightarrow e^{-ix} f(x) = \mathscr{C}$, $\mathscr{C} \in \mathbb{C}$. We obtain $f(x) = \mathscr{C}e^{ix}$, $\mathscr{C} \in \mathbb{C}$. Since $f(0) = 0 \Rightarrow \mathscr{C} = 0 \Rightarrow f(x) = 0$, for all $x \in \mathbb{R}$. ∎

(b) [127] **de Moivre's Formula** If $x \in \mathbb{R}$ and $n \in \mathbb{Z}$, then

$$(\cos x + i \sin x)^n = \cos(nx) + i \sin(nx).$$

   Since the formula is true for $n = 0$, we consider that $n \neq 0$. Let $f : \mathbb{R} \to \mathbb{C}$, $f(x) = \cos x + i \sin x$. We need to prove that $f^n(x) = f(nx)$ or, equivalently, $f(nx) f^n(-x) = 1$. Using that $f'(x) = if(x)$, we have

$$\left[f(nx) f^n(-x) \right]' = n f^{n-1}(-x) \left[f'(nx) f(-x) - f(nx) f'(-x) \right]$$

$$= i \cdot n \cdot f^{n-1}(-x) \left[f(nx) f(-x) - f(nx) f(-x) \right]$$

$$= 0.$$

(continued)

It follows that $f(nx)f^n(-x) = \mathscr{C}$, $\mathscr{C} \in \mathbb{C}$, for all $x \in \mathbb{R}$. When $x = 0 \Rightarrow$ $\mathscr{C} = f^{n+1}(0) = 1$. Therefore, $f(nx)f^n(-x) = 1$, for all $x \in \mathbb{R}$.   ∎

4.2   Integral Equations

4.11. Find all continuous functions $f : \mathbb{R} \to \mathbb{R}$ which satisfy the equation

$$f(x) = x + \int_0^x e^{-t} f(x - t)\, dt, \quad \forall x \in \mathbb{R}.$$

4.12. Let $a \in \mathbb{R}$. Find all continuous functions $f : \mathbb{R} \to \mathbb{R}$ which satisfy the equation:

(a) $f(x) = a + \displaystyle\int_0^x \sin t f(x - t)\, dt, \quad \forall x \in \mathbb{R}$;

(b) $f(x) = a - \displaystyle\int_0^x \sin t f(x - t)\, dt, \quad \forall x \in \mathbb{R}$.

4.13. Find all continuous functions $f : \mathbb{R} \to \mathbb{R}$ which satisfy the equation:

(a) $f(-x) = 1 + \displaystyle\int_0^x e^{-t} f(x - t)\, dt, \quad \forall x \in \mathbb{R}$;

(b) $f(-x) = x + \displaystyle\int_0^x e^{-t} f(x - t)\, dt, \quad \forall x \in \mathbb{R}$;

(c) $f(-x) = \sin x + \displaystyle\int_0^x e^{-t} f(x - t)\, dt, \quad \forall x \in \mathbb{R}$.

4.14. Find all continuous functions $f : \mathbb{R} \to \mathbb{R}$ which satisfy the equation:

(a) [70] $f(-x) = 1 + \displaystyle\int_0^x \sin t\, f(x - t)\, dt, \quad \forall x \in \mathbb{R}$;

(b) [78] $f(-x) = x + \displaystyle\int_0^x \sin t\, f(x - t)\, dt, \quad \forall x \in \mathbb{R}$.

4.15. Let $a \in \mathbb{R}$. Find all continuous functions $f : \mathbb{R} \to \mathbb{R}$ which satisfy the equation:

(a) $f(-x) = a - \displaystyle\int_0^x \sin t\, f(x - t)\, dt, \quad \forall x \in \mathbb{R}$;

(b) $f(-x) = x - \displaystyle\int_0^x \sin t\, f(x - t)\, dt, \quad \forall x \in \mathbb{R}$.

4.16. Let $a \in \mathbb{R}$. Find all continuous functions $f : \mathbb{R} \to \mathbb{R}$ which satisfy the equation

$$f(x) = a - \int_0^x \cos t\, f(x - t)\, dt, \quad \forall x \in \mathbb{R}.$$

4.17. Let $a \in \mathbb{R}$. Find all continuous functions $f : \mathbb{R} \to \mathbb{R}$ which satisfy the equation:

(a) $f(x) - \displaystyle\int_0^x t f(x - t)\, dt = a, \quad \forall x \in \mathbb{R}$;

(b) $f(x) + \displaystyle\int_0^x t f(x - t)\, dt = a, \quad \forall x \in \mathbb{R}$;

(c) $f(-x) + \displaystyle\int_0^x t f(x - t)\, dt = a, \quad \forall x \in \mathbb{R}$.

4.3   Differential Equations

4.18. *A problem of Vasile Pop.*

(a) Find all differentiable functions $f : \mathbb{R} \to \mathbb{R}$ which satisfy the equation

$$f'(x) - 2x f(-x) = x, \quad \text{for all } x \in \mathbb{R}.$$

(b) Find all differentiable functions $f : \mathbb{R} \to \mathbb{R}$ which satisfy the equation

$$f'(x) + 2x f(-x) = x, \quad \text{for all } x \in \mathbb{R}.$$

4.19. [52] Find all differentiable functions $f : \mathbb{R} \to \mathbb{R}$ which satisfy the equation

$$xf'(x) + f(-x) = x^2, \quad \text{for all } x \in \mathbb{R}.$$

4.20. [38] Let $k \in \mathbb{N}$. Find all differentiable functions $f : \mathbb{R} \to \mathbb{R}$ which satisfy the equation

$$xf'(x) + kf(-x) = x^2, \quad \text{for all } x \in \mathbb{R}.$$

4.21. *A Cauchy problem.*

(a) Find all differentiable functions $f : (0, \infty) \to (0, \infty)$, with $f(1) = \sqrt{2}$, such that

$$f'\left(\frac{1}{x}\right) = \frac{1}{f(x)}, \quad \forall x > 0.$$

(b) Find all differentiable functions $f : (0, \infty) \to (0, \infty)$, with $f(1) = 2$, such that

$$f'\left(\frac{1}{x}\right) = \frac{1}{f(x)}, \quad \forall x > 0.$$

4.22. Let $a, b > 0$. Find all differentiable functions $f : (0, \infty) \to (0, \infty)$ such that[1]

$$f'\left(\frac{a}{x}\right) = \frac{bx}{f(x)}, \quad \forall x > 0.$$

4.23. [A. Sîntămărian, 2016] Find all differentiable functions $f : [-1, 1] \to (0, \infty)$ such that

$$f'(\sin x) = \frac{\sin x}{f(\cos x)}, \quad \forall x \in \mathbb{R}.$$

4.24. [A. Sîntămărian, 2016] Find all differentiable functions $f : \left(0, \frac{\pi}{2}\right) \to (0, \infty)$ such that

$$f'(\arctan x) = \frac{1}{f\left(\arctan \frac{1}{x}\right)}, \quad \forall x > 0.$$

---

[1]For $b = 1$ we obtain problem B–3, Putnam 2005.

4.25. Find all continuous functions $f : \mathbb{R} \to \mathbb{R}$ which satisfy the equation

$$\int_{x-y}^{x+y} f(t) \mathrm{d}t = f(x) f(y), \quad \forall x, y \in \mathbb{R}.$$

4.26.

(a) Find all nonconstant differentiable functions $f : \mathbb{R} \to \mathbb{R}$ which satisfy the equation

$$f(x+y) - f(x-y) = 2 f'(x) f(y), \quad \forall x, y \in \mathbb{R}.$$

(b) Find all nonconstant differentiable functions $f : \mathbb{R} \to \mathbb{R}$ which satisfy the equation

$$f(x+y) - f(x-y) = 2 f'(x) f'(y), \quad \forall x, y \in \mathbb{R}.$$

4.27. A Differential Equation Used for the Calculation of a Series

(a) Find all differentiable functions $f : \mathbb{R} \to \mathbb{R}$ which satisfy the equation

$$f'(x) - f(x) = x \, e^x, \quad \forall x \in \mathbb{R}.$$

(b) Calculate

$$\sum_{n=1}^{\infty} n \left(e^x - 1 - \frac{x}{1!} - \frac{x^2}{2!} - \cdots - \frac{x^n}{n!} \right).$$

4.28. Other Differential Equations Used for the Calculation of Some Series
    Let $k \geq 0$ be an integer.

(a) Find all differentiable functions $f : \mathbb{R} \to \mathbb{R}$ such that

(continued)

$$f'(x) - f(x) = \frac{x^k}{k!} e^x, \quad \forall x \in \mathbb{R}.$$

(b) Prove that

$$\sum_{n=k}^{\infty} C_n^k \left(e^x - 1 - \frac{x}{1!} - \frac{x^2}{2!} - \cdots - \frac{x^n}{n!} \right) = \frac{x^{k+1}}{(k+1)!} e^x.$$

(c) Find all differentiable functions $f : \mathbb{R} \to \mathbb{R}$ such that

$$f'(x) - f(x) = \frac{x^k}{k!} (x + k) e^x, \quad \forall x \in \mathbb{R}.$$

(d) Prove that

$$\sum_{n=k}^{\infty} n C_n^k \left(e^x - 1 - \frac{x}{1!} - \frac{x^2}{2!} - \cdots - \frac{x^n}{n!} \right)$$

$$= \begin{cases} \dfrac{x^2}{2} e^x & \text{if } k = 0 \\[2mm] \left(\dfrac{x^{k+1}}{(k-1)!(k+1)} + \dfrac{x^{k+2}}{k!(k+2)} \right) e^x & \text{if } k \geq 1. \end{cases}$$

4.29. Let $x \in \mathbb{R}$.

(a) Prove that

$$\sum_{n=1}^{\infty} \left(\frac{x^n}{n!} - \frac{x^{n+1}}{(n+1)!} + \frac{x^{n+2}}{(n+2)!} - \cdots \right) = \sinh x.$$

In particular, when $x = 1$ we obtain the series

$$\sum_{n=1}^{\infty} \left(\frac{1}{n!} - \frac{1}{(n+1)!} + \frac{1}{(n+2)!} - \cdots \right) = \sinh 1.$$

(b) Prove that

$$\sum_{n=1}^{\infty} \left(\frac{x^n}{n!} - 2\frac{x^{n+1}}{(n+1)!} + 3\frac{x^{n+2}}{(n+2)!} - \cdots \right) = \frac{1}{2} \left(\sinh x + x e^{-x} \right).$$

In particular, when $x = 1$ we obtain the series

$$\sum_{n=1}^{\infty} \left(\frac{1}{n!} - \frac{2}{(n+1)!} + \frac{3}{(n+2)!} - \cdots \right) = \frac{\cosh 1}{2}.$$

4.30. Let $x \in \mathbb{R}$. Prove that

$$\sum_{n=1}^{\infty} n \left(\frac{x^n}{n!} - \frac{x^{n+1}}{(n+1)!} + \frac{x^{n+2}}{(n+2)!} - \cdots \right) = \frac{\sinh x}{2} + \frac{x e^x}{2}.$$

In particular, when $x = 1$ we obtain the series

$$\sum_{n=1}^{\infty} n \left(\frac{1}{n!} - \frac{1}{(n+1)!} + \frac{1}{(n+2)!} - \cdots \right) = \frac{3e - e^{-1}}{4}.$$

4.31. [54] **The Behavior of an Exponential Series**
Let $k \in \mathbb{N}$. Prove that

$$\lim_{x \to \infty} e^{-x} \sum_{n=k}^{\infty} (-1)^n C_n^k \left(e^x - 1 - \frac{x}{1!} - \frac{x^2}{2!} - \cdots - \frac{x^n}{n!} \right) = \frac{(-1)^k}{2^{k+1}}.$$

4.32. Let $x \in \mathbb{R}$. Prove that:

(a) $\displaystyle\sum_{n=1}^{\infty} \left\lfloor \frac{n}{2} \right\rfloor \left(e^x - 1 - \frac{x}{1!} - \frac{x^2}{2!} - \cdots - \frac{x^n}{n!} \right) = \frac{1}{4} \sinh x + \frac{1}{4}(x^2 - x)e^x;$

(b) $\displaystyle\sum_{n=1}^{\infty} (-1)^n \left\lfloor \frac{n}{2} \right\rfloor \left(e^x - 1 - \frac{x}{1!} - \frac{x^2}{2!} - \cdots - \frac{x^n}{n!} \right)$
$\qquad = \frac{1}{2} x \cosh x - \frac{1}{2} \sinh x;$

(c) $\displaystyle\sum_{n=1}^{\infty} \left\lfloor (-1)^n \frac{n}{2} \right\rfloor \left(e^x - 1 - \frac{x}{1!} - \frac{x^2}{2!} - \cdots - \frac{x^n}{n!} \right) = -\frac{x}{2} \sinh x;$

(d) $\displaystyle\sum_{n=1}^{\infty} (-1)^{\lfloor \frac{n}{2} \rfloor} \left\lfloor \frac{n}{2} \right\rfloor \left(e^x - 1 - \frac{x}{1!} - \frac{x^2}{2!} - \cdots - \frac{x^n}{n!} \right)$

(continued)

$$= \frac{1}{2} \left(\sin x + \cos x + x \sin x - e^x \right);$$

(e) $\displaystyle\sum_{n=1}^{\infty} (-1)^{\lfloor \frac{n}{2} \rfloor} n \left(e^x - 1 - \frac{x}{1!} - \frac{x^2}{2!} - \cdots - \frac{x^n}{n!} \right)$

$$= \left(x + \frac{1}{2} \right) \sin x + \frac{\cos x - e^x}{2}.$$

4.4 Higher Order Derivatives

4.33. Higher Order Derivatives Let $a, b \in \mathbb{R}$, with $a \neq 0$. Prove that:

(a) $f : \mathbb{R} \setminus \left\{ -\frac{b}{a} \right\} \to \mathbb{R}, \ f(x) = \frac{1}{ax+b}, \quad f^{(n)}(x) = (-1)^n \frac{a^n n!}{(ax+b)^{n+1}}, \ n \geq 0;$

(b) $f : \left(-\frac{b}{|a|}, \infty \right) \to \mathbb{R}, \ f(x) = \ln(|a|x + b),$

$$f^{(n)}(x) = (-1)^{n-1} \frac{|a|^n (n-1)!}{(|a|x+b)^n}, \quad n \in \mathbb{N};$$

(c) $f : \mathbb{R} \to \mathbb{R}, \ f(x) = \sin(ax + b),$
$f^{(n)}(x) = a^n \sin \left(ax + b + \frac{n\pi}{2} \right), \ n \geq 0;$

(d) $f : \mathbb{R} \to \mathbb{R}, \ f(x) = \cos(ax + b),$
$f^{(n)}(x) = a^n \cos \left(ax + b + \frac{n\pi}{2} \right), \ n \geq 0;$

(e) $f : \mathbb{R} \to \mathbb{R}, \ f(x) = e^{ax+b}, \quad f^{(n)}(x) = a^n e^{ax+b}, \ n \geq 0.$

4.34. Divertimento Prove that:

(a) $\sin x$ is not a polynomial function;
(b) e^x is not a polynomial function;
(c) e^x is not a rational function.

4.35. Let f be a polynomial with real coefficients such that $f(x) \geq 0$, for all $x \in \mathbb{R}$. Let $g(x) = f(x) + f'(x) + f''(x) + \cdots$. Prove that $g(x) \geq 0$, for all $x \in \mathbb{R}$.

An Inequality with Derivatives of a Polynomial Function

Let p be a polynomial with real coefficients and real roots. The following inequality holds true:

$$p'^2(x) \geq p(x)p''(x), \quad \forall x \in \mathbb{R}.$$

Proof. Let $x_1, x_2, \ldots, x_n$ be the distinct real roots of p with multiplicities $k_1, k_2, \ldots, k_n$ respectively. If $x = x_i$ is a root of p, then the inequality is true. We assume that x is not a root of p.

We have

$$\frac{p'(x)}{p(x)} = \sum_{i=1}^{n} \frac{k_i}{x - x_i}$$

and it follows by differentiation that

$$\frac{p''(x)p(x) - p'^2(x)}{p^2(x)} = -\sum_{i=1}^{n} \frac{k_i}{(x - x_i)^2},$$

and the inequality to prove follows. ∎

4.36. Special Differential Equations

(a) Find all functions $f \in C^\infty((-R, R), \mathbb{R})$, $R \in (0, \infty]$, such that

$$f(x) = \sum_{n=1}^{\infty} f^{(n)}(x), \quad \text{for all} \quad x \in (-R, R).$$

(b) Find all functions $f \in C^\infty((-R, R), \mathbb{R})$, $R \in (0, \infty]$, such that

$$f(x) = \sum_{n=1}^{\infty} (-1)^n f^{(n)}(x), \quad \text{for all} \quad x \in (-R, R).$$

(c) Let $a \in \mathbb{R}^*$. Find all functions $f \in C^\infty((-R, R), \mathbb{R})$, $R \in (0, \infty]$, such that

$$f(x) = \sum_{n=1}^{\infty} a^n f^{(n)}(x), \quad \text{for all} \quad x \in (-R, R).$$

(continued)

(d) Find all functions $f \in C^{\infty}((-R, R), \mathbb{R})$, $R \in (0, \infty]$, such that

$$f(x) = \sum_{n=1}^{\infty} n f^{(n)}(x), \quad \text{for all} \quad x \in (-R, R).$$

(e) Find all functions $f \in C^{\infty}((-R, R), \mathbb{R})$, $R \in (0, \infty]$, such that

$$f(x) = \sum_{n=1}^{\infty} f^{(2n-1)}(x), \quad \text{for all} \quad x \in (-R, R).$$

(f) Find all functions $f \in C^{\infty}((-R, R), \mathbb{R})$, $R \in (0, \infty]$, such that

$$f(x) = \sum_{n=1}^{\infty} f^{(2n)}(x), \quad \text{for all} \quad x \in (-R, R).$$

(g) Find all functions $f \in C^{\infty}((-R, R), \mathbb{R})$, $R \in (0, \infty]$, such that

$$f(-x) = \sum_{n=1}^{\infty} f^{(2n)}(x), \quad \text{for all} \quad x \in (-R, R).$$

4.37. Calculate the derivative of order n of the following functions:

(a) $f : \mathbb{R} \setminus \{-1\} \to \mathbb{R}, \quad f(x) = \frac{3x}{1+x};$

(b) $f : \mathbb{R} \setminus \left\{-3, \frac{1}{2}\right\} \to \mathbb{R}, \quad f(x) = \frac{1}{2x^2+5x-3};$

(c) $f : (0, \infty) \to \mathbb{R}, \quad f(x) = \frac{1+x}{\sqrt{x}};$

(d) $f : \mathbb{R}^* \to \mathbb{R}, \quad f(x) = \frac{1+x}{\sqrt[3]{x}};$

(e) $f : \mathbb{R} \to \mathbb{R}, \quad f(x) = \sin^2 x;$

(f) $f : \mathbb{R} \to \mathbb{R}, \quad f(x) = \cos^2 x;$

(g) $f : \mathbb{R} \to \mathbb{R}, \quad f(x) = \sin^3 x;$

(h) $f : \mathbb{R} \to \mathbb{R}, \quad f(x) = \cos^3 x;$

(i) $f : \mathbb{R} \to \mathbb{R}, \quad f(x) = \sinh x;$

(j) $f : \mathbb{R} \to \mathbb{R}, \quad f(x) = \cosh x.$

4.38. Let $f : \mathbb{R} \to \mathbb{R}$ be the function defined by $f(x) = \sin^4 x + \cos^4 x$. Prove that

$$f^{(n)}(x) = 4^{n-1} \cos\left(4x + \frac{n\pi}{2}\right), \quad n \in \mathbb{N}.$$

4.39. [M. Țena, 2011] Let $f, g : \mathbb{R} \to \mathbb{R}$, $f(x) = e^x \sin x$, $g(x) = e^x \cos x$. Prove that, for all $n \geq 0$ an integer and $x \in \mathbb{R}$, we have

$$\left(f^{(n)}(x) \right)^2 + \left(g^{(n)}(x) \right)^2 = 2^n e^{2x}.$$

Remark 4.1. If $f(x) = e^{-x} \sin x$ and $g(x) = e^{-x} \cos x$, then for all $n \geq 0$ an integer and $x \in \mathbb{R}$, we have

$$\left(f^{(n)}(x) \right)^2 + \left(g^{(n)}(x) \right)^2 = 2^n e^{-2x}.$$

4.40. *Series with derivatives.* Let $k \in \mathbb{N}$ and $f : \mathbb{R} \to \mathbb{R}$, $f(x) = \frac{1}{1+x^2}$.

(a) [5] Prove that

$$\sum_{n=0}^{\infty} \left(\frac{f^{(4n)}(1)}{(4n)!} + \frac{f^{(2n)}(1)}{(2n)!} \right) = 1.$$

(b) [34] Prove that

$$\sum_{n=0}^{\infty} \frac{f^{(2kn)}(1)}{(2kn)!} = 2^{k-1} \frac{2^k - \cos \frac{k\pi}{2} + \sin \frac{k\pi}{2}}{2^{2k} - 2^{k+1} \cos \frac{k\pi}{2} + 1}.$$

> **Leibniz's Formula** Let $n \geq 0$ be an integer, $a, b \in \overline{\mathbb{R}}$, $a < b$, and let $f, g : (a, b) \to \mathbb{R}$ be n times differentiable functions on (a, b). Then
>
> $$(f \cdot g)^{(n)}(x) = \sum_{k=0}^{n} C_n^k f^{(n-k)}(x) g^{(k)}(x), \quad x \in (a, b).$$

4.41. Calculate the derivative of order n of the following functions:

(a) $f : \mathbb{R} \to \mathbb{R}$, $\quad f(x) = (x^2 - 3x - 1)e^{2x}$;
(b) $f : \mathbb{R} \to \mathbb{R}$, $\quad f(x) = \frac{x-1}{e^x}$;
(c) $f : \mathbb{R} \to \mathbb{R}$, $\quad f(x) = e^{ax} \cos(bx + c)$, $a, b, c \in \mathbb{R}$.

4.42. Let $n \geq 0$ be an integer and $f, g : \mathbb{R} \to \mathbb{R}$, $f(x) = x \sin x$, $g(x) = x \cos x$. Solve the equation

$$f^{(2n)}(x) + g^{(2n+1)}(x) = (-1)^n \sin x.$$

4.43. Let $f : (-1, \infty) \to \mathbb{R}$, $f(x) = \ln^2(x + 1)$. Prove that, $\forall n \in \mathbb{N}$, $n \geq 2$, we have

$$f^{(n)}(x) = 2(-1)^{n-1}(n-1)! \left[\ln(x + 1) - \left(1 + \frac{1}{2} + \frac{1}{3} + \cdots + \frac{1}{n-1} \right) \right] \frac{1}{(x + 1)^n}.$$

4.44. Let $f : (0, \infty) \to \mathbb{R}$, $f(x) = \frac{\ln x}{x}$. Prove that

$$f^{(n)}(x) = (-1)^{n-1} \frac{n!}{x^{n+1}} \left(1 + \frac{1}{2} + \frac{1}{3} + \cdots + \frac{1}{n} - \ln x \right), \quad n \in \mathbb{N}.$$

4.45. Let $n \in \mathbb{N}$ and $f : (0, \infty) \to \mathbb{R}$, $f(x) = x^n \ln x$. Prove that

$$f^{(n)}(x) = n! \left(1 + \frac{1}{2} + \frac{1}{3} + \cdots + \frac{1}{n} + \ln x \right).$$

4.46. Let $f : (0, \infty) \to \mathbb{R}$, $f(x) = \frac{(1+x)^2 - \ln x}{x}$. Prove that

$$f^{(n)}(x) = (-1)^n \frac{n!}{x^{n+1}} \left(1 + 1 + \frac{1}{2} + \frac{1}{3} + \cdots + \frac{1}{n} - \ln x \right), \quad n \geq 2.$$

4.47. Let $n \in \mathbb{N}$ and $f : \mathbb{R}^* \to \mathbb{R}$, $f(x) = x^{n-1} e^{\frac{1}{x}}$. Prove that

$$f^{(n)}(x) = (-1)^n \frac{f(x)}{x^{2n}}.$$

4.48. Let $f : [-1, 1] \to \mathbb{R}$, $f(x) = \arcsin^2 x$. Calculate $f^{(n)}(0)$.

4.49. Let $f, g, h : \mathbb{R} \to \mathbb{R}$ be the functions

$$f(x) = \frac{1}{1 + x^2}, \quad g(x) = \arctan x, \quad h(x) = g^2(x).$$

Calculate $f^{(n)}(0)$, $g^{(n)}(0)$, and $h^{(n)}(0)$.

4.50. Let $f : \mathbb{R} \setminus \{-1\} \to \mathbb{R}$, $f(x) = \frac{1}{1+x^3}$. Calculate $f^{(n)}(0)$.

4.51. Let $f : \mathbb{R} \to \mathbb{R}$, $f(x) = \frac{1}{x^2+x+1}$. Calculate $f^{(n)}\left(-\frac{1}{2}\right)$.

4.52. Let $f : (-\infty, 1) \to \mathbb{R}$, $f(x) = \ln(1 - x^3)$. Calculate $f^{(n)}(0)$.

4.53. Limits of nth Order Derivatives Prove that:

- $$\lim_{n \to \infty} n \left(\frac{e^x - e^a}{x - a} \right)^{(n)} = e^x, \quad x, a \in \mathbb{R}, \ x \neq a;$$

- $$\lim_{n \to \infty} (-1)^{n-1} n \left(\frac{\sin x}{x} \right)^{(2n+1)} = \frac{\sin x}{2}, \quad x \in \mathbb{R}^*;$$

- $$\lim_{n \to \infty} (-1)^n n \left(\frac{\sin x}{x} \right)^{(2n)} = \frac{\cos x}{2}, \quad x \in \mathbb{R}^*;$$

- $$\lim_{n \to \infty} (-1)^n n \left(\frac{1 - \cos x}{x} \right)^{(2n)} = \frac{\sin x}{2}, \quad x \in \mathbb{R}^*;$$

- $$\lim_{n \to \infty} (-1)^{n-1} n \left(\frac{1 - \cos x}{x} \right)^{(2n-1)} = \frac{\cos x}{2}, \quad x \in \mathbb{R}^*;$$

- $$\lim_{n \to \infty} (-1)^n \frac{(1 + x)^n}{(n - 1)!} \left(\frac{\ln(1 + x)}{x} \right)^{(n)} = 1, \quad x \in (-1, 1), \ x \neq 0.$$

Remark 4.2. The case when $a = 0$ in the first limit is problem 4, part (b), section A, given in 2018 by M. Ivan at Traian Lalescu National Contest for University Students held every year in Romania [110].

The Generalized Leibniz Formula Let $a, b \in \mathbb{R}$, $a < b$, $m \geq 1$ an integer and let $f_1, f_2, \ldots, f_m : (a, b) \to \mathbb{R}$ be n times differentiable functions on (a, b). Then

$$(f_1 f_2 \cdots f_m)^{(n)} (x) = \sum_{k_1 + k_2 + \cdots + k_m = n} \frac{n!}{k_1! \, k_2! \cdots k_m!} f_1^{(k_1)}(x) f_2^{(k_2)}(x) \cdots f_m^{(k_m)}(x),$$

where the sum is over all m-tuples $(k_1, k_2, \ldots, k_m)$ of integers $k_1, k_2, \ldots, k_m \geq 0$ such that $k_1 + k_2 + \cdots + k_m = n$.

In particular, when $m = 2$ we obtain the classical Leibniz formula

$$(f_1 \cdot f_2)^{(n)}(x) = \sum_{k_1 + k_2 = n} \frac{n!}{k_1! k_2!} f_1^{(k_1)}(x) f_2^{(k_2)}(x) = \sum_{k_1 = 0}^{n} C_n^{k_1} f_1^{(k_1)}(x) f_2^{(n - k_1)}(x).$$

4.54. [69] Let $n \in \mathbb{N}$ and $f_n : \mathbb{R} \to \mathbb{R}$, $f_n(x) = \sin x \sin(2x) \cdots \sin(nx)$. Prove that $f_n^{(n)}(0) = (n!)^2$.

4.55. Let $n \in \mathbb{N}$ and $g_n : \mathbb{R} \to \mathbb{R}$, $g_n(x) = \cos x \cos(2x) \cdots \cos(nx)$. Calculate $g_n^{(n)}(0)$.

4.56. Let $n \in \mathbb{N}$ and $f_n : \mathbb{R} \to \mathbb{R}$, $f_n(x) = \sinh x \sinh(2x) \cdots \sinh(nx)$. Calculate $f_n^{(n)}(0)$.

4.57. Let $n \in \mathbb{N}$ and $g_n : \mathbb{R} \to \mathbb{R}$, $f_n(x) = \cosh x \cosh(2x) \cdots \cosh(nx)$. Calculate $g_n^{(n)}(0)$.

4.5   Taylor's Formula

Taylor's Formula Let $I \subseteq \mathbb{R}$ be an open interval and $f : I \to \mathbb{R}$ be an $n+1$ times differentiable function on I. Then, for all $x_0, x \in I$, with $x \neq x_0$, there exists $\theta_{x_0,x} \in (0, 1)$ such that the *Taylor formula or order n corresponding to the function f and the point x_0* holds true

$$f(x) = (T_n f)(x) + (R_n f)(x),$$

where

$$(T_n f)(x) = \sum_{k=0}^{n} \frac{f^{(k)}(x_0)}{k!}(x - x_0)^k$$

is the *Taylor polynomial of degree n corresponding to the function f and the point x_0*, and

$$(R_n f)(x) = \frac{f^{(n+1)}(x_0 + \theta_{x_0,x}(x - x_0))}{(n + 1)!}(x - x_0)^{n+1}$$

is the *remainder of order n corresponding to the function f and the point x_0* (the remainder in the form of Lagrange).

We mention three particular cases of Taylor's formula.

I. If $x_0 = 0 \in I$, then we obtain the *Maclaurin formula of order n corresponding to the function f*

$$f(x) = (M_n f)(x) + (R_n f)(x),$$

where

(continued)

$$(M_n f)(x) = \sum_{k=0}^{n} \frac{f^{(k)}(0)}{k!} x^k$$

is the *Maclaurin polynomial of degree n corresponding to the function f*, and

$$(R_n f)(x) = \frac{f^{(n+1)}(\theta_x x)}{(n+1)!} x^{n+1}$$

is the *remainder of order n corresponding to the function f*.

II. If $f : \mathbb{R} \to \mathbb{R}$ is a polynomial function of degree n and $x_0 \in \mathbb{R}$, then

$$f(x) = (T_n f)(x), \quad \forall x \in \mathbb{R}.$$

III. If $n = 0$, then we obtain *Lagrange's Mean Value Theorem*

$$f(x) = f(x_0) + f'(x_0 + \theta_{x_0,x}(x - x_0))(x - x_0).$$

Taylor's Formula with the Remainder in Integral Form

Let I be an open interval and let $f \in C^{n+1}(I, \mathbb{R})$. Then, for all $x_0, x \in I$, we have

$$f(x) = \sum_{k=0}^{n} \frac{f^{(k)}(x_0)}{k!} (x - x_0)^k + \int_{x_0}^{x} \frac{(x - t)^n}{n!} f^{(n+1)}(t) dt.$$

4.58. Write the Taylor formula of order n corresponding to the function f and the point x_0 in the following cases:

(a) $f : (-3, \infty) \to \mathbb{R}, \quad f(x) = \frac{1}{x+3}, \quad x_0 = -1;$
(b) $f : \mathbb{R} \to \mathbb{R}, \quad f(x) = e^{2x-1}, \quad x_0 = 2;$
(c) $f : \mathbb{R} \to \mathbb{R}, \quad f(x) = \sin(3x), \quad x_0 = \frac{\pi}{2};$
(d) $f : \mathbb{R} \to \mathbb{R}, \quad f(x) = \cosh(x - 1), \quad x_0 = 1.$

4.59. Write the following functions in terms of powers of x:

(a) $f : \mathbb{R} \to \mathbb{R}, \quad f(x) = (x - 1)^3 (x + 5)^2;$
(b) $f : \mathbb{R} \to \mathbb{R}, \quad f(x) = (x + 1)^2 (x - 9)(x + 2),$

without expanding the parenthesis.

4.60. Write the function in terms of powers of $x - 1$

$$f : \mathbb{R} \to \mathbb{R}, \quad f(x) = x^5 - 5x^4 + 10x^3 - 7x^2 - x + 3.$$

4.61. Write the function in terms of powers of $x + 1$

$$f : \mathbb{R} \to \mathbb{R}, \quad f(x) = x^5 + 5x^4 + 3x^3 - 11x^2 - 16x - 7.$$

4.62. Write the Maclaurin polynomial of degree n corresponding to the function f

(a) $f : (-1, \infty) \to \mathbb{R}, \quad f(x) = \frac{1}{x^2+4x+3};$
(b) $f : [-1, \infty) \to \mathbb{R}, \quad f(x) = \sqrt{x + 1};$
(c) $f : (-\infty, 1) \to \mathbb{R}, \quad f(x) = \ln \frac{x-1}{x-2}.$

4.63. The Maclaurin Formula of Order n for Elementary Functions
Write the Maclaurin formula of order n corresponding to the function f

(a) $f : \mathbb{R} \to \mathbb{R}, \quad f(x) = e^x;$
(b) $f : \mathbb{R} \to \mathbb{R}, \quad f(x) = \sin x;$
(c) $f : \mathbb{R} \to \mathbb{R}, \quad f(x) = \cos x;$
(d) $f : (-1, \infty) \to \mathbb{R}, \quad f(x) = (1 + x)^\alpha, \quad \alpha \in \mathbb{R};$
(e) $f : (-1, \infty) \to \mathbb{R}, \quad f(x) = \ln(1 + x);$
(f) $f : \mathbb{R} \to \mathbb{R}, \quad f(x) = \sinh x;$
(g) $f : \mathbb{R} \to \mathbb{R}, \quad f(x) = \cosh x.$

4.64. Let $n \in \mathbb{N}$. Prove that

$$1 + \frac{x}{1!} + \frac{x^2}{2!} + \cdots + \frac{x^{2n}}{(2n)!} > 0, \quad \forall x \in \mathbb{R}.$$

4.65. *Irrational numbers.* Prove that e, e^{-1}, $e^{\sqrt{2}}$, $\sin 1$, and $\cos 1$ are irrational numbers.

4.66. (a) Prove that

$$\sum_{n=1}^{\infty} \frac{n^3}{n!} = 5e.$$

(b) Let f be a function which has a Maclaurin series expansion with radius of convergence $R > 1$. Prove that

$$\sum_{n=1}^{\infty} \frac{n^3}{n!} f^{(n)}(0) = f'(1) + 3f''(1) + f'''(1).$$

4.67. Let $f : \mathbb{R} \to \mathbb{R}$ be a twice derivable function, with f, f', and f'' continuous on $\mathbb{R}$. If $f(0) = 1$, $f'(0) = 0$, and $f''(0) = p$, then prove that

$$\lim_{n \to \infty} \left(f\left(\frac{x}{\sqrt{n}} \right) \right)^n = e^{p \frac{x^2}{2}}.$$

4.6   Series with the Maclaurin Remainder of a Function f

4.68. [19 April 2019]

(a) Prove that

$$\sum_{n=1}^{\infty} \left(\frac{x^n}{n} + \frac{x^{n+1}}{n+1} + \frac{x^{n+2}}{n+2} + \cdots \right) = \frac{x}{1-x}, \quad x \in [-1, 1).$$

(b) Let f be a function which has the Maclaurin series expansion $f(x) = \sum_{n=0}^{\infty} \frac{f^{(n)}(0)}{n!} x^n$, $|x| < R$. Prove that

$$\sum_{n=1}^{\infty} \left(\frac{f^{(n)}(0)}{n!} x^n + \frac{f^{(n+1)}(0)}{(n+1)!} x^{n+1} + \cdots \right) = xf'(x), \quad |x| < R.$$

(c) Let f be a function which has the Maclaurin series expansion $f(x) = \sum_{n=0}^{\infty} \frac{f^{(n)}(0)}{n!} x^n$, $|x| < R$. Prove that

$$\sum_{n=0}^{\infty} a^n \left(\frac{f^{(n)}(0)}{n!} x^n + \frac{f^{(n+1)}(0)}{(n+1)!} x^{n+1} + \cdots \right)$$

$$= \frac{f(x) - af(ax)}{1-a}, \quad a \neq 1, \ |ax| < R.$$

4.69. Let $x \in \mathbb{R}$. Prove that:

(a) $\displaystyle\int_0^x \frac{(x-t)^n}{n!} e^t \, dt = e^x - 1 - \frac{x}{1!} - \frac{x^2}{2!} - \cdots - \frac{x^n}{n!}, \quad \forall n \geq 0;$

(b) $\displaystyle\sum_{n=0}^{\infty} \left(e^x - 1 - \frac{x}{1!} - \frac{x^2}{2!} - \cdots - \frac{x^n}{n!} \right) = xe^x;$

(c) $\displaystyle\sum_{n=0}^{\infty} (-1)^n \left(e^x - 1 - \frac{x}{1!} - \frac{x^2}{2!} - \cdots - \frac{x^n}{n!} \right) = \sinh x.$

4.70. Series with the Maclaurin Remainder of a Function f

Let f be a function which has the Maclaurin series expansion with radius of convergence $R \in (0, \infty]$. Prove that, for $|x| < R$, we have

(a) $\displaystyle\sum_{n=1}^{\infty} \left(f(x) - f(0) - \frac{f'(0)}{1!} x - \frac{f''(0)}{2!} x^2 - \cdots - \frac{f^{(n)}(0)}{n!} x^n \right)$
$$= xf'(x) - f(x) + f(0);$$

(b) $\displaystyle\sum_{n=1}^{\infty} n \left(f(x) - f(0) - \frac{f'(0)}{1!} x - \frac{f''(0)}{2!} x^2 - \cdots - \frac{f^{(n)}(0)}{n!} x^n \right) = \frac{x^2}{2} f''(x);$

(c) $\displaystyle\sum_{n=1}^{\infty} n^2 \left(f(x) - f(0) - \frac{f'(0)}{1!} x - \frac{f''(0)}{2!} x^2 - \cdots - \frac{f^{(n)}(0)}{n!} x^n \right)$
$$= \frac{x^3}{3} f'''(x) + \frac{x^2}{2} f''(x).$$

In particular, when $f(x) = \mathrm{Li}_k(x)$, $k \in \mathbb{N}$, $k \geq 2$ and $x \in [-1, 1]$, we obtain the polylogarithm series

(a) $\displaystyle\sum_{n=1}^{\infty} \left(\mathrm{Li}_k(x) - x - \frac{x^2}{2^k} - \cdots - \frac{x^n}{n^k} \right) = \mathrm{Li}_{k-1}(x) - \mathrm{Li}_k(x);$

(b) $\displaystyle\sum_{n=1}^{\infty} n \left(\mathrm{Li}_k(x) - x - \frac{x^2}{2^k} - \cdots - \frac{x^n}{n^k} \right) = \frac{\mathrm{Li}_{k-2}(x)}{2} - \frac{\mathrm{Li}_{k-1}(x)}{2};$

(continued)

(c) $\displaystyle\sum_{n=1}^{\infty} n^2 \left(\text{Li}_k(x) - x - \frac{x^2}{2^k} - \cdots - \frac{x^n}{n^k} \right)$

$\displaystyle = \frac{\text{Li}_{k-3}(x)}{3} - \frac{\text{Li}_{k-2}(x)}{2} + \frac{\text{Li}_{k-1}(x)}{6}.$

When $x = 1$ the above formulae become

(a) $\displaystyle\sum_{n=1}^{\infty} \left(\zeta(k) - 1 - \frac{1}{2^k} - \cdots - \frac{1}{n^k} \right) = \zeta(k-1) - \zeta(k), \quad k > 2;$

(b) $\displaystyle\sum_{n=1}^{\infty} n \left(\zeta(k) - 1 - \frac{1}{2^k} - \cdots - \frac{1}{n^k} \right) = \frac{\zeta(k-2)}{2} - \frac{\zeta(k-1)}{2}, \quad k > 3;$

(c) $\displaystyle\sum_{n=1}^{\infty} n^2 \left(\zeta(k) - 1 - \frac{1}{2^k} - \cdots - \frac{1}{n^k} \right)$

$\displaystyle = \frac{\zeta(k-3)}{3} - \frac{\zeta(k-2)}{2} + \frac{\zeta(k-1)}{6}, \quad k > 4$

4.71.

(a) Prove that

$$(-1) \cdot 1^2 + (-1)^2 \cdot 2^2 + \cdots + (-1)^n \cdot n^2 = (-1)^n \frac{n(n+1)}{2}, \quad n \geq 1.$$

(b) Prove that

$$\sum_{n=1}^{\infty} (-1)^{n-1} n^2 \left(e^x - 1 - \frac{x}{1!} - \frac{x^2}{2!} - \cdots - \frac{x^n}{n!} \right) = \frac{x^2}{2} e^{-x}, \quad x \in \mathbb{R}.$$

4.72. Let $x \in \mathbb{R}$. Prove that:

(a) $\displaystyle\sum_{n=0}^{\infty} \cos \frac{n\pi}{2} \left(e^x - 1 - \frac{x}{1!} - \frac{x^2}{2!} - \cdots - \frac{x^n}{n!} \right) = \frac{1}{2} \left(\sin x - \cos x + e^x \right);$

(continued)

(b) $\displaystyle\sum_{n=1}^{\infty} \sin \frac{n\pi}{2} \left(e^x - 1 - \frac{x}{1!} - \frac{x^2}{2!} - \cdots - \frac{x^n}{n!} \right) = \frac{1}{2} \left(-\sin x - \cos x + e^x \right).$

A generalization Let f be a function which has the Maclaurin series expansion with radius of convergence R, and let $x \in \mathbb{R}$, $|x| < R$. Prove that

$$\sum_{n=0}^{\infty} \cos \frac{n\pi}{2} \left(f(x) - f(0) - \frac{f'(0)}{1!}x - \frac{f''(0)}{2!}x^2 - \cdots - \frac{f^{(n)}(0)}{n!}x^n \right)$$

$$= \frac{f(x)}{2} + \frac{f(xi) - f(-xi)}{4i} - \frac{f(xi) + f(-xi)}{4}$$

and

$$\sum_{n=1}^{\infty} \sin \frac{n\pi}{2} \left(f(x) - f(0) - \frac{f'(0)}{1!}x - \frac{f''(0)}{2!}x^2 - \cdots - \frac{f^{(n)}(0)}{n!}x^n \right)$$

$$= \frac{f(x)}{2} - \frac{f(xi) - f(-xi)}{4i} - \frac{f(xi) + f(-xi)}{4}.$$

We record, as a gem, the proof of an amazing series formula involving the Maclaurin remainder of a function f.

A Series Formula with the Maclaurin Remainder of a Function f

Let f be a function which has the Maclaurin series expansion with radius of convergence $R = \infty$. Then

$$\sum_{n=0}^{\infty} 2^n \left(f(x) - f(0) - \frac{f'(0)}{1!}x - \frac{f''(0)}{2!}x^2 - \cdots - \frac{f^{(n)}(0)}{n!}x^n \right) = f(2x) - f(x),$$

for all $x \in \mathbb{R}$.

(continued)

Proof. We have

$$S(x) = \sum_{n=0}^{\infty} 2^n \left(f(x) - f(0) - \frac{f'(0)}{1!}x - \frac{f''(0)}{2!}x^2 - \cdots - \frac{f^{(n)}(0)}{n!}x^n \right)$$

$$= f(x) - f(0)$$

$$+ \sum_{n=1}^{\infty} 2^n \left(f(x) - f(0) - \frac{f'(0)}{1!}x - \frac{f''(0)}{2!}x^2 - \cdots - \frac{f^{(n)}(0)}{n!}x^n \right)$$

$$= f(x) - f(0)$$

$$+ \sum_{m=0}^{\infty} 2^{m+1} \left(f(x) - f(0) - \frac{f'(0)}{1!}x - \frac{f''(0)}{2!}x^2 - \cdots - \frac{f^{(m+1)}(0)}{(m+1)!}x^{m+1} \right)$$

$$= f(x) - f(0)$$

$$+ 2 \sum_{m=0}^{\infty} 2^m \left(f(x) - f(0) - \frac{f'(0)}{1!}x - \frac{f''(0)}{2!}x^2 - \cdots - \frac{f^{(m)}(0)}{m!}x^m \right)$$

$$- \sum_{m=0}^{\infty} \frac{f^{(m+1)}(0)}{(m+1)!}(2x)^{m+1}$$

$$= f(x) - f(0) + 2S(x) - (f(2x) - f(0)),$$

and it follows that $S(x) = f(2x) - f(x)$. ∎

4.73. Let $x \in \mathbb{R}$. Prove that the following series are absolutely convergent and show that:

(a) [44] $\displaystyle\sum_{n=0}^{\infty} 3^n \left(\sin x - x + \frac{1}{3!}x^3 - \cdots - \frac{\sin \frac{n\pi}{2}}{n!}x^n \right) = \sin x \cos(2x);$

(b) [41] $\displaystyle\sum_{n=0}^{\infty} 3^n \left(\cos x - 1 + \frac{1}{2!}x^2 - \cdots - \frac{\cos \frac{n\pi}{2}}{n!}x^n \right) = -\sin x \sin(2x).$

4.74. Let f be a function which has the Maclaurin series expansion with radius of convergence R and let $k \geq 0$. Prove that

$$\sum_{n=k}^{\infty} C_n^k f^{(n)}(0) \left(e^x - 1 - \frac{x}{1!} - \frac{x^2}{2!} - \cdots - \frac{x^n}{n!} \right) = \frac{1}{k!} \int_0^x e^{x-t} t^k f^{(k)}(t) dt, \quad |x| < R.$$

4.75. Let f be a function which has the Maclaurin series expansion with radius of convergence R. Calculate

$$\sum_{n=1}^{\infty} (-1)^n \left(f(x) - f(0) - \frac{f'(0)}{1!} x - \frac{f''(0)}{2!} x^2 - \cdots - \frac{f^{(n)}(0)}{n!} x^n \right), \quad |x| < R.$$

4.76. Let $x \in \mathbb{R}$. Prove that:

(a) [86] $\displaystyle\sum_{n=1}^{\infty} (-1)^{\lfloor \frac{n}{2} \rfloor} \left(e^x - 1 - \frac{x}{1!} - \frac{x^2}{2!} - \cdots - \frac{x^n}{n!} \right) = 1 - \cos x;$

(b) $\displaystyle\sum_{n=1}^{\infty} (-1)^{\lfloor \frac{n}{2} \rfloor} \left(e^x - 1 - \frac{x}{1!} - \frac{x^2}{2!} - \cdots - \frac{x^{n-1}}{(n-1)!} \right) = \sin x.$

A generalization Let f be a function which has the Maclaurin series expansion with radius of convergence R. Prove that, for $|x| < R$, we have

$$\sum_{n=1}^{\infty} (-1)^{\lfloor \frac{n}{2} \rfloor} \left(f(x) - f(0) - \frac{f'(0)}{1!} x - \frac{f''(0)}{2!} x^2 - \cdots - \frac{f^{(n)}(0)}{n!} x^n \right)$$

$$= f(0) - \frac{f(ix) + f(-ix)}{2}$$

and

$$\sum_{n=1}^{\infty} (-1)^{\lfloor \frac{n}{2} \rfloor} \left(f(x) - f(0) - \frac{f'(0)}{1!} x - \frac{f''(0)}{2!} x^2 - \cdots - \frac{f^{(n-1)}(0)}{(n-1)!} x^{n-1} \right)$$

$$= \frac{f(ix) - f(-ix)}{2i}.$$

4.77. A Spectacular Maclaurin Series

(a) Let $k, n \in \mathbb{N}$, $n \geq k$. Prove that

$$C_k^k + C_{k+1}^k + \cdots + C_n^k = C_{n+1}^{k+1}.$$

(b) Let $k \geq 0$ be an integer and f be a function which has the Maclaurin series expansion with radius of convergence $R \in (0, \infty]$. Prove that, for $|x| < R$, we have

$$\sum_{n=k}^{\infty} C_n^k \left(f(x) - f(0) - \frac{f'(0)}{1!}x - \frac{f''(0)}{2!}x^2 - \cdots - \frac{f^{(n)}(0)}{n!}x^n \right)$$

$$= \frac{f^{(k+1)}(x)}{(k+1)!} x^{k+1}.$$

4.78. Let $n \in \mathbb{N}$, $T_n(x) = \sum_{k=1}^{n}(-1)^{k-1}\frac{x^{2k-1}}{(2k-1)!}$ be the Maclaurin polynomial of degree $2n-1$ corresponding to the sine function and let

$$I_n = \int_0^{\infty} \frac{T_n(x) - \sin x}{x^{2n+1}}dx.$$

Prove that:

(a) $I_n = -\frac{1}{2n(2n-1)} I_{n-1}, n \geq 1$;

(b) $I_n = \frac{(-1)^{n-1}}{2(2n)!}\pi$.

4.79. [*O. Furdui, Problem 3, Traian Lalescu National Contest for University Students, 2015*][2]

Let $n \geq 0$ be an integer, $T_{2n}(x) = \sum_{k=1}^{n+1}(-1)^{k-1}\frac{x^{2k-2}}{(2k-2)!}$ be the Maclaurin polynomial of degree $2n$ corresponding to the cosine function and let

$$I_n = \int_0^{\infty} \frac{T_{2n}(x) - \cos x}{x^{2n+2}}dx.$$

---

[2]This problem was given in 2015 at Section B of Traian Lalescu National Contest for University Students, the 8th edition, Braşov, Romania.

(a) Prove that $I_n = -\frac{1}{2n(2n+1)} I_{n-1}, n \geq 1$.

(b) Calculate I_n.

4.7 Series with Fractional Part Function

4.80. Prove that

$$\lim_{n \to \infty} \{n!e\} = 0 \quad \text{and} \quad \lim_{n \to \infty} n\{n!e\} = 1,$$

where $\{a\}$ is the fractional part of the real number a.

An Invitation to the Calculation of Series with the Floor and the Fractional Part Function

4.81. [59] Prove that:

(a) $\displaystyle\sum_{n=1}^{\infty} \frac{\{n!e\}}{n!} = \sum_{n=1}^{\infty} \left(e - 1 - \frac{1}{1!} - \frac{1}{2!} - \cdots - \frac{1}{n!} \right) = 1;$

(b) $\displaystyle\sum_{n=1}^{\infty} \frac{\{(2n)!e\}}{(2n)!} = \sum_{n=1}^{\infty} \left(e - 1 - \frac{1}{1!} - \frac{1}{2!} - \cdots - \frac{1}{(2n)!} \right) = 1 - \frac{\cosh 1}{2};$

(c) $\displaystyle\sum_{n=1}^{\infty} \frac{\{(2n-1)!e\}}{(2n-1)!} = \sum_{n=1}^{\infty} \left(e - 1 - \frac{1}{1!} - \frac{1}{2!} - \cdots - \frac{1}{(2n-1)!} \right) = \frac{\cosh 1}{2};$

(d) $\displaystyle\sum_{n=1}^{\infty} (-1)^n \frac{\{n!e\}}{n!} = \sum_{n=1}^{\infty} (-1)^n \left(e - 1 - \frac{1}{1!} - \frac{1}{2!} - \cdots - \frac{1}{n!} \right) = 1 - \cosh 1;$

(e) $\displaystyle\sum_{n=1}^{\infty} \frac{\{n!e\}}{n!} x^n = \frac{e^x - ex}{x - 1} + 1, \text{ for } x \neq 1;$

(f) $\displaystyle\sum_{n=1}^{\infty} \frac{\lfloor n!e \rfloor}{n!} x^n = \frac{e^x}{1 - x} - 1, \text{ for } x \in (-1, 1);$

(continued)

(g) $\displaystyle\sum_{n=1}^{\infty}\left\{\frac{n!}{e}\right\}\frac{1}{n!} = \frac{e}{2} - \frac{3}{2e};$

(h) $\displaystyle\sum_{n=1}^{\infty}\left\{\frac{n!}{e}\right\}\frac{x^n}{n!} = \frac{x}{e(1-x)} + \frac{e^x}{2} + \frac{x+1}{x-1}\cdot\frac{e^{-x}}{2}, \text{ for } x \neq 1;$

(i) $\displaystyle\sum_{n=1}^{\infty}\left\lfloor\frac{n!}{e}\right\rfloor\frac{x^n}{n!} = -\frac{1}{2}\left(e^x + \frac{x+1}{x-1}e^{-x}\right), \text{ for } x \in [-1, 1);$

(j) $\displaystyle\sum_{n=1}^{\infty}(-1)^n\{n!e\} = 1 - e + \ln 2 + \int_0^1 e^x \ln(2-x)\,dx;$

(k) $\displaystyle\sum_{n=1}^{\infty}\frac{\{n!e\}}{n} = \int_0^1 \frac{e^x - 1}{x}\,dx;$

(l) $\displaystyle\sum_{n=1}^{\infty}\left(\{n!e\} - \frac{1}{n+1}\right) = 2 - e + \int_0^1 \frac{e^x - 1}{r}\,dx;$

(m) $\displaystyle\sum_{n=1}^{\infty}\frac{\{(2n-1)!\sinh 1\}}{(2n-1)!} = \frac{1}{2e};$

(n) $\displaystyle\sum_{n=1}^{\infty}\frac{\{(2n)!\cosh 1\}}{(2n)!} = 1 - \cosh 1 + \frac{\sinh 1}{2};$

(o) $\displaystyle\sum_{n=1}^{\infty}\frac{\{(2n)!\cos 1\}}{(2n)!} = \frac{\cosh 1}{2} - \frac{\cos 1}{2} - \frac{\sin 1}{2};$

(p) $\displaystyle\sum_{n=1}^{\infty}\frac{\{(2n-1)!\sin 1\}}{(2n-1)!} = \frac{\cos 1}{2} + \frac{\sinh 1}{2}.$

4.82. [59] Let $k \geq 0$ be an integer. Prove that

$$\sum_{n=1}^{\infty}\frac{\{(n+k)!e\}}{n!} = \frac{e}{k+1} - k!\left(e - 1 - \frac{1}{1!} - \cdots - \frac{1}{k!}\right).$$

4.8   Extrema of One Variable Functions

The Local Extremum Points of a One Variable Function
Let $I \subseteq \mathbb{R}$ be an open interval, $n \in \mathbb{N}$, $n \geq 2$, $f : I \to \mathbb{R}$ a function n times differentiable on I and $x_0 \in I$ such that

$$f'(x_0) = f''(x_0) = \ldots = f^{(n-1)}(x_0) = 0, \quad f^{(n)}(x_0) \neq 0.$$

(a) If n is even, then x_0 is a point of local extremum of f as follows: a local minimum point of f if $f^{(n)}(x_0) > 0$, and a local maximum point of f if $f^{(n)}(x_0) < 0$.

(b) If n is odd, then x_0 is not a local extremum point of f, it is an inflection point.

4.83. Find:

(a) the minimum and the maximum values of the function

$$f : \mathbb{R} \to \mathbb{R}, \quad f(x) = 8 \sin x + 15 \cos x;$$

(b) the maximum value of the function $f : \mathbb{R} \to \mathbb{R}$,

$$f(x) = \sinh x - \cosh^2 x - \frac{1}{2};$$

(c) the minimum value of the function $f : (-\infty, 0) \to \mathbb{R}$,

$$f(x) = \arctan^2 x - \frac{\pi}{2} \arctan \frac{1}{x};$$

(d) the minimum value of the function $f : \mathbb{R} \to \mathbb{R}$,

$$f(x) = x(x + 5)(x + 10)(x + 15).$$

4.84. Find the minimum and the maximum values of the function $f : \mathbb{R} \to \mathbb{R}$, $f(x) = x^5 - x$, on the set $C = \{ x \in \mathbb{R} \mid x^4 + 4 \leq 5x^2 \}$.

4.85. Find the local extremum points and the local extremum values of the following functions:

(a) $f : \mathbb{R} \to \mathbb{R}, \quad f(x) = \frac{x}{x^2+1}$;

(b) $f : \mathbb{R} \to \mathbb{R}, \quad f(x) = \frac{x^2}{x^4+1}$;

(c) $f : \mathbb{R} \to \mathbb{R}, \quad f(x) = 3x^5 - 15x^4 - 25x^3 + 1$;

(d) $f : \mathbb{R} \to \mathbb{R}, \quad f(x) = x^6 - 16x^3 - 1$;

(e) $f : \mathbb{R} \to \mathbb{R}, \quad f(x) = xe^{-x^2}$;

(f) $f : \mathbb{R} \to \mathbb{R}, \quad f(x) = x^2 e^{-x^2}$;

(g) $f : \mathbb{R} \to \mathbb{R}, \quad f(x) = \cosh(3x)$;

(h) $f : (0, \infty) \to \mathbb{R}, \quad f(x) = x \ln^2 x$;

(i) $f : (0, \infty) \to \mathbb{R}, \quad f(x) = x^x$;

(j) $f : \mathbb{R} \to \mathbb{R}, \quad f(x) = e^x \cos x$;

(k) $f : \mathbb{R} \to \mathbb{R}, \quad f(x) = \arctan\left(xe^x\right)$;

(l) $f : \mathbb{R} \to \mathbb{R}, \quad f(x) = 2 \sin x + x^2 + \pi x$;

(m) $f : \mathbb{R} \to \mathbb{R}, \quad f(x) = 2 \sin x - x^2 - \pi x$.

4.86. Find the distance from the point $(0, 33)$ to the parabola $x = 4y^2$.

4.87. Let $b \in \mathbb{R}$. Find the distance from the point $(0, b)$ to the parabola $x^2 = 4y$.

4.88. [A. Doboşan, 1978] Let $a, b, c \in \mathbb{R}$, with $ab < 0$. If m and M are the local
minimum and the local maximum values of $f : \mathbb{R} \to \mathbb{R}, f(x) = ax^5 + bx^3 + c$,
then $M + m = 2c$ and $Mm = \frac{108b^5}{3125a^3} + c^2$.

We record, as a gem, the proof of a conditional inequality in two variables.

A Gem with a Conditional Inequality

If $x, y \geq 0$, such that $x + y = 2$, then $x^2 + xy + y^2 \geq 3$.

Proof. Let $x = 1 + \alpha$ and $y = 1 - \alpha, \alpha \in [0, 1]$. We have

$$x^2 + xy + y^2 = (1 + \alpha)^2 + 1 - \alpha^2 + (1 - \alpha)^2 = 3 + \alpha^2 \geq 3,$$

with equality when $x = y = 1$.   ∎

4.89. Let $n \in \mathbb{N}$ and $x, y \geq 0$, with $x + y = 2$. Prove that $x^n - xy + y^n \geq 1$.

4.90. Open problem. Let $a, b \in \mathbb{R}$, $a^2 + b^2 \neq 0$. Find:

- $\max_{x \in \mathbb{R}} \{\sin(ax) + \sin(bx)\}$;
- $\max_{x \in \mathbb{R}} \{\sin(ax) + \cos(bx)\}$.

An Invitation to Maclaurin Inequalities

I. Let $k \in \mathbb{N}$ and let f be a function which has the Maclaurin series expansion with radius of convergence $R \in \{1, \infty\}$. If $f^{(n)}(0) \geq 0$, $\forall n \geq 0$, then

$$f\left(\frac{x^{2k+2}}{2k+2}\right) - 2f\left(\frac{x^{2k+1}}{2k+1}\right) + f\left(\frac{x^{2k}}{2k}\right) \geq 0, \quad \forall x \in (-R, R).$$

Proof. A calculation shows that

$$f\left(\frac{x^{2k+2}}{2k+2}\right) - 2f\left(\frac{x^{2k+1}}{2k+1}\right) + f\left(\frac{x^{2k}}{2k}\right)$$

$$= \sum_{n=0}^{\infty} \frac{f^{(n)}(0)}{n!} x^{2kn} \left[\frac{x^{2n}}{(2k+2)^n} - 2\frac{x^n}{(2k+1)^n} + \frac{1}{(2k)^n}\right].$$

We denote $x^n = t$ and observe that $\frac{t^2}{(2k+2)^n} - \frac{2t}{(2k+1)^n} + \frac{1}{(2k)^n} > 0, \forall t \in \mathbb{R}$. This follows since the discriminant of the quadratic function $\Delta < 0$. ∎

The case $k = 1$. If f is a function which has the Maclaurin series expansion with radius of convergence $R \in \{1, \infty\}$ and $f^{(n)}(0) \geq 0$, $\forall n \geq 0$, then

$$f\left(\frac{x^4}{4}\right) - 2f\left(\frac{x^3}{3}\right) + f\left(\frac{x^2}{2}\right) \geq 0, \quad \forall x \in (-R, R).$$

Applications. The following inequalities hold:

(a) $e^{\frac{x^4}{4}} - 2e^{\frac{x^3}{3}} + e^{\frac{x^2}{2}} \geq 0$, $\forall x \in \mathbb{R}$;

(b) $\cosh\left(\frac{x^4}{4}\right) - 2\cosh\left(\frac{x^3}{3}\right) + \cosh\left(\frac{x^2}{2}\right) \geq 0$, $\forall x \in \mathbb{R}$;

(continued)

(c) $\dfrac{2}{4 - x^4} - \dfrac{3}{3 - x^3} + \dfrac{1}{2 - x^2} \geq 0,\ \forall x \in [-1, 1]$;

(d) $\dfrac{2^{2n-1}}{(4 - x^4)^n} - \dfrac{3^n}{(3 - x^3)^n} + \dfrac{2^{n-1}}{(2 - x^2)^n} \geq 0,\ \forall x \in [-1, 1],\ n \in \mathbb{N}$;

(e) $\left(1 - \dfrac{x^4}{4}\right)\left(1 - \dfrac{x^2}{2}\right) \leq \left(1 - \dfrac{x^3}{3}\right)^2,\ \forall x \in [-1, 1]$;

(f) $\left(1 - \dfrac{x^{2k+2}}{2k + 2}\right)\left(1 - \dfrac{x^{2k}}{2k}\right) \leq \left(1 - \dfrac{x^{2k+1}}{2k + 1}\right)^2,\ \forall x \in [-1, 1],\ k \in \mathbb{N}$.

II. Let f be a function which has the Maclaurin series expansion with radius of convergence $R \in \{1, \infty\}$. If $f^{(n)}(0) \geq 0,\ \forall n \geq 0$, then

$$f\left(x^6\right) - 4f\left(x^5\right) + 6f\left(x^4\right) - 4f\left(x^3\right) + f\left(x^2\right) \geq 0, \quad \forall x \in (-R, R).$$

Proof. A calculation shows that

$$f\left(x^6\right) - 4f\left(x^5\right) + 6f\left(x^4\right) - 4f\left(x^3\right) + f\left(x^2\right)$$

$$= \sum_{n=0}^{\infty} \frac{f^{(n)}(0)}{n!} x^{2n} \left(x^n - 1\right)^4 \geq 0,$$

with equality when $x = 0$ or $x = 1$. ∎

Applications. We consider the particular cases when f is e^x, $\cosh x$, $\frac{1}{1-x}$, $\ln(1 - x)$, and we have that the following inequalities hold:

(g) [V. Brayman, 2020] $e^{x^6} - 4e^{x^5} + 6e^{x^4} - 4e^{x^3} + e^{x^2} \geq 0,\ \forall x \in \mathbb{R}$;

(h) $\cosh(x^6) - 4\cosh(x^5) + 6\cosh(x^4) - 4\cosh(x^3) + \cosh(x^2) \geq 0,\ \forall x \in \mathbb{R}$;

(i) $\dfrac{1}{1 - x^6} - \dfrac{4}{1 - x^5} + \dfrac{6}{1 - x^4} - \dfrac{4}{1 - x^3} + \dfrac{1}{1 - x^2} \geq 0,\ \forall x \in (-1, 1)$;

(j) $(1 - x^2)(1 - x^4)^6(1 - x^6) \leq (1 - x^3)^4(1 - x^5)^4,\ \forall x \in [-1, 1]$.

A *challenge.* Prove that the inequality (j) holds for all $x \in \mathbb{R}$.

Partial Derivatives and Applications

5

If I were again beginning my studies, I would follow
the advice of Plato and start with mathematics.

Galileo Galilei (1564–1642)

This chapter collects problems on partial derivatives and their applications, the Jacobian and the Hessian matrices, differential operators, the chain rule, homogeneous functions and Euler's identity, Taylor's formula for functions of two variables, as well as extrema of functions of several variables.

5.1 Partial Derivatives, the Jacobian and the Hessian Matrices, Differential Operators

5.1. Calculate the first order partial derivatives of the following functions:

(a) $f(x, y) = \sin(\cos(xy^2 + 2x - y + 10))$;
(b) $f(x, y) = \ln(\sin^2(xy) + 1)$;
(c) $f(x, y, z) = \ln(x^y y^z z^x)$, $x, y, z > 0$;
(d) $f(x, y, z) = e^{xyz} \cos^2(x - yz)$;
(e) $f(x, y, z) = (xy)^z$, $x, y > 0$;
(f) $f(x, y, z) = x^{yz}$, $x > 0$;
(g) $f(x, y, z) = x^{y^z}$, $x, y > 0$.

5.2. Calculate the first and second order partial derivatives of the following functions:

(a) $f(x, y) = \ln(xy + 1)$, $xy + 1 > 0$;
(b) $f(x, y) = (x - y) \cos(xy)$;
(c) $f(x, y) = (x^2 + y^2) \arctan \frac{x}{y}$, $y \neq 0$;

© The Author(s), under exclusive license to Springer Nature Switzerland AG 2021
A. Sîntămărian, O. Furdui, *Sharpening Mathematical Analysis Skills*, Problem Books
in Mathematics, https://doi.org/10.1007/978-3-030-77139-3_5

(d) $f(x, y, z) = e^{x^2 y} \sin(x - z)$;

(e) $f(x, y, z) = \frac{ye^{-z}}{xy-1}$, $xy - 1 \neq 0$;

(f) $f(x, y, z) = \frac{x-y}{xz-1}$, $xz - 1 \neq 0$;

(g) $f(x, y, z) = \frac{xz}{yz-1}$, $yz - 1 \neq 0$;

(h) $f(x, y, z) = \frac{\sin(x-y)}{\cos(y-z)}$, $y - z \neq \frac{\pi}{2} + k\pi$, $k \in \mathbb{Z}$.

5.3.   [3, p. 225], [96, p. 98, 99], [143, p. 72] Let $f : \mathbb{R}^2 \to \mathbb{R}$ be the function defined by

$$f(x, y) = \begin{cases} xy\frac{x^2-y^2}{x^2+y^2} & \text{if } (x, y) \neq (0, 0) \\ 0 & \text{if } (x, y) = (0, 0). \end{cases}$$

Prove that:

(a) f is of class C^1;

(b) $\frac{\partial^2 f}{\partial y \partial x}(0, 0) = -1$, $\frac{\partial^2 f}{\partial x \partial y}(0, 0) = 1$;

(c) f does not satisfy the conditions of Schwarz's Theorem.

5.4.   [143, p. 73] Let $f : \mathbb{R}^2 \to \mathbb{R}$ be the function defined by

$$f(x, y) = \begin{cases} x^2 \arctan\frac{y}{x} - y^2 \arctan\frac{x}{y} & \text{if } xy \neq 0 \\ 0 & \text{if } xy = 0. \end{cases}$$

Prove that:

(a) f is of class C^1;

(b) $\frac{\partial^2 f}{\partial y \partial x}(0, 0) = -1$, $\frac{\partial^2 f}{\partial x \partial y}(0, 0) = 1$;

(c) f does not satisfy the conditions of Schwarz's Theorem.

5.5.   Let $A = \{(x, y) \in \mathbb{R}^2 \mid y \neq 0\}$ and $f : A \to \mathbb{R}$, $f(x, y) = xy - xe^{\frac{x}{y}}$. Prove that, for all $(x, y) \in A$, the following equalities hold:

(a) $x\frac{\partial f}{\partial x}(x, y) + y\frac{\partial f}{\partial y}(x, y) = xy + f(x, y)$;

(b) $x^2\frac{\partial^2 f}{\partial x^2}(x, y) + 2xy\frac{\partial^2 f}{\partial y \partial x}(x, y) + y^2\frac{\partial^2 f}{\partial y^2}(x, y) = 2xy$.

5.6.   Let $A = \{(x, y) \in \mathbb{R}^2 \mid xy \neq 0\}$ and $f : A \to \mathbb{R}$, $f(x, y) = \frac{x}{y^3}e^{-\frac{y}{x}}$. Calculate $L(f) = x\frac{\partial f}{\partial x} + y\frac{\partial f}{\partial y}$ and $L(L(f))$.

5.7. Let $A = \{(x, y, z) \in \mathbb{R}^3 \mid x - yz \neq 0\}$ and $f : A \to \mathbb{R}$, $f(x, y, z) = \frac{xy-z}{x-yz}$. Prove that, for all $(x, y, z) \in A$, the following equality holds:

$$(x - yz)^2 \frac{\partial^2 f}{\partial z \partial y}(x, y, z) - 2xf(x, y, z) = 0.$$

5.8. Let $A = \{(x, y, z) \in \mathbb{R}^3 \mid x^3 + y^3 + z^3 - 3xyz > 0\}$ and $f, g, h : A \to \mathbb{R}$ be the functions defined by

$$f(x, y, z) = \ln(x^3 + y^3 + z^3 - 3xyz),$$
$$g(x, y, z) = f^2(x, y, z),$$
$$h(x, y, z) = f^3(x, y, z).$$

Prove that, for all $(x, y, z) \in A$, the following equalities hold:

(a) $f'_x(x, y, z) + f'_y(x, y, z) + f'_z(x, y, z) = \frac{3}{x+y+z}$;

(b) $f'^3_x(x, y, z) + f'^3_y(x, y, z) + f'^3_z(x, y, z) - 3f'_x(x, y, z)f'_y(x, y, z)f'_z(x, y, z)$
$$- \frac{27}{x^3+y^3+z^3-3xyz},$$

(c) $f'''_{x^3}(x, y, z) + f'''_{y^3}(x, y, z) + f'''_{z^3}(x, y, z) - 3f'''_{xyz}(x, y, z) = 0$;

(d) $g'''_{x^3}(x, y, z) + g'''_{y^3}(x, y, z) + g'''_{z^3}(x, y, z) - 3g'''_{xyz}(x, y, z) = 0$;

(e) $h'''_{x^3}(x, y, z) + h'''_{y^3}(x, y, z) + h'''_{z^3}(x, y, z) - 3h'''_{xyz}(x, y, z) = \frac{162}{x^3+y^3+z^3-3xyz}$.

Remark 5.1. If $n \in \mathbb{N}$ and $\varphi : A \to \mathbb{R}$, $\varphi(x, y, z) = f^n(x, y, z)$, then

$$\varphi'''_{x^3}(x, y, z) + \varphi'''_{y^3}(x, y, z) + \varphi'''_{z^3}(x, y, z) - 3\varphi'''_{xyz}(x, y, z) = \frac{27n(n-1)(n-2)f^{n-3}(x, y, z)}{x^3 + y^3 + z^3 - 3xyz}.$$

5.9. Let $n \geq 0$ and $f : \mathbb{R}^2 \to \mathbb{R}$, $f(x, y) = (x^n + y^n)e^{x+y}$. Calculate $\frac{\partial^{2n} f}{\partial y^n \partial x^n}(x, y)$.

5.10. Let $n \geq 0$ and $f : \mathbb{R}^2 \to \mathbb{R}$, $f(x, y) = \frac{xy}{(1+x^2)^2(1+y^2)^2}$. Calculate $\frac{\partial^{2n} f}{\partial y^n \partial x^n}(0, 0)$.

5.11. Let $A = \{(x, y) \in \mathbb{R}^2 \mid xy + 1 \neq 0\}$ and $f : A \to \mathbb{R}$, $f(x, y) = \frac{1}{xy+1}$.

(a) Prove that

$$\frac{\partial^{2n} f}{\partial y^n \partial x^n}(0, 0) = (-1)^n (n!)^2, \quad n \geq 0.$$

(b) Let $A = \{(x, y, z) \in \mathbb{R}^2 \mid xyz + 1 \neq 0\}$ and $f : A \to \mathbb{R}$, $f(x, y, z) = \frac{1}{xyz+1}$. Prove that

$$\frac{\partial^{3n} f}{\partial z^n \partial y^n \partial x^n}(0, 0, 0) = (-1)^n (n!)^3, \quad n \geq 0.$$

Remark 5.2. Let $k \in \mathbb{N}$, $k \geq 2$, $A = \{(x_1, x_2, \ldots, x_k) \in \mathbb{R}^k \mid x_1 x_2 \cdots x_k + 1 \neq 0\}$, and

$$f : A \to \mathbb{R}, \quad f(x_1, x_2, \ldots, x_k) = \frac{1}{x_1 x_2 \cdots x_k + 1}.$$

The following equality holds

$$\frac{\partial^{kn} f}{\partial x_k^n \partial x_{k-1}^n \cdots \partial x_1^n}(0, 0, \ldots, 0) = (-1)^n (n!)^k, \quad n \geq 0.$$

5.12. Let $f, g : \mathbb{R}^2 \to \mathbb{R}$, $f(x, y) = e^{x+y} \sin(x + y)$, $g(x, y) = e^{x+y} \cos(x + y)$. Prove that, for all $n \geq 0$ and $(x, y) \in \mathbb{R}^2$, we have:

(a) $\left(\frac{\partial^n f}{\partial x^n}(x, y)\right)^2 + \left(\frac{\partial^n f}{\partial y^n}(x, y)\right)^2 + \left(\frac{\partial^n g}{\partial x^n}(x, y)\right)^2 + \left(\frac{\partial^n g}{\partial y^n}(x, y)\right)^2 = 2^{n+1} e^{2(x+y)}$;

(b) $\left(\frac{\partial^{2n} f}{\partial y^n \partial x^n}(x, y)\right)^2 + \left(\frac{\partial^{2n} g}{\partial y^n \partial x^n}(x, y)\right)^2 = 2^{2n} e^{2(x+y)}$.

Let $A \subseteq \mathbb{R}^n$, $a \in \text{int}(A)$, and $f : A \to \mathbb{R}^m$, $f = (f_1, f_2, \ldots, f_m)$ be a function that has first order partial derivatives at a. The following matrix having m rows and n columns

$$J(f)(a) = \begin{pmatrix} \dfrac{\partial f_1}{\partial x_1}(a) & \dfrac{\partial f_1}{\partial x_2}(a) & \ldots & \dfrac{\partial f_1}{\partial x_n}(a) \\ \dfrac{\partial f_2}{\partial x_1}(a) & \dfrac{\partial f_2}{\partial x_2}(a) & \ldots & \dfrac{\partial f_2}{\partial x_n}(a) \\ \ldots & \ldots & \ldots & \ldots \\ \dfrac{\partial f_m}{\partial x_1}(a) & \dfrac{\partial f_m}{\partial x_2}(a) & \ldots & \dfrac{\partial f_m}{\partial x_n}(a) \end{pmatrix}$$

is called the *Jacobian matrix* of f at a.

If $m = n$, then the determinant of the Jacobian matrix of f is called *functional determinant* or the *Jacobian*

$$\frac{D(f_1, f_2, \ldots, f_n)}{D(x_1, x_2, \ldots, x_n)}(a) = \begin{vmatrix} \dfrac{\partial f_1}{\partial x_1}(a) & \dfrac{\partial f_1}{\partial x_2}(a) & \ldots & \dfrac{\partial f_1}{\partial x_n}(a) \\ \dfrac{\partial f_2}{\partial x_1}(a) & \dfrac{\partial f_2}{\partial x_2}(a) & \ldots & \dfrac{\partial f_2}{\partial x_n}(a) \\ \ldots & \ldots & \ldots & \ldots \\ \dfrac{\partial f_n}{\partial x_1}(a) & \dfrac{\partial f_n}{\partial x_2}(a) & \ldots & \dfrac{\partial f_n}{\partial x_n}(a) \end{vmatrix}.$$

(continued)

Let $A \subseteq \mathbb{R}^n$, $a \in \text{int}(A)$, and $f : A \rightarrow \mathbb{R}$ be a function that has second order partial derivatives at a. The following square matrix of order n

$$H(f)(a) = \begin{pmatrix} \dfrac{\partial^2 f}{\partial x_1^2}(a) & \dfrac{\partial^2 f}{\partial x_2 \partial x_1}(a) & \cdots & \dfrac{\partial^2 f}{\partial x_n \partial x_1}(a) \\ \dfrac{\partial^2 f}{\partial x_1 \partial x_2}(a) & \dfrac{\partial^2 f}{\partial x_2^2}(a) & \cdots & \dfrac{\partial^2 f}{\partial x_n \partial x_2}(a) \\ \cdots & \cdots & \cdots & \cdots \\ \dfrac{\partial^2 f}{\partial x_1 \partial x_n}(a) & \dfrac{\partial^2 f}{\partial x_2 \partial x_n}(a) & \cdots & \dfrac{\partial^2 f}{\partial x_n^2}(a) \end{pmatrix}$$

is called the *Hessian matrix* of f at a.

5.13. Let $f : \mathbb{R}^2 \rightarrow \mathbb{R}^3$, $f(x, y) = (x - \cos y, \arctan(xy), \sin(\pi x - y))$. Determine the matrix $J(f)(1, 0)$.

5.14. Let $f : \mathbb{R}^3 \rightarrow \mathbb{R}^2$, $f(x, y, z) = \left(\ln\left(e^{xy-z} + 1\right), (x^2 + 1)^{yz}\right)$. Determine the matrix $J(f)(-1, 1, -1)$.

5.15. *Polar coordinates in plane.* Calculate $\frac{D(x,y)}{D(\rho,\theta)}$ if

$$x = \rho \cos \theta, \quad y = \rho \sin \theta.$$

5.16. *Polar coordinates in space.* Calculate $\frac{D(x,y,z)}{D(\rho,\varphi,\theta)}$ if

$$x = \rho \sin \varphi \cos \theta, \quad y = \rho \sin \varphi \sin \theta, \quad z = \rho \cos \varphi.$$

5.17. Calculate $\frac{D(x,y)}{D(u,v)}$ if $x = \dfrac{u}{\sqrt{1-u^2-v^2}}$, $y = \dfrac{v}{\sqrt{1-u^2-v^2}}$, $u^2 + v^2 < 1$.

5.18. Calculate $\frac{D(x,y,z)}{D(u,v,w)}$ if

$$x = u^2 + v^2 + w^2, \quad y = u^3 + v^3 + w^3, \quad z = u^4 + v^4 + w^4.$$

5.19. Let $f : (0, \infty) \times \mathbb{R} \rightarrow \mathbb{R}$, $f(x, y) = x^y$. Calculate $H(f)(1, -1)$.

5.20. Let $A = \{(x, y, z) \in \mathbb{R}^3 \mid z - x \neq 0\}$ and $f : A \to \mathbb{R}$, $f(x, y, z) = \frac{y-z}{z-x}$. Calculate $H(f)(1, -1, 0)$.

Let $\emptyset \neq A \subseteq \mathbb{R}^3$ be an open set. We consider the following sets of functions:

$\mathscr{F} = \{ f \mid f : A \to \mathbb{R} \}$

$\mathscr{F}_1 = \{ f \mid f : A \to \mathbb{R} \text{ having first order partial derivatives on } A \}$

$\mathscr{F}_2 = \{ f \mid f : A \to \mathbb{R} \text{ having second order partial derivatives on } A \}$

$\mathscr{V} = \{ F \mid F : A \to \mathbb{R}^3 \}$

$\mathscr{V}_1 = \{ F \mid F = (P, Q, R) : A \to \mathbb{R}^3, \text{ with } P, Q, R \text{ having first order partial}$

derivatives on $A\}$.

We define the following differential operators:

the gradient $\nabla : \mathscr{F}_1 \to \mathscr{V}$, $\quad \nabla f = \left(\dfrac{\partial f}{\partial x}, \dfrac{\partial f}{\partial y}, \dfrac{\partial f}{\partial z} \right)$;

the divergence $\mathrm{div} : \mathscr{V}_1 \to \mathscr{F}$, $\quad \mathrm{div}\, F = \dfrac{\partial P}{\partial x} + \dfrac{\partial Q}{\partial y} + \dfrac{\partial R}{\partial z}$, $\quad F = (P, Q, R)$;

the rotation $\mathrm{rot} : \mathscr{V}_1 \to \mathscr{V}$, $\quad \mathrm{rot}\, F = \left(\dfrac{\partial R}{\partial y} - \dfrac{\partial Q}{\partial z}, \dfrac{\partial P}{\partial z} - \dfrac{\partial R}{\partial x}, \dfrac{\partial Q}{\partial x} - \dfrac{\partial P}{\partial y} \right)$;

the Laplacian $\Delta : \mathscr{F}_2 \to \mathscr{F}$, $\quad \Delta f = \dfrac{\partial^2 f}{\partial x^2} + \dfrac{\partial^2 f}{\partial y^2} + \dfrac{\partial^2 f}{\partial z^2}$.

The rotation is also known as *the curl*.

Remark 5.3. Let $A \subseteq \mathbb{R}^n$, $a \in \mathrm{int}\,(A)$, and $f : A \to \mathbb{R}$ having first order partial derivatives at a. The gradient of the function f at a is defined by

$$\nabla f(a) = \left(\frac{\partial f}{\partial x_1}(a), \frac{\partial f}{\partial x_2}(a), \dots, \frac{\partial f}{\partial x_n}(a) \right).$$

If f has second order partial derivatives at a, then

$$H(f)(a) = J(\nabla f)(a).$$

(continued)

Also, the Laplacian of f at a is given by

$$\Delta f(a) = \frac{\partial^2 f}{\partial x_1^2}(a) + \frac{\partial^2 f}{\partial x_2^2}(a) + \cdots + \frac{\partial^2 f}{\partial x_n^2}(a).$$

5.21. Let $f : \mathbb{R}^3 \to \mathbb{R}, f(x, y, z) = \arctan(xyz)$. Calculate ∇f and $\Delta f = \mathrm{div}(\nabla f)$.

5.22. Let $f : \mathbb{R}^3 \setminus \{(0, 0, 0)\} \to \mathbb{R}$, $f(x, y, z) = \ln \frac{1}{\sqrt{x^2+y^2+z^2}}$. Calculate ∇f and $\Delta f = \mathrm{div}(\nabla f)$.

5.23. Let $f \in C^2\left(\mathbb{R}^3, \mathbb{R}\right)$. Prove that $\mathrm{rot}(\nabla f) = \overline{0}$.

5.24. Let $f : \mathbb{R}^3 \to \mathbb{R}$ be the function defined by

$$f(x, y, z) = \frac{1}{3}(x^3 + y^3 + z^3) + \frac{3}{2}(x^2 + y^2 + z^2) + 9(x + y + z).$$

Find all points $(x, y, z) \in \mathbb{R}^3$ such that $\nabla f(x, y, z) = (y^3, z^3, x^3)$.

5.25. Let $f : \mathbb{R}^3 \to \mathbb{R}$, $f(x, y, z) = \frac{1}{2}(x+y+z)^2+xyz$. Find all points $(x, y, z) \in (-1, \infty)^3$ such that $\nabla f(x, y, z) = (x + 3, y + 8, z + 15)$.

5.26. Let $F : \mathbb{R}^3 \to \mathbb{R}^3$, $F(x, y, z) = (x - y + z, xy - e^z, \arctan(x - yz))$. Calculate $\mathrm{div}\, F$ and $\mathrm{rot}\, F$.

5.27. Let $F : \mathbb{R}^3 \to \mathbb{R}^3$, $F(x, y, z) = (x \cos(yz), ye^{xz}, \sin(xyz))$. Calculate $\mathrm{div}\, F$ and $\mathrm{rot}\, F$.

5.2 The Chain Rule

The Chain Rule

I. Let $\emptyset \neq A \subseteq \mathbb{R}^n$ be an open set, $J \subseteq \mathbb{R}$ an open interval, $f \in C^1(A, \mathbb{R})$ and $u_i : J \to \mathbb{R}, i \in \{1, \ldots, n\}$, derivable functions on J such that $(u_1(x), \ldots, u_n(x)) \in A$, for all $x \in J$. Then the function $F : J \to \mathbb{R}$, $F(x) = f(u_1(x), \ldots, u_n(x))$ is derivable on J and we have

(continued)

$$F'(x) = \sum_{i=1}^{n} f'_{u_i}(u_1(x), \ldots, u_n(x)) \cdot u'_i(x), \quad x \in J.$$

Case $n = 2$ Let $\emptyset \neq A \subseteq \mathbb{R}^2$ be an open set, $J \subseteq \mathbb{R}$ an open interval, $f \in C^1(A, \mathbb{R})$ and $u, v : J \to \mathbb{R}$ derivable functions on J such that $(u(x), v(x)) \in A$, for all $x \in J$. Then the function $F : J \to \mathbb{R}$, $F(x) = f(u(x), v(x))$ is derivable on J and we have

$$F'(x) = f'_u(u(x), v(x)) \cdot u'(x) + f'_v(u(x), v(x)) \cdot v'(x), \quad x \in J.$$

II. Let $I \subseteq \mathbb{R}$ be an open interval, $\emptyset \neq B \subseteq \mathbb{R}^m$ an open set, $f : I \to \mathbb{R}$ a derivable function on I, and $u : B \to \mathbb{R}$ a function having first order partial derivatives on B such that $u(x_1, \ldots, x_m) \in I$, for all $(x_1, \ldots, x_m) \in B$. Then the function $F : B \to \mathbb{R}$, $F(x_1, \ldots, x_m) = f(u(x_1, \ldots, x_m))$ has first order partial derivatives on B and we have, for $j \in \{1, \ldots, m\}$,

$$F'_{x_j}(x_1, \ldots, x_m) = f'(u(x_1, \ldots, x_m)) \cdot u'_{x_j}(x_1, \ldots, x_m), \quad (x_1, \ldots, x_m) \in B.$$

Case $m = 2$ Let $I \subseteq \mathbb{R}$ be an open interval, $\emptyset \neq B \subseteq \mathbb{R}^2$ an open set, $f : I \to \mathbb{R}$ a derivable function on I, and $u : B \to \mathbb{R}$ a function having first order partial derivatives on B such that $u(x, y) \in I$, for all $(x, y) \in B$. Then the function $F : B \to \mathbb{R}$, $F(x, y) = f(u(x, y))$ has first order partial derivatives on B and we have

$$F'_x(x, y) = f'(u(x, y)) \cdot u'_x(x, y), \quad (x, y) \in B,$$
$$F'_y(x, y) = f'(u(x, y)) \cdot u'_y(x, y), \quad (x, y) \in B.$$

III. Let $\emptyset \neq A \subseteq \mathbb{R}^n$ and $\emptyset \neq B \subseteq \mathbb{R}^m$ be open sets, $f \in C^1(A, \mathbb{R})$ and $u_1, \ldots, u_n : B \to \mathbb{R}$ functions having first order partial derivatives on B such that $(u_1(x), \ldots, u_n(x)) \in A$, for all $x = (x_1, \ldots, x_m) \in B$. Then the function $F : B \to \mathbb{R}$, $F(x) = f(u_1(x), \ldots, u_n(x))$ has first order partial derivatives on B and we have, for $j \in \{1, \ldots, m\}$,

$$F'_{x_j}(x) = \sum_{i=1}^{n} f'_{u_i}(u_1(x), \ldots, u_n(x)) \cdot (u_i)'_{x_j}(x), \quad x = (x_1, \ldots, x_m) \in B.$$

Case $m = n = 2$ Let $\emptyset \neq A, B \subseteq \mathbb{R}^2$ be open sets, $f \in C^1(A, \mathbb{R})$ and $u, v : B \to \mathbb{R}$ functions having first order partial derivatives on B such that $(u(x, y), v(x, y)) \in A$, for all $(x, y) \in B$. Then the function $F : B \to \mathbb{R}$,

(continued)

$F(x, y) = f(u(x, y), v(x, y))$ has first order partial derivatives on B and we have

$$F_x'(x, y) = f_u'(u(x, y), v(x, y)) \cdot u_x'(x, y) + f_v'(u(x, y), v(x, y)) \cdot v_x'(x, y),$$
$$F_y'(x, y) = f_u'(u(x, y), v(x, y)) \cdot u_y'(x, y) + f_v'(u(x, y), v(x, y)) \cdot v_y'(x, y),$$

for all $(x, y) \in B$.

Now we give some examples of how to apply the chain rule.

1. Calculate F''' if $F(x) = f(2x - 1, -x)$, where $f = f(u, v)$ is a function of class C^3.

   Let $u(x) = 2x - 1$ and $v(x) = -x$. Then $F(x) = f(u(x), v(x))$. We have

   $$F'(x) = 2 f_u'(u(x), v(x)) - f_v'(u(x), v(x)).$$

   We calculate

   $$F''(x) = 2[f_{u^2}''(u(x), v(x)) \cdot 2 + f_{uv}''(u(x), v(x)) \cdot (-1)]$$
   $$-[f_{vu}''(u(x), v(x)) \cdot 2 + f_{v^2}''(u(x), v(x)) \cdot (-1)]$$
   $$= 4 f_{u^2}''(u(x), v(x)) - 4 f_{uv}''(u(x), v(x)) + f_{v^2}''(u(x), v(x)).$$

   We derive F'' and we obtain

   $$F'''(x) = 4[f_{u^3}'''(u(x), v(x)) \cdot 2 + f_{u^2 v}'''(u(x), v(x)) \cdot (-1)]$$
   $$-4[f_{uvu}'''(u(x), v(x)) \cdot 2 + f_{uv^2}'''(u(x), v(x)) \cdot (-1)]$$
   $$+[f_{v^2 u}'''(u(x), v(x)) \cdot 2 + f_{v^3}'''(u(x), v(x)) \cdot (-1)]$$
   $$= 8 f_{u^3}'''(u(x), v(x)) - 12 f_{u^2 v}'''(u(x), v(x))$$
   $$+6 f_{uv^2}'''(u(x), v(x)) - f_{v^3}'''(u(x), v(x)).$$

2. Calculate $F_{x^2 y}'''$, F_{xyz}''', $F_{y^2 z}'''$, and F_{yz^2}''' if $F(x, y, z) = f(x - yz)$, where $f = f(u)$ is a three times derivable function.

   We denote $u(x, y, z) = x - yz$ and we have $F(x, y, z) = f(u(x, y, z))$.

   To determine $F_{x^2 y}'''$, first we calculate $F_x'(x, y, z) = f'(u(x, y, z))$ and then $F_{x^2}''(x, y, z) = f''(u(x, y, z))$, and taking the derivative of F_{x^2}'' with respect to y we obtain that

   $$F_{x^2 y}'''(x, y, z) = -z f'''(u(x, y, z)).$$

For calculating F'''_{xyz} we take the derivative of F'_x with respect to y and we have that

$$F''_{xy}(x, y, z) = -zf''(u(x, y, z)),$$

and by taking the derivative of the preceding equality with respect to z we obtain that

$$F'''_{xyz}(x, y, z) = -f''(u(x, y, z)) - zf'''(u(x, y, z)) \cdot (-y)$$
$$= yzf'''(u(x, y, z)) - f''(u(x, y, z)).$$

For determining F'''_{y^2z} we calculate one at a time:

$$F'_y(x, y, z) = -zf'(u(x, y, z)),$$

$$F''_{y^2}(x, y, z) = (F'_y)'_y(x, y, z) = -zf''(u(x, y, z)) \cdot (-z) = z^2 f''(u(x, y, z)),$$

$$F'''_{y^2z}(x, y, z) = (F''_{y^2})'_z(x, y, z) = 2zf''(u(x, y, z)) + z^2 f'''(u(x, y, z)) \cdot (-y)$$

$$= -yz^2 f'''(u(x, y, z)) + 2zf''(u(x, y, z)).$$

Now we calculate F'''_{yz^2} and we have

$$F''_{yz}(x, y, z) = (F'_y)'_z(x, y, z) = -f'(u(x, y, z)) - zf''(u(x, y, z)) \cdot (-y)$$
$$= yzf''(u(x, y, z)) - f'(u(x, y, z)),$$

$$F'''_{yz^2}(x, y, z) = (F''_{yz})'_z(x, y, z)$$

$$= yf''(u(x, y, z)) + yzf'''(u(x, y, z)) \cdot (-y) - f''(u(x, y, z)) \cdot (-y)$$

$$= -y^2zf'''(u(x, y, z)) + 2yf''(u(x, y, z)).$$

3. Calculate F^{IV}_{xyzt}, if $F(x, y, z, t) = f(x - t, yz, y - t)$, where $f = f(u, v, w)$ is a function of class C^4.

Let $u(x, y, z, t) = x - t$, $v(x, y, z, t) = yz$ and $w(x, y, z, t) = y - t$. Then we have

$$F(x, y, z, t) = f(u(x, y, z, t), v(x, y, z, t), w(x, y, z, t)).$$

First we calculate

$$F'_x(x, y, z, t) = f'_u(u(x, y, z, t), v(x, y, z, t), w(x, y, z, t)).$$

Next, for simplifying the writing, we shall drop out the argument of the functions and we have

$$F'_x = f'_u,$$

$$F''_{xy} = (F'_x)'_y = f''_{uv} \cdot z + f''_{uw} = zf''_{uv} + f''_{uw},$$

$$F'''_{xyz} = (F''_{xy})'_z = f''_{uv} + zf'''_{uv^2} \cdot y + f'''_{uwv} \cdot y = yzf'''_{uv^2} + yf'''_{uvw} + f''_{uv},$$

$$F^{IV}_{xyzt} = (F'''_{xyz})'_t$$

$$= yz[f^{IV}_{uv^2u} \cdot (-1) + f^{IV}_{uv^2w} \cdot (-1)] + y[f^{IV}_{uvwu} \cdot (-1) + f^{IV}_{uvw^2} \cdot (-1)]$$

$$+ [f'''_{uvu} \cdot (-1) + f'''_{uvw} \cdot (-1)]$$

$$= -yzf^{IV}_{u^2v^2} - yf^{IV}_{u^2vw} - yzf^{IV}_{uv^2w} - yf^{IV}_{uvw^2} - f'''_{u^2v} - f'''_{uvw}.$$

5.28. Calculate F' and F'' if:

(a) $F(x) = f(\sin x, \arctan x);$
(b) $F(x) = f(\ln(x^2 + 1), \cos x),$

where $f = f(u, v)$ is a function of class C^2.

5.29. Calculate F''' if:

(a) $F(x) = f(-x, 3x);$
(b) $F(x) = f(x, x^2),$

where $f = f(u, v)$ is a function of class C^3.

5.30. Calculate F' and F'' if:

(a) $F(x) = f(e^x, \ln(x + \sqrt{x^2 + 1}), x^2);$
(b) $F(x) = f(2^x, \arccot x, e^{x^2}),$

where $f = f(u, v, w)$ is a function of class C^2.

5.31. Calculate F''_{x^2}, F''_{xy}, and F''_{y^2} if:

(a) $F(x, y) = f(x^2 - 2xy + 3y);$
(b) $F(x, y) = f(x \sin y - y \cos x),$

where $f = f(u)$ is a twice derivable function.

5.32. Calculate the derivatives written to the right hand side of the function:

(a) $F(x, y, z) = f(x - y - z), \quad F'''_{xyz}, \; F'''_{yz^2};$
(b) $F(x, y, z) = f(xyz), \quad F'''_{x^3}, \; F'''_{x^2z};$

(c) $F(x, y, z) = f(x^2 - xy + z)$, F'''_{xy^2}, F'''_{xyz};

(d) $F(x, y, z, t) = f(xyt - z + t)$, F'''_{xyt}, F'''_{xzt}, F'''_{xt^2}, F'''_{yzt}, F^{IV}_{xyzt},

where $f = f(u)$ is a three times derivable function for parts (a), (b), and (c), and a four times derivable function for part (d).

5.33. Calculate F''_{x^2}, F''_{xy}, and F''_{y^2} if:

(a) $F(x, y) = f(xy - x + y, x \arctan y)$;

(b) $F(x, y) = f(\sin(x - y), xy - \ln(y^2 + 1))$,

where $f = f(u, v)$ is a function of class C^2.

5.34. Calculate the derivatives written to the right hand side of the function:

(a) $F(x, y, z) = f(x - yz, x - y)$, F''_{y^2}, F''_{yz}, F'''_{xyz};

(b) $F(x, y, z) = f(x - z, y - z)$, F''_{xz}, F''_{yz}, F'''_{xyz}, F'''_{xz^2};

(c) $F(x, y, z) = f(xy - z, xz - y)$, F''_{xy}, F''_{yz}, F'''_{xyz}, F'''_{y^2z};

(d) $F(x, y, z, t) = f(x - t, yz - t)$, F''_{yt}, F''_{zt}, F'''_{xyz}, F'''_{xyt}, F'''_{xzt}, F'''_{xt^2}, F^{IV}_{xyzt},

where $f = f(u, v)$ is a function of class C^3 for parts (a), (b), and (c), and of class C^4 for part (d).

5.35. Calculate the derivatives written to the right hand side of the function:

(a) $F(x, y, z) = f(x - y, y - z, xy - z)$, F''_{xy}, F''_{xz}, F''_{yz}, F'''_{xyz}, F'''_{yz^2};

(b) $F(x, y, z) = f(y, x - z, x - yz)$, F''_{xy}, F''_{xz}, F''_{yz}, F''_{z^2}, F'''_{xy^2}, F'''_{xyz}, F'''_{y^2z};

(c) $F(x, y, z) = f(z, x - y - z, x)$, F''_{xy}, F''_{xz}, F''_{yz}, F''_{z^2}, F'''_{xyz}, F'''_{xz^2}, $F^{IV}_{xy^3}$;

(d) $F(x, y, z, t) = f(xt, y - z, x - t)$, F''_{xy}, F''_{yt}, F''_{zt}, F''_{t^2}, F'''_{xyz}, F'''_{xyt}, F'''_{yzt}, F'''_{t^3}, F^{IV}_{xyzt},

where $f = f(u, v, w)$ is a function of class C^3 for parts (a) and (b), and of class C^4 for parts (c) and (d).

5.36. Let $a, b \in \mathbb{R}$, $f : \mathbb{R} \to \mathbb{R}$ be a nonconstant twice derivable function on $\mathbb{R}$ and $g : \mathbb{R}^2 \to \mathbb{R}$ be the function defined by $g(x, y) = e^{f(ax + by)}$. Find the relation that has to be satisfied by the numbers a and b so that, for all $(x, y) \in \mathbb{R}^2$, to have

$$\frac{\partial^2 g}{\partial x^2}(x, y) - 2\frac{\partial^2 g}{\partial y \partial x}(x, y) + \frac{\partial^2 g}{\partial y^2}(x, y) = 0.$$

5.37. Let $a, b, x_0 \in \mathbb{R}$, $f : \mathbb{R}^2 \to \mathbb{R}$ be a function of class C^2, and $g : \mathbb{R} \to \mathbb{R}$ be the function defined by $g(x) = e^{f(ax,bx)}$. Knowing that the second order partial derivatives of f at (ax_0, bx_0) are equal to each other to a nonzero real number, find the relation that has to be satisfied by the numbers a and b such that $g''(x_0)g(x_0) = g'^2(x_0)$.

5.3 Homogeneous Functions. Euler's Identity

Let $\alpha \in \mathbb{R}$ and $A \subseteq \mathbb{R}^n$ be a *cone*, i.e. a set with the property that, for all $x \in A$ and $t \in (0, \infty)$, we have $tx \in A$. A function $f : A \to \mathbb{R}$ is called *homogeneous of degree α* if

$$f(tx) = t^\alpha f(x), \quad \text{for all } x \in A \text{ and } t \in (0, \infty).$$

Euler's identity. Let $\alpha \in \mathbb{R}$, $A \subseteq \mathbb{R}^n$ be an open set, which is also a cone, and let $f \in C^1(A, \mathbb{R})$. The function f is homogeneous of degree α if and only if

$$\sum_{i=1}^n x_i \frac{\partial f}{\partial x_i}(x_1, \ldots, x_n) = \alpha f(x_1, \ldots, x_n), \quad \text{for all } (x_1, \ldots, x_n) \in A.$$

Euler's identity can be written using the gradient of the function f as follows:

$$x \cdot \nabla f(x) = \alpha f(x), \quad \text{for all } x \in A,$$

where "$\cdot$" denotes the inner product of vectors in $\mathbb{R}^n$.

5.38. Prove that the following functions are homogeneous in two different ways (using the definition and Euler's identity):

(a) $f(x, y, z) = \frac{1}{\sqrt{x+y+z}}$, $x + y + z > 0$;

(b) $f(x, y, z) = \frac{x+y+z}{xy+yz+zx}$, $xy + yz + zx \neq 0$;

(c) $f(x, y, z) = xe^{\frac{yz}{x^2}} + ye^{\frac{xz}{y^2}} + ze^{\frac{xy}{z^2}}$, $x, y, z \neq 0$;

(d) $f(x, y, z) = \frac{1}{x^2} \sin \frac{y}{z} + \frac{1}{y^2} \sin \frac{z}{x} + \frac{1}{z^2} \sin \frac{x}{y}$, $x, y, z \neq 0$;

(e) $f(x, y, z) = \frac{xy+z^2}{x-y}$, $x - y \neq 0$.

5.39. Let $\alpha \in \mathbb{R}$, $A \subseteq \mathbb{R}^2$ be an open set, which is also a cone, and $f \in C^3(A, \mathbb{R})$ be a homogeneous function of degree α. Prove that:

(a) $\frac{\partial f}{\partial x}$ and $\frac{\partial f}{\partial y}$ are homogeneous functions of degree $\alpha - 1$.

(b) The following equality holds:

$$x^2 \frac{\partial^2 f}{\partial x^2}(x, y) + 2xy \frac{\partial^2 f}{\partial y \partial x}(x, y) + y^2 \frac{\partial^2 f}{\partial y^2}(x, y) = \alpha(\alpha - 1) f(x, y),$$

for all $(x, y) \in A$.

(c) The following equality holds:

$$x^3 \frac{\partial^3 f}{\partial x^3}(x, y) + 3x^2 y \frac{\partial^3 f}{\partial y \partial x^2}(x, y) + 3xy^2 \frac{\partial^3 f}{\partial y^2 \partial x}(x, y) + y^3 \frac{\partial^3 f}{\partial y^3}(x, y)$$

$$= \alpha(\alpha - 1)(\alpha - 2) f(x, y),$$

for all $(x, y) \in A$.

Remark 5.4. More generally, if $n \in \mathbb{N}$ and $f \in C^n(A, \mathbb{R})$ is a homogeneous function of degree α, then the following equality holds:

$$\left(x \frac{\partial}{\partial x} + y \frac{\partial}{\partial y} \right)^{(n)} f(x, y) = \alpha(\alpha - 1) \cdots (\alpha - n + 1) f(x, y),$$

for all $(x, y) \in A$, where "(n)" means "symbolic power (symbolic order of derivability)", i.e. we have

$$\sum_{k=0}^{n} C_n^k x^{n-k} y^k \frac{\partial^n f}{\partial y^k \partial x^{n-k}}(x, y) = \alpha(\alpha - 1) \cdots (\alpha - n + 1) f(x, y),$$

for all $(x, y) \in A$.

5.4   Taylor's Formula for Real Functions of Two Real Variables

Taylor's Formula for Real Functions of Two Real Variables

Let $\emptyset \neq A \subseteq \mathbb{R}^2$ be a convex open set and $f \in C^{n+1}(A, \mathbb{R})$. Then $\forall (x_0, y_0), (x, y) \in A, \exists (\theta_1, \theta_2) \in (0, 1)^2$ such that *Taylor's formula of order n corresponding to the function f and the point (x_0, y_0) holds true*

(continued)

$$f(x, y) = (T_n f)(x, y) + (R_n f)(x, y),$$

where

$$(T_n f)(x, y) = f(x_0, y_0) + \sum_{k=1}^{n} \frac{1}{k!} \left((x - x_0)\frac{\partial}{\partial x} + (y - y_0)\frac{\partial}{\partial y} \right)^{(k)} f(x_0, y_0)$$

is the *Taylor polynomial of degree n corresponding to the function f and the point* (x_0, y_0), and

$$(R_n f)(x, y) = \frac{1}{(n+1)!} \left((x - x_0)\frac{\partial}{\partial x} + (y - y_0)\frac{\partial}{\partial y} \right)^{(n+1)} f(c_1, c_2),$$

with $(c_1, c_2) = (x_0 + \theta_1(x - x_0), y_0 + \theta_2(y - y_0))$, is the *remainder of order n corresponding to the function f and the point* (x_0, y_0) (the remainder in the Lagrange form).

Here "(k)" means "symbolic power (symbolic order of derivability)", i.e. we have

$$\left((x - x_0)\frac{\partial}{\partial x} + (y - y_0)\frac{\partial}{\partial y} \right)^{(k)} f(x_0, y_0)$$

$$= \sum_{i=0}^{k} C_k^i \frac{\partial^k f}{\partial y^i \partial x^{k-i}} (x_0, y_0)(x - x_0)^{k-i}(y - y_0)^i.$$

For example, the Taylor polynomial of degree 3 corresponding to the function f and the point (x_0, y_0) is given by

$$(T_3 f)(x, y)$$

$$= f(x_0, y_0) + \frac{1}{1!} \left[\frac{\partial f}{\partial x}(x_0, y_0)(x - x_0) + \frac{\partial f}{\partial y}(x_0, y_0)(y - y_0) \right]$$

$$+ \frac{1}{2!} \left[\frac{\partial^2 f}{\partial x^2}(x_0, y_0)(x - x_0)^2 + 2\frac{\partial^2 f}{\partial y \partial x}(x_0, y_0)(x - x_0)(y - y_0) \right.$$

$$\left. + \frac{\partial^2 f}{\partial y^2}(x_0, y_0)(y - y_0)^2 \right]$$

$$+ \frac{1}{3!} \left[\frac{\partial^3 f}{\partial x^3}(x_0, y_0)(x - x_0)^3 + 3\frac{\partial^3 f}{\partial y \partial x^2}(x_0, y_0)(x - x_0)^2(y - y_0) \right.$$

(continued)

$$+3\frac{\partial^3 f}{\partial y^2 \partial x}(x_0, y_0)(x - x_0)(y - y_0)^2 + \frac{\partial^3 f}{\partial y^3}(x_0, y_0)(y - y_0)^3 \Bigg].$$

We mention the following three particular cases of Taylor's formula.

I. When $(x_0, y_0) = (0, 0) \in A$, then we obtain the *Maclaurin formula of order n corresponding to the function* f

$$f(x, y) = (M_n f)(x, y) + (R_n f)(x, y),$$

where

$$(M_n f)(x, y) = f(0, 0) + \sum_{k=1}^{n} \frac{1}{k!}\left(x\frac{\partial}{\partial x} + y\frac{\partial}{\partial y}\right)^{(k)} f(0, 0)$$

is the *Maclaurin polynomial of degree n corresponding to the function* f, and

$$(R_n f)(x, y) = \frac{1}{(n+1)!}\left(x\frac{\partial}{\partial x} + y\frac{\partial}{\partial y}\right)^{(n+1)} f(\theta_1 x, \theta_2 y)$$

is the *remainder of order n corresponding to the function* f.

II. When $f : \mathbb{R}^2 \to \mathbb{R}$ is a polynomial function of degree n and $(x_0, y_0) \in \mathbb{R}^2$, then we have $f(x, y) = (T_n f)(x, y)$, for all $(x, y) \in \mathbb{R}^2$.

III. When $n = 0$, we obtain the *Lagrange Mean Value Theorem for a real function of two real variables*

$$f(x, y) = f(x_0, y_0) + \frac{\partial f}{\partial x}(c_1, c_2)(x - x_0) + \frac{\partial f}{\partial y}(c_1, c_2)(y - y_0),$$

where $(c_1, c_2) = (x_0 + \theta_1(x - x_0), y_0 + \theta_2(y - y_0))$.

The extension of the previous formulae to functions of more than two variables is left to the interested reader.

5.40. Write the following functions in terms of powers of x and y, without expanding the parenthesis:

(a) $f : \mathbb{R}^2 \to \mathbb{R}$, $f(x, y) = (x + 2)(y - 1)^2$;
(b) $f : \mathbb{R}^2 \to \mathbb{R}$, $f(x, y) = (x - 1)^2(y + 1)$.

5.41. Write in terms of powers of $x - 1$ and $y + 1$ the following function

$$f : \mathbb{R}^2 \to \mathbb{R}, \quad f(x, y) = x^3 + x^2 y + x^2 - 3xy - 3x + 2y + 8.$$

5.42. Write in terms of powers of $x + 1$ and $y - 2$ the following function

$$f : \mathbb{R}^2 \to \mathbb{R}, \quad f(x, y) = x^2 y + 4y^3 - 2x^2 + 4xy - 20y^2 - 8x + 35y - 14.$$

5.43. Find the Maclaurin polynomial of degree n corresponding to the function f in the following cases:

(a) $f : \mathbb{R}^2 \to \mathbb{R}, \ f(x, y) = \sin(xy), \ n = 2$;
(b) $f : \mathbb{R}^2 \to \mathbb{R}, \ f(x, y) = e^x \cos y, \ n = 3$;
(c) $f : \mathbb{R}^2 \to \mathbb{R}, \ f(x, y) = e^{xy}, \ n = 4$.

5.44. Find the Taylor polynomial of degree 2 corresponding to the function f and the point (x_0, y_0), if:

(a) $f : \mathbb{R}^2 \to \mathbb{R}, \ f(x, y) = e^{x^2 + xy}, \quad (x_0, y_0) = (1, -1)$;
(b) $f : \mathbb{R}^2 \to \mathbb{R}, \ f(x, y) = \cos(xy - 1), \ (x_0, y_0) = (1, 1)$.

5.45. Using the Taylor polynomial of degree 2 approximate $\sqrt{1,01} \cdot \sqrt[3]{0,97}$.

5.46. Using the Taylor polynomial of degree 2 approximate $0,98^{3,01}$.

5.47. Let $n \geq 0$ be an integer. Find the coefficient of $x^n y^n$ in the Maclaurin polynomial of degree $2n$ corresponding to the function

$$f(x, y) = \frac{3}{(3x + 1)(5y + 3)}.$$

5.48. Let $n \geq 0$ be an integer. Find the coefficient of $(x - 1)^n (y - 1)^n$ in the Taylor polynomial of degree $2n$ corresponding to the function $f(x, y) = \frac{2^{2n+1}}{y^n(xy+1)}$ and the point $(1, 1)$.

5.49. Find the coefficient of $x^n y^n$ in the Maclaurin series expansion of the function

$$f(x, y) = \frac{1}{(1 - x)(1 - x - y)}.$$

5.5   The Differential of Several Real Variable Functions

Let $A \subseteq \mathbb{R}^n$, $x^0 \in \text{int}\,(A)$, and $f : A \to \mathbb{R}$ a function k times differentiable at x^0.

The differential of order k of the function f at x^0 is defined by
$\mathrm{d}^k f(x^0) : \mathbb{R}^n \to \mathbb{R}$,

$$\mathrm{d}^k f(x^0)(\mathrm{d}x) = \left(\mathrm{d}x_1 \frac{\partial}{\partial x_1} + \cdots + \mathrm{d}x_n \frac{\partial}{\partial x_n}\right)^{(k)} f(x^0), \quad \mathrm{d}x = (\mathrm{d}x_1, \ldots, \mathrm{d}x_n) \in \mathbb{R}^n,$$

where "(k)" means "symbolic power (symbolic order of derivability)."

We mention the differential expressions of orders one and two of a function of two and three real variables.

Let $A \subseteq \mathbb{R}^2$, $(x_0, y_0) \in \text{int}\,(A)$, and $f : A \to \mathbb{R}$ a function twice differentiable at (x_0, y_0).

The differential of order one of the function f at (x_0, y_0)

$$\mathrm{d}\,f(x_0, y_0) : \mathbb{R}^2 \to \mathbb{R}, \quad \mathrm{d}\,f(x_0, y_0)(\mathrm{d}x, \mathrm{d}y) = \frac{\partial f}{\partial x}(x_0, y_0)\,\mathrm{d}x + \frac{\partial f}{\partial y}(x_0, y_0)\,\mathrm{d}y.$$

The differential of order two of the function f at (x_0, y_0)
$\mathrm{d}^2 f(x_0, y_0) : \mathbb{R}^2 \to \mathbb{R}$,

$$\mathrm{d}^2 f(x_0, y_0)(\mathrm{d}x, \mathrm{d}y) = \frac{\partial^2 f}{\partial x^2}(x_0, y_0)\,\mathrm{d}x^2 + 2\frac{\partial^2 f}{\partial y \partial x}(x_0, y_0)\,\mathrm{d}x\,\mathrm{d}y + \frac{\partial^2 f}{\partial y^2}(x_0, y_0)\,\mathrm{d}y^2.$$

Let $A \subseteq \mathbb{R}^3$, $(x_0, y_0, z_0) \in \text{int}\,(A)$, and $f : A \to \mathbb{R}$ a function twice differentiable at (x_0, y_0, z_0).

The differential of order one of the function f at (x_0, y_0, z_0)
$\mathrm{d}\,f(x_0, y_0, z_0) : \mathbb{R}^3 \to \mathbb{R}$,

$$\mathrm{d}\,f(x_0, y_0, z_0)(\mathrm{d}x, \mathrm{d}y, \mathrm{d}z) = \frac{\partial f}{\partial x}(x_0, y_0, z_0)\,\mathrm{d}x + \frac{\partial f}{\partial y}(x_0, y_0, z_0)\,\mathrm{d}y + \frac{\partial f}{\partial z}(x_0, y_0, z_0)\,\mathrm{d}z.$$

The differential of order two of the function f at (x_0, y_0, z_0)
$\mathrm{d}^2 f(x_0, y_0, z_0) : \mathbb{R}^3 \to \mathbb{R}$,

$$\mathrm{d}^2 f(x_0, y_0, z_0)(\mathrm{d}x, \mathrm{d}y, \mathrm{d}z) = \frac{\partial^2 f}{\partial x^2}(x_0, y_0, z_0)\,\mathrm{d}x^2 + \frac{\partial^2 f}{\partial y^2}(x_0, y_0, z_0)\,\mathrm{d}y^2$$

(continued)

$$+\frac{\partial^2 f}{\partial z^2}(x_0, y_0, z_0)\, dz^2 + 2\frac{\partial^2 f}{\partial y \partial x}(x_0, y_0, z_0)\, dx\, dy$$

$$+2\frac{\partial^2 f}{\partial z \partial y}(x_0, y_0, z_0)\, dy\, dz + 2\frac{\partial^2 f}{\partial z \partial x}(x_0, y_0, z_0)\, dx\, dz.$$

5.50. Let $f : \mathbb{R}^2 \to \mathbb{R}$, $f(x, y) = e^{x^2+y} + \sin(xy^2)$. Calculate

$$df(0, 1)(1, 2) \quad \text{and} \quad d^2 f(0, 1)(1, 2).$$

5.51. Let $f : \mathbb{R}^2 \to \mathbb{R}$, $f(x, y) = \arctan(xy) + e^{x^2-y^2}$. Calculate

$$df(1, 0)(2, 1) \quad \text{and} \quad d^2 f(1, 0)(2, 1).$$

5.52. Let $f : \mathbb{R}^3 \to \mathbb{R}$, $f(x, y, z) = \sin(xyz)$. Calculate

$$df(\pi, -1, 1)(-1, 2, 1) \quad \text{and} \quad d^2 f(\pi, -1, 1)(-1, 2, 1).$$

5.53. Let $f : \mathbb{R}^3 \to \mathbb{R}$, $f(x, y, z) = \cos(xy + yz + xz)$. Calculate

$$df\left(\frac{\pi}{2}, 1, 0\right)(1, -2, 2) \quad \text{and} \quad d^2 f\left(\frac{\pi}{2}, 1, 0\right)(1, -2, 2).$$

5.54. Let $f : (0, \infty)^3 \to \mathbb{R}$, $f(x, y, z) = \ln(x^x y^y z^z)$. Calculate

$$d^4 f(2, 1, 2)(-1, 1, 1).$$

Let $A \subseteq \mathbb{R}^n$, $x^0 \in \text{int}\,(A)$, and $f : A \to \mathbb{R}$ be a differentiable function at x^0. Then the function f is *derivable at x^0 by the direction* $v \in \mathbb{R}^n \setminus \{(0, \dots, 0)\}$ and this derivative is equal to the differential of f at x^0 evaluated at the unit vector of the direction, i.e.

$$\frac{df}{dv}(x^0) = df(x^0)\left(\frac{v}{\|v\|}\right).$$

5.55. Let $f : \mathbb{R}^2 \to \mathbb{R}$ be a differentiable function and let $(x_0, y_0) \in \mathbb{R}^2$. Knowing that the derivative of f at (x_0, y_0) by the direction of a vector parallel to the bisector of the first quadrant is -5, and the derivative of f at (x_0, y_0) by the direction of a vector parallel to the bisector of the second quadrant is 9, calculate $\frac{\partial f}{\partial x}(x_0, y_0) \cdot \frac{\partial f}{\partial y}(x_0, y_0)$.

5.56. Let $f : \mathbb{R}^3 \to \mathbb{R}$, $f(x, y, z) = \frac{x}{y^2+1} + \frac{y}{z^2+1} + \frac{z}{x^2+1}$. Calculate the derivatives of f at $(1, 1, -1)$ and $(-1, 1, -1)$ by the direction of a vector parallel to the normal to the plane $x - 2y + 2z + 3 = 0$.

5.6　　Extrema of Several Real Variable Functions

Local Extremum Points of Several Real Variable Functions
Let $A \subseteq \mathbb{R}^n$ be an open set, $f \in C^2(A, \mathbb{R})$, and $x^0 \in A$ a critical point of f.

(a) If $d^2 f(x^0)$ is *positive definite*, then x^0 is a point of *local minimum* of f, and if $d^2 f(x^0)$ is *negative definite*, then x^0 is a point of *local maximum* of f.
(b) If $d^2 f(x^0)$ is *indefinite*, then x^0 is *not a point of local extremum* of f, it is a *saddle point*.

Remark 5.5. The quadratic form $d^2 f(x^0) : \mathbb{R}^n \to \mathbb{R}$ is:

(i)   *positive definite* if $d^2 f(x^0)(h) > 0$, for all $h \in \mathbb{R}^n \setminus \{(0, \ldots, 0)\}$;
(ii) *negative definite* if $d^2 f(x^0)(h) < 0$, for all $h \in \mathbb{R}^n \setminus \{(0, \ldots, 0)\}$;
(iii) *indefinite* if there exist $h_1, h_2 \in \mathbb{R}^n$ such that

$$d^2 f(x^0)(h_1) \cdot d^2 f(x^0)(h_2) < 0.$$

Let

$$d_1 = \frac{\partial^2 f}{\partial x_1^2}(x^0)$$

$$d_2 = \begin{vmatrix} \dfrac{\partial^2 f}{\partial x_1^2}(x^0) & \dfrac{\partial^2 f}{\partial x_2 \partial x_1}(x^0) \\ \dfrac{\partial^2 f}{\partial x_2 \partial x_1}(x^0) & \dfrac{\partial^2 f}{\partial x_2^2}(x^0) \end{vmatrix}$$

$$d_3 = \begin{vmatrix} \dfrac{\partial^2 f}{\partial x_1^2}(x^0) & \dfrac{\partial^2 f}{\partial x_2 \partial x_1}(x^0) & \dfrac{\partial^2 f}{\partial x_3 \partial x_1}(x^0) \\ \dfrac{\partial^2 f}{\partial x_2 \partial x_1}(x^0) & \dfrac{\partial^2 f}{\partial x_2^2}(x^0) & \dfrac{\partial^2 f}{\partial x_3 \partial x_2}(x^0) \\ \dfrac{\partial^2 f}{\partial x_3 \partial x_1}(x^0) & \dfrac{\partial^2 f}{\partial x_3 \partial x_2}(x^0) & \dfrac{\partial^2 f}{\partial x_3^2}(x^0) \end{vmatrix}$$

(continued)

$$\cdots$$

$$d_n = \det(H(f)(x^0)).$$

Sylvester's Criterion

1. $d^2 f(x^0)$ is *positive definite* if and only if $d_k > 0$, for all $k \in \{1, 2, \ldots, n\}$;
2. $d^2 f(x^0)$ is *negative definite* if and only if $(-1)^k d_k > 0$, for all $k \in \{1, 2, \ldots, n\}$.

5.57. *A particular case of Huygens' problem.* Prove that the function

$$f : (0, \infty)^2 \to \mathbb{R}, \quad f(x, y) = \frac{xy}{(x+1)(y+1)(x+y)},$$

has a global maximum and determine this maximum.

5.58. Find the local extremum points and the local extremum values of the following functions:

(a) $f : (0, \infty)^2 \to \mathbb{R}, \ f(x, y) = \frac{x}{y} + \frac{y}{x}$;

(b) $f : \mathbb{R}^2 \to \mathbb{R}, \ f(x, y) = (x^2 + 1)(y^2 + 1)(xy + 1)$;

(c) $f : \mathbb{R}^2 \to \mathbb{R}, \ f(x, y) = xy\,e^{-2x-y}$;

(d) $f : \{(x, y) \in \mathbb{R}^2 \mid x \neq 0\} \to \mathbb{R}, \ f(x, y) = -\frac{y}{x}e^{-y} + \frac{1}{ex} - x^2 - 4x$;

(e) $f : (0, \pi)^2 \to \mathbb{R}, \ f(x, y) = \cos x + \cos y + \cos(x + y)$;

(f) $f : (0, \infty)^3 \to \mathbb{R}, \ f(x, y, z) = \frac{x}{y} + \frac{y}{z} + \frac{z}{x}$;

(g) $f : \mathbb{R}^3 \to \mathbb{R}, \ f(x, y, z) = \arctan(x^2 + 2x) + y^2 + yz + z^2$;

(h) $f : (0, \infty)^3 \to \mathbb{R}, \ f(x, y, z) = xyz + \frac{1}{x} + \frac{1}{y} + \frac{1}{z}$;

(i) $f : \mathbb{R}^3 \to \mathbb{R}, \ f(x, y, z) = (x - z)e^{x - y^2 - z}$;

(j) $f : \left(-\frac{\pi}{2}, \frac{\pi}{2}\right)^3 \to \mathbb{R}, \ f(x, y, z) = \cos x + \cos y + \cos z - \cos(x + y + z)$.

5.59. Let $k, l \in \mathbb{R}$. Find the local extremum points and the local extremum values of the function $f : (0, \infty)^2 \to \mathbb{R}, \ f(x, y) = xy + \frac{k}{x} + \frac{l}{y}$.

5.60. The Minimum Value of Two Integrals

(a) Calculate

$$\min_{a, b \in \mathbb{R}} \int_0^1 (x^2 - a - bx)^2 \mathrm{d}x.$$

(continued)

(b) Calculate

$$\min_{a, b, c \in \mathbb{R}} \int_{-1}^{1} (x^3 - a - bx - cx^2)^2 dx.$$

5.61. Find the maximum value of the function $f : \mathbb{R}^2 \to \mathbb{R}$, $f(x, y) = xy$, on the set $C = \{(x, y) \in (0, \infty)^2 \mid 9x^2 + 23xy + 16y^2 = 141\}$.

5.62. Find the minimum value of the function $f : \mathbb{R}^2 \to \mathbb{R}$, $f(x, y) = x + y^4$, on the set $C = \{(x, y) \in (0, \infty)^2 \mid xy = 4\}$.

5.63. Find the local extremum points of the function $f : \mathbb{R}^2 \to \mathbb{R}$, $f(x, y) = x^2 + 2y^2$, relative to the set $C = \{(x, y) \in \mathbb{R}^2 \mid 2x + 4y = 3\}$.

5.64. Find the minimum and maximum values of the function $f : \mathbb{R}^2 \to \mathbb{R}$, $f(x, y) = x - y$, relative to the set $C = \{(x, y) \in \mathbb{R}^2 \mid x^2 + y^2 = 2\}$.

5.65. Find the minimum and maximum values of the function $f : \mathbb{R}^2 \to \mathbb{R}$, $f(x, y) = xy$, relative to the set $C = \{(x, y) \in \mathbb{R}^2 \mid x^2 + y^2 = 1\}$.

5.66. Find the local extremum points of the function $f : \mathbb{R}^2 \to \mathbb{R}$, $f(x, y) = xy$, relative to the set $C = \{(x, y) \in \mathbb{R}^2 \mid x + 2y = 4\}$.

5.67. Prove that, of all rectangles of constant perimeter, the square has the maximum area.

5.68.

(a) Prove that, of all right trapezoids of constant area, the square has the minimum perimeter.
(b) Prove that, of all right trapezoids of constant perimeter, the square has the maximum area.

5.69. Find the maximum value of the function $f : [0, \infty)^3 \to \mathbb{R}$, $f(x, y, z) = \sin x + \sin y + \sin z$, on the set $C = \{(x, y, z) \in [0, \infty)^3 \mid x + y + z = \pi\}$.

5.70. Find the local extremum points of the function $f : \mathbb{R}^3 \to \mathbb{R}$, $f(x, y, z) = x + y^2 + z^3$, relative to the set $C = \{(x, y, z) \in \mathbb{R}^3 \mid x + 2y + 3z = a\}$, $a \in \mathbb{R}$.

5.71. Find the local extremum points of the function $f : (0, \infty)^3 \to \mathbb{R}$, $f(x, y, z) = xyz^2$, relative to the set $C = \{(x, y, z) \in (0, \infty)^3 \mid x + y + 2z = a\}$, $a > 0$.

5.72. Find the local extremum points of the function $f : \mathbb{R}^3 \to \mathbb{R}$, $f(x, y, z) = x(y + z) + 3yz$, relative to the set $C = \{(x, y, z) \in \mathbb{R}^3 \mid xyz = 3\}$.

5.73. Find the minimum and maximum values of the function $f : \mathbb{R}^3 \to \mathbb{R}$, $f(x, y, z) = x + y + z$, on the set

$$C = \left\{(x, y, z) \in \mathbb{R}^3 \mid x - 2y + z = 2, \ x^2 + y^2 + z^2 = 1\right\}.$$

5.74. Find the local extremum points of the function $f : \mathbb{R}^3 \to \mathbb{R}$, $f(x, y, z) = xyz$, relative to the set $C = \{(x, y, z) \in \mathbb{R}^3 \mid x + y + z = 4, \ xy + yz + xz = 5\}$.

5.75. Find the minimum and maximum values of the function $f : \mathbb{R}^3 \to \mathbb{R}$, $f(x, y, z) = 1 - x - y - z$, on the set $C = \{(x, y, z) \in \mathbb{R}^3 \mid x^2 + y^2 + z^2 \leq 1\}$.

5.76. Find the minimum and maximum values of the function $f : \mathbb{R}^3 \to \mathbb{R}$, $f(x, y, z) = 2xy + yz + xz$, on the set $C = \{(x, y, z) \in \mathbb{R}^3 \mid x^2 + y^2 \leq z \leq 1\}$.

5.77. [93] *A discrete extremum.* The positive integers x, y, z satisfy the equality $xy + z = 160$. Find the minimum value of the expression $x + yz$.

5.78. Prove that

$$\left|(x + y)e^{-x^2 - y^2}\right| \leq \frac{1}{\sqrt{e}}, \quad \text{for all } x, y \in \mathbb{R}.$$

5.79. The Extremum of Some Linear Functions Over Conics

(a) Find the minimum and maximum values of $f(x, y) = x + y$ on the ellipse
$\mathscr{E} = \{(x, y) \in \mathbb{R}^2 \mid 3x^2 - 2xy + 3y^2 + 4x + 4y - 4 = 0\}$.
(b) Find the local extreme values of $f(x, y) = 2x + y$ on the hyperbola
$\mathscr{H} = \{(x, y) \in \mathbb{R}^2 \mid 3x^2 + 10xy + 3y^2 - 16x - 16y - 16 = 0\}$.
(c) Find the local extreme values of $f(x, y) = 2x - y$ on the parabola
$\mathscr{P} = \{(x, y) \in \mathbb{R}^2 \mid 9x^2 + 24xy + 16y^2 - 40x + 30y = 0\}$.

A gem of geometry, analysis, and linear algebra.

We find the minimum and maximum values of the function $f(x, y) = x^2 + y^2 - 16x - 10y$ relative to the set $D = \{(x, y) \in \mathbb{R}^2 \mid x^2 + y^2 + 2y \leq 0\}$.

Solution. We notice that the set D is the disk with center $C(0, -1)$ and radius 1. Let $M(x, y) \in D$ and let A be the point whose coordinates are $(8, 5)$. We have that $f(x, y) = (x - 8)^2 + (y - 5)^2 - 89 = AM^2 - 89$. Thus, the minimum and maximum values of f are obtained when AM is minimum and maximum respectively.

Let $\mathscr{C} = \partial D$ and let E and F be the points where the line AC intersects the circle $\mathscr{C}$. A calculation shows that $E\left(\frac{4}{5}, -\frac{2}{5}\right)$ and $F\left(-\frac{4}{5}, -\frac{8}{5}\right)$. It follows that $\min f(x, y) = AE^2 - 89 = -8$ and $\max f(x, y) = AF^2 - 89 = 32$. ∎

A challenge. Find the minimum and maximum values of the function $f(x, y, z) = x^2 + y^2 + z^2 - 4x - 6y - 14z$ relative to the set $B = \{(x, y, z) \in \mathbb{R}^3 \mid x^2 + y^2 + z^2 - 2z \leq 0\}$.

5.80. Let P_2 be the set of polynomials of degree at most 2, with real coefficients, and let $J : P_2 \to \mathbb{R}$ be the function defined by

$$J(f) = \int_0^1 f^2(x)\,dx.$$

(a) [15, Problem 2.2.19 (Sp80), p. 35] If $Q = \{f \in P_2 \mid f(1) = 1\}$, then prove that J attains its minimum over Q and determine $f \in Q$ for which this minimum is attained.
(b) If $Q = \{f \in P_2 \mid f'(1) = 1\}$, then prove that J attains its minimum over Q and determine $f \in Q$ for which this minimum is attained.
(c) If $Q = \{f \in P_2 \mid f''(1) = 1\}$, then prove that J attains its minimum over Q and determine $f \in Q$ for which this minimum is attained.

Implicit Functions

6

The problems included in this chapter are about studying the existence and unicity of implicit functions of one or more variables defined by an equation or by a system of equations.

6.1 Implicit Functions of One Real Variable Defined by an Equation

The Theorem of Existence and Unicity of Implicit Functions of One Real Variable Defined by an Equation

Let $\emptyset \neq A$, $B \subseteq \mathbb{R}$ be open sets, $F : A \times B \to \mathbb{R}$, and $x_0 \in A$, $y_0 \in B$. If

(i) $F(x_0, y_0) = 0$,
(ii) $F \in C^1(A \times B, \mathbb{R})$,
(iii) $F_y'(x_0, y_0) \neq 0$,

then there exists $U \in \mathscr{V}(x_0)$ open, with $U \subseteq A$, there exists $V \in \mathscr{V}(y_0)$ open, with $V \subseteq B$, and there exists a unique function $y : U \to V$ such that:

(continued)

© The Author(s), under exclusive license to Springer Nature Switzerland AG 2021   177
A. Sîntămărian, O. Furdui, *Sharpening Mathematical Analysis Skills*, Problem Books
in Mathematics, https://doi.org/10.1007/978-3-030-77139-3_6

1° $F(x, y(x)) = 0$, for all $x \in U$,

2° $y(x_0) = y_0$,

3° $y \in C^1(U, V)$ and $y'(x) = -\dfrac{F'_x(x, y(x))}{F'_y(x, y(x))}$, for all $x \in U$.

Remark 6.1. If in condition (ii) we have $F \in C^n(A \times B, \mathbb{R})$, $n \geq 2$, then in conclusion 3° we can write $y \in C^n(U, V)$.

6.1. Prove that the equation $x - y^2 + e^{xy} = 0$ defines y as an implicit function of x in the neighborhood of $(0, 1)$ and calculate $y'(0)$, $y''(0)$ and $y'''(0)$.

6.2. Prove that the equation $x^5 + xy^2 - \sin y = 0$ defines y as an implicit function of x in the neighborhood of $(0, \pi)$. Calculate $y'(0)$ and $y''(0)$.

6.3. Prove that the equation $x^2 - xy + 2y^2 + x - y + \cos(\pi y) = 0$ defines y as an implicit function of x in the neighborhood of $(0, 1)$ and calculate $y'(0)$, $y''(0)$ and $y'''(0)$.

6.4. Prove that the equation $\cos(x - y) + x \sin y - y \cos x + \pi + 1 = 0$ defines y as an implicit function of x in the neighborhood of $(0, \pi)$. Calculate $y'(0)$ and $y''(0)$.

6.5. Prove that the equation $(x^2 + y^2) \arctan \frac{x}{y} + x - y + 1 = 0$ defines y as an implicit function of x in the neighborhood of $(0, 1)$ and calculate $y'(0)$ and $y''(0)$.

6.6. Prove that the equation $x^y = y^x$ defines y as an implicit function of x in the neighborhood of $(1, 1)$ and calculate $y'(1)$ and $y''(1)$.

6.7. Prove that the sum of the lengths of the segments determined by the origin and the intersection points with the coordinate axes of a tangent to the curve of equation $\sqrt{x} + \sqrt{y} = \sqrt{c}$, $c > 0$, is equal to c. (See Fig. 6.1.)

Remark 6.2. The curve $\sqrt{x} + \sqrt{y} = \sqrt{c}$, $c > 0$, is in fact a part of a parabola (prove it!). In general, the curves defined implicitly by the equation $\frac{x^\alpha}{a^\alpha} + \frac{y^\alpha}{b^\alpha} = 1$, where a and b are positive numbers and $\alpha \in \mathbb{R}$, are called *Lamé's curves* or *superellipses*.

6.8. *Folium of Descartes.* Let us consider the equation $x^3 + y^3 - 3axy = 0$, $a > 0$, and let $\alpha = \frac{3a}{2}$. Prove that the preceding equation defines y as an implicit function of x in the neighborhood of (α, α) and calculate $y'(\alpha)$, $y''(\alpha)$, and $y'''(\alpha)$. (See Fig. 6.2.)

Fig. 6.1 The tangent to
Lamé's curve
$\sqrt{x} + \sqrt{y} = \sqrt{c}, \; c > 0$

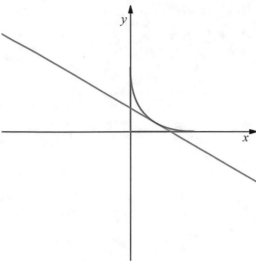

Fig. 6.2 Folium of Descartes
$x^3 + y^3 - 3axy = 0, \; a > 0$

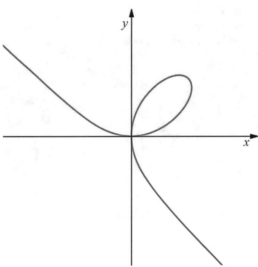

6.9. *Lemniscate of Bernoulli.* Let us consider the equation $(x^2 + y^2)^2 - 2a^2xy = 0$,
$a > 0$, and let $\alpha = \frac{a\sqrt{2}}{2}$. Prove that the preceding equation defines y as an implicit
function of x in the neighborhood of (α, α) and calculate $y'(\alpha)$ and $y''(\alpha)$. (See
Fig. 6.3.)

6.10. *Lemniscate of Bernoulli.* Determine the point, in the first quadrant, located on
the lemniscate of the equation $(x^2 + y^2)^2 - 2a^2(x^2 - y^2) = 0$, $a > 0$, where the
tangent is parallel to the x axis. (See Fig. 6.4.)

Fig. 6.3 Lemniscate of
Bernoulli
$(x^2 + y^2)^2 - 2a^2xy = 0,$
$a > 0$

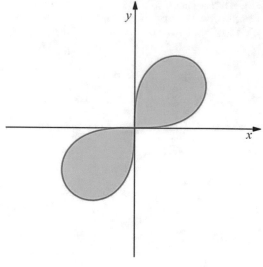

Fig. 6.4 Lemniscate of
Bernoulli
$(x^2 + y^2)^2 - 2a^2(x^2 - y^2) = 0,$
$a > 0$

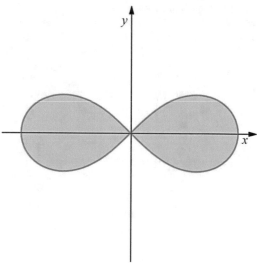

6.11. *Astroid.* Prove that the tangents to the astroid of equation $x^{\frac{2}{3}} + y^{\frac{2}{3}} = a^{\frac{2}{3}}$, $a > 0$, at the points of abscissa $\frac{a\sqrt{2}}{4}$ are perpendicular. (See Fig. 6.5.)

6.12. *Cardioid.* Prove that the tangents to the cardioid of equation $(x^2 + y^2 + ax)^2 = a^2(x^2 + y^2)$, $a > 0$, at the points of abscissa 0 and ordinate different from 0 are perpendicular. (See Fig. 6.6.)

Fig. 6.5 Astroid
$x^{\frac{2}{3}} + y^{\frac{2}{3}} = a^{\frac{2}{3}}$, $a > 0$,
and the two tangents

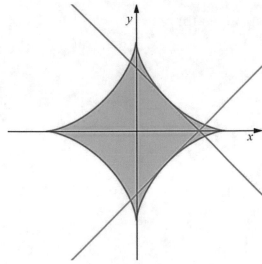

Fig. 6.6 Cardioid
$(x^2+y^2+ax)^2 = a^2(x^2+y^2)$,
$a > 0$, and the two tangents

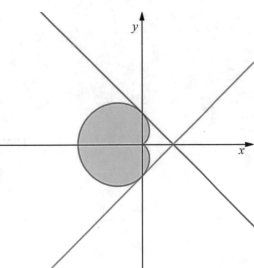

6.13. *Heart.* Let us consider the equation $(x^2 + y^2 - 1)^2 - x^2y^3 = 0$.

(a) Prove that the preceding equation defines y as an implicit function of x in the neighborhood of $(1, 1)$ and calculate $y'(1)$ and $y''(1)$.
(b) Prove that the preceding equation defines y as an implicit function of x in the neighborhood of $(-1, 1)$ and calculate $y'(-1)$ and $y''(-1)$.

(See Fig. 6.7.)

Fig. 6.7 Heart
(Eugène Beutel, 1909)
$(x^2 + y^2 - 1)^3 - x^2 y^3 = 0$

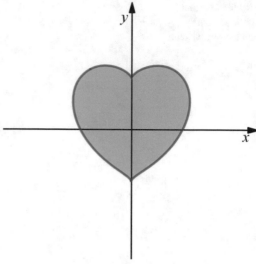

Fig. 6.8 Four leaf clover
$(x^2 + y^2 - 1)^3 - x^2 y^2 = 0$

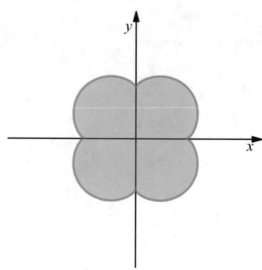

6.14. *Four leaf clover.* Prove that the equation $(x^2 + y^2 - 1)^3 - x^2 y^2 = 0$ defines y as an implicit function of x in the neighborhood of $(1, 1)$ and calculate $y'(1)$ and $y''(1)$. (See Fig. 6.8.)

6.2   Implicit Functions of Two Real Variables Defined by an Equation

The Theorem of Existence and Unicity of Implicit Functions of Two Variables Defined by an Equation

Let $\emptyset \neq A \subseteq \mathbb{R}^2$ be open set, $\emptyset \neq B \subseteq \mathbb{R}$ be open set, $F : A \times B \to \mathbb{R}$, and $(x_0, y_0) \in A$, $z_0 \in B$. If

(i) $F(x_0, y_0, z_0) = 0$,
(ii) $F \in C^1(A \times B, \mathbb{R})$,
(iii) $F'_z(x_0, y_0, z_0) \neq 0$,

then there exists $U \in \mathcal{V}((x_0, y_0))$ open, with $U \subseteq A$, there exists $V \in \mathcal{V}(z_0)$ open, with $V \subseteq B$, and there exists a unique function $z : U \to V$ such that:

$1°$ $F(x, y, z(x, y)) = 0$, for all $(x, y) \in U$,
$2°$ $z(x_0, y_0) = z_0$,
$3°$ $z \in C^1(U, V)$, and

$$z'_x(x, y) = -\frac{F'_x(x, y, z(x, y))}{F'_z(x, y, z(x, y))}, \quad z'_y(x, y) = -\frac{F'_y(x, y, z(x, y))}{F'_z(x, y, z(x, y))},$$

for all $(x, y) \in U$.

Remark 6.3. If in condition (ii) we have $F \in C^n(A \times B, \mathbb{R})$, $n \geq 2$, then in conclusion $3°$ we can write $z \in C^n(U, V)$.

6.15. Prove that the equation $xz - z^2 + y - \cos z = 0$ defines z as an implicit function of x and y in the neighborhood of $(-1, 1, 0)$ and calculate $z'_x(-1, 1)$, $z'_y(-1, 1)$, $z''_{x^2}(-1, 1)$, $z''_{xy}(-1, 1)$, and $z''_{y^2}(-1, 1)$.

6.16. Prove that the equation $(2x - y)e^z + 3xy - z = 0$ defines z as an implicit function of x and y in the neighborhood of $(1, -1, 0)$ and calculate $z'_x(1, -1)$, $z'_y(1, -1)$, $z''_{x^2}(1, -1)$, $z''_{xy}(1, -1)$, and $z''_{y^2}(1, -1)$.

6.17. Prove that the equation $xy \arctan z + 2x^2 - y + z = 0$ defines z as an implicit function of x and y in the neighborhood of $(-1, 2, 0)$ and calculate $z'_x(-1, 2)$, $z'_y(-1, 2)$, $z''_{x^2}(-1, 2)$, $z''_{xy}(-1, 2)$, and $z''_{y^2}(-1, 2)$.

6.3   Implicit Functions of One Real Variable Defined by a System of Equations

The Theorem of Existence and Unicity of Implicit Functions of One Real Variable Defined by a System of Equations
Let $\emptyset \neq A, B, C \subseteq \mathbb{R}$ be open sets, $F, G : A \times B \times C \to \mathbb{R}$ and $x_0 \in A$, $y_0 \in B, z_0 \in C$. If

(i) $F(x_0, y_0, z_0) = 0, G(x_0, y_0, z_0) = 0$,
(ii) $F, G \in C^1(A \times B \times C, \mathbb{R})$,
(iii) $\begin{vmatrix} F'_y & F'_z \\ G'_y & G'_z \end{vmatrix}_{(x_0,y_0,z_0)} \neq 0$,

then there exists $U \in \mathcal{V}(x_0)$ open, with $U \subseteq A$, there exists $V \in \mathcal{V}(y_0)$ open, with $V \subseteq B$, there exists $W \in \mathcal{V}(z_0)$ open, with $W \subseteq C$, and there exist a unique function $y : U \to V$ and a unique function $z : U \to W$ such that:

$1°$ $F(x, y(x), z(x)) = 0, G(x, y(x), z(x)) = 0$, for all $x \in U$,
$2°$ $y(x_0) = y_0, z(x_0) = z_0$,
$3°$ $y \in C^1(U, V), z \in C^1(U, W)$ and

$$y'(x) = -\frac{\begin{vmatrix} F'_x & F'_z \\ G'_x & G'_z \end{vmatrix}_{(x,y(x),z(x))}}{\begin{vmatrix} F'_y & F'_z \\ G'_y & G'_z \end{vmatrix}_{(x,y(x),z(x))}}, \quad z'(x) = -\frac{\begin{vmatrix} F'_y & F'_x \\ G'_y & G'_x \end{vmatrix}_{(x,y(x),z(x))}}{\begin{vmatrix} F'_y & F'_z \\ G'_y & G'_z \end{vmatrix}_{(x,y(x),z(x))}},$$

for all $x \in U$.

Remark 6.4. If in condition (ii) we have $F, G \in C^n(A \times B \times C, \mathbb{R}), n \geq 2$, then in conclusion $3°$ we can write $y \in C^n(U, V), z \in C^n(U, W)$.

6.18. Prove that the system of equations

$$\begin{cases} z^2 - 2x + y + e^y = 0 \\ x^3 - y^2 + z = 0 \end{cases}$$

defines y and z as implicit functions of x in the neighborhood of $(1, 0, -1)$ and calculate $y'(1), z'(1), y''(1)$, and $z''(1)$.

6.19. Prove that the system of equations

$$\begin{cases} x^8 - xy + yz = 1 \\ -x^2z^2 + 10x + 5e^y = 14 \end{cases}$$

defines y and z as implicit functions of x in the neighborhood of $(1, 0, -1)$ and calculate $y'(1)$, $z'(1)$, $y''(1)$, and $z''(1)$.

6.20. Prove that the system of equations

$$\begin{cases} x \sin(yz) - \cos z + y = 0 \\ 2e^{x+y} + xy - z^2 = 1 \end{cases}$$

defines y and z as implicit functions of x in the neighborhood of $(-1, 1, 0)$ and calculate $y'(-1)$, $z'(-1)$, $y''(-1)$, and $z''(-1)$.

6.4   Implicit Functions of Two Real Variables Defined by a System of Equations

The Theorem of Existence and Unicity of Implicit Functions of Two Real Variables Defined by a System of Equations

Let $\emptyset \neq A \subseteq \mathbb{R}^2$ be open set, $\emptyset \neq B, C \subseteq \mathbb{R}$ open sets, $F, G : A \times B \times C \to \mathbb{R}$ and $(x_0, y_0) \in A$, $u_0 \in B$, $v_0 \in C$. If

(i) $F(x_0, y_0, u_0, v_0) = 0$, $G(x_0, y_0, u_0, v_0) = 0$,

(ii) $F, G \in C^1(A \times B \times C, \mathbb{R})$,

(iii) $\begin{vmatrix} F'_u & F'_v \\ G'_u & G'_v \end{vmatrix}_{(x_0, y_0, u_0, v_0)} \neq 0$,

then there exists $W \in \mathscr{V}((x_0, y_0))$ open, with $W \subseteq A$, there exists $U \in \mathscr{V}(u_0)$ open, with $U \subseteq B$, there exists $V \in \mathscr{V}(v_0)$ open, with $V \subseteq C$, and there exist a unique function $u : W \to U$ and a unique function $v : W \to V$ such that:

$1°$ $F(x, y, u(x, y), v(x, y)) = 0$ and $G(x, y, u(x, y), v(x, y)) = 0$, for all $(x, y) \in W$,

$2°$ $u(x_0, y_0) = u_0$, $v(x_0, y_0) = v_0$,

$3°$ $u \in C^1(W, U)$, $v \in C^1(W, V)$, and

(continued)

$$u'_x(x, y) = - \frac{\begin{vmatrix} F'_x & F'_v \\ G'_x & G'_v \end{vmatrix}_{(x,y,u(x,y),v(x,y))}}{\begin{vmatrix} F'_u & F'_v \\ G'_u & G'_v \end{vmatrix}_{(x,y,u(x,y),v(x,y))}}, \quad u'_y(x, y) = - \frac{\begin{vmatrix} F'_y & F'_v \\ G'_y & G'_v \end{vmatrix}_{(x,y,u(x,y),v(x,y))}}{\begin{vmatrix} F'_u & F'_v \\ G'_u & G'_v \end{vmatrix}_{(x,y,u(x,y),v(x,y))}},$$

$$v'_x(x, y) = - \frac{\begin{vmatrix} F'_u & F'_x \\ G'_u & G'_x \end{vmatrix}_{(x,y,u(x,y),v(x,y))}}{\begin{vmatrix} F'_u & F'_v \\ G'_u & G'_v \end{vmatrix}_{(x,y,u(x,y),v(x,y))}}, \quad v'_y(x, y) = - \frac{\begin{vmatrix} F'_u & F'_y \\ G'_u & G'_y \end{vmatrix}_{(x,y,u(x,y),v(x,y))}}{\begin{vmatrix} F'_u & F'_v \\ G'_u & G'_v \end{vmatrix}_{(x,y,u(x,y),v(x,y))}},$$

for all $(x, y) \in W$.

Remark 6.5. If in condition (ii) we have $F, G \in C^n(A \times B \times C, \mathbb{R})$, $n \geq 2$, then in conclusion 3° we can write $u \in C^n(W, U)$, $v \in C^n(W, V)$.

6.21. Prove that the system of equations

$$\begin{cases} xy + xv - uv + e^u = \pi \\ xu + x - y - \cos v = 3 \end{cases}$$

defines u and v as implicit functions of x and y in the neighborhood of $(1, -1, 0, \pi)$. Calculate $u'_x(1, -1)$, $u'_y(1, -1)$, $v'_x(1, -1)$, and $v'_y(1, -1)$.

6.22. Prove that the system of equations

$$\begin{cases} x \sin u - y \cos v - x + y = 1 \\ e^{2x+y} - \sin(u - v) = 1 \end{cases}$$

defines u and v as implicit functions of x and y in the neighborhood of $(-1, 2, \pi, 0)$ and calculate $u'_x(-1, 2)$, $u'_y(-1, 2)$, $v'_x(-1, 2)$, and $v'_y(-1, 2)$.

Challenges, Gems, and Mathematical Beauties 7

> *In mathematics the art of proposing a question*
> *must be held of higher value than solving it.*
>
> Georg Ferdinand Cantor (1845–1918)

This chapter contains challenging problems on various topics discussed in the previous chapters of this book. A special attention is given to the calculation of series involving harmonic numbers and tails of Riemann zeta function values. The calculation of these series reduces to the evaluation of linear or nonlinear Euler sums. A series involving a single harmonic number is called a linear Euler sum, while a series involving a product of at least two harmonic numbers is called a nonlinear Euler sum. The study of these series has been initiated by the famous mathematicians Leonhard Euler and Christian Goldbach in the mid-eighteenth century. Most of the series in this chapter are new and we have included them here because they are simply unusual sums containing combinations of harmonic numbers and tails of zeta function values. These series, linear or quadratic, alternating or not, are evaluated in terms of zeta function, logarithms, and polylogarithm of special values. Proofs of special results of analysis, which are called gems, challenges, and several open problems are scattered throughout the chapter.

7.1   Limits of Sequences

7.1. Calculate:

(a) $\lim\limits_{n\to\infty} \left(2\sqrt[n]{x} - 1\right)^n$, $x > 0$;

(b) $\lim\limits_{n\to\infty} \left(\sqrt[n]{x} + \sqrt[n]{y} - 1\right)^n$, $x, y > 0$;

(c) $\lim\limits_{x\to 0^+} (2x^x - 1)^{\frac{1}{x}}$.

© The Author(s), under exclusive license to Springer Nature Switzerland AG 2021
A. Sîntămărian, O. Furdui, *Sharpening Mathematical Analysis Skills*, Problem Books
in Mathematics, https://doi.org/10.1007/978-3-030-77139-3_7

A Gem of Mathematical Analysis

We prove that the sequence $\left(\left\{1 + \frac{1}{2} + \cdots + \frac{1}{n}\right\}\right)_{n \geq 1}$ diverges.

By way of contradiction, we assume that the sequence converges. This implies that there exists $a \in \mathbb{R}$ such that

$$\lim_{n \to \infty} \left\{1 + \frac{1}{2} + \cdots + \frac{1}{n}\right\} = a.$$

Equivalently

$$\lim_{n \to \infty} \left(1 + \frac{1}{2} + \cdots + \frac{1}{n} - \left\lfloor 1 + \frac{1}{2} + \cdots + \frac{1}{n} \right\rfloor\right) = a.$$

We also have

$$\lim_{n \to \infty} \left(1 + \frac{1}{2} + \cdots + \frac{1}{2n} - \left\lfloor 1 + \frac{1}{2} + \cdots + \frac{1}{2n} \right\rfloor\right) = a.$$

Subtracting the previous two limits we obtain that

$$\lim_{n \to \infty} \left(\frac{1}{n+1} + \cdots + \frac{1}{2n} - \left\lfloor 1 + \frac{1}{2} + \cdots + \frac{1}{2n} \right\rfloor + \left\lfloor 1 + \frac{1}{2} + \cdots + \frac{1}{n} \right\rfloor\right) = 0.$$

It follows that

$$\lim_{n \to \infty} \left(\left\lfloor 1 + \frac{1}{2} + \cdots + \frac{1}{2n} \right\rfloor - \left\lfloor 1 + \frac{1}{2} + \cdots + \frac{1}{n} \right\rfloor\right) = \ln 2,$$

which is a contradiction, since a sequence of integers cannot converge to an irrational number. ∎

In fact, it can be proved that, this is left as *a challenge* to the interested reader, the sequence $\left(\left\{1 + \frac{1}{2} + \cdots + \frac{1}{n}\right\}\right)_{n \geq 1}$ *is dense* in $[0, 1]$.

7.2. Study the convergence of the sequence $(\{\ln n\})_{n \geq 1}$, where $\{a\}$ denotes the fractional part of a.

7.3. Let $k \in \mathbb{N}$. Calculate

$$\lim_{n \to \infty} \sqrt[n]{\frac{(kn)!}{(n!)^k}}.$$

7.4. Calculate $\lim_{n \to \infty} (e^{H_n - \gamma} - n)$.

7.5. [A. Boțan, 2016] Prove that:

(a) $\ln x \le x - 1 \le \ln x + \ln^2 x$, $\forall x \in [1, 2)$;

(b) $\lim\limits_{n \to \infty} \left(\sqrt[n]{1} + \sqrt[n]{2} + \cdots + \sqrt[n]{n} - n - \ln n \right) = -1.$

7.6. Let $a, b \in \mathbb{R}$, with $a - b = \frac{\gamma}{2}$. Calculate

$$\lim_{n \to \infty} \left(2 e^{1 + \frac{1}{2} + \cdots + \frac{1}{n} - a} - \sqrt{n}\, e^{1 + \frac{1}{3} + \cdots + \frac{1}{2n-1} - b} \right).$$

7.7. Calculate

$$\lim_{n \to \infty} \sqrt[n]{(\sin 1)^{2n} + (\sin 2)^{2n} + \cdots + (\sin n)^{2n}}.$$

7.8. [M. Ivan, 2019]

(a) Calculate

$$\lim_{n \to \infty} \sqrt[n]{2^{n \sin 1} + 2^{n \sin 2} + \cdots + 2^{n \sin n}}.$$

(b) Let $(x_n)_{n \in \mathbb{N}}$ be a bounded sequence of real numbers. Calculate

$$\lim_{n \to \infty} \sqrt[n]{2^{n x_1} + 2^{n x_2} + \cdots + 2^{n x_n}}.$$

7.9. Trigonometric Limits Calculate:

(a) $\lim\limits_{n \to \infty} \sqrt[n]{|\sin 1 \sin 2 \cdots \sin n|}$;

(b) $\lim\limits_{n \to \infty} \sqrt[n]{\sin^2 n \sin^2 (n+1) \cdots \sin^2 (3n)}$;

(c) $\lim\limits_{n \to \infty} \sqrt[n]{(a \pm \cos 1)(a \pm \cos 2) \cdots (a \pm \cos n)}$, $a \ge 1$;

(d) $\lim\limits_{n \to \infty} \sqrt[n]{(a \sin 1 + b \cos 1)^2 (a \sin 2 + b \cos 2)^2 \cdots (a \sin n + b \cos n)^2}$,
$a, b \in \mathbb{R}$.

7.10. Prove that

$$\lim_{n \to \infty} \left(\frac{1 + \frac{1}{3} + \cdots + \frac{1}{2n-1}}{\frac{1}{2} + \frac{1}{4} + \cdots + \frac{1}{2n}} \right)^{\ln n} = 4.$$

7.11. Let $a, b \in (0, \infty)$ and $\alpha, \beta \in (0, 1)$. Prove that

$$\lim_{n \to \infty} \frac{\left(\dfrac{1}{a^\alpha} + \dfrac{1}{(a+1)^\alpha} + \dfrac{1}{(a+2)^\alpha} + \cdots + \dfrac{1}{(a+n-1)^\alpha} \right)^{\beta-1}}{\left(\dfrac{1}{b^\beta} + \dfrac{1}{(b+1)^\beta} + \dfrac{1}{(b+2)^\beta} + \cdots + \dfrac{1}{(b+n-1)^\beta} \right)^{\alpha-1}} = \frac{(1-\beta)^{\alpha-1}}{(1-\alpha)^{\beta-1}}.$$

7.12. A. Exponential Limits

Let $x > 0$. Prove that:

(a) $\displaystyle \lim_{n \to \infty} n \sqrt[n]{e^x - 1 - \frac{x}{1!} - \cdots - \frac{x^n}{n!}} = ex;$

(b) $\displaystyle \lim_{n \to \infty} n \left(\sqrt[2n]{n^3} \cdot n \sqrt[n]{e^x - 1 - \frac{x}{1!} - \cdots - \frac{x^n}{n!}} - ex \right) = ex \ln \frac{x}{\sqrt{2\pi}};$

(c) $\displaystyle \lim_{n \to \infty} n \sqrt[n]{\sinh x - \frac{x}{1!} - \cdots - \frac{x^{2n-1}}{(2n-1)!}} = \frac{ex}{2};$

(d) $\displaystyle \lim_{n \to \infty} n \sqrt[n]{\cosh x - 1 - \frac{x^2}{2!} - \cdots - \frac{x^{2n}}{(2n)!}} = \frac{ex}{2};$

(e) *A challenge.*

$$\lim_{n \to \infty} \left[\left(e^x - 1 - \frac{x}{1!} - \cdots - \frac{x^{n+1}}{(n+1)!} \right)^{-\frac{1}{n+1}} \right.$$

$$\left. - \left(e^x - 1 - \frac{x}{1!} - \cdots - \frac{x^n}{n!} \right)^{-\frac{1}{n}} \right] = \frac{1}{ex}.$$

B. Polylogarithm Limits

Let $k \in \mathbb{N}$ and $x \in [0, 1)$. Prove that:

(a) $\displaystyle \lim_{n \to \infty} \sqrt[n]{\mathrm{Li}_k(x) - x - \frac{x^2}{2^k} - \cdots - \frac{x^n}{n^k}} = x;$

(b) $\displaystyle \lim_{n \to \infty} n \left(\sqrt[n]{n^k} \cdot \sqrt[n]{\mathrm{Li}_k(x) - x - \frac{x^2}{2^k} - \cdots - \frac{x^n}{n^k}} - x \right) = x \ln \frac{x}{1-x}.$

We mention that $\mathrm{Li}_1(x) = -\ln(1-x)$.

7.13. Prove that

$$\lim_{n \to \infty} \left[\sum_{k=1}^{n} \left(\frac{1}{k^2} + \frac{1}{(k+1)^2} + \cdots \right) - \ln n \right] = 1 + \gamma.$$

7.14.

(a) Let $n \in \mathbb{N}$. Prove that the equation $x^n + x^{n-1} + \cdots + x - 1 = 0$ has a unique positive solution x_n, with $x_n \leq 1$.

(b) Prove that the sequence $(x_n)_{n \geq 1}$ is decreasing.

(c) Prove that $\lim\limits_{n \to \infty} x_n = \frac{1}{2}$.

(d) Prove that $\lim\limits_{n \to \infty} 2^n \left(x_n - \frac{1}{2} - \frac{1}{2^{n+2}} \right) = 0$.

7.15. Prove that:

(a) [142] $\lim\limits_{n \to \infty} \sum\limits_{k=1}^{n-1} \dfrac{n^2}{k^2(n-k)^2} = \dfrac{\pi^2}{3}$;

(b) $\lim\limits_{n \to \infty} \sum\limits_{k=1}^{n-1} \dfrac{n^3}{k^3(n-k)^3} = 2\zeta(3)$.

7.16. Taylorian Limits
Let $n \geq 0$ be an integer. Calculate:

(a) $\lim\limits_{x \to 0} \left((n+1)! \dfrac{e^x - 1 - \frac{x}{1!} - \frac{x^2}{2!} - \cdots - \frac{x^n}{n!}}{x^{n+1}} \right)^{\frac{1}{x}}$;

(b) $\lim\limits_{x \to 0} \left((n+1) \dfrac{\ln \frac{1}{1-x} - x - \frac{x^2}{2} - \cdots - \frac{x^n}{n}}{x^{n+1}} \right)^{\frac{1}{x}}$.

7.17.

(a) Let $x \in \mathbb{R}$. Calculate $\lim\limits_{n \to \infty} n\left(\cos^n \frac{x}{n} - 1\right)$.

(b) Let $f \in C^2(\mathbb{R}, \mathbb{R})$, with $f(0) = 1$. Prove that $\lim\limits_{n \to \infty} f^n\left(\frac{x}{n}\right) = e^{f'(0)x}$ and calculate

$$\lim_{n \to \infty} n\left(f^n\left(\frac{x}{n}\right) - e^{f'(0)x} \right).$$

7.18. Let $f : [0, 1] \to \mathbb{R}$ be a continuous function. Calculate:

(a) $\lim\limits_{n \to \infty} n^2 \displaystyle\int_0^1 \left(\sum\limits_{k=n}^{\infty} \frac{x^k}{k^2} \right) \left(\sum\limits_{i=n}^{\infty} \frac{x^i}{i} \right) f(x)\,dx$;

(continued)

(b) $\displaystyle \lim_{n \to \infty} n^3 \int_0^1 \left(\sum_{k=n}^{\infty} \frac{x^k}{k^2} \right)^2 f(x) dx.$

7.19. [137], [128, problems 59–61, pp. 19, 20], [130, Proposition 2]

Let $k \geq 0$ be an integer, $p, q, s \in \mathbb{N}$, with $p \geq 2$, and $a, r \in (0, +\infty)$. We consider the sequence $(a_n)_{n \in \mathbb{N}}$ defined by $a_n = a + (n-1)r$.

Prove that:

(i) $\displaystyle \lim_{n \to \infty} \frac{a_{qn+k+1} a_{qn+k+1+p} \cdots a_{qn+k+1+s(n-1)p}}{a_{qn+k} a_{qn+k+p} \cdots a_{qn+k+s(n-1)p}} = \sqrt[p]{\frac{ps+q}{q}};$

(ii) $\displaystyle \lim_{n \to \infty} \sqrt[n]{\frac{a_{qn+k} a_{qn+k+p} \cdots a_{qn+k+s(n-1)p}}{(n!)^s}} = \left(\sqrt[p]{\frac{ps+q}{q}} \right)^q [(ps+q)r]^s;$

(iii) $\displaystyle \lim_{n \to \infty} \frac{\sqrt[n]{a_{qn+k} a_{qn+k+p} \cdots a_{qn+k+s(n-1)p}}}{n^s}$

$= \left(\sqrt[p]{\frac{ps+q}{q}} \right)^q \left[\frac{(ps+q)r}{e} \right]^s.$

7.20. [131, Problem 3] Evaluate

$$\lim_{n \to \infty} \frac{5^{5n} \left(C_{2n}^n \right)^3}{C_{10n}^{5n} C_{5n}^n C_{4n}^{2n}}.$$

7.21. [131, Problem 4] Let $p, \alpha, \beta \in \mathbb{N}$, with $p \geq 2$ and $\alpha < \beta$. We consider the sequences $(x_n)_{n \in \mathbb{N}}$ and $(y_n)_{n \in \mathbb{N}}$, with $x_1 \neq 0$ and $y_1 \neq 0$, defined by the recurrence relations

$$n x_{n+1} = \left(n + \frac{1}{p} \right) x_n \quad \text{and} \quad n y_{n+1} = (n+1) y_n.$$

Evaluate $\displaystyle \lim_{n \to \infty} \frac{y_{\alpha n}(x_{\alpha n} + x_{\alpha n+1} + \cdots + x_{\beta n})}{x_{\alpha n}(y_{\alpha n} + y_{\alpha n+1} + \cdots + y_{\beta n})}.$

7.22. [131, Problem 5] Let $\lambda, \mu, \alpha, \beta \in \mathbb{N}$, with $\lambda \neq \mu$ and $\alpha < \beta$. We consider the sequences $(x_n)_{n \in \mathbb{N}}$ and $(y_n)_{n \in \mathbb{N}}$, with $x_1 \neq 0$ and $y_1 \neq 0$, defined by the recurrence relations

$$nx_{n+1} = \left(n + \frac{1}{\lambda}\right)x_n \quad \text{and} \quad ny_{n+1} = \left(n + \frac{1}{\mu}\right)y_n.$$

Evaluate $\displaystyle\lim_{n\to\infty} \frac{y_{\alpha n}(x_{\alpha n} + x_{\alpha n+1} + \cdots + x_{\beta n})}{x_{\alpha n}(y_{\alpha n} + y_{\alpha n+1} + \cdots + y_{\beta n})}.$

7.23. [133] Let $p, a, b \in \mathbb{N}$, with $a < b$. We consider the sequence $(x_n)_{n \geq 1}$ defined by the recurrence $nx_{n+1} = (n + 1/p)x_n$ and an initial condition $x_1 \neq 0$. Evaluate

$$\lim_{n\to\infty} \frac{x_{an} + x_{an+1} + \cdots + x_{bn}}{nx_{an}}.$$

7.24. [138] Let $p \geq 2$ be an integer and let $(x_n)_{n \in \mathbb{N}}$ be the sequence defined by

$$x_n = \prod_{j=0}^{n-1} \frac{n + jp + 1}{n + jp}.$$

Assuming that $\displaystyle\lim_{n\to\infty} x_n$ exists, find its value.

Remark 7.1. The limit in Problem 7.24 is a particular case of the limit in Problem 1.16 ($q = 1$) and also a particular case of part (i) in Problem 7.19 ($k = 0$, $q = s = 1$). Different solutions are given there, without the assumption that the limit exists, while here we present a telescoping way for finding the limit (see the solution).

7.25. [135] Let $p \geq 2$ be an integer. Determine the limit

$$\lim_{n\to\infty} \sum_{k=1}^{\infty} \frac{\sqrt[p]{n}}{\sum_{j=1}^{p} \sqrt[p]{k^j(n+k)^{p-j+1}}}.$$

7.26. [136] Let $(S_n)_{n \in \mathbb{N}}$ be the sequence defined by $S_n = \sum_{k=1}^{n}(-1)^{k-1}\left(\frac{1}{k} - \ln\frac{k+1}{k}\right)$.

Calculate:

(a) $\displaystyle\lim_{n\to\infty} n^2\left|S_n - \ln\frac{4}{\pi}\right|$;

(b) $\displaystyle\lim_{n\to\infty} n^3\left|S_n - \frac{(-1)^{n-1}}{4n^2} - \ln\frac{4}{\pi}\right|$.

7.2　　Limits of Integrals

7.27.

(a) Calculate

$$\lim_{n\to\infty}\int_0^2 \frac{x^n}{1+x^n}dx.$$

(b) Let $f : [0, \infty) \to \mathbb{R}$ be a continuous function such that $\lim_{x\to\infty} f(x) = f(\infty)$ exists and is finite. Prove that

$$\lim_{n\to\infty}\int_0^2 f(x^n)dx = f(0) + f(\infty).$$

7.28. Let $a \in [0, 1)$ and let $f : [a, 1] \to \mathbb{R}$ be a continuous function. Prove that

$$\lim_{x\to\infty} x \int_a^1 y^x f(y)dy = f(1).$$

7.29. Calculate

$$\lim_{x\to\infty} x \int_0^1 (t^2 - t + 1)^x dt.$$

7.30. [O. Furdui, 30 December 2016]
　　Calculate

$$\lim_{x\to\infty} x \int_0^1 t^{tx} dt.$$

7.31. [89] Let $a > 1$. Calculate

$$\lim_{x\to\infty} x \int_0^1 a^{t(t-1)x} dt.$$

7.32. Calculate

$$\lim_{n\to\infty} \sqrt[n]{\int_0^\infty \{x\}^n \, e^{-x} dx},$$

where $\{x\}$ denotes the fractional part of x.

7.33. Calculate

$$\lim_{n\to\infty} \frac{1}{n} \sqrt[n]{\int_0^\infty \lfloor x \rfloor^n \, e^{-x} dx},$$

where $\lfloor x \rfloor$ denotes the floor of x.

7.34. Let $a, b > 0$. Calculate

$$\lim_{n\to\infty} n^2 \int_0^1 x^n \left(\sqrt[n]{ax+b} - 1 \right) dx.$$

7.35. Prove that

$$\lim_{n\to\infty} n^2 \left(\int_0^1 \sqrt[n]{1+x^n} dx - 1 \right) = \frac{\pi^2}{12}.$$

7.36. Let $a, b \ge 0$.

(a) Calculate

$$\lim_{n\to\infty} \sqrt{n} \int_0^{\frac{\pi}{2}} \sqrt{a \sin^{2n} x + b \cos^{2n} x} \, dx.$$

(b) Let $f : \left[0, \frac{\pi}{2}\right] \to \mathbb{R}$ be a continuous function. Calculate

$$\lim_{n\to\infty} \sqrt{n} \int_0^{\frac{\pi}{2}} \sqrt{a \sin^{2n} x + b \cos^{2n} x} \, f(x) \, dx.$$

7.37. Let $a, b \in \mathbb{R}$, $a < b$, and let $f : [a, b] \to \mathbb{R}$ be a Riemann integrable function. Calculate

$$\lim_{n\to\infty} \int_a^b \frac{f(x)}{1 + \cos x \cos(x+1) \cdots \cos(x+n)} \, dx.$$

7.38. Limits of Integrals with Exponential and Trigonometric Functions

(a) [75] Let $k > 0$ be a real number. Calculate

$$\lim_{n \to \infty} n^k \int_0^\infty \frac{|\sin x|^k}{e^{(n+1)x} - e^{nx}} dx.$$

(b) [79] Let $k > -1$ be a real number. Calculate

$$\lim_{n \to \infty} n^{k+1} \int_0^\infty \frac{x^k \sin x}{e^{(n+1)x} - e^{nx}} dx.$$

7.39. Let $f : [0, \infty) \to \mathbb{R}$ be continuous at 0 and bounded. Prove that

$$\lim_{n \to \infty} n \int_0^\infty \left(\frac{\sqrt{x^2 + 1} - x}{\sqrt{x^2 + 1} + x} \right)^n f(x) \, dx = \frac{f(0)}{2}.$$

7.40.

(a) Calculate

$$\lim_{n \to \infty} \frac{1}{n} \int_0^n |\sin x| \, dx.$$

(b) Calculate

$$\lim_{n \to \infty} \frac{1}{n} \int_0^n |\sin x| \cos^2 x \, dx.$$

7.41. Let $f : [0, 1] \to \mathbb{R}$ be a Riemann integrable function.

(a) Prove that

$$\lim_{n \to \infty} \int_0^1 f(|\sin nx|) dx = \frac{1}{\pi} \int_0^\pi f(\sin x) dx.$$

(b) Prove that

$$\lim_{n \to \infty} \int_0^1 f(\sin nx) dx = \frac{1}{2\pi} \int_0^{2\pi} f(\sin x) dx.$$

7.42. Let $k, m \geq 0$. Prove that

$$\lim_{n \to \infty} \left(\int_0^1 \sqrt[n]{x^k (1-x)^m} \, dx \right)^n = e^{-k-m}.$$

An interesting identity

$$\sum_{i=1}^n \sum_{j=1}^n \min\{i, j\} = \sum_{i=1}^n i^2 = \frac{n(n+1)(2n+1)}{6}, \quad n \in \mathbb{N}.$$

7.43. Let $k \geq 2$ be an integer. Calculate

$$\lim_{n \to \infty} \int_1^k \frac{\{x^n\}}{x} \, dx,$$

where $\{a\}$ denotes the fractional part of a.

7.44. Let $k \in \mathbb{N}$. Calculate

$$\lim_{n \to \infty} \int_1^2 \left\{ nx^k \right\} dx,$$

where $\{a\}$ denotes the fractional part of a.

7.3   Convergence and Evaluation of Series

7.45. Study the convergence of the series $\displaystyle\sum_{n=2}^{\infty} \left(\sqrt[n]{n} - 1 \right)^\alpha \ln^\beta n$, $\alpha, \beta > 0$.

7.46. Study the convergence of the series $\displaystyle\sum_{n=1}^{\infty} \sin n \sin \frac{1}{n}$.

7.47. *A trigonometric series*

(a) Prove that $|\sin x| \leq |x|$, for all $x \in \mathbb{R}$.

(b) Prove that the series $\displaystyle\sum_{n=1}^{\infty} \sin\left(\pi (2 + \sqrt{3})^n \right)$ converges absolutely.

7.48. Study the convergence of the series $\displaystyle\sum_{n=2}^{\infty}\left(1 - n^{\frac{\sin n}{n}}\right)$.

7.49. Let $f : [0, 1] \to \mathbb{R}$ be a continuous function. Study the convergence of the series

$$\sum_{n=1}^{\infty}\int_0^1 \frac{x^n}{1 + x + x^2 + \cdots + x^n} f(x)\mathrm{d}x.$$

7.50. Study the convergence of the series:

(a) $\displaystyle\sum_{n=1}^{\infty}\left(\frac{\pi}{2} - \int_0^n \frac{\sin^2 x}{x^2}\mathrm{d}x\right)$;

(b) $\displaystyle\sum_{n=1}^{\infty}\left(\ln 2 - \int_0^n \frac{\sin^4 x}{x^3}\mathrm{d}x\right)$.

7.51. Let $p \geq 2$ be an integer. Prove that:

(a) $\dfrac{1}{(n + 1)\sqrt[p]{n}} < \dfrac{p}{\sqrt[p]{n}} - \dfrac{p}{\sqrt[p]{n + 1}}$, $\forall n \in \mathbb{N}$;

(b) $\displaystyle\sum_{n=1}^{\infty}\frac{1}{(n + 1)\sqrt[p]{n}} < p.$

See also Problem 2.24.

7.52. Calculate

$$\int_0^{\infty} \{x\}\,\mathrm{e}^{-x}\mathrm{d}x \qquad \text{and} \qquad \int_0^{\infty} x\mathrm{e}^{-\lfloor x \rfloor}\mathrm{d}x.$$

7.53. Calculate

$$\sum_{n=1}^{\infty}\frac{1}{(n - 1)! + n!} \qquad \text{and} \qquad \sum_{n=0}^{\infty}\frac{(-1)^n}{(n + (-1)^n)!}.$$

7.54. Calculate

$$\sum_{n=1}^{\infty}\frac{n}{(n + 1)!} \qquad \text{and} \qquad \sum_{n=1}^{\infty}\frac{n}{(n + (-1)^n)!}.$$

7.55. Let $k \in \mathbb{N}$. Prove that

$$\sum_{n=1}^{\infty} \frac{1}{(n+k-1)(n+k)(n+k)!} = 1 + \frac{1}{1!} + \frac{1}{2!} + \cdots + \frac{1}{k!} + \frac{1}{k \cdot k!} - e.$$

Remark 7.2. When $k = 1$ we obtain Problem 1.3.

7.56. Prove that

$$\sum_{n=2}^{\infty} \frac{1}{(-1)^n n + 1} = \ln 2 - 1 \quad \text{and} \quad \sum_{n=2}^{\infty} \frac{1}{(-1)^n n - 1} = \ln 2 + \frac{1}{2}.$$

7.57. Prove that

$$\sum_{n=1}^{\infty} \frac{(-1)^{\lfloor \frac{n}{2} \rfloor}}{n} = \frac{\pi}{4} - \frac{1}{2} \ln 2 \quad \text{and} \quad \sum_{n=1}^{\infty} \frac{(-1)^{\lfloor \frac{n}{3} \rfloor}}{n} = \frac{2\pi}{3\sqrt{3}} - \frac{1}{3} \ln 2.$$

7.58. A Wallis Series Prove that

$$\sum_{n=1}^{\infty} (-1)^{\lfloor \frac{n}{2} \rfloor} \ln\left(1 + \frac{1}{n}\right) = \ln \frac{4}{\pi},$$

where $\lfloor x \rfloor$ denotes the floor of x.

7.59.

(a) Let $n, k \in \mathbb{N}$. Prove that

$$\int_0^1 (1-x)^n x^{k-1} dx = \frac{n!}{k(k+1)(k+2) \cdots (k+n)}.$$

(b) Prove that

$$\sum_{k=1}^{\infty} \frac{1}{k(k+1)(k+2) \cdots (k+n)} = \frac{1}{n \cdot n!}.$$

See also Problem 2.19.

7.4   Harmonic Series

7.60. A Famous Series of Euler Revisited
  Prove, by calculating in two different ways the integral

$$\int_0^1 \frac{\ln x \ln(1-x)}{x(1-x)}\,dx,$$

that

$$\sum_{n=1}^{\infty} \frac{H_n}{n^2} = 2\zeta(3).$$

7.61.

(a) Let $k \geq 1$ be an integer. Prove that

$$\sum_{n=1}^{\infty} \frac{1}{n(n+k)} = \frac{H_k}{k}.$$

(b) *Euler's series.* Prove, by using part (a), that

$$\sum_{k=1}^{\infty} \frac{H_k}{k^2} = 2\zeta(3).$$

(c) *A telescoping term.* Prove that

$$\frac{H_n H_{n+1}}{n(n+1)} = \frac{H_n^2}{n} - \frac{H_{n+1}^2}{n+1} + \frac{H_n}{n} - \frac{H_{n+1}}{n+1} + \frac{1}{(n+1)^2} + \frac{H_{n+1}}{(n+1)^2}, \quad n \geq 1.$$

(d) [25] *A nonlinear Euler sum.* Prove that

$$\sum_{n=1}^{\infty} \frac{H_n H_{n+1}}{n(n+1)} = \zeta(2) + 2\zeta(3).$$

(e) Prove that

(continued)

$$\sum_{n=1}^{\infty} \frac{H_n^2}{n(n+1)} = 3\zeta(3).$$

7.62. [O. Furdui, 16 July 2018]

(a) Prove that

$$\sum_{k=1}^{n} H_k^2 = (n+1)H_n^2 - (2n+1)H_n + 2n, \quad n \in \mathbb{N}.$$

(b) [65] Prove that

$$\sum_{n=1}^{\infty} H_n^2 \left(\zeta(2) - 1 - \frac{1}{2^2} - \cdots - \frac{1}{n^2} - \frac{1}{n} \right) = 2 - \zeta(2) - 2\zeta(3).$$

(c) Prove that

$$\sum_{n=1}^{\infty} H_n^2 \left(\zeta(3) - 1 - \frac{1}{2^3} - \cdots - \frac{1}{n^3} \right) = 2\zeta(2) - 4\zeta(3) + 3\zeta(4).$$

7.63. *Two series and Apéry's constant.*
  Prove that:

(a) [26, p. 142] $\displaystyle\sum_{n=1}^{\infty} \frac{1}{n} \left(\zeta(2) - \frac{1}{1^2} - \frac{1}{2^2} - \cdots - \frac{1}{n^2} \right) = \zeta(3);$

(b) $\displaystyle\sum_{n=1}^{\infty} \frac{1}{n} \left(\frac{1}{n^2} + \frac{1}{(n+1)^2} + \frac{1}{(n+2)^2} + \cdots \right) = 2\zeta(3).$

A Famous Series of Goldbach Revisited
We prove that

$$\sum_{n=1}^{\infty} \frac{H_n}{n^3} = \frac{\pi^4}{72}.$$

(continued)

We have

$$S = \sum_{n=1}^{\infty} \frac{H_n}{n^3} = \sum_{n=1}^{\infty} \frac{1}{n^2} \sum_{k=1}^{\infty} \frac{1}{k(n+k)} = \sum_{n=1}^{\infty} \sum_{k=1}^{\infty} \frac{1}{n^2 k(n+k)}.$$

It follows, based on symmetry reasons, that $S = \sum\limits_{n=1}^{\infty} \sum\limits_{k=1}^{\infty} \frac{1}{nk^2(n+k)}$ and this shows

$$2S = \sum_{n=1}^{\infty} \sum_{k=1}^{\infty} \left(\frac{1}{n^2 k(n+k)} + \frac{1}{nk^2(n+k)} \right) = \sum_{n=1}^{\infty} \sum_{k=1}^{\infty} \frac{1}{n^2 k^2} = \sum_{n=1}^{\infty} \frac{1}{n^2} \sum_{k=1}^{\infty} \frac{1}{k^2} = \frac{\pi^4}{36}.$$

7.64. Prove that

$$\sum_{n=1}^{\infty} H_n \left(\frac{\pi^4}{72} - \frac{H_1}{1^3} - \frac{H_2}{2^3} - \cdots - \frac{H_n}{n^3} \right) = \frac{17}{4} \zeta(4) - 2\zeta(3).$$

7.65. *A challenge.* Prove that

$$\sum_{n=1}^{\infty} (-1)^n H_n \left(2\zeta(3) - \frac{H_1}{1^2} - \frac{H_2}{2^2} - \cdots - \frac{H_n}{n^2} \right)$$

$$= \frac{\ln^4 2}{48} - \frac{\pi^2 \ln^2 2}{48} - \frac{\pi^4}{1440} - \frac{7}{8} \zeta(3) \ln 2 + \frac{1}{2} \mathrm{Li}_4 \left(\frac{1}{2} \right).$$

This result, which answers a question of O. Furdui and A. Sîntămărian [60], was communicated to the authors by M. Levy on 24 November, 2019. The second open problem in [60], of calculating the series

$$\sum_{n=1}^{\infty} (-1)^n H_n \left(\frac{\pi^4}{72} - \frac{H_1}{1^3} - \frac{H_2}{2^3} - \cdots - \frac{H_n}{n^3} \right),$$

was solved by C.I. Vălean and M. Levy on December 2019 in [147].

7.66. **Computing** $\mathrm{Li}_2\left(\frac{1}{2}\right)$ **and** $\sum_{n=1}^{\infty}(-1)^n\frac{H_n}{n}$

(a) Prove the identity

$$\int_0^1 x^{n-1}\ln(1-x)dx = -\frac{H_n}{n}, \quad n \geq 1.$$

(b) *An alternating harmonic series.* Prove that

$$\sum_{n=1}^{\infty}(-1)^{n-1}\frac{H_n}{n} = \frac{\pi^2}{12} - \frac{\ln^2 2}{2}.$$

We solve this part of the problem here. We have

$$\int_0^1 \ln\left(\frac{1}{1-x}\right)\frac{dx}{1+x} = \int_0^1 \ln\left(\frac{1}{1-x}\right)\left(1-x+x^2-x^3+\cdots\right)dx$$

$$= 1 - \frac{H_2}{2} + \frac{H_3}{3} - \cdots.$$

A calculation, based on the substitution $x = \frac{1-t}{1+t}$, shows that

$$\int_0^1 \ln\left(\frac{1}{1-x}\right)\frac{dx}{1+x} = \frac{\pi^2}{12} - \frac{\ln^2 2}{2},$$

and we obtain, probably a new method, that

$$\sum_{n=1}^{\infty}(-1)^{n-1}\frac{H_n}{n} = \frac{\pi^2}{12} - \frac{\ln^2 2}{2}.$$

(c) Justify the steps in *Legendre's formula*

$$\sum_{n=1}^{\infty}\frac{1}{n^2 2^n} = \int_0^1 \ln\left(\frac{1}{x}\right)\frac{dx}{2-x}$$

$$= \int_0^1 \ln\left(\frac{1}{1-x}\right)\left(1-x+x^2-\cdots\right)dx$$

$$= 1 - \frac{H_2}{2} + \frac{H_3}{3} - \cdots$$

$$= \frac{\pi^2}{12} - \frac{\ln^2 2}{2}.$$

(continued)

Remark 7.3.

(i) We mention the equality

$$\operatorname{Li}_2\left(\frac{1}{2}\right) = \sum_{n=1}^{\infty} \frac{1}{n^2 2^n} = \frac{\pi^2}{12} - \frac{\ln^2 2}{2}.$$

See also Problems 3.64 and 7.83 for the calculation of $\operatorname{Li}_2\left(\frac{1}{2}\right)$.

(ii) *The calculation of a class of log integrals.*

   We record the calculation of a class of logarithmic integrals. We have

$$\int_0^1 \left(\frac{\ln(1-x)}{1+x}\right)^2 dx = \int_0^1 \left(\frac{\ln t}{2-t}\right)^2 dt$$

$$= \sum_{n=0}^{\infty} \frac{n+1}{2^{n+2}} \int_0^1 t^n \ln^2 t\, dt$$

$$= \sum_{n=0}^{\infty} \frac{1}{2^{n+1}(n+1)^2}$$

$$\overset{7.66\,(c)}{=} \frac{\pi^2}{12} - \frac{\ln^2 2}{2}.$$

More generally, one can prove that, for $k \geq 2$, we have

$$\int_0^1 \left(\frac{\ln(1-x)}{1+x}\right)^k dx = \frac{k}{2^{k-1}} \sum_{i=1}^{k-1} (-1)^{i-1} s(k-1,i) \operatorname{Li}_{k+1-i}\left(\frac{1}{2}\right),$$

where $s(k-1,i)$ are the *Stirling numbers of the first kind*.

7.67. A Gem with Two Series
Prove that

$$\sum_{n=1}^{\infty} (-1)^{n-1} \frac{H_n^- - H_n}{n} = \ln^2 2 \quad \text{and} \quad \sum_{n=1}^{\infty} (-1)^{n-1} \frac{H_n^- + H_n}{n} = \zeta(2).$$

7.68. Prove that:

(a) $\displaystyle\sum_{n=1}^{\infty} H_n \left(\ln\frac{1}{2} + 1 - \frac{1}{2} + \cdots + \frac{(-1)^{n-1}}{n} \right) = \frac{1 - \ln 2}{2}$;

(b) $\displaystyle\sum_{n=1}^{\infty} \frac{H_n}{n+1} \left(\ln\frac{1}{2} + 1 - \frac{1}{2} + \cdots + \frac{(-1)^{n-1}}{n} \right) = \frac{\zeta(3)}{8} - \frac{\ln^3 2}{6}$.

7.69. Intriguing Series with $H_{2n} - H_n - \ln 2$
Prove that:

(a) [68] $\displaystyle\sum_{n=1}^{\infty} (-1)^n n \left(\frac{1}{n+1} + \frac{1}{n+2} + \cdots + \frac{1}{2n} - \ln 2 + \frac{1}{4n} \right)$
$= \frac{\ln 2 - 1}{8}$;

(b) $\displaystyle\sum_{n=1}^{\infty} (-1)^n n^2 \left(\frac{1}{n+1} + \frac{1}{n+2} + \cdots + \frac{1}{2n} - \ln 2 + \frac{1}{4n} - \frac{1}{16n^2} \right)$
$= \frac{\pi - 3}{32}$.

7.70. A Challenging Series with $H_{2n} - H_n - \ln 2$ and the Tail of $\zeta(2)$

(a) Prove that

$$\int_0^1 \frac{x^{2n}}{1+x}\,dx = H_n - H_{2n} + \ln 2, \quad n \geq 1.$$

(b) Prove that

$$\sum_{n=1}^{\infty} \left(\zeta(2) - 1 - \frac{1}{2^2} - \cdots - \frac{1}{n^2} \right)(H_{2n} - H_n - \ln 2)$$

$$= -\frac{1}{2}\zeta(2) - \frac{7}{8}\zeta(3) + \zeta(2)\ln 2 + \ln^2 2.$$

7.71. *The Glaisher–Kinkelin constant A* is defined by the formula

$$A = \lim_{n \to \infty} n^{-\frac{n^2}{2} - \frac{n}{2} - \frac{1}{12}} e^{\frac{n^2}{4}} \prod_{k=1}^{n} k^k = 1.282427129100622\ldots.$$

Prove that

$$\sum_{n=1}^{\infty} n \left(H_n - \ln n - \gamma - \frac{1}{2n} + \frac{1}{12n^2} \right) = \frac{5}{24} - \ln A + \frac{\gamma}{12}.$$

7.5   Series with Factorials

7.72. A Mathematical Jewel I

(a) Check the identity

$$\frac{(m-1)!}{(n+m)!} = \frac{(m-1)!}{n(n+m-1)!} - \frac{m!}{n(n+m)!}, \quad m, n \geq 1.$$

(b) [149, B. Cloitre, 9 December 2004, entry 8]

Prove that

$$\sum_{n=1}^{\infty} \sum_{m=1}^{\infty} \frac{(n-1)!(m-1)!}{(n+m)!} = \zeta(2).$$

7.73. [48] A Mathematical Jewel II

(a) Let $a \in [-1, 1]$. Prove that

$$\sum_{i=1}^{\infty} \sum_{j=1}^{\infty} \frac{(i-1)!(j-1)!}{(i+j)!} a^{i+j} = 2\text{Li}_2 \left(\frac{a}{2-a} \right) - 2\text{Li}_2 \left(-\frac{a}{2-a} \right) - 2\text{Li}_2(a),$$

where Li_2 is the Dilogarithm function defined in Problems 3.4 and 7.83.

(continued)

(b) *A double sum with factorials.*
   Prove that

$$\sum_{i=1}^{\infty}\sum_{j=1}^{\infty}\frac{(i-1)!(j-1)!}{(i+j)!}=\zeta(2).$$

A Mathematical Jewel III

We record as a gem the third proof, the first two proofs being given in Problems 7.72 and 7.73, of the following remarkable formula due to B. Cloitre, involving a double series with factorials

$$\sum_{i=1}^{\infty}\sum_{j=1}^{\infty}\frac{(i-1)!(j-1)!}{(i+j)!}=\zeta(2).$$

We have

$$\sum_{i=1}^{\infty}\sum_{j=1}^{\infty}\frac{(i-1)!(j-1)!}{(i+j)!}=\sum_{i=1}^{\infty}(i-1)!\sum_{j=1}^{\infty}\frac{1}{j(j+1)\cdots(j+i)}$$

$$=\sum_{i=1}^{\infty}\frac{(i-1)!}{i\cdot i!}$$

$$=\sum_{i=1}^{\infty}\frac{1}{i^2}$$

$$=\zeta(2).$$

We used in our calculations the following formula due to M. Andreoli [2]

$$\sum_{k=1}^{\infty}\frac{1}{k(k+1)\cdots(k+n)}=\frac{1}{n\cdot n!},\quad n\in\mathbb{N},$$

whose solution is given in detail in [26, Problem 3.10, p. 140]. For another proof of the preceding formula see the solution of Problem 7.59. ∎

Remark 7.4. It is worth mentioning the fabulous series with four factorials

$$\sum_{i=1}^{\infty}\sum_{j=1}^{\infty}\sum_{k=1}^{\infty} \frac{(i-1)!(j-1)!(k-1)!}{(i+j+k)!} = \frac{13}{4}\zeta(3) - \frac{\pi^2}{2}\ln 2$$

whose proof, which is challenging, is left to the interested reader.

7.74. *Amazing series with three factorials.*
   Prove that:

(a) *An alternating series* $\displaystyle\sum_{i=1}^{\infty}\sum_{j=1}^{\infty}(-1)^{i-1}\frac{(i-1)!(j-1)!}{(i+j)!} = \frac{\zeta(2)}{2}$;

(b) *A $\zeta(2)$ series* $\displaystyle\sum_{i=1}^{\infty}\sum_{j=1}^{\infty}\left((-1)^{i-1}+(-1)^{j-1}\right)\frac{(i-1)!(j-1)!}{(i+j)!} = \zeta(2)$;

(c) $\displaystyle\sum_{i=1}^{\infty}\sum_{j=1}^{\infty}\frac{(i-1)!(j-1)!}{(i+j+1)!} = 2-\zeta(2)$;

(d) $\displaystyle\sum_{i=1}^{\infty}\sum_{j=1}^{\infty}\frac{(i-1)!(j-1)!}{(i+j+2)!} = \frac{\zeta(2)}{2}-\frac{3}{4}$.

7.6   Series of Functions

7.75. *A uniformly continuous function.*
   Prove that the series $\sum_{n=1}^{\infty}\frac{\cos(n^4 x)}{n^2}$ defines a uniformly continuous function on $\mathbb{R}$.

7.76. Let $(a_n)_{n\geq 0}$ be a sequence of real numbers such that $\lim_{n\to\infty} a_n = l$. Prove that

$$\lim_{x\to\infty} e^{-x}\sum_{n=0}^{\infty} a_n\frac{x^n}{n!} = l.$$

7.77.

(a) Let $(a_n)_{n\geq 1}$ be a sequence of positive real numbers. Prove that

$$\lim_{x \to \infty} \left(\sum_{n=1}^{\infty} a_n^x \right)^{\frac{1}{x}} = \sup_{n \geq 1} a_n.$$

(b) *Application.* Calculate $\lim\limits_{x \to \infty} \zeta(x)^{\frac{1}{x}}$.

7.78. [O. Furdui, M. Ivan, A. Sîntămărian, 28 December 2016]

(a) Prove that

$$\sum_{n=1}^{\infty} \left(\frac{x}{n} \right)^n = x \int_0^1 \frac{1}{t^{xt}} dt, \quad x \in \mathbb{R}.$$

(b) [56] Prove that

$$\lim_{x \to \infty} \left(\sum_{n=1}^{\infty} \left(\frac{x}{n} \right)^n \right)^{\frac{1}{x}} = e^{e^{-1}}.$$

7.79. [O. Furdui, 29 December 2016]

(a) Prove that

$$\sum_{n=1}^{\infty} (-1)^{n-1} \left(\frac{x}{n} \right)^n = x \int_0^1 t^{xt} dt, \quad x \in \mathbb{R}.$$

(b) Prove that

$$\lim_{x \to \infty} \sum_{n=1}^{\infty} (-1)^{n-1} \left(\frac{x}{n} \right)^n = 1.$$

7.80. Find the convergence set and the sum of the power series

$$\sum_{n=2}^{\infty} \frac{x^n}{(-1)^n n + 1}.$$

7.81. Prove that

$$\sum_{n=1}^{\infty} H_n \left(\frac{1}{1-x} - 1 - x - x^2 - \cdots - x^n \right) = -\frac{x}{(1-x)^2} \ln(1-x), \quad x \in (-1, 1).$$

7.82. Prove that

$$\sum_{n=0}^{\infty}\sum_{m=0}^{\infty}\left(e^x - 1 - \frac{x}{1!} - \frac{x^2}{2!} - \cdots - \frac{x^{n+m}}{(n+m)!}\right) = \frac{x^2 + 2x}{2}e^x, \quad x \in \mathbb{R}.$$

7.83. The Dilogarithm function Li_2 is the special function defined by

$$\text{Li}_2(z) = \sum_{n=1}^{\infty} \frac{z^n}{n^2} = -\int_0^z \frac{\ln(1-t)}{t}\,dt, \quad |z| \le 1.$$

(a) Prove that $\text{Li}_2(z) + \text{Li}_2(1-z) = \frac{\pi^2}{6} - \ln z \ln(1-z)$.

In particular, $\text{Li}_2\left(\frac{1}{2}\right) = \frac{\pi^2}{12} - \frac{\ln^2 2}{2}$ (see also Problems 3.64 and 7.66).

(b) *Application.* Calculate

$$\sum_{n=1}^{\infty} \frac{\sin^{2n}\theta + \cos^{2n}\theta}{n^2}, \quad \theta \in \mathbb{R}.$$

7.84. [62] Let $k \in \mathbb{N}$. Find the convergence set and the sum of the power series

$$\sum_{n=k}^{\infty} C_n^k \left(e - 1 - \frac{1}{1!} - \frac{1}{2!} - \cdots - \frac{1}{n!}\right) x^n.$$

Remark 7.5. More generally, it can be proved that, if f has a Maclaurin series with radius of convergence $R > 1$, then

$$\sum_{n=k}^{\infty} C_n^k \left(e - 1 - \frac{1}{1!} - \frac{1}{2!} - \cdots - \frac{1}{n!}\right) f^{(n)}(0) = \frac{1}{k!}\int_0^1 e^{1-t}t^k f^{(k)}(t)\,dt.$$

See also Problem 4.74.

7.85. Prove that

$$\sum_{n=0}^{\infty} n! \left(e - 1 - \frac{1}{1!} - \frac{1}{2!} - \cdots - \frac{1}{n!}\right)^2 = e \sum_{n=1}^{\infty} \frac{1}{n \cdot n!}.$$

7.86. Harmonic–Trigonometric Series
    Prove that

(continued)

$$\sum_{n=1}^{\infty} \frac{H_n}{n} x^n = \mathrm{Li}_2(x) + \frac{1}{2} \ln^2(1 - x), \quad -1 \le x < 1. \tag{7.1}$$

Remark 7.6. Formula (7.1), which is the generating function for the sequence $\left(\frac{H_n}{n}\right)_{n \ge 1}$, is also true for complex numbers z, with $|z| \le 1$, $z \ne 1$.

Replacing x by $e^{\frac{\pi i}{k}}$, $k \in \mathbb{N}$, in formula (7.1) one obtains the *harmonic–cosine series formula*

$$\sum_{n=1}^{\infty} \frac{H_n}{n} \cos \frac{n\pi}{k} = \frac{\pi^2}{24} \cdot \frac{k^2 - 6k + 3}{k^2} + \frac{1}{2} \ln^2\left(2 \sin \frac{\pi}{2k}\right).$$

In particular, when k equals 1, 2, or 3, we obtain the following formulae:

- $$\sum_{n=1}^{\infty} (-1)^n \frac{H_n}{n} = \frac{\ln^2 2}{2} - \frac{\pi^2}{12};$$
- $$\sum_{n=1}^{\infty} (-1)^n \frac{H_{2n}}{n} = \frac{\ln^2 2}{4} - \frac{5\pi^2}{48};$$
- $$\sum_{n=1}^{\infty} \frac{H_n}{n} \cos \frac{n\pi}{3} = -\frac{\pi^2}{36}.$$

Replacing x by e^{ix} and $-e^{ix}$ in formula (7.1), we obtain the *amazing Fourier series formulae*:

- $$\sum_{n=1}^{\infty} \frac{H_n}{n} \cos(nx) = \frac{\pi^2}{24} - \frac{\pi x}{4} + \frac{x^2}{8} + \frac{1}{2} \ln^2\left(2 \sin \frac{x}{2}\right), \quad x \in [0, 2\pi];$$
- $$\sum_{n=1}^{\infty} (-1)^n \frac{H_n}{n} \cos(nx) = \frac{x^2}{8} - \frac{\pi^2}{12} + \frac{1}{2} \ln^2\left(2 \cos \frac{x}{2}\right), \quad x \in [-\pi, \pi].$$

7.87. (a) **The Generating Function of** $\frac{1}{n}\left(\zeta(2) - 1 - \frac{1}{2^2} - \cdots - \frac{1}{n^2}\right)$

Prove that

$$\sum_{n=1}^{\infty} \left(\zeta(2) - 1 - \frac{1}{2^2} - \cdots - \frac{1}{n^2}\right) \frac{x^n}{n}$$

$$= \ln(1 - x)\left[\mathrm{Li}_2(x) - \zeta(2)\right] - \mathrm{Li}_3(x) + \int_0^x \frac{\ln^2(1 - t)}{t} dt, \quad x \in [-1, 1].$$

(continued)

In particular, for $x = 1$ we have

$$\sum_{n=1}^{\infty} \frac{\zeta(2) - 1 - \frac{1}{2^2} - \cdots - \frac{1}{n^2}}{n} = \zeta(3)$$

and when $x = -1$ we obtain, see also part (a) of Problem 3.81, that

$$\sum_{n=1}^{\infty} (-1)^n \frac{\zeta(2) - 1 - \frac{1}{2^2} - \cdots - \frac{1}{n^2}}{n} = \zeta(3) - \frac{\pi^2}{4} \ln 2.$$

It follows, from the previous formula, that

$$\sum_{n=1}^{\infty} (-1)^n \frac{1 + \frac{1}{2^2} + \cdots + \frac{1}{n^2}}{n} = \zeta(2) \frac{\ln 2}{2} - \zeta(3).$$

(b) **The Generating Function of** $\frac{1}{n} \left(\zeta(3) - 1 - \frac{1}{2^3} - \cdots - \frac{1}{n^3} \right)$
Prove that

$$\sum_{n=1}^{\infty} \left(\zeta(3) - 1 - \frac{1}{2^3} - \cdots - \frac{1}{n^3} \right) \frac{x^n}{n}$$

$$= \ln(1 - x) \left[\mathrm{Li}_3(x) - \zeta(3) \right] - \mathrm{Li}_4(x) + \frac{\mathrm{Li}_2^2(x)}{2}, \quad x \in [-1, 1].$$

In particular, for $x = 1$ we have

$$\sum_{n=1}^{\infty} \frac{\zeta(3) - 1 - \frac{1}{2^3} - \cdots - \frac{1}{n^3}}{n} = \frac{\zeta(4)}{4}$$

and when $x = -1$ we obtain, see also part (b) of Problem 3.81, that

$$\sum_{n=1}^{\infty} (-1)^n \frac{\zeta(3) - 1 - \frac{1}{2^3} - \cdots - \frac{1}{n^3}}{n} = \frac{19\pi^4}{1440} - \frac{7}{4} \zeta(3) \ln 2.$$

It follows, from the previous formula, that

$$\sum_{n=1}^{\infty} (-1)^n \frac{1 + \frac{1}{2^3} + \cdots + \frac{1}{n^3}}{n} = \frac{3}{4} \zeta(3) \ln 2 - \frac{19}{16} \zeta(4).$$

7.88. A Gem with a Logarithmic Series

Let $x \in [-1, 1)$. Prove that

$$\sum_{n=1}^{\infty} \left[\ln^2(1-x) - \left(x + \frac{x^2}{2} + \cdots + \frac{x^n}{n} \right)^2 \right] = \frac{x+1}{x-1} \ln(1-x^2) - \ln^2(1-x).$$

In particular, when $x = -1$ we obtain the amazing logarithmic series

$$\sum_{n=1}^{\infty} \left[\left((-1) + \frac{(-1)^2}{2} + \cdots + \frac{(-1)^n}{n} \right)^2 - \ln^2 2 \right] = \ln^2 2.$$

7.7 Pearls of Series with Tails of Zeta Function Values

7.89. Two Superb Series with the Tail of $\zeta(2)$

Prove that:

(a) [61] $\displaystyle\sum_{n=1}^{\infty} \left[n \left(\zeta(2) - 1 - \frac{1}{2^2} - \cdots - \frac{1}{n^2} \right) - 1 + \frac{1}{2n} \right] = \frac{1}{4}$;

(b) $\displaystyle\sum_{n=1}^{\infty} (-1)^n \left[n^2 \left(\zeta(2) - 1 - \frac{1}{2^2} - \cdots - \frac{1}{n^2} \right) - n + \frac{1}{2} \right] = \frac{1 - \ln 4}{4}$.

7.90. Series with $\zeta(2) - 1 - \frac{1}{2^2} - \cdots - \frac{1}{n^2} - \frac{1}{n}$

(a) Prove that

$$\sum_{n=1}^{\infty} H_n \left(\zeta(2) - 1 - \frac{1}{2^2} - \cdots - \frac{1}{n^2} - \frac{1}{n} \right) = -1.$$

(b) Prove that

$$\sum_{n=1}^{\infty} \left(1 + \frac{1}{2^2} + \cdots + \frac{1}{n^2} \right) \left(\zeta(2) - 1 - \frac{1}{2^2} - \cdots - \frac{1}{n^2} - \frac{1}{n} \right) = \zeta(2) - 2\zeta(3).$$

(continued)

(c) *Open problem* Calculate

$$\sum_{n=1}^{\infty} H_{2n}\left(\zeta(2)-1-\frac{1}{2^2}-\cdots-\frac{1}{n^2}-\frac{1}{n}\right).$$

7.91. Prove that

$$\sum_{n=1}^{\infty}\left(1+\frac{1}{2^2}+\cdots+\frac{1}{n^2}\right)\left(\zeta(3)-1-\frac{1}{2^3}-\cdots-\frac{1}{n^3}\right)=\frac{\zeta(4)}{2}.$$

See the connection of this exercise with part (b) of Problem 2.56.

Open Problem Since

$$H_n\left(1+\frac{1}{2^2}+\cdots+\frac{1}{n^2}\right)\left(\zeta(3)-1-\frac{1}{2^3}-\cdots-\frac{1}{n^3}\right)\sim\frac{\zeta(2)}{2}\cdot\frac{H_n}{n^2},$$

it is natural to consider as an open problem the calculation of the series

$$\sum_{n=1}^{\infty} H_n\left(1+\frac{1}{2^2}+\cdots+\frac{1}{n^2}\right)\left(\zeta(3)-1-\frac{1}{2^3}-\cdots-\frac{1}{n^3}\right).$$

7.92. Prove that:

(a) [85] $\displaystyle\sum_{n=1}^{\infty} n\left(\zeta(3)-1-\frac{1}{2^3}-\cdots-\frac{1}{n^3}-\frac{1}{2n^2}\right)=\frac{1}{4}-\frac{\pi^2}{12}$;

(b) $\displaystyle\sum_{n=1}^{\infty} n^2\left(\zeta(4)-1-\frac{1}{2^4}-\cdots-\frac{1}{n^4}-\frac{1}{3n^3}\right)=\frac{1}{9}+\frac{\zeta(3)}{6}-\frac{\zeta(2)}{2}$;

(c) [76] $\displaystyle\sum_{n=1}^{\infty} n^3\left(\zeta(5)-1-\frac{1}{2^5}-\cdots-\frac{1}{n^5}-\frac{1}{4n^4}\right)=\frac{1}{16}+\frac{\zeta(3)}{4}-\frac{\zeta(2)}{2}$.

7.93. Exotic Series Prove that:

(a) $\displaystyle\sum_{n=1}^{\infty} \left[n^2 \left(\zeta(3) - 1 - \frac{1}{2^3} - \cdots - \frac{1}{n^3} \right) - \frac{1}{2} + \frac{1}{2n} \right] = \frac{2\zeta(2) + 1}{12}$;

(b) $\displaystyle\sum_{n=1}^{\infty} \left[n^3 \left(\zeta(4) - 1 - \frac{1}{2^4} - \cdots - \frac{1}{n^4} \right) - \frac{1}{3} + \frac{1}{2n} \right] = \frac{6\zeta(2) + 1}{24}$.

A challenge. If $x > 1$, one can prove that

$$\lim_{n \to \infty} n^2 \left[n^{x-1} \left(\zeta(x) - \sum_{k=1}^{n} \frac{1}{k^x} \right) - \frac{1}{x-1} + \frac{1}{2n} \right] = \frac{x}{12},$$

which shows that $n^{x-1} \left(\zeta(x) - \sum_{k=1}^{n} \frac{1}{k^x} \right) - \frac{1}{x-1} + \frac{1}{2n} \sim \frac{1}{n^2}$.

It is natural to consider, as a challenge, the calculation of the series

$$\sum_{n=1}^{\infty} \left[n^{k-1} \left(\zeta(k) - 1 - \frac{1}{2^k} - \cdots - \frac{1}{n^k} \right) - \frac{1}{k-1} + \frac{1}{2n} \right],$$

where $k \in \mathbb{N}$, $k \geq 5$.

7.94. Prove that:

(a) $\displaystyle\sum_{k=1}^{n} k H_k = \frac{n(n+1)}{2} \left(H_{n+1} - \frac{1}{2} \right)$, $n \geq 1$;

(b) $\displaystyle\sum_{n=1}^{\infty} n H_n \left(\zeta(2) - 1 - \frac{1}{2^2} - \cdots - \frac{1}{n^2} - \frac{1}{n} + \frac{1}{2n^2} \right) = \frac{3}{8}$;

(c) $\displaystyle\sum_{n=1}^{\infty} n H_n \left(\zeta(3) - 1 - \frac{1}{2^3} - \cdots - \frac{1}{n^3} - \frac{1}{2n^2} \right) = \frac{\zeta(2)}{4} - \zeta(3) - \frac{1}{8}$;

(d) $\displaystyle\sum_{n=1}^{\infty} n H_n \left(\zeta(4) - 1 - \frac{1}{2^4} - \cdots - \frac{1}{n^4} \right) = -\frac{\zeta(2)}{4} + \frac{5}{4}\zeta(3) - \frac{5}{8}\zeta(4)$.

7.95.

(a) Prove that, for all $n \geq 1$, the following identity holds:

$$1^2 - 2^2 + 3^2 - \cdots + (-1)^{n+1}n^2 = (-1)^{n+1}\frac{n(n+1)}{2}.$$

(b) Prove that

$$\sum_{n=1}^{\infty}(-1)^n \left[n^2 \left(\zeta(3) - 1 - \frac{1}{2^3} - \cdots - \frac{1}{n^3} \right) - \frac{1}{2} \right] = \frac{1}{4} + \frac{\ln 2}{2} - \frac{\zeta(2)}{4}.$$

(c) Prove that

$$\sum_{n=1}^{\infty}(-1)^n n^2 \left(\zeta(4) - 1 - \frac{1}{2^4} - \cdots - \frac{1}{n^4} \right) = \frac{\zeta(2)}{4} - \frac{3\zeta(3)}{8}.$$

The next problems in this section are about the calculation of exotic series involving the product of tails of Riemann zeta function values. These exercises have been motivated by problems 3.22 and 3.45 in [26] and by the results in the papers [27, 28, 58, 92].

7.96. Fascinating Series Prove that:

(a) $\displaystyle\sum_{n=1}^{\infty}\frac{1}{n}\left[\zeta^2(2) - \left(1 + \frac{1}{2^2} + \cdots + \frac{1}{n^2} \right)^2 \right] = 9\zeta(5) - 3\zeta(2)\zeta(3);$

(b) $\displaystyle\sum_{n=1}^{\infty}\left[\zeta^2(2) - \left(1 + \frac{1}{2^2} + \cdots + \frac{1}{n^2} \right)^2 - \frac{2\zeta(2)}{n} \right]$
$= 2\zeta(2) - 3\zeta(3) - \frac{5}{2}\zeta(4);$

(c) H_n *enters the scene.*

$$\sum_{n=1}^{\infty} H_n \left[\zeta^2(2) - \left(1 + \frac{1}{2^2} + \cdots + \frac{1}{n^2} \right)^2 - \frac{2\zeta(2)}{n} \right]$$

$$= -2\zeta(2) + 3\zeta(3) - \frac{19}{4}\zeta(4);$$

(continued)

(d) *A challenge.*

$$\sum_{n=1}^{\infty} (-1)^n \left[\zeta^2(2) - \left(1 + \frac{1}{2^2} + \cdots + \frac{1}{n^2} \right)^2 \right]$$

$$= -\frac{38}{9} \zeta(4) + 4 \mathrm{Li}_4 \left(\frac{1}{2} \right) + \frac{7}{2} \zeta(3) \ln 2 - \zeta(2) \ln^2 2 + \frac{\ln^4 2}{6};$$

(e) *Open problem.* Since $\zeta^2(2) - \left(1 + \frac{1}{2^2} + \cdots + \frac{1}{n^2} \right)^2 \sim \frac{2\zeta(2)}{n}$, it is natural to consider as an open problem the calculation of the series

$$\sum_{n=1}^{\infty} \left[\zeta^2(2) - \left(1 + \frac{1}{2^2} + \cdots + \frac{1}{n^2} \right)^2 \right]^2.$$

7.97. Prove that

$$\sum_{n=1}^{\infty} \left[\zeta^2(3) - \left(1 + \frac{1}{2^3} + \cdots + \frac{1}{n^3} \right)^2 \right] = 10\zeta(5) - 4\zeta(2)\zeta(3) - \zeta^2(3).$$

We mention that, for an integer $k \geq 3$, the following formula holds:

$$\sum_{n=1}^{\infty} \left[\zeta^2(k) - \left(1 + \frac{1}{2^k} + \cdots + \frac{1}{n^k} \right)^2 \right]$$

$$= 2 \sum_{n=1}^{\infty} \frac{1}{n^{k-1}} \left(1 + \frac{1}{2^k} + \cdots + \frac{1}{n^k} \right) - \zeta^2(k) - \zeta(2k - 1).$$

The last series, which is an Euler series of weight $2k - 1$, can be evaluated in terms of zeta values [22, Theorem 3.1].

7.98. [61] **Pearls of Quadratic Series with the Tail of $\zeta(2)$**
Prove that:

(a) *An alternating sum.*

$$\sum_{n=1}^{\infty} (-1)^n \left[n^2 \left(\zeta(2) - 1 - \frac{1}{2^2} - \cdots - \frac{1}{n^2} \right)^2 - 1 \right] = \frac{1}{2} + \frac{\zeta(2)}{2} + \frac{5}{8}\zeta(3) - \frac{\pi^2 \ln 2}{4};$$

(continued)

(b) $\displaystyle\sum_{n=1}^{\infty}\left[n\left(\zeta(2)-1-\frac{1}{2^2}-\cdots-\frac{1}{n^2}\right)^2-\frac{1}{n}\right]=\frac{3}{2}-\frac{\zeta(2)}{2}-\frac{3}{2}\zeta(3);$

(c) $\displaystyle\sum_{n=1}^{\infty}\left[n^2\left(\zeta(2)-1-\frac{1}{2^2}-\cdots-\frac{1}{n^2}\right)^2-1+\frac{1}{n}\right]=-\frac{2}{3}+\frac{\zeta(2)}{2}+\frac{\zeta(3)}{2};$

(d) *A quadratic series and Apéry's constant.*

$$\sum_{n=1}^{\infty}\left[(n^2+n)\left(\zeta(2)-1-\frac{1}{2^2}-\cdots-\frac{1}{n^2}\right)^2-1\right]=\frac{5}{6}-\zeta(3).$$

7.99. Prove that:

(a) [27] $\displaystyle\sum_{n=1}^{\infty}\frac{1}{n}\left(\zeta(2)-1-\frac{1}{2^2}-\cdots-\frac{1}{n^2}\right)^2=5\zeta(2)\zeta(3)-9\zeta(5);$

(b) $\displaystyle\sum_{n=1}^{\infty}\frac{1}{n+1}\left(\zeta(2)-1-\frac{1}{2^2}-\cdots-\frac{1}{n^2}\right)^2=\zeta(2)\zeta(3)-\frac{5}{2}\zeta(4)+\zeta(5);$

(c) $\displaystyle\sum_{n=1}^{\infty}\frac{1}{n^2}\left(\zeta(2)-1-\frac{1}{2^2}-\cdots-\frac{1}{n^2}\right)^2=\zeta^2(3)-\frac{23}{24}\zeta(6);$

(d) $\displaystyle\sum_{n=1}^{\infty}\frac{1}{(n+1)^2}\left(\zeta(2)-1-\frac{1}{2^2}-\cdots-\frac{1}{n^2}\right)^2=-\zeta^2(3)+\frac{101}{24}\zeta(6)$
$$-\frac{5}{2}\zeta(4).$$

7.100. Quadratic Series with Tail of $\zeta(3)$
Prove that:

(a) $\displaystyle\sum_{n=1}^{\infty}n\left(\zeta(3)-1-\frac{1}{2^3}-\cdots-\frac{1}{n^3}\right)^2=\frac{3}{4}\zeta(4)-3\zeta(2)\zeta(3)+5\zeta(5);$

(b) $\displaystyle\sum_{n=1}^{\infty}n^2\left(\zeta(3)-1-\frac{1}{2^3}-\cdots-\frac{1}{n^3}\right)^2=\frac{2}{3}\zeta(2)-\frac{\zeta(3)}{3}-\frac{3}{4}\zeta(4)-\frac{5}{3}\zeta(5)$
$$+\zeta(2)\zeta(3);$$

(continued)

(c) *A fabulous series.*

$$\sum_{n=1}^{\infty} \left[n^3 \left(\zeta(3) - 1 - \frac{1}{2^3} - \cdots - \frac{1}{n^3} \right)^2 - \frac{1}{4n} \right] = -\zeta(2) + \frac{\zeta(3)}{2} + \frac{3}{8}\zeta(4) + \frac{3}{16}.$$

7.101. A Quadratic Series with the Tail of $\zeta(4)$

Prove that

$$\sum_{n=1}^{\infty} n^3 \left(\zeta(4) - 1 - \frac{1}{2^4} - \cdots - \frac{1}{n^4} \right)^2$$

$$= \frac{1}{4} \left[2\zeta(3) - \zeta(4) - 10\zeta(5) - \frac{5}{3}\zeta(6) + 2\zeta^2(3) + 4\zeta(2)\zeta(3) \right].$$

7.102.

(a) Prove that

$$\sum_{n=1}^{\infty} n \left(\zeta(2) - 1 - \frac{1}{2^2} - \cdots - \frac{1}{n^2} \right) \left(\zeta(3) - 1 - \frac{1}{2^3} - \cdots - \frac{1}{n^3} \right)$$

$$= \frac{1}{2} \left(\zeta(2) + \zeta(3) - 2\zeta(4) \right).$$

(b) Prove that

$$\sum_{n=1}^{\infty} n \left(\zeta(2) - 1 - \frac{1}{2^2} - \cdots - \frac{1}{n^2} \right) \left(\zeta(4) - 1 - \frac{1}{2^4} - \cdots - \frac{1}{n^4} \right)$$

$$= \frac{3}{2}\zeta(2)\zeta(3) - \frac{15}{4}\zeta(5) + \frac{3}{8}\zeta(4) + \frac{1}{2}\zeta(3).$$

A challenge. Prove that

$$\sum_{n=1}^{\infty} n \left(\frac{1}{n^2} + \frac{1}{(n+1)^2} + \cdots \right) \left(\frac{1}{n^3} + \frac{1}{(n+1)^3} + \cdots \right) = \frac{1}{2}\zeta(2) + \frac{1}{2}\zeta(3) + \zeta(4).$$

7.103. *Open problem.* Calculate:

(a) $\displaystyle\sum_{n=1}^{\infty}(-1)^n n\left(\zeta(2)-1-\frac{1}{2^2}-\cdots-\frac{1}{n^2}\right)\left(\zeta(3)-1-\frac{1}{2^3}-\cdots-\frac{1}{n^3}\right);$

(b) $\displaystyle\sum_{n=1}^{\infty}(-1)^n n\left(\zeta(2)-1-\frac{1}{2^2}-\cdots-\frac{1}{n^2}\right)\left(\zeta(4)-1-\frac{1}{2^4}-\cdots-\frac{1}{n^4}\right);$

(c) $\displaystyle\sum_{n=1}^{\infty}(-1)^n n\left(\zeta(k)-1-\frac{1}{2^k}-\cdots-\frac{1}{n^k}\right)\left(\zeta(m)-1-\frac{1}{2^m}-\cdots-\frac{1}{n^m}\right),$
where $k, m \geq 2$.

7.104. (a) *A challenge.* Prove that

$$\sum_{n=1}^{\infty} H_n\left(\zeta(2)-1-\frac{1}{2^2}-\cdots-\frac{1}{n^2}\right)\left(\zeta(3)-1-\frac{1}{2^3}-\cdots-\frac{1}{n^3}\right)$$

$$= -2\zeta(4) - \frac{3}{2}\zeta(5) + 2\zeta(2)\zeta(3).$$

Open problems. Calculate:

(b) $\displaystyle\sum_{n=1}^{\infty} H_n\left(\zeta(2)-1-\frac{1}{2^2}-\cdots-\frac{1}{n^2}\right)\left(\zeta(4)-1-\frac{1}{2^4}-\cdots-\frac{1}{n^4}\right);$

(c) $\displaystyle\sum_{n=1}^{\infty}(-1)^n H_n\left(\zeta(2)-1-\frac{1}{2^2}-\cdots-\frac{1}{n^2}\right)\left(\zeta(3)-1-\frac{1}{2^3}-\cdots-\frac{1}{n^3}\right);$

(d) $\displaystyle\sum_{n=1}^{\infty}(-1)^n H_n\left(\zeta(2)-1-\frac{1}{2^2}-\cdots-\frac{1}{n^2}\right)\left(\zeta(4)-1-\frac{1}{2^4}-\cdots-\frac{1}{n^4}\right).$

(e) More generally, calculate in closed form the harmonic series

$$\sum_{n=1}^{\infty} H_n\left(\zeta(k)-1-\frac{1}{2^k}-\cdots-\frac{1}{n^k}\right)\left(\zeta(m)-1-\frac{1}{2^m}-\cdots-\frac{1}{n^m}\right)$$

and

$$\sum_{n=1}^{\infty}(-1)^n H_n\left(\zeta(k)-1-\frac{1}{2^k}-\cdots-\frac{1}{n^k}\right)\left(\zeta(m)-1-\frac{1}{2^m}-\cdots-\frac{1}{n^m}\right),$$

for integers $k, m > 1$.

A Remarkable Quadratic Series
We prove that

$$\sum_{n=1}^{\infty} \left(\frac{1}{n^2} + \frac{1}{(n+1)^2} + \cdots \right)^2 = 3\zeta(3). \tag{7.2}$$

Solution I. Using Abel's summation formula, with $a_n = 1$ and $b_n = \left(\frac{1}{n^2} + \frac{1}{(n+1)^2} + \cdots \right)^2$, we have that

$$\sum_{n=1}^{\infty} \left(\frac{1}{n^2} + \frac{1}{(n+1)^2} + \cdots \right)^2 = \sum_{n=1}^{\infty} \frac{1}{n} \left(\frac{1}{n^2} + \frac{2}{(n+1)^2} + \cdots \right)$$

$$= 2 \sum_{n=1}^{\infty} \frac{1}{n} \left(\frac{1}{n^2} + \frac{1}{(n+1)^2} + \cdots \right) - \zeta(3)$$

$$\stackrel{(*)}{=} 2 \sum_{n=1}^{\infty} \frac{H_n}{n^2} - \zeta(3)$$

$$= 3\zeta(3).$$

We used at step $(*)$ Abel's summation formula with $a_n = \frac{1}{n}$ and $b_n = \frac{1}{n^2} + \frac{1}{(n+1)^2} + \cdots$. ∎

Solution II. We have

$$\sum_{n=1}^{\infty} \left(\frac{1}{n^2} + \frac{1}{(n+1)^2} + \cdots \right)^2 = \zeta^2(2) + \sum_{n=2}^{\infty} \left(\frac{1}{n^2} + \frac{1}{(n+1)^2} + \cdots \right)^2$$

$$= \frac{5}{2}\zeta(4) + \sum_{m=1}^{\infty} \left(\frac{1}{(m+1)^2} + \frac{1}{(m+2)^2} + \cdots \right)^2$$

$$= \frac{5}{2}\zeta(4) + \sum_{m=1}^{\infty} \left(\zeta(2) - 1 - \frac{1}{2^2} - \cdots - \frac{1}{m^2} \right)^2$$

$$\stackrel{(**)}{=} \frac{5}{2}\zeta(4) + 3\zeta(3) - \frac{5}{2}\zeta(4)$$

$$= 3\zeta(3),$$

where the equality $(**)$ follows based on problem 3.22 in [26]. ∎

7.105.

(a) Prove that

$$\sum_{n=1}^{\infty} \frac{1}{n^2} \left(\frac{1}{n^2} + \frac{1}{(n+1)^2} + \cdots \right) = \frac{7}{4}\zeta(4).$$

(b) *A generalization.* Let $k > 1$. Prove that

$$\sum_{n=1}^{\infty} \frac{1}{n^k} \left(\frac{1}{n^k} + \frac{1}{(n+1)^k} + \cdots \right) = \frac{\zeta^2(k)+\zeta(2k)}{2}.$$

Remark 7.7. We mention that, for $k > 1$, the following equality holds:

$$\sum_{n=1}^{\infty} \frac{1 + \frac{1}{2^k} + \cdots + \frac{1}{n^k}}{n^k} = \sum_{n=1}^{\infty} \frac{1}{n^k} \left(\frac{1}{n^k} + \frac{1}{(n+1)^k} + \cdots \right) = \frac{\zeta^2(k) + \zeta(2k)}{2}.$$

7.106. Prove that

$$\sum_{n=1}^{\infty} \left(\frac{1}{n^2} + \frac{1}{(n+1)^2} + \cdots \right) \left(\frac{1}{(n+1)^2} + \frac{1}{(n+2)^2} + \cdots \right) = 3\zeta(3) - \frac{7}{4}\zeta(4).$$

7.107. Series with Tails of $\zeta(k)$

(a) Prove that

$$\sum_{n=1}^{\infty} \frac{1}{n^2} \left(\frac{1}{n^2} + \frac{1}{(n+1)^2} + \cdots \right) \left(\frac{1}{(n+1)^2} + \frac{1}{(n+2)^2} + \cdots \right) = \frac{9}{8}\zeta(6).$$

(b) *A generalization.* Let $k > 1$. Prove that

$$\sum_{n=1}^{\infty} \frac{1}{n^k} \left(\frac{1}{n^k} + \frac{1}{(n+1)^k} + \cdots \right) \left(\frac{1}{(n+1)^k} + \frac{1}{(n+2)^k} + \cdots \right) = \frac{\zeta^3(k) - \zeta(3k)}{3}.$$

A Gem with a Cubic Series

We prove that

$$\sum_{n=1}^{\infty} (2n-1) \left(\frac{1}{n^2} + \frac{1}{(n+1)^2} + \cdots \right)^3 = 9\zeta(3) - \frac{17}{4}\zeta(4).$$

Solution. We apply Abel's summation formula with $a_n = 2n - 1$ and $b_n = \left(\frac{1}{n^2} + \frac{1}{(n+1)^2} + \cdots \right)^3$. A calculation shows that

$$b_n - b_{n+1} = \frac{1}{n^2} \left[\left(\frac{1}{n^2} + \frac{1}{(n+1)^2} + \cdots \right)^2 \right.$$

$$+ \left(\frac{1}{n^2} + \frac{1}{(n+1)^2} + \cdots \right) \left(\frac{1}{(n+1)^2} + \frac{1}{(n+2)^2} + \cdots \right)$$

$$\left. + \left(\frac{1}{(n+1)^2} + \frac{1}{(n+2)^2} + \cdots \right)^2 \right]$$

and it follows

$$\sum_{n=1}^{\infty} (2n-1) \left(\frac{1}{n^2} + \frac{1}{(n+1)^2} + \cdots \right)^3$$

$$= \sum_{n=1}^{\infty} \left(\frac{1}{n^2} + \frac{1}{(n+1)^2} + \cdots \right)^2$$

$$+ \sum_{n=1}^{\infty} \left(\frac{1}{n^2} + \frac{1}{(n+1)^2} + \cdots \right) \left(\frac{1}{(n+1)^2} + \frac{1}{(n+2)^2} + \cdots \right)$$

$$+ \sum_{n=1}^{\infty} \left(\frac{1}{(n+1)^2} + \frac{1}{(n+2)^2} + \cdots \right)^2$$

$$= 3\zeta(3) + 3\zeta(3) - \frac{7}{4}\zeta(4) + 3\zeta(3) - \frac{5}{2}\zeta(4)$$

$$= 9\zeta(3) - \frac{17}{4}\zeta(4).$$

(continued)

We used in our calculations formula (7.2), Problem 7.106, and the series $\sum_{n=1}^{\infty} \left(\frac{1}{(n+1)^2} + \frac{1}{(n+2)^2} + \cdots \right)^2 = 3\zeta(3) - \frac{5}{2}\zeta(4)$, which follows from the second solution of formula (7.2). ∎

Remark 7.8. We mention that the *cubic series*

$$\sum_{n=1}^{\infty} n \left(\frac{1}{n^2} + \frac{1}{(n+1)^2} + \cdots \right)^3 = \frac{9}{2}\zeta(3) - \frac{17}{8}\zeta(4) - \frac{25}{4}\zeta(5) + \frac{9}{2}\zeta(2)\zeta(3)$$

was calculated in [27].

Subtracting the previous two cubic series one obtains the remarkable formula

$$\sum_{n=1}^{\infty} \left(\frac{1}{n^2} + \frac{1}{(n+1)^2} + \cdots \right)^3 = 9\zeta(2)\zeta(3) - \frac{25}{2}\zeta(5),$$

communicated to the second author, without proof, in 2014 by C.I. Vălean.

7.108. Prove that:

(a) $\displaystyle\sum_{n=1}^{\infty} \left[\left(\frac{1}{n^2} + \frac{1}{(n+1)^2} + \cdots \right) - \frac{1}{n} \right]^2 = \zeta(2) - \zeta(3);$

(b) $\displaystyle\sum_{n=1}^{\infty} n \left[\left(\frac{1}{n^2} + \frac{1}{(n+1)^2} + \cdots \right) - \frac{1}{n} \right]^2 = \frac{3}{2}\zeta(3) - \frac{1}{2}\zeta(2) - \frac{1}{2}.$

Part (b) of the problem is related to part (a) of Problem 2.82.

7.109. Prove that:

(a) [74] $\displaystyle\sum_{n=1}^{\infty} (2n-1) \left(\frac{1}{n^3} + \frac{1}{(n+1)^3} + \cdots \right)^2 = \frac{\pi^4}{60};$

(b) [77] $\displaystyle\sum_{n=1}^{\infty} (2n-1) \left[\left(\frac{1}{n^2} + \frac{1}{(n+1)^2} + \cdots \right)^2 - \frac{1}{n^2} \right] = 3.$

7.110.

(a) **A Quadratic Series, $\zeta(2)$ and $\zeta(3)$**
Prove that

$$\sum_{n=1}^{\infty}(3n^2 - 3n + 1)\left(\frac{1}{n^3} + \frac{1}{(n+1)^3} + \cdots\right)^2 = 2\zeta(2) - \zeta(3).$$

(b) **A Quadratic Series, $\zeta(3)$ and $\zeta(4)$**
Prove that

$$\sum_{n=1}^{\infty}(4n^3 - 6n^2 + 4n - 1)\left(\frac{1}{n^4} + \frac{1}{(n+1)^4} + \cdots\right)^2 = 2\zeta(3) - \zeta(4).$$

7.111. Two Harmonic Series with the Tail of $\zeta(3)$
Prove that:

(a) $\displaystyle\sum_{n=1}^{\infty} H_n\left(\frac{1}{n^3} + \frac{1}{(n+1)^3} + \cdots\right) = 2\zeta(3) - \zeta(2) + \frac{5}{4}\zeta(4);$

(b) $\displaystyle\sum_{n=1}^{\infty} n\left(\frac{1}{n^3} + \frac{1}{(n+1)^3} + \cdots\right)^2 = \frac{3}{4}\zeta(4) - 5\zeta(5) + 3\zeta(2)\zeta(3).$

7.112. Prove that

$$\sum_{n=1}^{\infty} \frac{H_n}{n}\left(\frac{1}{n^2} + \frac{1}{(n+1)^2} + \cdots\right) = 3\zeta(4).$$

Remark 7.9. For a similar exercise see problem 3.62 in [26].

7.113. Prove that:

(a) $\displaystyle\sum_{n=1}^{\infty} \frac{H_n}{n(n+1)}\left(\zeta(2) - 1 - \frac{1}{2^2} - \cdots - \frac{1}{n^2}\right) = \frac{\zeta(4)}{2};$

(continued)

(b) $\displaystyle\sum_{n=1}^{\infty} \frac{H_n}{n(n+1)} \left(\zeta(3) - 1 - \frac{1}{2^3} - \cdots - \frac{1}{n^3} \right) = -\frac{15}{2}\zeta(5) + 4\zeta(2)\zeta(3).$

7.114. A Challenging Alternating Zeta Series

(a) Prove that

$$\zeta(2) - 1 - \frac{1}{2^2} - \cdots - \frac{1}{n^2} = -\int_0^1 \frac{x^n}{1-x} \ln x \, dx, \quad n \geq 1.$$

(b) Prove that

$$\sum_{n=1}^{\infty} \frac{(-1)^n}{n} \left(\zeta(2) - 1 - \frac{1}{2^2} - \cdots - \frac{1}{n^2} \right) = \zeta(3) - \frac{\pi^2}{4}\ln 2. \qquad (7.3)$$

Remark 7.10. This problem was motivated by the calculation, and this is a very challenging problem, of the quadratic alternating series

$$\sum_{n=1}^{\infty} (-1)^n n \left(\frac{1}{n^2} + \frac{1}{(n+1)^2} + \cdots \right)^2$$

$$= -\frac{49}{16}\zeta(4) + \frac{5}{8}\zeta(3) - \frac{\pi^2}{4}\ln 2 + \frac{7}{4}\ln 2\, \zeta(3) + \frac{\ln^4 2}{12} + 2\mathrm{Li}_4\left(\frac{1}{2}\right) - \frac{\ln^2 2\, \zeta(2)}{2}.$$

From part (b) of the problem and the previous formula one obtains the closely related series

$$\sum_{n=1}^{\infty} (-1)^n n \left(\zeta(2) - 1 - \frac{1}{2^2} - \cdots - \frac{1}{n^2} \right)^2$$

$$= -\frac{49}{16}\zeta(4) + \frac{7}{4}\ln 2\, \zeta(3) + \frac{\ln^4 2}{12} + 2\mathrm{Li}_4\left(\frac{1}{2}\right) - \frac{\zeta(2)\ln^2 2}{2} - \frac{5}{8}\zeta(3) + \frac{\pi^2}{4}\ln 2.$$

It is worth mentioning that the non-alternating series (see also part (a) of Problem 7.63)

$$\sum_{n=1}^{\infty} \frac{1}{n} \left(\zeta(2) - 1 - \frac{1}{2^2} - \cdots - \frac{1}{n^2} \right) = \zeta(3) \qquad (7.4)$$

(continued)

appears in [26, problem 3.20, p. 142].

Adding and subtracting series (7.3) and (7.4) we obtain the following interesting formulae:

- $$\sum_{n=1}^{\infty} \frac{1}{n} \left(\zeta(2) - 1 - \frac{1}{2^2} - \cdots - \frac{1}{(2n)^2} \right) = 2\zeta(3) - \frac{\pi^2}{4} \ln 2;$$

- $$\sum_{n=1}^{\infty} \frac{1}{2n-1} \left(\zeta(2) - 1 - \frac{1}{2^2} - \cdots - \frac{1}{(2n-1)^2} \right) = \frac{\pi^2}{8} \ln 2.$$

7.115. Prove that

$$\sum_{n=1}^{\infty} (-1)^n H_n \left(\zeta(2) - 1 - \frac{1}{2^2} - \cdots - \frac{1}{n^2} \right) = \frac{7\zeta(3)}{16} - \frac{\pi^2 \ln 2}{8}.$$

A challenge. Calculate

$$\sum_{n=1}^{\infty} (-1)^n H_n^2 \left(\zeta(2) - 1 - \frac{1}{2^2} - \cdots - \frac{1}{n^2} \right).$$

7.116. A Series with the nth Harmonic Number and Tail of $\zeta(3)$

Prove that:

(a) $\zeta(3) - 1 - \dfrac{1}{2^3} - \cdots - \dfrac{1}{n^3} = \dfrac{1}{2} \displaystyle\int_0^1 \dfrac{x^n}{1-x} \ln^2 x \, dx, \quad n \geq 1;$

(b) [145] $\displaystyle\sum_{n=1}^{\infty} \dfrac{H_n}{n} \left(\zeta(3) - 1 - \dfrac{1}{2^3} - \cdots - \dfrac{1}{n^3} \right) = 2\zeta(2)\zeta(3) - \dfrac{7}{2}\zeta(5);$

(c) *Challenges.* Calculate

$$\sum_{n=1}^{\infty} (-1)^n H_n \left(\zeta(3) - 1 - \frac{1}{2^3} - \cdots - \frac{1}{n^3} \right)$$

and

$$\sum_{n=1}^{\infty} (-1)^n H_n^2 \left(\zeta(3) - 1 - \frac{1}{2^3} - \cdots - \frac{1}{n^3} \right).$$

(continued)

Remark 7.11. We mention that the problem involving the calculation of the series with the nth harmonic number and the tail of $\zeta(2)$

$$\sum_{n=1}^{\infty} \frac{H_n}{n}\left(\zeta(2) - 1 - \frac{1}{2^2} - \cdots - \frac{1}{n^2}\right) = \frac{7}{4}\zeta(4)$$

can be found in [26, problem 3.62, p. 149] and [28].

7.117. Various Series with the Tail of $\zeta(2)$

Open problem. Calculate, in terms of well-known constants, the series:

(a) $\displaystyle\sum_{n=1}^{\infty}(-1)^n H_n\left(\zeta(2) - 1 - \frac{1}{2^2} - \cdots - \frac{1}{n^2}\right)^2$;

(b) $\displaystyle\sum_{n=1}^{\infty} H_n^2\left(\zeta(2) - 1 - \frac{1}{2^2} - \cdots - \frac{1}{n^2}\right)^2$;

(c) $\displaystyle\sum_{n=1}^{\infty}(-1)^n H_n^2\left(\zeta(2) - 1 - \frac{1}{2^2} - \cdots - \frac{1}{n^2}\right)^2$.

Remark 7.12. This problem is motivated by the evaluation of the *quadratic harmonic series*

$$\sum_{n=1}^{\infty} H_n\left(\zeta(2) - 1 - \frac{1}{2^2} - \cdots - \frac{1}{n^2}\right)^2 = \frac{19}{4}\zeta(4) - 3\zeta(3),$$

which is calculated in [28].

7.118. Alternating Quadratic Zeta Series

Open problem. Let $k \in \mathbb{N}$, $k \geq 2$, and $i = 0, 1, \ldots, 2k - 3$.

(a) Calculate

$$\sum_{n=1}^{\infty}(-1)^n n^i\left(\frac{1}{n^k} + \frac{1}{(n+1)^k} + \cdots\right)^2.$$

(continued)

(b) Calculate

$$\sum_{n=1}^{\infty}(-1)^n n^i \left(\zeta(k) - 1 - \frac{1}{2^k} - \cdots - \frac{1}{n^k}\right)^2.$$

The series in part (b), with $i = 0$ and $k = 2$, was evaluated by C.I. Vălean. When $i = 1$ and $k = 2$, the series is given in Problem 7.114.

7.119. A Mosaic of Series with $\frac{\pi^2}{8} - 1 - \frac{1}{3^2} - \cdots - \frac{1}{(2n-1)^2}$

(a) Prove that

$$\sum_{n=1}^{\infty}\left(\frac{\pi^2}{8} - 1 - \frac{1}{3^2} - \cdots - \frac{1}{(2n-1)^2} - \frac{1}{4n}\right) = \frac{\ln 2}{2} - \frac{\pi^2}{16} + \frac{1}{4}.$$

A generalization. **Let** $a > b > 0$. **Then**

$$\sum_{n=1}^{\infty}\left(\frac{1}{a^2}\zeta\left(2, \frac{a-b}{a}\right) - \frac{1}{(a-b)^2} - \frac{1}{(2a-b)^2} - \cdots - \frac{1}{(na-b)^2} - \frac{1}{a^2 n}\right)$$
$$= \frac{1}{a^2}\left[1 - \gamma - \psi\left(1 - \frac{b}{a}\right) - \frac{a-b}{a}\zeta\left(2, \frac{a-b}{a}\right)\right],$$

where ψ is the Digamma function, $\psi(x) = -\frac{1}{x} - \gamma - \sum_{n=1}^{\infty}\left(\frac{1}{x+n} - \frac{1}{n}\right)$, $x \neq 0, -1, -2 \ldots$ and $\zeta(2, a) = \sum_{n=0}^{\infty}\frac{1}{(n+a)^2}$, $a \neq 0, -1, -2 \ldots$, denotes the Hurwitz zeta function.

(b) Prove that

$$\sum_{n=1}^{\infty} n\left(\frac{\pi^2}{8} - 1 - \frac{1}{3^2} - \cdots - \frac{1}{(2n-1)^2} - \frac{1}{4n}\right) = \frac{1}{8}\left(1 - \frac{\pi^2}{8}\right).$$

(c) Prove that

$$\sum_{n=1}^{\infty}(-1)^n n\left(\frac{\pi^2}{8} - 1 - \frac{1}{3^2} - \cdots - \frac{1}{(2n-1)^2} - \frac{1}{4n}\right) = \frac{1}{8} + \frac{\pi}{16} - \frac{\pi^2}{32}.$$

(continued)

(d) Prove that

$$\sum_{n=1}^{\infty} \frac{1}{2n-1} \left(\frac{\pi^2}{8} - 1 - \frac{1}{3^2} - \cdots - \frac{1}{(2n-1)^2} \right) = \frac{\pi^2 \ln 2}{8} - \frac{7\zeta(3)}{16}.$$

(e) Prove that

$$\sum_{n=1}^{\infty} \frac{1}{(2n-1)^2} \left(\frac{\pi^2}{8} - 1 - \frac{1}{3^2} - \cdots - \frac{1}{(2n-1)^2} \right) = \frac{\pi^4}{384}.$$

This also implies that

$$\sum_{n=1}^{\infty} \frac{1}{(2n-1)^2} \left(1 + \frac{1}{3^2} + \cdots + \frac{1}{(2n-1)^2} \right) = \frac{5\pi^4}{384}.$$

(f) Prove that

$$\sum_{n=1}^{\infty} \left[\frac{\pi^4}{64} - \left(1 + \frac{1}{3^2} + \cdots + \frac{1}{(2n-1)^2} \right)^2 - \frac{\pi^2}{16n} \right] = \frac{\pi^2}{16} - \frac{\pi^4}{128}.$$

(g) **A Masterpiece of Quadratic Series**
   Prove that

$$\sum_{n=1}^{\infty} \left(\frac{\pi^2}{8} - 1 - \frac{1}{3^2} - \cdots - \frac{1}{(2n-1)^2} \right)^2 = \frac{\pi^2 \ln 2}{8} - \frac{\pi^4}{128}.$$

7.120. Open Problem. Calculate:

(a) $\displaystyle\sum_{n=1}^{\infty} (-1)^n \left(\frac{\pi^2}{8} - 1 - \frac{1}{3^2} - \cdots - \frac{1}{(2n-1)^2} \right)^2$;

(b) $\displaystyle\sum_{n=1}^{\infty} (-1)^n n \left(\frac{\pi^2}{8} - 1 - \frac{1}{3^2} - \cdots - \frac{1}{(2n-1)^2} \right)^2$;

(c) $\displaystyle\sum_{n=1}^{\infty} \frac{1}{n} \left(\frac{\pi^2}{8} - 1 - \frac{1}{3^2} - \cdots - \frac{1}{(2n-1)^2} \right)^2$;

(continued)

(d) $\displaystyle\sum_{n=1}^{\infty} \frac{(-1)^n}{n} \left(\frac{\pi^2}{8} - 1 - \frac{1}{3^2} - \cdots - \frac{1}{(2n-1)^2} \right)^2$;

(e) $\displaystyle\sum_{n=1}^{\infty} H_n \left(\frac{\pi^2}{8} - 1 - \frac{1}{3^2} - \cdots - \frac{1}{(2n-1)^2} \right)^2$;

(f) $\displaystyle\sum_{n=1}^{\infty} (-1)^n H_n \left(\frac{\pi^2}{8} - 1 - \frac{1}{3^2} - \cdots - \frac{1}{(2n-1)^2} \right)^2$;

(g) $\displaystyle\sum_{n=1}^{\infty} \left(\frac{\pi^2}{8} - 1 - \frac{1}{3^2} - \cdots - \frac{1}{(2n-1)^2} \right) \left(\frac{\pi^2}{24} - \frac{1}{2^2} - \frac{1}{4^2} - \cdots - \frac{1}{(2n)^2} \right)$;

(h) *Special quadratic sums.*

Let $k \geq 2$ be an integer. Calculate, in closed form, the series

$$\sum_{n=1}^{\infty} \left(\zeta(k) \frac{2^k - 1}{2^k} - 1 - \frac{1}{3^k} - \cdots - \frac{1}{(2n-1)^k} \right)^2$$

and

$$\sum_{n=1}^{\infty} (-1)^n \left(\zeta(k) \frac{2^k - 1}{2^k} - 1 - \frac{1}{3^k} - \cdots - \frac{1}{(2n-1)^k} \right)^2 .$$

7.8   Exotic Zeta Series

7.121. Prove that:

(a) $\displaystyle\sum_{n=0}^{\infty} \left(\frac{1}{(6n+1)^2} + \frac{1}{(6n+5)^2} \right) = \frac{\pi^2}{9}$;

(b) $\displaystyle\sum_{n=0}^{\infty} \left(\frac{1}{(6n+1)^3} + \frac{1}{(6n+5)^3} \right) = \frac{91}{108} \zeta(3)$;

(c) $\displaystyle\sum_{n=0}^{\infty} \left(\frac{1}{(6n+1)^k} + \frac{1}{(6n+5)^k} \right) = \left(1 - \frac{1}{2^k} \right) \left(1 - \frac{1}{3^k} \right) \zeta(k), \quad k > 1$;

(d) $\displaystyle\sum_{n=0}^{\infty} (-1)^n \left(\frac{1}{6n+1} + \frac{1}{6n+5} \right) = \frac{\pi}{3}$;

(continued)

(e) $\displaystyle\sum_{n=0}^{\infty}(-1)^n\left(\frac{1}{(6n+1)^2}+\frac{1}{(6n+5)^2}\right)=\frac{10}{9}G,$

where G denotes *Catalan's constant*;

(f) $\displaystyle\sum_{n=0}^{\infty}(-1)^n\left(\frac{1}{(6n+1)^3}+\frac{1}{(6n+5)^3}\right)=\frac{7}{216}\pi^3;$

(g) $\displaystyle\sum_{n=0}^{\infty}(-1)^n\left(\frac{1}{(6n+1)^k}+\frac{1}{(6n+5)^k}\right)=\left(1+\frac{1}{3^k}\right)\beta(k),\ \ k>0,$

where β denotes the *Dirichlet beta* function.

7.122. Prove that:

(a) $\displaystyle\sum_{n=2}^{\infty}\left(\frac{n(n+1)}{2}-2\zeta(3)-3\zeta(4)-\cdots-n\zeta(n+1)\right)=2\zeta(3)-1;$

(b) $\displaystyle\sum_{n=2}^{\infty}(-1)^n\left(\frac{n(n+1)}{2}-2\zeta(3)-3\zeta(4)-\cdots-n\zeta(n+1)\right)=\frac{3}{8};$

(c) $\displaystyle\sum_{n=1}^{\infty}\left(\frac{3}{4}+n-\zeta(2)-\zeta(4)-\cdots-\zeta(2n)\right)=\frac{\pi^2}{12}-\frac{11}{16};$

(d) $\displaystyle\sum_{n=1}^{\infty}(-1)^n\left(\frac{3}{4}+n-\zeta(2)-\zeta(4)-\cdots-\zeta(2n)\right)=\frac{\pi}{4}\coth\pi-\frac{7}{8};$

(e) $\displaystyle\sum_{n=1}^{\infty}\left(\frac{1}{4}+n-\zeta(3)-\zeta(5)-\cdots-\zeta(2n+1)\right)=\frac{1}{16};$

(f) $\displaystyle\sum_{n=2}^{\infty}\left(\ln 2-\frac{\zeta(2)}{2^2}-\frac{\zeta(3)}{2^3}\cdots-\frac{\zeta(n)}{2^n}\right)=\frac{\pi^2}{8}-\ln 2;$

(g) $\displaystyle\sum_{n=2}^{\infty}(-1)^n\left(H_n-\gamma-\frac{\zeta(2)}{2}-\frac{\zeta(3)}{3}-\cdots-\frac{\zeta(n)}{n}\right)=1-\gamma-\frac{\ln 2}{2}.$

7.123. Prove that

$$\sum_{n=1}^{\infty}\left(\frac{1}{2}-\zeta(2)-\zeta(4)-\cdots-\zeta(2n)+\zeta(3)+\zeta(5)+\cdots+\zeta(2n+1)\right)=\frac{\pi^2}{12}-\frac{3}{4}.$$

7.9   Special Differential Equations

7.124.

(a) Let $k \in \mathbb{R}^*$. Find all differentiable functions $f : \mathbb{R} \to \mathbb{R}$ that satisfy the equation

$$f(x)f'(-x) = k, \quad \forall x \in \mathbb{R}.$$

(b) Let k be an even positive integer. Find all differentiable functions $f : \mathbb{R} \to \mathbb{R}$ that satisfy the equation

$$f(x)f'(-x) = x^k, \quad \forall x \in \mathbb{R}.$$

7.125.

(a) Find all differentiable functions $f : \mathbb{R} \to \mathbb{R}$ that satisfy the equation

$$f'(x) = f(x)f(-x), \quad \forall x \in \mathbb{R}.$$

(b) Find all differentiable functions $f : \mathbb{R} \to \mathbb{R}$ that satisfy the equation

$$f'(x) = f^2(x)f(-x), \quad \forall x \in \mathbb{R}.$$

7.126.

(a) Find all differentiable functions $f : \mathbb{R} \to \mathbb{R}$, with $f(0) = 1$, which satisfy the equation

$$f'(x) = f^2(-x)f(x), \quad \forall x \in \mathbb{R}.$$

(b) Find all differentiable functions $f : \mathbb{R} \to \mathbb{R}$, with $f(0) = 1$, which satisfy the equation

$$f'(x) = x^2 f^2(-x)f(x), \quad \forall x \in \mathbb{R}.$$

7.127. Find all differentiable functions $f : \mathbb{R} \to \mathbb{R}$ that satisfy the equation

$$f'(x) = f^3(-x)f(x), \quad \forall x \in \mathbb{R}.$$

7.128. Find all continuous functions $f : \mathbb{R} \to \mathbb{R}$ such that

$$\frac{1}{y-x} \int_x^y f(t)dt = \frac{f(x)+f(y)}{2}, \quad \forall x, y \in \mathbb{R}, \ x \neq y.$$

7.129. Find all differentiable functions $f : \mathbb{R} \to \mathbb{R}$ such that

$$\frac{f(x) - f(y)}{x - y} = \frac{f'(x) + f'(y)}{2}, \quad \forall x, y \in \mathbb{R}, \ x \neq y.$$

7.130. Find all differentiable functions $f : \mathbb{R} \to \mathbb{R}$ such that

$$\frac{f(x) - f(y)}{x - y} = f'\left(\frac{x + y}{2}\right), \quad \forall x, y \in \mathbb{R}, \ x \neq y.$$

7.131. Find all differentiable functions $f : \mathbb{R} \to \mathbb{R}$ such that

$$\frac{f(x) - f(y)}{x - y} = f'\left(\frac{x - y}{2}\right), \quad \forall x, y \in \mathbb{R}, \ x \neq y.$$

7.132. [87] Find all continuous functions $f : \mathbb{R} \to \mathbb{R}$ such that

$$\int_{-x}^{0} f(t)\mathrm{d}t + \int_{0}^{x} t f(x - t)\mathrm{d}t = x, \quad \forall x \in \mathbb{R}.$$

7.10   Inequalities

Two New Proofs of Nesbitt's Inequality
Method I. Using the $AM - GM$ inequality.
   If a, b, c are positive real numbers, then

$$\frac{a}{b + c} + \frac{b}{a + c} + \frac{c}{a + b} \geq \frac{3}{2}.$$

   A new proof of this inequality is based on a method that combines integral calculus with the $AM - GM$ inequality. The inequality is equivalent to

$$\frac{1}{b + c} + \frac{1}{a + c} + \frac{1}{a + b} \geq \frac{9}{2(a + b + c)}.$$

We have that

(continued)

$$\frac{1}{b+c} + \frac{1}{a+c} + \frac{1}{a+b} = \int_0^1 \left(x^{b+c-1} + x^{a+c-1} + x^{a+b-1} \right) dx$$

$$= \int_0^1 \frac{x^{b+c} + x^{a+c} + x^{a+b}}{x} dx$$

$$\overset{(*)}{\geq} 3 \int_0^1 \frac{\sqrt[3]{x^{2(a+b+c)}}}{x} dx$$

$$= \frac{9}{2(a+b+c)}.$$

We used at step $(*)$ the $AM - GM$ inequality. ∎

Generalization. Let $n \in \mathbb{N}$, $n \geq 2$, and let $x_1, x_2, \ldots, x_n$ be positive real numbers. The following inequality holds true:

$$\frac{x_1}{x_2+x_3 + \cdots + x_n} + \frac{x_2}{x_1+x_3 + \cdots + x_n} + \cdots + \frac{x_n}{x_1+x_2 + \cdots + x_{n-1}} \geq \frac{n}{n-1}.$$

The inequality is equivalent to

$$\frac{1}{S - x_1} + \frac{1}{S - x_2} + \cdots + \frac{1}{S - x_n} \geq \frac{n^2}{(n-1)S},$$

where $S = x_1 + x_2 + \cdots + x_n$.

We have that

$$\frac{1}{S - x_1} + \frac{1}{S - x_2} + \cdots + \frac{1}{S - x_n} = \int_0^1 \left(t^{S-x_1-1} + \cdots + t^{S-x_n-1} \right) dt$$

$$\overset{(*)}{\geq} n \int_0^1 t^{\frac{n-1}{n}S-1} dt$$

$$= \frac{n^2}{(n-1)S}.$$

We used at step $(*)$ the $AM - GM$ inequality. ∎

Method II. Using the convexity of the exponential function.

Let $S = a + b + c$. We have

(continued)

$$\frac{a}{b+c} + \frac{b}{a+c} + \frac{c}{a+b} = \int_0^\infty \left(ae^{-(b+c)x} + be^{-(a+c)x} + ce^{-(a+b)x} \right) dx$$

$$= S \int_0^\infty \left(\frac{a}{S}e^{-(b+c)x} + \frac{b}{S}e^{-(a+c)x} + \frac{c}{S}e^{-(a+b)x} \right) dx$$

$$\overset{(**)}{\geq} S \int_0^\infty e^{-\frac{a(b+c)+b(a+c)+c(a+b)}{S}x} dx$$

$$= \frac{S^2}{2(ab+bc+ac)}$$

$$\geq \frac{3}{2}.$$

We used at step (**) the convexity of the exponential function.   ∎

7.133. Prove that if $a_1, a_2, \ldots, a_n \in \mathbb{R}^*$, then

$$\sum_{i=1}^n \sum_{j=1}^n \frac{a_i a_j}{a_i^2 + a_j^2} \geq 0.$$

Remark 7.13. More generally, if $m, k \in \mathbb{N}$ and $a_1, a_2, \ldots, a_n \in \mathbb{R}^*$, then

$$\sum_{i=1}^n \sum_{j=1}^n \frac{a_i^m a_j^m}{a_i^k + a_j^k} \geq 0.$$

7.134. Prove that if $a_1, a_2, \ldots, a_n \in \mathbb{R}$, then

$$\sum_{i=1}^n \sum_{j=1}^n \frac{ij}{i+j-1}a_i a_j \geq \left(\sum_{i=1}^n a_i \right)^2.$$

7.11   Fabulous Integrals

7.135. A Fabulous Logarithmic Integral
Prove that

(continued)

$$\int_0^1 \frac{\ln(x^a + (1-x)^a)}{x}\,dx = \frac{\pi^2}{12}\left(\frac{1}{a} - a\right), \quad a > 0.$$

Remark 7.14. The following particular cases are worth mentioning:

- $a = \sqrt{2} - 1 \Rightarrow \int_0^1 \dfrac{\ln(x^{\sqrt{2}-1} + (1-x)^{\sqrt{2}-1})}{x}\,dx = \zeta(2);$

- $a = \sqrt{37} - 6 \Rightarrow \int_0^1 \dfrac{\ln(x^{\sqrt{37}-6} + (1-x)^{\sqrt{37}-6})}{x}\,dx = \pi^2;$

- $a = \sqrt{k^2 + 1} - k,\, k \in \mathbb{N},$ the integral equals $\frac{k\pi^2}{6}.$

7.136. [36] *The Stirling numbers of the first kind* $s(n, k)$ *are defined by the generating function*

$$z(z+1)(z+2)\cdots(z+n-1) = \sum_{k=0}^{n}(-1)^{n+k}s(n,k)z^k.$$

We have that $s(0, 0) = 1$, $s(n, 0) = 0$, for all $n \in \mathbb{N}$ and $s(n, n) = 1$.
Let $n \geq 1$ be an integer. Prove that

$$\int_0^1 \left(\frac{\ln(1-x)}{x}\right)^n dx = n\sum_{k=0}^{n-1}(-1)^{k-1}s(n-1, k)\zeta(n+1-k).$$

7.137. A Harmonic Integral
Prove that

$$\int_1^\infty \frac{H_{\lfloor x \rfloor}}{x^2}\,dx = \zeta(2),$$

where H_n denotes the nth harmonic number and $\lfloor x \rfloor$ is the floor of x.

7.138. [83] Let $m \geq 0$ be an integer. Prove that

(continued)

$$\int_0^1 \frac{x^m}{\left\lfloor \frac{1}{x} \right\rfloor} dx = \frac{\zeta(2) + \zeta(3) + \cdots + \zeta(m+2)}{m+1} - 1,$$

where $\lfloor x \rfloor$ denotes the floor of x.

7.139. Sophomore's Dream for Integrals of the Form $\int_1^\infty f(x)dx$

Prove that:

(a) $\displaystyle\int_1^\infty \left(e - 1 - \frac{1}{1!} - \cdots - \frac{1}{\lfloor x \rfloor !} \right) dx = 1;$

(b) $\displaystyle\int_1^\infty \{x\} \left(e - 1 - \frac{1}{1!} - \cdots - \frac{1}{\lfloor x \rfloor !} \right) dx = \frac{1}{2};$

(c) $\displaystyle\int_1^\infty x \left(e - 1 - \frac{1}{1!} - \cdots - \frac{1}{\lfloor x \rfloor !} \right) dx = \frac{e+1}{2};$

(d) $\displaystyle\int_1^\infty \frac{\{e \lfloor x \rfloor !\}}{\lfloor x \rfloor !} dx = 1;$

(e) $\displaystyle\int_1^\infty \left(\sinh 1 - 1 - \frac{1}{3!} - \cdots - \frac{1}{\lfloor 2x - 1 \rfloor !} \right) dx = \frac{1}{2e};$

(f) $\displaystyle\int_1^\infty (-1)^{\lfloor x \rfloor} \left(e - 1 - \frac{1}{1!} - \cdots - \frac{1}{\lfloor x \rfloor !} \right) dx = 1 - \cosh 1.$

We record as gems the proofs of two famous harmonic series $\sum_{n=1}^\infty \frac{H_n}{n^2} = 2\zeta(3)$ and $\sum_{n=1}^\infty \frac{H_n}{n^3} = \frac{\pi^4}{72}$. The first series is a result of Leonhard Euler and the second series is attributed to the German mathematician Christian Goldbach.

A Polylogarithm Integral and a Famous Series of Euler

We prove that

$$2\zeta(3) = \int_0^1 \frac{\zeta(2) - \text{Li}_2(x)}{1-x} dx = \sum_{n=1}^\infty \frac{H_n}{n^2}.$$

We calculate the preceding integral by two different methods. We have

(continued)

$$\frac{\zeta(2) - \mathrm{Li}_2(x)}{1 - x} = \sum_{n=1}^{\infty} \frac{1 + x + \cdots + x^{n-1}}{n^2}$$

and it follows that

$$\int_0^1 \frac{\zeta(2) - \mathrm{Li}_2(x)}{1 - x} \mathrm{d}x = \int_0^1 \sum_{n=1}^{\infty} \frac{1 + x + \cdots + x^{n-1}}{n^2} \mathrm{d}x$$

$$= \sum_{n=1}^{\infty} \frac{1}{n^2} \int_0^1 (1 + x + x^2 + \cdots + x^{n-1}) \mathrm{d}x$$

$$= \sum_{n=1}^{\infty} \frac{H_n}{n^2}.$$

We calculate the integral by parts and we have

$$\int_0^1 \frac{\zeta(2) - \mathrm{Li}_2(x)}{1 - x} \mathrm{d}x = -(\zeta(2) - \mathrm{Li}_2(x)) \ln(1 - x)\Big|_0^1 + \int_0^1 \frac{\ln^2(1 - x)}{x} \mathrm{d}x$$

$$= \int_0^1 \frac{\ln^2 x}{1 - x} \mathrm{d}x$$

$$= \int_0^1 \ln^2 x \left(\sum_{n=0}^{\infty} x^n \right) \mathrm{d}x$$

$$= \sum_{n=0}^{\infty} \int_0^1 x^n \ln^2 x \, \mathrm{d}x$$

$$= \sum_{n=0}^{\infty} \frac{2}{(n + 1)^3}$$

$$= 2\zeta(3).$$

The Euler series formula is proved.   ∎

A Polylogarithm Integral and a Famous Series of Goldbach
We prove that

(continued)

$$\frac{\pi^4}{72} = \int_0^1 \frac{\zeta(3) - \text{Li}_3(x)}{1 - x} \, dx = \sum_{n=1}^{\infty} \frac{H_n}{n^3}.$$

We calculate the preceding integral by two different methods.
We have

$$\frac{\zeta(3) - \text{Li}_3(x)}{1 - x} = \sum_{n=1}^{\infty} \frac{1 + x + \cdots + x^{n-1}}{n^3}$$

and it follows that

$$\int_0^1 \frac{\zeta(3) - \text{Li}_3(x)}{1 - x} dx = \int_0^1 \sum_{n=1}^{\infty} \frac{1 + x + \cdots + x^{n-1}}{n^3} dx$$

$$= \sum_{n=1}^{\infty} \frac{1}{n^3} \int_0^1 (1 + x + x^2 + \cdots + x^{n-1}) dx$$

$$= \sum_{n=1}^{\infty} \frac{H_n}{n^3}.$$

We calculate the integral by parts and we have

$$\int_0^1 \frac{\zeta(3) - \text{Li}_3(x)}{1 - x} dx = -(\zeta(3) - \text{Li}_3(x)) \ln(1 - x) \Big|_0^1 - \int_0^1 \text{Li}_2(x) \frac{\ln(1 - x)}{x} dx$$

$$= -\int_0^1 \text{Li}_2(x) \frac{\ln(1 - x)}{x} dx$$

$$= \frac{\text{Li}_2^2(x)}{2} \Big|_0^1$$

$$= \frac{\zeta^2(2)}{2}$$

$$= \frac{\pi^4}{72}.$$

The Goldbach series formula is proved.   ∎

7.140. Euler and Goldbach Related Series

(a) *An Euler related series.*
Prove that

$$\sum_{n=1}^{\infty} \frac{H_{2n}}{n^2} = \int_0^1 \frac{\zeta(2) - \text{Li}_2(x^2)}{1 - x} dx = \frac{11}{4}\zeta(3).$$

See also Problem 3.79.
More generally, one can prove that if $k \in \mathbb{N}$, then

$$\sum_{n=1}^{\infty} \frac{H_{kn}}{n^2} = k\int_0^1 \frac{\ln(1 - x)\ln(1 - x^k)}{x}dx$$

$$= (k + 1)\zeta(3) - \frac{k}{2}\int_0^1 \frac{\ln^2(1 + x + \cdots + x^{k-1})}{x}dx.$$

We leave, as *a challenge*, the calculation of the previous logarithmic integrals that lead to the evaluation of the series $\sum_{n=1}^{\infty} \frac{H_{kn}}{n^2}$.

(b) Prove, using the identity $O_n = H_{2n} - \frac{1}{2}H_n$, that

$$\sum_{n=1}^{\infty} \frac{O_n}{n^2} = \frac{7}{4}\zeta(3).$$

See also part (c) of Problem 3.65.

(c) *A Goldbach related series.*

A challenge. Prove that

$$\sum_{n=1}^{\infty} \frac{H_{2n}}{n^3} = -6\zeta(4) + 7\zeta(3)\ln 2 - 2\zeta(2)\ln^2 2 + \frac{1}{3}\ln^4 2 + 8\text{Li}_4\left(\frac{1}{2}\right).$$

7.141. A Series with Double Factorials and $H_{2n} - H_n - \ln 2$

(a) Prove that

$$\int_0^{\frac{\pi}{2}} \sin^{2n} x \ln \sin x \, dx = \frac{\pi}{2} \cdot \frac{(2n - 1)!!}{(2n)!!} (H_{2n} - H_n - \ln 2), \quad n \geq 1.$$

(continued)

(b) Prove that

$$\sum_{n=1}^{\infty} \frac{(2n-1)!!}{(2n)!!} \left(H_{2n} - H_n - \ln 2\right) = \ln 2 - 1.$$

For other series involving the term $H_{2n} - H_n - \ln 2$, see also Problems 7.69 and 7.70.

7.142. A Series with Double Factorials and $H_{2n+1} - H_n - \ln 2$

(a) Prove that

$$\int_0^{\frac{\pi}{2}} \sin^{2n+1} x \ln \sin x \, \mathrm{d}x = \frac{(2n)!!}{(2n+1)!!} \left(\ln 2 + H_n - H_{2n+1}\right), \quad n \geq 0.$$

(b) Prove that

$$\sum_{n=0}^{\infty} \frac{(2n)!!}{(2n+1)!!} \left(H_{2n+1} - H_n - \ln 2\right) = \ln 2,$$

with the convention $H_0 = 0$.

7.143. [O. Furdui, Problem 4, SEEMOUS 2018, Iaşi, Romania]

(a) Let $f : \mathbb{R} \to \mathbb{R}$ be a polynomial function. Prove that

$$\int_0^{\infty} \mathrm{e}^{-x} f(x) \mathrm{d}x = f(0) + f'(0) + f''(0) + \cdots .$$

(b) Let f be a function that has a Maclaurin series expansion with radius of convergence $R = \infty$. Prove that if $\sum_{n=0}^{\infty} f^{(n)}(0)$ converges absolutely, then $\int_0^{\infty} \mathrm{e}^{-x} f(x) \mathrm{d}x$ converges and

$$\sum_{n=0}^{\infty} f^{(n)}(0) = \int_0^{\infty} \mathrm{e}^{-x} f(x) \mathrm{d}x.$$

7.144. [O. Furdui, 23 May 2018]

(a) Let f be a function that has a Maclaurin series expansion with radius of convergence $R = \infty$. Prove that if $\sum\limits_{n=0}^{\infty} f^{(2n)}(0)$ converges absolutely, then $\int_0^{\infty} e^{-x}(f(x) + f(-x))\mathrm{d}x$ converges and

$$\sum_{n=0}^{\infty} f^{(2n)}(0) = \frac{1}{2} \int_0^{\infty} e^{-x}(f(x) + f(-x))\mathrm{d}x.$$

(b) Let f be a function that has a Maclaurin series expansion with radius of convergence $R = \infty$. Prove that if $\sum\limits_{n=1}^{\infty} f^{(2n-1)}(0)$ converges absolutely, then $\int_0^{\infty} e^{-x}(f(x) - f(-x))\mathrm{d}x$ converges and

$$\sum_{n=1}^{\infty} f^{(2n-1)}(0) = \frac{1}{2} \int_0^{\infty} e^{-x}(f(x) - f(-x))\mathrm{d}x.$$

An Artistry of Quadratic Series: Two New Proofs of Sandham–Yeung Series

8

> *Truth is ever to be found in simplicity, and not in the multiplicity and confusion of things.*
>
> Sir Isaac Newton (1642–1727)

In this chapter we give two new proofs of the following remarkable series formula:

$$\sum_{n=1}^{\infty} \frac{H_n^2}{n^2} = \frac{17}{4}\zeta(4), \tag{8.1}$$

where $H_n = 1 + \frac{1}{2} + \cdots + \frac{1}{n}$ denotes the nth harmonic number.

This formula has an interesting history. It was the first quadratic series introduced in the literature by H.F. Sandham, in 1948, as a problem in the American Mathematical Monthly [125]. Apparently, the series went unnoticed. D. Castellanos recorded it in his survey article [11, p. 86], attributed it rightly to Sandham, but with a wrong entry in the bibliography. P.J. De Doelder [14] evaluated the associated series $\sum_{n=1}^{\infty} \frac{H_n^2}{(n+1)^2} = \frac{11\pi^4}{360}$ without any reference to Sandham's series. In April 1993, the series was discovered numerically by Enrico Au-Yeung, an undergraduate student in the Faculty of Mathematics in Waterloo, and proved rigorously by D. Borwein and J. Borwein in [8], who used Parseval's theorem to prove it. Formula (8.1) was rediscovered by P. Freitas as Proposition A.1 in the appendix section of [23]. Freitas proved it by calculating a double integral involving a logarithmic function. This formula was revived and brought into light by C.I. Vălean and O. Furdui [146], who proved it by calculating a special integral involving a quadratic logarithmic function. The series has also appeared as a problem in [26, Problem 3.70, p. 150] and [122, Problem 2.6.1, p. 110]. It is clear that this remarkable quadratic harmonic series has attracted lots of attention lately and has become a classic in the theory of nonlinear harmonic series.

© The Author(s), under exclusive license to Springer Nature Switzerland AG 2021 245
A. Sîntămărian, O. Furdui, *Sharpening Mathematical Analysis Skills*, Problem Books in Mathematics, https://doi.org/10.1007/978-3-030-77139-3_8

8.1   The First Proof

The first proof of formula (8.1) is based on calculating the series

$$\sum_{n=1}^{\infty} \frac{1}{n}\left(2\zeta(3) - \frac{H_1}{1^2} - \frac{H_2}{2^2} - \cdots - \frac{H_n}{n^2}\right) = \frac{\pi^4}{30}$$

in two different ways [60].

We have, since $\sum_{n=1}^{\infty} \frac{H_n}{n^2} = 2\zeta(3)$ (see part (b) of Problem 7.61), that

$$\sum_{n=1}^{\infty} \frac{1}{n}\left(2\zeta(3) - \frac{H_1}{1^2} - \frac{H_2}{2^2} - \cdots - \frac{H_n}{n^2}\right) = \sum_{n=1}^{\infty}\sum_{m=1}^{\infty} \frac{H_{n+m}}{n(n+m)^2}.$$

It follows, based on symmetry reasons, that

$$\sum_{n=1}^{\infty}\sum_{m=1}^{\infty} \frac{H_{n+m}}{n(n+m)^2} = \sum_{n=1}^{\infty}\sum_{m=1}^{\infty} \frac{H_{n+m}}{m(n+m)^2},$$

and this implies

$$\sum_{n=1}^{\infty}\sum_{m=1}^{\infty} \frac{H_{n+m}}{n(n+m)^2} = \frac{1}{2}\sum_{n=1}^{\infty}\sum_{m=1}^{\infty}\left(\frac{H_{n+m}}{m(n+m)^2} + \frac{H_{n+m}}{n(n+m)^2}\right) = \frac{1}{2}\sum_{n=1}^{\infty}\sum_{m=1}^{\infty} \frac{H_{n+m}}{nm(n+m)}.$$

Therefore,

$$\sum_{n=1}^{\infty} \frac{1}{n}\left(2\zeta(3) - \frac{H_1}{1^2} - \frac{H_2}{2^2} - \cdots - \frac{H_n}{n^2}\right) = \frac{1}{2}\sum_{n=1}^{\infty}\sum_{m=1}^{\infty} \frac{H_{n+m}}{nm(n+m)}. \qquad (8.2)$$

Using (8.2) and the identity $\sum_{k=1}^{\infty} \frac{1}{k(k+n)} = \frac{H_n}{n}$ (see part (a) of Problem 7.61), we have

$$\sum_{n=1}^{\infty} \frac{1}{n}\left(2\zeta(3) - \frac{H_1}{1^2} - \frac{H_2}{2^2} - \cdots - \frac{H_n}{n^2}\right) = \frac{1}{2}\sum_{n=1}^{\infty}\sum_{m=1}^{\infty}\sum_{k=1}^{\infty} \frac{1}{nmk(k+n+m)}$$

$$= \frac{1}{2}\sum_{n=1}^{\infty}\sum_{m=1}^{\infty}\sum_{k=1}^{\infty} \frac{1}{nmk}\int_0^1 x^{n+m+k-1}dx$$

$$= \frac{1}{2}\int_0^1 \left(\sum_{k=1}^{\infty}\frac{x^{k-1}}{k}\sum_{n=1}^{\infty}\frac{x^n}{n}\sum_{m=1}^{\infty}\frac{x^m}{m}\right)dx$$

$$= -\frac{1}{2}\int_0^1 \frac{\ln^3(1-x)}{x}dx$$

$$= -\frac{1}{2}\int_0^1 \frac{\ln^3 x}{1-x}dx$$

$$= -\frac{1}{2}\int_0^1 \ln^3 x \left(\sum_{i=0}^{\infty}x^i\right)dx$$

$$= -\frac{1}{2}\sum_{i=0}^{\infty}\int_0^1 x^i \ln^3 x\, dx$$

$$= 3\sum_{i=0}^{\infty}\frac{1}{(i+1)^4}$$

$$= 3\zeta(4).$$

Now we calculate the series by using Abel's summation formula with

$$a_n = \frac{1}{n} \quad \text{and} \quad b_n = 2\zeta(3) - \frac{H_1}{1^2} - \frac{H_2}{2^2} - \cdots - \frac{H_n}{n^2},$$

and we have that

$$\sum_{n=1}^{\infty} \frac{1}{n} \left(2\zeta(3) - \frac{H_1}{1^2} - \frac{H_2}{2^2} - \cdots - \frac{H_n}{n^2} \right) = \lim_{n\to\infty} H_n \left(2\zeta(3) - \frac{H_1}{1^2} - \cdots - \frac{H_{n+1}}{(n+1)^2} \right)$$

$$+ \sum_{n=1}^{\infty} H_n \frac{H_{n+1}}{(n+1)^2}$$

$$= \sum_{n=1}^{\infty} \frac{\left(H_{n+1} - \frac{1}{n+1} \right) H_{n+1}}{(n+1)^2}$$

$$= \sum_{n=1}^{\infty} \left(\frac{H_{n+1}}{n+1} \right)^2 - \sum_{n=1}^{\infty} \frac{H_{n+1}}{(n+1)^3}$$

$$= \sum_{n=1}^{\infty} \left(\frac{H_n}{n} \right)^2 - \sum_{n=1}^{\infty} \frac{H_n}{n^3}$$

$$= \sum_{n=1}^{\infty} \left(\frac{H_n}{n} \right)^2 - \frac{\pi^4}{72},$$

since

$$\lim_{n\to\infty} H_n \left(2\zeta(3) - \frac{H_1}{1^2} - \frac{H_2}{2^2} - \cdots - \frac{H_{n+1}}{(n+1)^2} \right) = 0$$

and $\sum_{n=1}^{\infty} \frac{H_n}{n^3} = \frac{\pi^4}{72}$ (see **A Famous Series of Goldbach Revisited**, after Problem 7.63).

It follows that

$$\sum_{n=1}^{\infty} \frac{H_n^2}{n^2} = 3\zeta(4) + \frac{5}{4}\zeta(4) = \frac{17}{4}\zeta(4).$$

8.2   The Second Proof

The second proof of formula (8.1) is based on calculating the series

$$\sum_{n=1}^{\infty} \frac{H_n}{n} \left(\frac{1}{n^2} + \frac{1}{(n+1)^2} + \cdots \right) = 3\zeta(4) \tag{8.3}$$

in two different ways.

One method for calculating the series (8.3) is given in the solution of Problem 7.112. Before we calculate the series (8.3) by another method we record a formula we need in our analysis.

Let $n \geq 1$ be an integer. The following formula holds true:

$$2 \sum_{k=1}^{n} \frac{H_k}{k} = H_n^2 + 1 + \frac{1}{2^2} + \cdots + \frac{1}{n^2}. \tag{8.4}$$

Formula (8.4) can be proved by mathematical induction or by an application of Abel's summation formula with $a_k = \frac{1}{k}$ and $b_k = H_k$.

Now we calculate the series (8.3) by Abel's summation formula with

$$a_n = \frac{H_n}{n} \quad \text{and} \quad b_n = \frac{1}{n^2} + \frac{1}{(n+1)^2} + \cdots$$

and we have that

$$\sum_{n=1}^{\infty} \frac{H_n}{n} \left(\frac{1}{n^2} + \frac{1}{(n+1)^2} + \cdots \right)$$

$$= \lim_{n \to \infty} \left(\frac{H_1}{1} + \frac{H_2}{2} + \cdots + \frac{H_n}{n} \right) \left(\frac{1}{(n+1)^2} + \frac{1}{(n+2)^2} + \cdots \right)$$

$$+ \sum_{n=1}^{\infty} \left(\frac{H_1}{1} + \frac{H_2}{2} + \cdots + \frac{H_n}{n} \right) \frac{1}{n^2}$$

$$= \sum_{n=1}^{\infty} \left(\frac{H_1}{1} + \frac{H_2}{2} + \cdots + \frac{H_n}{n} \right) \frac{1}{n^2}$$

$$\overset{(8.4)}{=} \frac{1}{2} \sum_{n=1}^{\infty} \frac{H_n^2}{n^2} + \frac{1}{2} \sum_{n=1}^{\infty} \frac{1 + \frac{1}{2^2} + \cdots + \frac{1}{n^2}}{n^2}$$

$$= \frac{1}{2} \sum_{n=1}^{\infty} \frac{H_n^2}{n^2} + \frac{7}{8} \zeta(4),$$

since (see Remark 7.7 with $k = 2$, after Problem 7.105)

$$\sum_{n=1}^{\infty} \frac{1 + \frac{1}{2^2} + \cdots + \frac{1}{n^2}}{n^2} = \frac{7}{4} \zeta(4).$$

It follows that $3\zeta(4) = \frac{1}{2} \sum_{n=1}^{\infty} \frac{H_n^2}{n^2} + \frac{7}{8}\zeta(4) \Rightarrow \sum_{n=1}^{\infty} \frac{H_n^2}{n^2} = \frac{17}{4}\zeta(4).$

Part II

Solutions

Sequences of Real Numbers

9

9.1   Limits of Sequences

1.1. Let $x_n = \frac{n!e^n}{n^{n+\frac{1}{2}}}$, $n \geq 1$. We have

$$\frac{x_{n+1}}{x_n} = \frac{e}{\left(1 + \frac{1}{n}\right)^{n+\frac{1}{2}}}, \quad n \geq 1.$$

One can prove that the function $f(x) = \left(1 + \frac{1}{x}\right)^{x+\frac{1}{2}}$ is strictly decreasing on $[1, \infty)$. Since $\lim\limits_{x \to \infty} f(x) = e$, we get that $f(x) > e$, for all $x \in [1, \infty)$. Therefore, $\frac{x_{n+1}}{x_n} < 1$, for all $n \geq 1$, and this implies that the sequence $(x_n)_{n \geq 1}$ is strictly decreasing.

1.2. Let $x_n = \sum\limits_{k=1}^{n} \frac{9k^2 + 12k + 5}{(3k+2)!}$, $n \geq 1$. We have

$$x_n = \sum_{k=1}^{n} \left[\frac{1}{(3k)!} + \frac{1}{(3k+1)!} + \frac{1}{(3k+2)!} \right] = \sum_{k=0}^{3n+2} \frac{1}{k!} - 1 - \frac{1}{1!} - \frac{1}{2!}, \quad n \geq 1.$$

Therefore, $\lim\limits_{n \to \infty} x_n = e - \frac{5}{2}$.

1.3. We have

$$x_n = \sum_{k=1}^{n} \frac{1}{k(k+1)(k+1)!} = 3 - \sum_{k=0}^{n+1} \frac{1}{k!} - \frac{1}{(n+1)(n+1)!}, \quad n \geq 1.$$

Therefore, $\lim\limits_{n \to \infty} x_n = 3 - e$.

© The Author(s), under exclusive license to Springer Nature Switzerland AG 2021   253
A. Sîntămărian, O. Furdui, *Sharpening Mathematical Analysis Skills*, Problem Books in Mathematics, https://doi.org/10.1007/978-3-030-77139-3_9

1.4. We have $x_n = \frac{1}{2}\left[\frac{3}{2} - \frac{2n+3}{(n+1)(n+2)}\right], n \geq 1$. It follows that

$$\lim_{n\to\infty}\left(2x_n - \frac{1}{2}\right)^n = \lim_{n\to\infty}\left[1 - \frac{2n+3}{(n+1)(n+2)}\right]^n = e^{-2}.$$

1.5. We have

$$\lim_{n\to\infty}\frac{\sqrt[k]{n^k+3} - \sqrt[k]{n^k+1}}{\sqrt[m]{n^m+2} - \sqrt[m]{n^m-5}} = \begin{cases} \infty & \text{if } k < m \\ \frac{2}{7} & \text{if } k = m \\ 0 & \text{if } k > m. \end{cases}$$

1.6. $a = \frac{3}{2}, b = \frac{9}{4}$, and $c = 0$.

1.7.

(a) The limit equals $-\frac{3}{2}$.

(b) Let $S_n = 1 + 2 + \cdots + n$, $n \geq 1$. We have $\sum_{k=1}^{2n}(-1)^k S_k = 2S_n, n \geq 1$. Thus,

$$\lim_{n\to\infty}\frac{\sum_{k=1}^{2n}(-1)^k S_k}{\sum_{k=1}^{2n+1}(-1)^{k-1}S_k} = \lim_{n\to\infty}\frac{2S_n}{-2S_n + S_{2n+1}}$$

$$= \lim_{n\to\infty}\frac{n(n+1)}{-n(n+1) + (2n+1)(n+1)} = \lim_{n\to\infty}\frac{n}{n+1} = 1.$$

(c) The limit equals 2.

(d) The limit equals 2.

1.8.

(a) Prove (see the solution of Problem 7.47) that $\left\{(2+\sqrt{3})^n\right\} = 1 - (2-\sqrt{3})^n$. The other parts of the problem can be solved similarly.

1.9. The limit equals 1. Observe that $\left(\sqrt[n]{n}\right)^{-1} \leq n^{\frac{\sin n}{n}} \leq \sqrt[n]{n}, \forall n \geq 1$.

1.10.

(a) Let $x_n = \frac{1}{(n+1)^2} + \frac{1}{(n+2)^2} + \cdots + \frac{1}{(2n)^2}, n \geq 1$. We have

$$\frac{n}{(2n)^2} \leq x_n \leq \frac{n}{(n+1)^2}, \quad n \geq 1.$$

It follows, based on the Squeeze Theorem, that $\lim\limits_{n\to\infty} x_n = 0$.

(b) Let $x_n = \dfrac{1}{\sqrt{n^2+1}} + \dfrac{1}{\sqrt{n^2+2}} + \cdots + \dfrac{1}{\sqrt{n^2+n}}, n \geq 1$. We have

$$\frac{n}{\sqrt{n^2 + n}} \leq x_n \leq \frac{n}{\sqrt{n^2 + 1}}, \quad n \geq 1.$$

It follows, based on the Squeeze Theorem, that $\lim\limits_{n\to\infty} x_n = 1$.

(c) Let $x_n = \sum\limits_{k=1}^{n} \dfrac{5k^3+3k^2+2k+1}{n^4+k+7}, n \geq 1$. We have

$$\sum_{k=1}^{n} \frac{5k^3 + 3k^2 + 2k + 1}{n^4 + n + 7} \leq x_n \leq \sum_{k=1}^{n} \frac{5k^3 + 3k^2 + 2k + 1}{n^4 + 8}, \quad n \geq 1.$$

It follows, based on the Squeeze Theorem, that $\lim\limits_{n\to\infty} x_n = \frac{5}{4}$.

(d) We mention that this limit is a particular case of a problem proposed by L. Pârşan in Gazeta Matematică [114].

Let $x_n = \dfrac{(2n + 1)(2n + 3) \cdots (4n + 1)}{(2n)(2n + 2) \cdots (4n)}, n \geq 1$. Since $n(n + 2) < (n + 1)^2$, we have

$$\frac{(4n + 1)^2}{(2n)(4n)} = \frac{(2n + 1)^2(2n + 3)^2 \cdots (4n + 1)^2}{(2n)(2n + 1)^2(2n + 3)^2 \cdots (4n - 1)^2(4n)} < x_n^2$$

$$< \frac{(2n + 1)(2n + 2)^2(2n + 4)^2 \cdots (4n)^2(4n + 1)}{(2n)^2(2n + 2)^2 \cdots (4n)^2}$$

$$= \frac{(2n + 1)(4n + 1)}{(2n)^2}, \quad n \geq 1.$$

It follows, based on the Squeeze Theorem, that $\lim\limits_{n\to\infty} x_n^2 = 2$, so $\lim\limits_{n\to\infty} x_n = \sqrt{2}$.
We notice that if we take $p = q = 2$ in Problem 1.16 or $k = 0$, $a = r = 1$, $p = q = 2$, and $s = 1$ in part (i) of Problem 7.19, then we get the limit in (d).

(e) Let $x_n = \dfrac{(n + 2)(n + 5) \cdots (4n - 1)}{(n + 1)(n + 4) \cdots (4n - 2)}, n \geq 1$. Since $n(n + 3)^2 < (n + 2)^3$ and $n^2(n + 3) < (n + 1)^3$, we have that

$$\frac{(4n - 1)^3}{(n + 1)(4n - 2)^2} = \frac{(n + 2)^3(n + 5)^3 \cdots (4n - 4)^3(4n - 1)^3}{(n + 1)(n + 2)^3(n + 5)^3 \cdots (4n - 4)^3(4n - 2)^2} < x_n^3$$

$$< \frac{(n + 2)^2(n + 4)^3(n + 7)^3 \cdots (4n - 2)^3(4n - 1)}{(n + 1)^3(n + 4)^3 \cdots (4n - 2)^3}$$

$$= \frac{(n + 2)^2(4n - 1)}{(n + 1)^3}, \quad n \geq 1.$$

It follows, based on the Squeeze Theorem, that $\lim_{n\to\infty} x_n^3 = 4$. Thus, $\lim_{n\to\infty} x_n = \sqrt[3]{4}$.

We remark that if we take $k = 1$, $a = r = 1$, $p = 3$, and $q = s = 1$ in part (i) of Problem 7.19, then we get the limit in (e).

(f) [128, problem 63, p. 20] We have

$$
\begin{aligned}
\frac{C_{4n}^{2n}}{4^n C_{2n}^n} &= \frac{1}{4^n} \cdot \frac{(4n)!}{[(2n)!]^2} \cdot \frac{(n!)^2}{(2n)!} \\
&= \frac{1}{4^n} \cdot \frac{(2n+1)(2n+2)\cdots(4n)}{[(n+1)(n+2)\cdots(2n)]^2} \\
&= \frac{(2n+1)(2n+2)\cdots(4n)}{[(2n+2)(2n+4)\cdots(4n)]^2} \\
&= \frac{(2n+1)(2n+3)\cdots(4n-1)}{(2n+2)(2n+4)\cdots(4n)}, \quad n \geq 1.
\end{aligned}
$$

Using part (d) of the problem, we get that $\lim_{n\to\infty} \dfrac{C_{4n}^{2n}}{4^n C_{2n}^n} = \dfrac{\sqrt{2}}{2}$.

(g) [128, problem 64, p. 20] We have, for $n \geq 1$, that

$$
\begin{aligned}
\frac{3^{3n}(C_{2n}^n)^2}{C_{3n}^n C_{6n}^{3n}} &= 3^{3n} \cdot \frac{[(2n)!]^2}{(n!)^4} \cdot \frac{n!(2n)!}{(3n)!} \cdot \frac{[(3n)!]^2}{(6n)!} \\
&= 3^{3n} \cdot \frac{[(n+1)(n+2)\cdots(2n)]^3}{(3n+1)(3n+2)\cdots(6n)} \\
&= \frac{[(3n+3)(3n+6)\cdots(6n)]^3}{(3n+1)(3n+2)\cdots(6n)} \\
&= \frac{(3n+3)(3n+6)\cdots(6n)}{(3n+1)(3n+4)\cdots(6n-2)} \cdot \frac{(3n+3)(3n+6)\cdots(6n)}{(3n+2)(3n+5)\cdots(6n-1)}.
\end{aligned}
$$

Using the idea in the solution of part (e), we get that

$$
\lim_{n\to\infty} \frac{(3n+3)(3n+6)\cdots(6n)}{(3n+1)(3n+4)\cdots(6n-2)} = \sqrt[3]{4}
$$

$$
\lim_{n\to\infty} \frac{(3n+3)(3n+6)\cdots(6n)}{(3n+2)(3n+5)\cdots(6n-1)} = \sqrt[3]{2}.
$$

Therefore,

$$
\lim_{n\to\infty} \frac{3^{3n}(C_{2n}^n)^2}{C_{3n}^n C_{6n}^{3n}} = \sqrt[3]{4} \cdot \sqrt[3]{2} = 2.
$$

(h) The limit equals 0. We have

$$\frac{1}{2^n} \sum_{k=0}^{n} \frac{C_n^k}{\sqrt{k+2}} = \frac{1}{2^n} \sum_{k=0}^{n} C_n^k \frac{\sqrt{k+2}}{k+2}$$

$$= \frac{1}{2^n} \sum_{k=0}^{n} \frac{(n+1)!}{(k+1)!(n-k)!} \cdot \frac{\sqrt{k+2}}{n+1} \cdot \frac{k+1}{k+2}$$

$$= \frac{1}{2^n} \sum_{k=0}^{n} C_{n+1}^{k+1} \frac{\sqrt{k+2}}{n+1} \cdot \frac{k+1}{k+2}$$

$$< \frac{\sqrt{n+2}}{n+1} \cdot \frac{1}{2^n} \sum_{k=0}^{n} C_{n+1}^{k+1}$$

$$= \frac{\sqrt{n+2}}{n+1} \cdot \frac{2^{n+1}-1}{2^n}.$$

Remark. More generally, if $\alpha \geq 0$, then

$$\lim_{n\to\infty} \frac{1}{2^n} \sum_{k=1}^{n} \frac{C_n^k}{\sqrt{k+\alpha}} = 0.$$

The case $\alpha = 0$ is a problem given at the entrance examination test of Technical University of Cluj-Napoca on July 2012.

1.11. For $\epsilon > 0$ there exists $n_0 \in \mathbb{N}$ such that $l-\epsilon < 2x_n+x_{2n} < l+\epsilon, \forall n \geq n_0$. Let $k \geq 1$ be an integer. Replacing n by $2n, 2^2n, \ldots, 2^{k-1}n$ in the previous inequalities, we obtain that

$$\sum_{i=0}^{k-1} \left(l - (-1)^i \epsilon\right) \left(-\frac{1}{2}\right)^i < 2x_n + \left(-\frac{1}{2}\right)^{k-1} x_{2^k n} < \sum_{i=0}^{k-1} \left(l + (-1)^i \epsilon\right) \left(-\frac{1}{2}\right)^i,$$

$\forall n \geq n_0$. This implies that

$$\frac{2l}{3} \left[1 - \left(-\frac{1}{2}\right)^k\right] - 2\epsilon \left(1 - \frac{1}{2^k}\right) < 2x_n + \left(-\frac{1}{2}\right)^{k-1} x_{2^k n}$$

$$< \frac{2l}{3} \left[1 - \left(-\frac{1}{2}\right)^k\right] + 2\epsilon \left(1 - \frac{1}{2^k}\right), \quad \forall n \geq n_0.$$

Passing to the limit as $k \to \infty$ in the previous inequalities, we have that $\frac{2l}{3} - 2\epsilon \le 2x_n \le \frac{2l}{3} + 2\epsilon$, $\forall n \ge n_0$. This implies that $\frac{l}{3} - \epsilon \le x_n \le \frac{l}{3} + \epsilon$, $\forall n \ge n_0 \Rightarrow \lim_{n \to \infty} x_n = \frac{l}{3}$.

1.12.

(a) Let $I_n = \int_0^1 \sqrt[n]{x^n + (1-x)^n}\, dx$. We have

$$I_n < \int_0^{\frac{1}{2}} \sqrt[n]{(1-x)^n + (1-x)^n}\, dx + \int_{\frac{1}{2}}^1 \sqrt[n]{x^n + x^n}\, dx = \frac{3}{4}\sqrt[n]{2}$$

and

$$I_n > \int_0^{\frac{1}{2}} (1-x)dx + \int_{\frac{1}{2}}^1 x\, dx = \frac{3}{4}.$$

1.13. For part (a) see the solution of Problem 1.12.
1.14. The limit equals $\frac{3}{2}$.
1.15.

(a) The limit equals $\sqrt{2}$. Observe that $x_n = k$, for $1 + 2 + \cdots + (k-1) + 1 \le n \le 1 + 2 + \cdots + k \Rightarrow x_n = k$, for $\frac{(k-1)k}{2} + 1 \le n \le \frac{k(k+1)}{2}$. If $n \ge 1$, then there exists a unique $k \ge 1$ such that $1 + 2 + \cdots + (k-1) + 1 \le n \le 1 + 2 + \cdots + k$ or $1 + 2 + \cdots + (k-1) < n \le 1 + 2 + \cdots + k$. It follows that $\frac{x_n}{\sqrt{n}} = \frac{k}{\sqrt{n}}$. We have

$$\frac{(k-1)k}{2} < n \le \frac{k(k+1)}{2} \quad \Rightarrow \quad \sqrt{\frac{2}{k(k+1)}} \le \frac{1}{\sqrt{n}} < \sqrt{\frac{2}{(k-1)k}}$$

and it follows that

$$k\sqrt{\frac{2}{k(k+1)}} \le \frac{k}{\sqrt{n}} < k\sqrt{\frac{2}{(k-1)k}}.$$

Thus, $\lim_{n \to \infty} \frac{x_n}{\sqrt{n}} = \lim_{n \to \infty} \frac{k}{\sqrt{n}} = \sqrt{2}$.

(b) Similarly, one can prove that the limit equals $\sqrt[3]{3}$.

1.16. *Solution I.* The first solution, based on inequalities, is due to the problem author, Alina Sîntămărian.

(continued)

One can prove that the following inequalities hold:

$$n(n + p)^{p-1} < (n + p - 1)^p$$

and

$$n^{p-1}(n + p) < (n + 1)^p,$$

for all $n \in \mathbb{N}$ and $p \in \mathbb{N}$, $p \geq 2$.
Let

$$x_n = \frac{(qn + 1)(qn + p + 1) \cdots (qn + (n - 1)p + 1)}{qn(qn + p) \cdots (qn + (n - 1)p)}, \quad n \geq 1.$$

We have

$$
\begin{aligned}
x_n^p &= \frac{(qn+1)^{p-1}\left(\prod_{k=1}^{n-1}(qn+(k-1)p+1)(qn+kp+1)^{p-1}\right)(qn+(n-1)p+1)}{(qn)^p(qn+p)^p \cdots (qn+(n-1)p)^p} \\
&< \frac{(qn+1)^{p-1}(qn+p)^p \cdots (qn+(n-1)p)^p(qn+(n-1)p+1)}{(qn)^p(qn+p)^p \cdots (qn+(n-1)p)^p} \\
&= \frac{(qn+1)^{p-1}(qn+(n-1)p+1)}{(qn)^p},
\end{aligned}
$$

for all $n \in \mathbb{N}$, $n \geq 2$.
Also,

$$
\begin{aligned}
x_n^p &= \frac{(qn+1)^p(qn+p+1)^p \cdots (qn+(n-1)p+1)^p}{(qn)\left(\prod_{k=1}^{n-1}(qn+(k-1)p)^{p-1}(qn+kp)\right)(qn+(n-1)p)^{p-1}} \\
&> \frac{(qn+1)^p(qn+p+1)^p \cdots (qn+(n-2)p+1)^p(qn+(n-1)p+1)^p}{(qn)(qn+1)^p(qn+p+1)^p \cdots (qn+(n-2)p+1)^p(qn+(n-1)p)^{p-1}} \\
&= \frac{(qn+(n-1)p+1)^p}{(qn)(qn+(n-1)p)^{p-1}},
\end{aligned}
$$

for all $n \in \mathbb{N}$, $n \geq 2$.
Therefore, we write

(continued)

$$\frac{(qn + (n-1)p + 1)^p}{(qn)(qn + (n-1)p)^{p-1}} < x_n^p < \frac{(qn + 1)^{p-1}(qn + (n-1)p + 1)}{(qn)^p},$$

for all $n \in \mathbb{N}, n \geq 2$.

It follows that $\lim\limits_{n \to \infty} x_n^p = \frac{p+q}{q}$, since

$$\lim_{n \to \infty} \frac{(qn + (n-1)p + 1)^p}{(qn)(qn + (n-1)p)^{p-1}} = \frac{p+q}{q}$$

and

$$\lim_{n \to \infty} \frac{(qn + 1)^{p-1}(qn + (n-1)p + 1)}{(qn)^p} = \frac{p+q}{q}.$$

Thus,

$$\lim_{n \to \infty} \left(\frac{qn + 1}{qn} \cdot \frac{qn + p + 1}{qn + p} \cdots \frac{qn + np + 1}{qn + np} \right) = \sqrt[p]{\frac{p+q}{q}}.$$

Solution II. The second solution, based on calculus, is due to Dorian Popa. This solution holds for $p, q > 0$.

Let $x_n = \dfrac{qn + 1}{qn} \cdot \dfrac{qn + p + 1}{qn + p} \cdots \dfrac{qn + np + 1}{qn + np}, n \geq 1$.

We have

$$\ln x_n = \sum_{k=0}^{n} \ln \frac{qn + kp + 1}{qn + kp} = \sum_{k=0}^{n} \frac{1}{qn + kp} \cdot \frac{\ln\left(1 + \frac{1}{qn+kp}\right)}{\frac{1}{qn+kp}},$$

for all $n \geq 1$.

Let $f : (0, +\infty) \to \mathbb{R}, f(x) = \frac{\ln(1+x)}{x}$. Since f is strictly decreasing on $(0, +\infty)$, we have that

$$\frac{\ln\left(1 + \frac{1}{qn}\right)}{\frac{1}{qn}} \leq \frac{\ln\left(1 + \frac{1}{qn+kp}\right)}{\frac{1}{qn+kp}} \leq \frac{\ln\left(1 + \frac{1}{qn+np}\right)}{\frac{1}{qn+np}},$$

for all $k \in \{0, 1, \ldots, n\}$ and $n \geq 1$.

Therefore, we have that

$$\frac{\ln\left(1 + \frac{1}{qn}\right)}{\frac{1}{qn}} \sum_{k=0}^{n} \frac{1}{qn + kp} \leq \ln x_n \leq \frac{\ln\left(1 + \frac{1}{qn+np}\right)}{\frac{1}{qn+np}} \sum_{k=0}^{n} \frac{1}{qn + kp},$$

(continued)

for all $n \geq 1$.

It follows that

$$\lim_{n\to\infty} \ln x_n = \frac{1}{p} \ln \frac{p+q}{q},$$

since

$$\lim_{n\to\infty} \frac{\ln\left(1+\frac{1}{qn}\right)}{\frac{1}{qn}} = 1, \qquad \lim_{n\to\infty} \frac{\ln\left(1+\frac{1}{qn+np}\right)}{\frac{1}{qn+np}} = 1$$

and

$$\lim_{n\to\infty} \sum_{k=0}^{n} \frac{1}{qn+kp} = \lim_{n\to\infty} \frac{1}{n} \sum_{k=0}^{n} \frac{1}{q+p\frac{k}{n}} = \int_0^1 \frac{1}{q+px}\, dx = \frac{1}{p} \ln \frac{q+p}{q}.$$

Thus, $\lim_{n\to\infty} x_n = \left(\frac{p+q}{q}\right)^{\frac{1}{p}}$.

1.17. Observe that $a_{n+2}a_{n+1} = a_{n+1}a_n + 1$, $\forall n \geq 1$. It follows that $a_{2k} = \frac{(2k-1)!!}{(2k-2)!!}$, $k \geq 2$, and $a_{2k+1} = \frac{(2k)!!}{(2k-1)!!}$, $k \geq 1$.

1.18. Prove by mathematical induction that $x_n = 2\cos\frac{\pi}{2^{n+1}}$, $\forall n \geq 1$. It follows that $\lim_{n\to\infty} 4^n (2 - x_n) = \frac{\pi^2}{4}$.

1.19. We consider the case when $x_0 \neq 1$. We have, based on the $AM - GM$ inequality, that $1 + x_n > 2\sqrt{x_n}$ and it follows $x_{n+1} > 1$, $\forall n \geq 0$. Also, $1 < x_{n+1} = \frac{1+x_n}{2\sqrt{x_n}} < \frac{1+x_n}{2} < x_n$, $\forall n \geq 0$. The sequence is strictly decreasing and bounded, since $1 < x_{n+1} < x_n < x_0$, $\forall n \geq 0$. Thus, the sequence converges. Let $l = \lim_{n\to\infty} x_n$. Since $1 < x_{n+1} < \frac{1+x_n}{2}$, $\forall n \geq 0$, we have that $1 \leq l \leq \frac{1+l}{2} \Rightarrow l = 1$.

1.20.

(a) By induction one can prove that $x_n \geq 1$, $\forall n \geq 1$. Using the $AM-GM$ inequality we have that $\sqrt{x_n} + \frac{1}{n+1} + 1 = x_{n+1} + 1 \geq 2\sqrt{x_{n+1}}$, $\forall n \geq 0$. It follows that $2\sqrt{x_{n+1}} \leq \frac{n+2}{n+1} + \sqrt{x_n} \Rightarrow 2^{n+1}\sqrt{x_{n+1}} \leq 2^n \cdot \frac{n+2}{n+1} + 2^n \sqrt{x_n}$, $\forall n \geq 0$. This implies that

$$1 \leq x_{n+1} \leq \left(\frac{1}{2^{n+1}} \sum_{k=0}^{n} 2^k \frac{k+2}{k+1} + \frac{\sqrt{x_0}}{2^{n+1}}\right)^2, \quad n \geq 1.$$

Passing to the limit, as $n \to \infty$, in the previous inequalities, we have that $\lim_{n\to\infty} x_n = 1$.

(b) Similarly, one can prove that $\lim_{n\to\infty} x_n = 1$.

1.21.

(a) e; (b) $\frac{1}{32}$; (c) $\frac{4}{e}$;

(d) Let $x_n = \frac{(2n)(2n+3)\cdots(8n-6)}{(n!)^2}$, $n \geq 1$. We have

$$
\frac{x_{n+1}}{x_n} = \frac{(2n+2)(2n+5)\cdots(8n-4)(8n-1)(8n+2)}{[(n+1)!]^2} \cdot \frac{(n!)^2}{(2n)(2n+3)\cdots(8n-6)}
$$

$$
= \frac{(2n+2)(2n+5)\cdots(8n-4)}{(2n)(2n+3)\cdots(8n-6)} \cdot \frac{(8n-1)(8n+2)}{(n+1)^2}, \quad n \geq 1.
$$

Using the idea from the solution of part (e) of Problem 1.10, we get that

$$
\lim_{n\to\infty} \frac{(2n+2)(2n+5)\cdots(8n-4)}{(2n)(2n+3)\cdots(8n-6)} = 2\sqrt[3]{2}.
$$

An application of Cauchy–d'Alembert's criterion shows that

$$
\lim_{n\to\infty} \sqrt[n]{x_n} = \lim_{n\to\infty} \frac{x_{n+1}}{x_n} = 2\sqrt[3]{2} \cdot 64 = 128\sqrt[3]{2}.
$$

(e) Let $x_n = \dfrac{(2n+1)(2n+4)\cdots(5n-2)}{n^n}$, $n \geq 1$. We have

$$
\frac{x_{n+1}}{x_n} = \frac{(2n+3)(2n+6)\cdots(5n)(5n+3)}{(n+1)^{n+1}} \cdot \frac{n^n}{(2n+1)(2n+4)\cdots(5n-2)}
$$

$$
= \frac{(2n+3)(2n+6)\cdots(5n)}{(2n+1)(2n+4)\cdots(5n-2)} \cdot \frac{5n+3}{n+1} \cdot \frac{1}{\left(1+\frac{1}{n}\right)^n}, \quad n \geq 1.
$$

Using the idea from the solution of part (e) of Problem 1.10, we get that

$$
\lim_{n\to\infty} \frac{(2n+3)(2n+6)\cdots(5n)}{(2n+1)(2n+4)\cdots(5n-2)} = \sqrt[3]{\frac{25}{4}}.
$$

It follows, based on Cauchy–d'Alembert's criterion, that

$$
\lim_{n\to\infty} \sqrt[n]{x_n} = \lim_{n\to\infty} \frac{x_{n+1}}{x_n} = \sqrt[3]{\frac{25}{4}} \cdot 5 \cdot \frac{1}{e} = \frac{5}{e}\sqrt[3]{\frac{25}{4}}.
$$

1.22. The limit equals 1. Use Cauchy–d'Alembert's criterion.

1.23. Solution due to D.M. Bătinețu-Giurgiu.

Let $L_n = \sqrt[n+1]{(n+1)!} - \sqrt[n]{n!}$, $n \geq 2$. We consider the sequence $(x_n)_{n \geq 2}$ defined by

$$x_n = \frac{\sqrt[n+1]{(n+1)!}}{\sqrt[n]{n!}} - 1, \quad n \geq 2.$$

We have

$$\lim_{n \to \infty} \frac{\sqrt[n+1]{(n+1)!}}{\sqrt[n]{n!}} = \lim_{n \to \infty} \left(\frac{\sqrt[n+1]{(n+1)!}}{n+1} \cdot \frac{n}{\sqrt[n]{n!}} \cdot \frac{n+1}{n} \right) = \frac{1}{e} \cdot e \cdot 1 = 1,$$

and this implies $\lim_{n \to \infty} x_n = 0$.

It follows that

$$\lim_{n \to \infty} L_n = \lim_{n \to \infty} x_n \sqrt[n]{n!} = \lim_{n \to \infty} \left(\frac{x_n}{\ln(1 + x_n)} \cdot \sqrt[n]{n!} \cdot \ln(1 + x_n) \right)$$

$$= \lim_{n \to \infty} \left(\frac{x_n}{\ln(1 + x_n)} \cdot \sqrt[n]{n!} \cdot \ln \frac{\sqrt[n+1]{(n+1)!}}{\sqrt[n]{n!}} \right)$$

$$= \lim_{n \to \infty} \left(\frac{x_n}{\ln(1 + x_n)} \cdot \frac{\sqrt[n]{n!}}{n} \cdot \ln \frac{n+1}{\sqrt[n+1]{(n+1)!}} \right)$$

$$= 1 \cdot \frac{1}{e} \cdot \ln e = \frac{1}{e}.$$

Remark. We mention that a generalization of Traian Lalescu's sequence involving the Gamma function was studied in [24].

1.24. Let $(a_n)_{n \geq 0}$ and $(b_n)_{n \geq 0}$ be the sequences defined by

$$a_n = C_n^0 + C_{n+1}^2 + C_{n+2}^4 + \cdots + C_{2n}^{2n} = \sum_{k=0}^{n} C_{n+k}^{2k}, \quad n \geq 0;$$

$$b_n = C_n^1 + C_{n+1}^3 + C_{n+2}^5 + \cdots + C_{2n-1}^{2n-1} = \sum_{k=1}^{n} C_{n+k-1}^{2k-1}, \quad n \geq 1.$$

Using the recurrence formula $C_n^k = C_{n-1}^k + C_{n-1}^{k-1}$, $0 < k < n$, we have

$$a_{n+1} = \sum_{k=0}^{n+1} C_{n+1+k}^{2k} = 2 + \sum_{k=1}^{n} C_{n+1+k}^{2k} = 2 + \sum_{k=1}^{n} \left(C_{n+k}^{2k} + C_{n+k}^{2k-1} \right)$$

$$= 1 + \sum_{k=1}^{n} C_{n+k}^{2k} + 1 + \sum_{k=1}^{n} C_{n+k}^{2k-1} = a_n + 1 + \sum_{k=1}^{n} \left(C_{n+k-1}^{2k-1} + C_{n+k-1}^{2k-2} \right)$$

$$= a_n + 1 + \sum_{k=1}^{n} C_{n+k-1}^{2k-1} + \sum_{k=1}^{n} C_{n+k-1}^{2k-2} = a_n + b_n + 1 + \sum_{p=0}^{n-1} C_{n+p}^{2p}$$

$$= 2a_n + b_n,$$

and

$$b_{n+1} = \sum_{k=1}^{n+1} C_{n+k}^{2k-1} = \sum_{k=1}^{n} C_{n+k}^{2k-1} + 1 = \sum_{k=1}^{n} \left(C_{n+k-1}^{2k-1} + C_{n+k-1}^{2k-2} \right) + 1$$

$$= \sum_{k=1}^{n} C_{n+k-1}^{2k-1} + \sum_{p=0}^{n-1} C_{n+p}^{2p} + 1$$

$$= b_n + a_n.$$

It follows that $a_{n+2} - 3a_{n+1} + a_n = 0$, with $a_0 = 1$ and $a_1 = 2$. We determine the sequence $(a_n)_{n \geq 0}$ using the method for solving sequences defined by recurrence relations of order two and we get that $a_n = F_{2n+1}$. This implies $b_n = a_{n+1} - 2a_n = F_{2n+3} - 2F_{2n+1} = F_{2n}$.

1.25.

(a) The limit equals 1. Let $f_n(x) = \underbrace{f \circ f \circ \cdots \circ f}_{n \text{ times}}(x)$. We have that $f_n(x) = $
$\frac{a_n x + b_n}{c_n x + d_n}$, where a_n, b_n, c_n, and d_n are sequences of real numbers defined by

$$\begin{pmatrix} a_n & b_n \\ c_n & d_n \end{pmatrix} = \begin{pmatrix} 3 & 2 \\ 2 & 3 \end{pmatrix}^n = \frac{1}{2} \begin{pmatrix} 5^n + 1 & 5^n - 1 \\ 5^n - 1 & 5^n + 1 \end{pmatrix}.$$

It follows that $f_n(x) = \frac{(5^n+1)x+5^n-1}{(5^n-1)x+5^n+1} \Rightarrow \lim\limits_{n \to \infty} f_n(x) = 1$.

(b) The limit equals $\frac{\sqrt{5}-1}{2}$. See the solution of part (a).

For information regarding sequences defined by homographic recurrence relations the reader is referred to [119, section 3.3, p. 129].

9.2 Applications of Stolz–Cesàro Theorem, the ∞/∞ and the 0/0 Cases

1.26. (a) 1; (b) $\frac{1}{p+1}$; (c) $\frac{2}{5}$; (d) $\frac{3}{4}$; (e) $\frac{1}{3}$; (f) 1; (g) 1; (h) $\frac{a}{2}$; (i) a.

1.27. Since $x_1 > 0$, we get that $x_n > 0$, for all $n \geq 1$. The inequality $\ln(1 + x) < x$, for $x > 0$, implies $x_{n+1} = \ln(1+x_n) < x_n$, for all $n \geq 1$. It follows that the sequence $(x_n)_{n\geq 1}$ is strictly decreasing and bounded, hence convergent. Let $l = \lim_{n\to\infty} x_n$. Passing to the limit in the recurrence relation we obtain $l = \ln(1 + l)$, and this implies $l = 0$.

Since $(x_n)_{n\geq 1}$ is strictly decreasing and $\lim_{n\to\infty} x_n = 0$, we have that the sequence $(1/x_n)_{n\geq 1}$ is strictly increasing and $\lim_{n\to\infty} \frac{1}{x_n} = \infty$. An application of Stolz–Cesàro's theorem shows that

$$\lim_{n\to\infty} nx_n = \lim_{n\to\infty} \frac{n+1-n}{\frac{1}{x_{n+1}} - \frac{1}{x_n}} = \lim_{n\to\infty} \frac{x_n x_{n+1}}{x_n - x_{n+1}}$$

$$= \lim_{n\to\infty} \frac{x_n \ln(1+x_n)}{x_n - \ln(1+x_n)} = \lim_{n\to\infty} \frac{\ln(1+x_n)}{x_n} \cdot \frac{x_n^2}{x_n - \ln(1+x_n)} = 2.$$

1.28. The sequence $(x_n)_{n\geq 1}$ is bounded because $\arctan x \in \left(-\frac{\pi}{2}, \frac{\pi}{2}\right)$, for all $x \in \mathbb{R}$. Since $x_1 > 0$, we have $x_n > 0$, for all $n \geq 1$. The inequality $\arctan x < x$, for $x > 0$, implies $x_{n+1} = \arctan x_n < x_n$, for all $n \geq 1$. It follows that the sequence $(x_n)_{n\geq 1}$ is strictly decreasing. The sequence $(x_n)_{n\geq 1}$ converges, since it is bounded and decreases. Let $l = \lim_{n\to\infty} x_n$. Passing to the limit in the recurrence relation we obtain $l = \arctan l$, and it follows $l = 0$.

To calculate $\lim_{n\to\infty} \sqrt{n} x_n$ we consider the sequence $(nx_n^2)_{n\geq 1}$. Since $(x_n)_{n\geq 1}$ is strictly decreasing and $\lim_{n\to\infty} x_n = 0$, we get that the sequence $\left(\frac{1}{x_n^2}\right)_{n\geq 1}$ is strictly increasing and $\lim_{n\to\infty} \frac{1}{x_n^2} = \infty$. We apply Stolz–Cesàro's theorem and we have

$$\lim_{n\to\infty} nx_n^2 = \lim_{n\to\infty} \frac{n+1-n}{\frac{1}{x_{n+1}^2} - \frac{1}{x_n^2}} = \lim_{n\to\infty} \frac{x_n^2 x_{n+1}^2}{x_n^2 - x_{n+1}^2}$$

$$= \lim_{n\to\infty} \frac{x_n^2 \arctan^2 x_n}{x_n^2 - \arctan^2 x_n} = \lim_{n\to\infty} \left[\left(\frac{\arctan x_n}{x_n}\right)^2 \cdot \frac{x_n^4}{x_n^2 - \arctan^2 x_n}\right] = \frac{3}{2}.$$

Thus, $\lim_{n\to\infty} \sqrt{n} x_n = \sqrt{\frac{3}{2}}$.

1.29. We have $x_n < x_n + e^{-x_n} = x_{n+1}$, for all $n \geq 1$, and this shows that the sequence $(x_n)_{n\geq 1}$ is strictly increasing. Also, we observe that the sequence $(x_n)_{n\geq 1}$ is unbounded. If $(x_n)_{n\geq 1}$ would be bounded, then the sequence would converge. Denoting by l its limit we get, from the recurrence relation, that $l = l + e^{-l}$, which is impossible. Since the sequence $(x_n)_{n\geq 1}$ is strictly increasing and unbounded, it follows that $\lim\limits_{n\to\infty} x_n = \infty$. We apply Stolz–Cesàro's theorem and we have

$$\lim_{n\to\infty}\frac{x_n}{\ln n} = \lim_{n\to\infty}\frac{x_{n+1}-x_n}{\ln(n+1)-\ln n} = \lim_{n\to\infty}\frac{1}{\frac{e^{x_n}}{n}\ln\left(1+\frac{1}{n}\right)^n} = 1,$$

since, based on the same theorem of Stolz–Cesàro, one has

$$\lim_{n\to\infty}\frac{e^{x_n}}{n} = \lim_{n\to\infty}\frac{e^{x_{n+1}}-e^{x_n}}{n+1-n} = \lim_{n\to\infty}\left(e^{x_n+e^{-x_n}}-e^{x_n}\right) = \lim_{n\to\infty}\frac{e^{\frac{1}{e^{x_n}}}-1}{\frac{1}{e^{x_n}}} = 1.$$

1.31. Let $(y_n)_{n\geq 1}$ be the sequence defined by

$$y_n = 1 + \frac{1}{2} + \frac{1}{3} + \cdots + \frac{1}{n} - \ln(n+1), \quad n \geq 1.$$

A calculation shows that:

$$x_{n+1} - x_n = \frac{1}{n+1} + \ln\left(1 - \frac{1}{n+1}\right) < 0, \quad n \geq 1;$$

$$y_{n+1} - y_n = \frac{1}{n+1} - \ln\left(1 + \frac{1}{n+1}\right) > 0, \quad n \geq 1;$$

$$x_n - y_n = \ln\frac{n+1}{n} > 0, \quad n \geq 1.$$

It follows $0 < 1 - \ln 2 = y_1 < y_2 < \ldots < y_n < x_n < x_{n-1} < \ldots < x_1 = 1$, which implies that $(x_n)_{n\geq 1}$ is bounded and being strictly decreasing it converges. We denote its limit by γ.

To calculate $\lim\limits_{n\to\infty} n(x_n - \gamma)$ we apply Stolz–Cesàro's theorem, the $0/0$ case, and we have

$$\lim_{n\to\infty} n(x_n-\gamma) = \lim_{n\to\infty}\frac{x_n-\gamma}{\frac{1}{n}} = \lim_{n\to\infty}\frac{x_{n+1}-x_n}{\frac{1}{n+1}-\frac{1}{n}} = \lim_{n\to\infty}\frac{\frac{1}{n+1}-\ln(n+1)+\ln n}{\frac{1}{n+1}-\frac{1}{n}} = \frac{1}{2}.$$

Remark. We mention that interesting properties of Euler's constant are studied in [129].

1.32.

(a) The limit equals 1. We use the following limit (see Problem 1.31):

$$\lim_{n\to\infty} n\left(1+\frac{1}{2}+\frac{1}{3}+\cdots+\frac{1}{n}-\ln n-\gamma\right) = \lim_{n\to\infty} n(H_n - \ln n - \gamma) = \frac{1}{2}.$$

We have

$$\lim_{n\to\infty} n\left(2H_n - H_{n^2} - \gamma\right) = \lim_{n\to\infty} n\left[2(H_n - \ln n - \gamma) - (H_{n^2} - \ln(n^2) - \gamma)\right]$$

$$= 2\lim_{n\to\infty} n(H_n - \ln n - \gamma) - \lim_{n\to\infty} n(H_{n^2} - \ln(n^2) - \gamma)$$

$$= 1.$$

(b) The limit equals $\frac{k}{2}$. See the solution of part (a) of the problem.

1.33. Use that $x = e^{\ln x}$, $\forall x > 0$.

$$\lim_{n\to\infty} \frac{1}{n}\left(1 + \frac{H_n}{n}\right)^n = e^\gamma, \quad \lim_{n\to\infty} \frac{1}{\sqrt{n}}\left(1 + \frac{O_n}{n}\right)^n = 2e^{\frac{\gamma}{2}}, \quad \lim_{n\to\infty} \frac{1}{n}\left(1 + \frac{2O_n}{n}\right)^n = 4e^\gamma.$$

1.34. $\frac{1}{2}e^{\frac{\gamma}{2}}$.

1.35. Let $(y_n)_{n\geq 1}$ be the sequence defined by

$$y_n = 1 + \frac{1}{\sqrt{2}} + \frac{1}{\sqrt{3}} + \cdots + \frac{1}{\sqrt{n}} - 2(\sqrt{n+1} - 1), \quad n \geq 1.$$

A calculation shows that:

$$x_{n+1} - x_n = \frac{1}{\sqrt{n+1}} - 2(\sqrt{n+1} - \sqrt{n}) < 0, \quad n \geq 1;$$

$$y_{n+1} - y_n = \frac{1}{\sqrt{n+1}} - 2(\sqrt{n+2} - \sqrt{n+1}) > 0, \quad n \geq 1;$$

$$x_n - y_n = 2(\sqrt{n+1} - \sqrt{n}) > 0, \quad n \geq 1.$$

It follows $0 < 3 - 2\sqrt{2} = y_1 < y_2 < \ldots < y_n < x_n < x_{n-1} < \ldots < x_1 = 1$, which implies that the sequence $(x_n)_{n\geq 1}$ is bounded and being strictly decreasing it converges. We denote its limit by $\mathscr{I}$.

To calculate $\lim_{n\to\infty} \sqrt{n}(x_n - \mathscr{I})$ we apply Stolz–Cesàro's theorem, the $0/0$ case, and we have

$$\lim_{n\to\infty} \sqrt{n}(x_n - \mathscr{I}) = \lim_{n\to\infty} \frac{x_n - \mathscr{I}}{\frac{1}{\sqrt{n}}} = \lim_{n\to\infty} \frac{x_{n+1} - x_n}{\frac{1}{\sqrt{n+1}} - \frac{1}{\sqrt{n}}}$$

$$= \lim_{n\to\infty} \frac{\frac{1}{\sqrt{n+1}} - 2(\sqrt{n+1} - \sqrt{n})}{\frac{1}{\sqrt{n+1}} - \frac{1}{\sqrt{n}}} = \frac{1}{2}.$$

1.36. 1.

1.37. $\frac{1}{3}$.

1.38. $e^{\frac{2}{\pi^4}}$.

1.39. (b) $-\frac{1}{2}$.

1.40. The limit equals 0. We have, based on part (a) of Problem 1.39, that

$$\lim_{n \to \infty} \left(1 + \frac{1}{2} + \cdots + \frac{1}{n}\right)\left(\zeta(k) - 1 - \frac{1}{2^k} - \cdots - \frac{1}{n^k}\right)$$

$$= \lim_{n \to \infty} \frac{1 + \frac{1}{2} + \cdots + \frac{1}{n}}{n^{k-1}} \cdot \lim_{n \to \infty} n^{k-1}\left(\zeta(k) - 1 - \frac{1}{2^k} - \cdots - \frac{1}{n^k}\right)$$

$$= 0.$$

1.41. Let $x > 1$ and $k \geq 1$. We have

$$\frac{1}{(x-1)k^{x-1}} = \int_k^\infty \frac{1}{t^x} dt = \sum_{i=k}^\infty \int_i^{i+1} \frac{1}{t^x} dt > \sum_{i=k}^\infty \frac{1}{(i+1)^x}$$

and it follows that

$$\zeta(x) - 1 - \frac{1}{2^x} - \cdots - \frac{1}{k^x} < \frac{1}{(x-1)k^{x-1}}. \qquad (9.1)$$

(a) We have, based on inequality (9.1) with $k = 1$, that $1 < \zeta(x) < \frac{x}{x-1}$, $\forall x > 1$. This implies that $\lim_{x \to \infty} \zeta(x) = 1$.

(b) Using inequality (9.1) with $k = 2$, we have $\frac{1}{2^x} < \zeta(x) - 1 < \frac{1}{2^x} + \frac{1}{(x-1)2^{x-1}} \Rightarrow$ $1 < 2^x(\zeta(x) - 1) < \frac{x+1}{x-1}$, $\forall x > 1$. This implies $\lim_{x \to \infty} 2^x(\zeta(x) - 1) = 1$. It follows that

$$\lim_{x \to \infty} \frac{\zeta(2x) - 1}{\zeta(x) - 1} = \lim_{x \to \infty} \frac{2^{2x}(\zeta(2x) - 1)}{2^x(\zeta(x) - 1)} \cdot \lim_{x \to \infty} \frac{1}{2^x} = 0.$$

(c) We have

$$\lim_{x \to \infty} \zeta(x)^{2^x} = \lim_{x \to \infty} (1 + \zeta(x) - 1)^{2^x} = e^{\lim_{x \to \infty} 2^x(\zeta(x)-1)} = e.$$

(d) We have, based on Lagrange Mean Value Theorem, that

$$\left(\frac{3}{2}\right)^x \left(\zeta(x)^{2^x} - e\right) = \left(\frac{3}{2}\right)^x \left(e^{2^x \ln \zeta(x)} - e\right) = \left(\frac{3}{2}\right)^x \left(2^x \ln \zeta(x) - 1\right) \cdot e^{\theta(x)},$$

where $\theta(x)$ is between 1 and $2^x \ln \zeta(x)$. Observe that $\lim_{x \to \infty} \theta(x) = 1$.

We are able to write

$$\left(\frac{3}{2}\right)^x \left(2^x \ln \zeta(x) - 1\right) = \left(\frac{3}{2}\right)^x \left[2^x \left(\ln \zeta(x) - \ln 1\right) - 1\right]$$

$$= \left(\frac{3}{2}\right)^x \left[2^x \left(\zeta(x) - 1\right) \cdot \frac{1}{\theta_1(x)} - 1\right],$$

where $1 < \theta_1(x) < \zeta(x)$. Observe that $\lim\limits_{x \to \infty} \theta_1(x) = 1$.

It follows that

$$\left(\frac{3}{2}\right)^x \left(2^x \ln \zeta(x) - 1\right) = \frac{1}{\theta_1(x)} \left\{ \left[2^x(\zeta(x) - 1) - 1\right] \left(\frac{3}{2}\right)^x + (1 - \theta_1(x)) \left(\frac{3}{2}\right)^x \right\}.$$

A calculation, based on inequality (9.1) with $k = 3$, shows that

$$\lim\limits_{x \to \infty} \left[2^x(\zeta(x) - 1) - 1\right] \left(\frac{3}{2}\right)^x = \lim\limits_{x \to \infty} \left(\zeta(x) - 1 - \frac{1}{2^x}\right) 3^x = 1.$$

The inequalities

$$0 < (\theta_1(x) - 1) \left(\frac{3}{2}\right)^x < (\zeta(x) - 1) \left(\frac{3}{2}\right)^x = 2^x (\zeta(x) - 1) \left(\frac{3}{4}\right)^x$$

imply that $\lim\limits_{x \to \infty} (\theta_1(x) - 1) \left(\frac{3}{2}\right)^x = 0$.

Thus, $\lim\limits_{x \to \infty} \left(\frac{3}{2}\right)^x (2^x \ln \zeta(x) - 1) = 1 \Rightarrow \lim\limits_{x \to \infty} \left(\frac{3}{2}\right)^x \left(\zeta(x)^{2^x} - e\right) = e$.

1.42. Both limits are equal to 1. The first limit can be calculated by applying the Stolz–Cesàro theorem, the $0/0$ case.

For the calculation of the second limit we consider separately the cases when $n = 2m$ and $n = 2m - 1$.

The case $n = 2m$. Let

$$x_m = \sum_{k=0}^{2m} (-1)^k \frac{1}{k!} - \frac{1}{e}.$$

We have

$$x_{m+1} - x_m = -\frac{1}{(2m)!(2m + 2)},$$

which shows the sequence $(x_m)_{m \geq 1}$ decreases and converges to 0. We apply Stolz–Cesàro's theorem, the $0/0$ case, and we have

$$\lim_{m\to\infty} (2m+1)! \left(\sum_{k=0}^{2m} (-1)^k \frac{1}{k!} - \frac{1}{e} \right) = \lim_{m\to\infty} \frac{x_{m+1} - x_m}{\frac{1}{(2m+3)!} - \frac{1}{(2m+1)!}}$$

$$= \lim_{m\to\infty} \frac{-\frac{1}{(2m)!(2m+2)}}{-\frac{4m^2+10m+5}{(2m+3)!}}$$

$$= 1.$$

The case $n = 2m - 1$. Let

$$y_m = \frac{1}{e} - \sum_{k=0}^{2m-1} (-1)^k \frac{1}{k!}.$$

We have

$$y_{m+1} - y_m = -\frac{1}{(2m-1)!(2m+1)},$$

which shows the sequence $(y_m)_{m\geq 1}$ decreases to 0. We apply Stolz–Cesàro's theorem, the $0/0$ case, and we have

$$\lim_{m\to\infty} (2m)! \left(\frac{1}{e} - \sum_{k=0}^{2m-1} (-1)^k \frac{1}{k!} \right) = \lim_{m\to\infty} \frac{y_{m+1} - y_m}{\frac{1}{(2m+2)!} - \frac{1}{(2m)!}}$$

$$= \lim_{m\to\infty} \frac{-\frac{1}{(2m-1)!(2m+1)}}{-\frac{4m^2+6m+1}{(2m+2)!}}$$

$$= 1.$$

9.3   Wolstenholme Sequences

1.43.

(a) This part follows from part (b) with $f(x) = x^s$. Two solutions of this problem, one elementary and the other one, which uses Lebesgue Dominated Convergence Theorem, are given in [98].

(b) If $f(1) > 1$, then $x_n > f^n(1) \Rightarrow \lim_{n\to\infty} x_n = \infty$.

   Let $f(1) \leq 1$. We have

(continued)

$$x_n = \sum_{k=0}^{n-1} f^n\left(1 - \frac{k}{n}\right) = \sum_{k=0}^{n-1} e^{n \ln f\left(1 - \frac{k}{n}\right)}.$$

We apply Lagrange Mean Value Theorem to the function $\ln f$ and we have that there exists $\theta \in \left(1 - \frac{k}{n}, 1\right)$ such that

$$\ln f(1) - \ln f\left(1 - \frac{k}{n}\right) = \frac{k}{n} \cdot \frac{f'(\theta)}{f(\theta)} \geq \frac{k}{n} \cdot \frac{f'(1)}{f(1)},$$

where the last inequality holds since the function $(\ln f)' = \frac{f'}{f}$ decreases. It follows that $\ln f\left(1 - \frac{k}{n}\right) \leq \ln f(1) - \frac{k}{n} \cdot \frac{f'(1)}{f(1)} \Rightarrow e^{n \ln f\left(1 - \frac{k}{n}\right)} \leq f^n(1) e^{-k \frac{f'(1)}{f(1)}}, \forall 1 \leq k \leq n$. Therefore,

$$0 < x_n < f^n(1) \sum_{k=0}^{n-1} e^{-k \frac{f'(1)}{f(1)}} = f^n(1) \frac{1 - e^{-n \frac{f'(1)}{f(1)}}}{1 - e^{-\frac{f'(1)}{f(1)}}}. \tag{9.2}$$

If $f(1) < 1$, then we have, based on inequality (9.2), that $\lim_{n \to \infty} x_n = 0$.

If $f(1) = 1$, then inequality (9.2) implies $x_n < \frac{1}{1 - e^{-f'(1)}}, \forall n \geq 1$, and this shows $\limsup x_n \leq \frac{1}{1 - e^{-f'(1)}}$.

Let $i \geq 1$ be a fixed integer. Then, for all $n \geq i$, we have

$$x_n \geq \sum_{k=0}^{i-1} e^{n \ln f\left(1 - \frac{k}{n}\right)}. \tag{9.3}$$

If k is fixed, then

$$\lim_{n \to \infty} n \ln f\left(1 - \frac{k}{n}\right) = -k \lim_{n \to \infty} \frac{\ln f\left(1 - \frac{k}{n}\right)}{f\left(1 - \frac{k}{n}\right) - 1} \cdot \frac{f\left(1 - \frac{k}{n}\right) - 1}{-\frac{k}{n}} = -kf'(1).$$

Passing to the limit as $n \to \infty$ in (9.3), we get

$$\liminf x_n \geq \sum_{k=0}^{i-1} e^{-kf'(1)} = \frac{1 - e^{-if'(1)}}{1 - e^{-f'(1)}}, \quad \forall i \geq 1.$$

Passing to the limit as $i \to \infty$ in the previous inequality, we have that $\liminf x_n \geq \frac{1}{1 - e^{-f'(1)}}$.

(continued)

For studying the sequence $(y_n)_{n\geq 1}$ see the solution of Problem 1.46, with $a = 1$.

1.44. The first limit is a particular case of Problem 1.46, with $f(x) = x$, $\alpha = 1$, $\beta = 0$. We calculate the second limit.

Let $x_n = \sum_{k=1}^{n} a^{k-n} \left(\frac{k}{n}\right)^k$.

We have

$$x_n = \sum_{k=1}^{n-\sqrt[3]{n}-1} a^{k-n} \left(\frac{k}{n}\right)^k + \sum_{k=n-\sqrt[3]{n}}^{n} a^{k-n} \left(\frac{k}{n}\right)^k = S_1(n) + S_2(n).$$

Observe that

$$S_1(n) = \sum_{k=1}^{n-\sqrt[3]{n}-1} a^{k-n} \left(\frac{k}{n}\right)^k < \sum_{k=1}^{n-\sqrt[3]{n}} a^{k-n} \left(\frac{k}{n}\right)^k < \sum_{k=1}^{n-\sqrt[3]{n}} \left(\frac{k}{n}\right)^k = \frac{1}{n} + \sum_{k=2}^{n-\sqrt[3]{n}} \left(\frac{k}{n}\right)^k.$$

Let $f : [2, n - \sqrt[3]{n}] \to \mathbb{R}$, $f(x) = \left(\frac{x}{n}\right)^x$, and $g(x) = \ln f(x) = x \ln x - x \ln n$. A calculation shows that g decreases on the interval $\left[2, \frac{n}{e}\right]$ and increases on the interval $\left[\frac{n}{e}, n - \sqrt[3]{n}\right]$, so it follows that

$$\max_{2 \leq k \leq n-\sqrt[3]{n}} \left(\frac{k}{n}\right)^k \leq \max\left\{\frac{4}{n^2}, \left(\frac{n - \sqrt[3]{n}}{n}\right)^{n-\sqrt[3]{n}}\right\}.$$

This implies that

$$S_1(n) < \frac{1}{n} + \max\left\{\frac{4}{n^2}, \left(\frac{n - \sqrt[3]{n}}{n}\right)^{n-\sqrt[3]{n}}\right\} (n - \sqrt[3]{n} - 1).$$

Since $\ln(1 - x) \leq -x, \forall x \in [0, 1)$, we have that $\ln\left(1 - \frac{\sqrt[3]{n}}{n}\right) \leq -\frac{\sqrt[3]{n}}{n} \Rightarrow$
$(n - \sqrt[3]{n}) \ln\left(1 - \frac{\sqrt[3]{n}}{n}\right) \leq -\sqrt[3]{n} + \frac{1}{\sqrt[3]{n}} \Rightarrow \left(\frac{n-\sqrt[3]{n}}{n}\right)^{n-\sqrt[3]{n}} \leq e^{-\sqrt[3]{n}+\frac{1}{\sqrt[3]{n}}}$.
This implies

$$0 \leq \lim_{n\to\infty} \left(\frac{n - \sqrt[3]{n}}{n}\right)^{n-\sqrt[3]{n}} (n - \sqrt[3]{n} - 1) \leq \lim_{n\to\infty} e^{-\sqrt[3]{n}+\frac{1}{\sqrt[3]{n}}} (n - \sqrt[3]{n} - 1) = 0$$

(continued)

and it follows that $\lim\limits_{n\to\infty}\left(\frac{n-\sqrt[3]{n}}{n}\right)^{n-\sqrt[3]{n}}(n-\sqrt[3]{n}-1)=0$.

Thus,

$$\lim_{n\to\infty}\max\left\{\frac{4}{n^2},\left(\frac{n-\sqrt[3]{n}}{n}\right)^{n-\sqrt[3]{n}}\right\}(n-\sqrt[3]{n}-1)=0$$

and this shows that $\lim\limits_{n\to\infty}S_1(n)=0$.

We can write

$$S_2(n)=\sum_{k=n-\sqrt[3]{n}}^{n}a^{k-n}\left(\frac{k}{n}\right)^k=\sum_{i=0}^{\sqrt[3]{n}}\frac{1}{a^i}\left(1-\frac{i}{n}\right)^{n-i}.$$

We have $(n-i)\ln\left(1-\frac{i}{n}\right)\leq-\frac{i}{n}(n-i)=-i+\frac{i^2}{n}\Rightarrow\left(1-\frac{i}{n}\right)^{n-i}\leq$ $e^{-i+\frac{i^2}{n}}\leq e^{-i}\cdot e^{\frac{1}{\sqrt[3]{n}}}$. It follows that

$$S_2(n)\leq e^{\frac{1}{\sqrt[3]{n}}}\sum_{i=0}^{\sqrt[3]{n}}\frac{1}{(ae)^i}<e^{\frac{1}{\sqrt[3]{n}}}\sum_{i=0}^{\infty}\frac{1}{(ae)^i}=e^{\frac{1}{\sqrt[3]{n}}}\frac{ae}{ae-1}.$$

The previous inequality implies that $\limsup S_2(n)\leq\frac{ae}{ae-1}$.

Let $k\geq1$ be a fixed integer. Then, for $\sqrt[3]{n}>k$, we have

$$S_2(n)>\sum_{i=0}^{k}\frac{1}{a^i}\left(1-\frac{i}{n}\right)^{n-i}.$$

It follows

$$\liminf S_2(n)\geq\sum_{i=0}^{k}\frac{1}{a^i}e^{-i}=\frac{1-\left(\frac{1}{ae}\right)^{k+1}}{1-\frac{1}{ae}}.$$

Passing to the limit, as $k\to\infty$, in the previous inequality, we have that $\liminf S_2(n)\geq\frac{ae}{ae-1}$.

1.45.

(a) *Solution I.* Use a similar technique to the one given in the solution of Problem 1.44.

Solution II. We use Tannery's Theorem for Series.

Tannery's Theorem for Series [106, p. 216] For each natural number n, let $\sum_{k=1}^{m_n} a_k(n)$ be a finite sum such that $\lim_{n\to\infty} m_n = \infty$. If, for each k, $\lim_{n\to\infty} a_k(n)$ exists, and there is a convergent series $\sum_{k=1}^{\infty} M_k$ of nonnegative real numbers such that $|a_k(n)| \le M_k$, for all $n \in \mathbb{N}$ and $1 \le k \le m_n$, then

$$\lim_{n\to\infty} \sum_{k=1}^{m_n} a_k(n) = \sum_{k=1}^{\infty} \lim_{n\to\infty} a_k(n),$$

that is, both sides are well defined (the limits and the sums converge) and are equal.

Let $k_0 \in \mathbb{N}$ be such that $sk_0 > 1$, i.e. $k_0 = \left\lfloor \frac{1}{s} \right\rfloor + 1$. We have

$$\sum_{k=1}^{n} \left(\frac{k}{n}\right)^{sk} = \sum_{k=1}^{k_0} \left(\frac{k}{n}\right)^{sk} + \sum_{k=k_0+1}^{n} \left(\frac{k}{n}\right)^{sk}.$$

It is easy to see that $\lim_{n\to\infty} \sum_{k=1}^{k_0} \left(\frac{k}{n}\right)^{sk} = 0$.

We are able to write

$$\sum_{k=k_0+1}^{n} \left(\frac{k}{n}\right)^{sk} = \sum_{i=0}^{n-k_0-1} \left(1 - \frac{i}{n}\right)^{s(n-i)} = \sum_{i=0}^{n-k_0} \left(1 - \frac{i}{n}\right)^{s(n-i)} - \left(\frac{k_0}{n}\right)^{sk_0}.$$

Now we observe that the function $f : [i, \infty) \to \mathbb{R}$, $f(x) = \left(1 - \frac{i}{x}\right)^{s(x-i)}$ is strictly decreasing on $[i, \infty)$. It follows that, for $n \ge k_0 + i$, we have

$$\left(1 - \frac{i}{n}\right)^{s(n-i)} \le \left(\frac{k_0}{i + k_0}\right)^{sk_0}.$$

(continued)

Since the series $\sum_{i=0}^{\infty} \frac{1}{(i+k_0)^{sk_0}}$ converges and $\lim_{n\to\infty} \left(1 - \frac{i}{n}\right)^{s(n-i)} = e^{-si}$,
we have, based on Tannery's Theorem for Series, that

$$\lim_{n\to\infty} \sum_{i=0}^{n-k_0-1} \left(1 - \frac{i}{n}\right)^{s(n-i)} = \sum_{i=0}^{\infty} e^{-si} = \frac{e^s}{e^s - 1}.$$

(b) Observe that

$$\sum_{k=1}^{n} \left(\frac{k}{n}\right)^{(a+b)n} \leq \sum_{k=1}^{n} \left(\frac{k}{n}\right)^{ak+bn} \leq \sum_{k=1}^{n} \left(\frac{k}{n}\right)^{(a+b)k}.$$

1.46. If $f(1) > 1$, then $x_n > f^{\alpha n+\beta}(1) \Rightarrow \lim_{n\to\infty} x_n = \infty$.
Let $f(1) \leq 1$. We have

$$x_n = \sum_{k=0}^{n-1} \frac{1}{a^k} f^{\alpha n+\beta} \left(1 - \frac{k}{n}\right) = \sum_{k=0}^{n-1} \frac{1}{a^k} e^{(\alpha n+\beta)\ln f\left(1-\frac{k}{n}\right)}.$$

We apply Lagrange Mean Value Theorem to the function $\ln f$ and we have
that there exists $\theta \in \left(1 - \frac{k}{n}, 1\right)$ such that

$$\ln f(1) - \ln f\left(1 - \frac{k}{n}\right) = \frac{k}{n} \cdot \frac{f'(\theta)}{f(\theta)} \geq \frac{k}{n} \cdot \frac{f'(1)}{f(1)},$$

where the last inequality holds since the function $(\ln f)' = \frac{f'}{f}$ decreases. It
follows that $\ln f\left(1 - \frac{k}{n}\right) \leq \ln f(1) - \frac{k}{n} \cdot \frac{f'(1)}{f(1)} \Rightarrow (\alpha n + \beta)\ln f\left(1 - \frac{k}{n}\right) \leq$
$(\alpha n + \beta)\ln f(1) - k \cdot \frac{\alpha n+\beta}{n} \cdot \frac{f'(1)}{f(1)} \leq (\alpha n + \beta)\ln f(1) - k \cdot \alpha \cdot \frac{f'(1)}{f(1)}$. This
implies that $e^{(\alpha n+\beta)\ln f\left(1-\frac{k}{n}\right)} \leq f^{\alpha n+\beta}(1)e^{-k\alpha \frac{f'(1)}{f(1)}}, \forall 1 \leq k \leq n$. Therefore,

$$0 < x_n < f^{\alpha n+\beta}(1) \sum_{k=0}^{n-1} \left(\frac{e^{-\alpha \frac{f'(1)}{f(1)}}}{a}\right)^k = f^{\alpha n+\beta}(1) \frac{1 - \left(\frac{e^{-\alpha \frac{f'(1)}{f(1)}}}{a}\right)^n}{1 - \frac{e^{-\alpha \frac{f'(1)}{f(1)}}}{a}}.$$

$$\tag{9.4}$$

If $f(1) < 1$, then we have, based on inequality (9.4), that $\lim_{n\to\infty} x_n = 0$.

(continued)

If $f(1) = 1$, then inequality (9.4) implies that $x_n < \dfrac{1}{1-\frac{e^{-\alpha f'(1)}}{a}}$, $\forall n \geq 1$,

and it follows $\limsup x_n \leq \dfrac{ae^{\alpha f'(1)}}{ae^{\alpha f'(1)}-1}$.

Let $i \geq 1$ be a fixed integer. Then, for all $n \geq i$, we have

$$x_n \geq \sum_{k=0}^{i-1} \frac{1}{a^k} e^{(\alpha n+\beta)\ln f\left(1-\frac{k}{n}\right)}. \tag{9.5}$$

If k is fixed, then

$$(\alpha n + \beta)\ln f\left(1 - \frac{k}{n}\right) = -k \cdot \frac{\ln f\left(1 - \frac{k}{n}\right)}{f\left(1 - \frac{k}{n}\right) - 1} \cdot \frac{f\left(1 - \frac{k}{n}\right) - 1}{-\frac{k}{n}} \cdot \frac{\alpha n + \beta}{n}$$

and we have that $\lim\limits_{n\to\infty} (\alpha n + \beta)\ln f\left(1 - \frac{k}{n}\right) = -k\alpha f'(1)$.

Passing to the limit as $n \to \infty$ in (9.5), we get that

$$\liminf x_n \geq \sum_{k=0}^{i-1} \left(\frac{e^{-\alpha f'(1)}}{a}\right)^k = \frac{1 - \left(\frac{e^{-\alpha f'(1)}}{a}\right)^i}{1 - \frac{e^{-\alpha f'(1)}}{a}}, \quad \forall\, i \geq 1.$$

We let $i \to \infty$ in the previous inequality and we obtain $\liminf x_n \geq \dfrac{ae^{\alpha f'(1)}}{ae^{\alpha f'(1)}-1}$.

9.4   Limits of Integrals

1.47.

(a) The case when $k = 1$ is trivial, so we consider only the case when $k \neq 1$.
Using the substitution $x = t^n$, we have

$$I_n = \int_0^1 \left(\frac{\sqrt[n]{x} + k - 1}{k}\right)^n dx = \frac{n}{k^n} \int_0^1 [t(t + k - 1)]^{n-1}(t + k - 1)dt.$$

We make the substitution $t(t + k - 1) = y \Rightarrow t = \dfrac{\sqrt{(k-1)^2+4y}-k+1}{2}$ and we get, after some calculations, that

$$I_n = \frac{n}{2k^n} \int_0^k y^{n-1} \left(1 + \frac{k-1}{\sqrt{(k-1)^2 + 4y}} \right) dy$$

$$= \frac{1}{2} + \frac{n(k-1)}{2k^n} \int_0^k \frac{y^{n-1}}{\sqrt{(k-1)^2 + 4y}} dy.$$

We integrate by parts and we obtain that

$$I_n = \frac{k}{k+1} + \frac{k-1}{k^n} \int_0^k \frac{y^n}{\left(\sqrt{(k-1)^2 + 4y}\right)^3} dy.$$

It follows $\lim\limits_{n \to \infty} I_n = \frac{k}{k+1}$, since

$$0 < \frac{1}{k^n} \int_0^k \frac{y^n}{\left(\sqrt{(k-1)^2 + 4y}\right)^3} dy < \frac{k}{(n+1)|k-1|^3}.$$

Now we prove that the second limit holds. We have

$$I_n - \frac{k}{k+1} = \frac{k-1}{k^n} \int_0^k \frac{y^n}{\left(\sqrt{(k-1)^2 + 4y}\right)^3} dy.$$

We integrate by parts and we get

$$I_n - \frac{k}{k+1} = \frac{k(k-1)}{(n+1)(k+1)^3} + \frac{k-1}{k^n} \cdot \frac{6}{n+1} \int_0^k \frac{y^{n+1}}{\left(\sqrt{(k-1)^2 + 4y}\right)^5} dy.$$

This implies that

$$\lim_{n \to \infty} n \left(I_n - \frac{k}{k+1} \right)$$

$$= \lim_{n \to \infty} \left[\frac{k(k-1)n}{(n+1)(k+1)^3} + \frac{k-1}{k^n} \cdot \frac{6n}{n+1} \int_0^k \frac{y^{n+1}}{\left(\sqrt{(k-1)^2 + 4y}\right)^5} dy \right]$$

$$= \frac{k(k-1)}{(k+1)^3},$$

since

$$0 < \frac{1}{k^n} \int_0^k \frac{y^{n+1}}{\left(\sqrt{(k-1)^2 + 4y}\right)^5} dy < \frac{k^2}{(n+2)|k-1|^5}.$$

(b) Using the substitution $x = y^n$, we have

$$I_n = \int_0^1 \left(\frac{k}{\sqrt[n]{x} + k - 1}\right)^n dx = nk^n \int_0^1 \left(\frac{y}{y + k - 1}\right)^{n-1} \frac{dy}{y + k - 1}.$$

We make the substitution $\frac{y}{y+k-1} = t \Rightarrow y = \frac{(k-1)t}{1-t}$ and we obtain, after some calculations, that

$$I_n = nk^n \int_0^{\frac{1}{k}} \frac{t^{n-1}}{1 - t} dt.$$

We integrate by parts and we get

$$I_n = \frac{k}{k-1} - k^n \int_0^{\frac{1}{k}} \frac{t^n}{(1-t)^2} dt.$$

It follows $\lim_{n \to \infty} I_n = \frac{k}{k-1}$, since

$$0 < k^n \int_0^{\frac{1}{k}} \frac{t^n}{(1-t)^2} dt < \frac{k^{n+2}}{(k-1)^2} \int_0^{\frac{1}{k}} t^n dt = \frac{k}{(n+1)(k-1)^2}.$$

Now we prove that the second limit holds. We have

$$\frac{k}{k-1} - I_n = k^n \int_0^{\frac{1}{k}} \frac{t^n}{(1-t)^2} dt.$$

We integrate by parts and we obtain

$$\frac{k}{k-1} - I_n = \frac{1}{n+1} \cdot \frac{k}{(k-1)^2} - \frac{2k^n}{n+1} \int_0^{\frac{1}{k}} \frac{t^{n+1}}{(1-t)^3} dt.$$

This implies that

$$\lim_{n \to \infty} n \left[\frac{k}{k-1} - \int_0^1 \left(\frac{k}{\sqrt[n]{x}+k-1} \right)^n dx \right]$$

$$= \lim_{n \to \infty} \left[\frac{n}{n+1} \cdot \frac{k}{(k-1)^2} - \frac{2k^n n}{n+1} \int_0^{\frac{1}{k}} \frac{t^{n+1}}{(1-t)^3} dt \right]$$

$$= \frac{k}{(k-1)^2},$$

since

$$0 < k^n \int_0^{\frac{1}{k}} \frac{t^{n+1}}{(1-t)^3} dt < \frac{k^{n+3}}{(k-1)^3} \int_0^{\frac{1}{k}} t^{n+1} dt = \frac{k}{(k-1)^3(n+2)}.$$

1.48.

(a) Using the substitution $x = t^n$, we have that

$$\frac{4^n}{\sqrt{n}} \int_0^1 \left(1 - \sqrt[n]{x} \right)^n dx = 4^n \sqrt{n} \int_0^1 (1-t)^n t^{n-1} dt = 4^n \sqrt{n} \cdot \frac{n!(n-1)!}{(2n)!}$$

and the limit follows based on Stirling's formula.

(b) Let $\epsilon > 0$. Since f is continuous at 0, there exists $\delta > 0$ such that $|f(x) - f(0)| < \epsilon$, for $|x| < \delta$. We have

$$\frac{4^n}{\sqrt{n}} \int_0^1 \left(1 - \sqrt[n]{x} \right)^n f(x) dx = \frac{4^n}{\sqrt{n}} \int_0^\delta \left(1 - \sqrt[n]{x} \right)^n (f(x) - f(0)) dx$$

$$+ \frac{4^n}{\sqrt{n}} \int_\delta^1 \left(1 - \sqrt[n]{x} \right)^n (f(x) - f(0)) dx$$

$$+ f(0) \frac{4^n}{\sqrt{n}} \int_0^1 \left(1 - \sqrt[n]{x} \right)^n dx.$$

$$(9.6)$$

We can write

$$\left| \frac{4^n}{\sqrt{n}} \int_0^\delta \left(1 - \sqrt[n]{x} \right)^n (f(x) - f(0)) dx \right| \le \epsilon \frac{4^n}{\sqrt{n}} \int_0^\delta \left(1 - \sqrt[n]{x} \right)^n dx$$

$$< \epsilon \frac{4^n}{\sqrt{n}} \int_0^1 \left(1 - \sqrt[n]{x} \right)^n dx$$

and it follows, based on part (a), that

$$\lim_{n\to\infty} \left| \frac{4^n}{\sqrt{n}} \int_0^\delta \left(1 - \sqrt[n]{x}\right)^n (f(x) - f(0))dx \right| \le \epsilon\sqrt{\pi}.$$

This implies, since $\epsilon > 0$ was arbitrary taken, that

$$\lim_{n\to\infty} \frac{4^n}{\sqrt{n}} \int_0^\delta \left(1 - \sqrt[n]{x}\right)^n (f(x) - f(0))dx = 0. \tag{9.7}$$

Since f is Riemann integrable, we have that f is bounded, i.e. there exists $M > 0$ such that $|f(x)| \le M$, $\forall x \in [0, 1]$. It follows

$$\left| \frac{4^n}{\sqrt{n}} \int_\delta^1 \left(1 - \sqrt[n]{x}\right)^n (f(x) - f(0))dx \right| \le 2M \frac{4^n}{\sqrt{n}} (1 - \delta)(1 - \sqrt[n]{\delta})^n.$$

Passing to the limit in the previous inequality we have, since $\lim_{n\to\infty} \frac{4^n}{\sqrt{n}}(1 - \sqrt[n]{\delta})^n = 0$, that

$$\lim_{n\to\infty} \frac{4^n}{\sqrt{n}} \int_\delta^1 \left(1 - \sqrt[n]{x}\right)^n (f(x) - f(0))dx = 0. \tag{9.8}$$

Combining (9.6), (9.7), (9.8), and part (a) of the problem, we obtain

$$\lim_{n\to\infty} \frac{4^n}{\sqrt{n}} \int_0^1 \left(1 - \sqrt[n]{x}\right)^n f(x)dx = \sqrt{\pi} f(0).$$

1.49. *Solution I.*

(a) The limit equals $1 + a$. Let $y_n = \sqrt[n]{\int_0^1 (1 + a_n x^n)^n \, dx}$. We have $y_n \le 1 + a_n$, $\forall n \ge 1$. Let $0 < \epsilon < 1$. We have

$$y_n \ge \sqrt[n]{\int_{\sqrt[n]{\epsilon}}^1 (1 + a_n x^n)^n \, dx} \ge \sqrt[n]{\int_{\sqrt[n]{\epsilon}}^1 (1 + a_n\epsilon)^n \, dx} = (1 + a_n\epsilon)\sqrt[n]{1 - \sqrt[n]{\epsilon}}.$$

Thus,

$$(1 + a_n\epsilon)\sqrt[n]{1 - \sqrt[n]{\epsilon}} \le y_n \le 1 + a_n, \quad \forall n \ge 1.$$

Passing to the limit in the previous inequalities we get, since $\lim_{n\to\infty} \sqrt[n]{1 - \sqrt[n]{\epsilon}} = 1$, that

$$1 + a\epsilon \le \lim_{n\to\infty} y_n \le 1 + a.$$

Since $\epsilon \in (0, 1)$ was arbitrary taken, we obtain, by letting $\epsilon \to 1$, that $\lim\limits_{n\to\infty} y_n = 1 + a$.

Solution II.

(a) Using integration by parts we have

$$\int_0^1 \left(1 + a_n x^n\right)^n dx = x \left(1 + a_n x^n\right)^n \Big|_0^1 - n^2 \int_0^1 a_n x^n \left(1 + a_n x^n\right)^{n-1} dx$$

$$= (1 + a_n)^n - n^2 \int_0^1 \left(1 + a_n x^n\right)^n dx + n^2 \int_0^1 \left(1 + a_n x^n\right)^{n-1} dx$$

$$\geq (1 + a_n)^n - n^2 \int_0^1 \left(1 + a_n x^n\right)^n dx.$$

This implies $\sqrt[n]{\int_0^1 (1 + a_n x^n)^n \, dx} \geq \dfrac{1+a_n}{\sqrt[n]{n^2+1}}$, $\forall n \geq 1$.

Since

$$\frac{1 + a_n}{\sqrt[n]{n^2 + 1}} \leq \sqrt[n]{\int_0^1 (1 + a_n x^n)^n \, dx} \leq 1 + a_n, \quad \forall n \geq 1,$$

we have, by passing to the limit as $n \to \infty$, that $\lim\limits_{n\to\infty} y_n = 1 + a$.

(b) The limit equals $1 + a$. We have

$$\sqrt[n]{\int_0^1 \left(1 + \frac{a_1 x + a_3 x^3 + \cdots + a_{2n-1} x^{2n-1}}{n}\right)^n dx} \leq 1 + \frac{a_1 + a_3 + \cdots + a_{2n-1}}{n}.$$

The $AM - GM$ inequality implies that

$$\sqrt[n]{\int_0^1 \left(1 + \frac{a_1 x + a_3 x^3 + \cdots + a_{2n-1} x^{2n-1}}{n}\right)^n dx} \geq \sqrt[n]{\int_0^1 \left(1 + \sqrt[n]{a_1 a_3 \cdots a_{2n-1}} x^n\right)^n dx}.$$

Thus,

$$\sqrt[n]{\int_0^1 \left(1 + \sqrt[n]{a_1 a_3 \cdots a_{2n-1}} x^n\right)^n dx} \leq \sqrt[n]{\int_0^1 \left(1 + \frac{a_1 x + a_3 x^3 + \cdots + a_{2n-1} x^{2n-1}}{n}\right)^n dx}$$

$$\leq 1 + \frac{a_1 + a_3 + \cdots + a_{2n-1}}{n}.$$

Since $\lim\limits_{n\to\infty} \sqrt[n]{a_1 a_3 \cdots a_{2n-1}} = a = \lim\limits_{n\to\infty} \dfrac{a_1 + a_3 + \cdots + a_{2n-1}}{n}$, we have, by part (a) and by passing to the limit as $n \to \infty$ in the previous inequalities, that the limit equals $1 + a$.

(c) The limit equals $1 + a$. See the solution of part (b).

1.51.

(a) Using the substitution $nx = y$, we have

$$\int_0^1 f(\{nx\})dx = \frac{1}{n}\int_0^n f(\{y\})\,dy$$

$$= \frac{1}{n}\sum_{k=0}^{n-1}\int_k^{k+1} f(y-k)\,dy$$

$$\overset{y-k=t}{=} \frac{1}{n}\sum_{k=0}^{n-1}\int_0^1 f(t)dt$$

$$= \int_0^1 f(t)dt.$$

(b) See the solution of part (a) of Problem 1.52.

1.52.

(a) Using the substitution $nx = y$, we have

$$\int_0^1 f(\lfloor nx\rfloor)\,g(\{nx\})\,dx = \frac{1}{n}\int_0^n f(\lfloor y\rfloor)\,g(\{y\})\,dy$$

$$= \frac{1}{n}\sum_{k=0}^{n-1}\int_k^{k+1} f(k)g(y-k)dy$$

$$\overset{y-k=t}{=} \frac{1}{n}\sum_{k=0}^{n-1} f(k)\int_0^1 g(t)dt.$$

(b) We apply Stolz–Cesàro's theorem, the ∞/∞ case, and we have, based on part (a) of the problem, that

$$\lim_{n\to\infty}\frac{1}{n^\alpha}\int_0^1 f(\lfloor nx\rfloor)\,g(\{nx\})\,dx$$

$$= \lim_{n\to\infty}\frac{f(0)+f(1)+\cdots+f(n-1)}{n^{1+\alpha}}\int_0^1 g(x)dx$$

$$= \lim_{n\to\infty}\frac{f(n)}{(1+n)^{1+\alpha}-n^{1+\alpha}}\int_0^1 g(x)dx$$

$$= \lim_{n\to\infty}\frac{f(n)}{n^\alpha}\cdot\lim_{n\to\infty}\frac{n^\alpha}{(1+n)^{1+\alpha}-n^{1+\alpha}}\int_0^1 g(x)dx$$

$$= \frac{L}{1+\alpha}\int_0^1 g(x)dx.$$

1.53. The limit equals $\int_0^1 f(x)dx$. Since $f = f^+ - f^-$, where f^+ and f^- are the positive and negative parts of f, without losing the generality, we consider that $f \geq 0$. Using the substitution $\frac{n}{x} = t$, we have

$$\int_0^1 f\left(\left\{\frac{n}{x}\right\}\right) dx = n \int_n^\infty \frac{f(\{t\})}{t^2} dt$$

$$= n \sum_{k=n}^\infty \int_k^{k+1} \frac{f(\{t\})}{t^2} dt$$

$$= n \sum_{k=n}^\infty \int_k^{k+1} \frac{f(t-k)}{t^2} dt$$

$$\overset{t-k=y}{=\!=\!=} n \sum_{k=n}^\infty \int_0^1 \frac{f(y)}{(k+y)^2} dy$$

$$= \int_0^1 f(y) \left(\sum_{k=n}^\infty \frac{n}{(k+y)^2}\right) dy.$$

Since

$$\sum_{k=n}^\infty \frac{n}{(k+1)^2} < \sum_{k=n}^\infty \frac{n}{(k+y)^2} < \sum_{k=n}^\infty \frac{n}{k^2},$$

it follows that

$$\sum_{k=n}^\infty \frac{n}{(k+1)^2} \int_0^1 f(y)dy < \int_0^1 f\left(\left\{\frac{n}{x}\right\}\right) dx < \sum_{k=n}^\infty \frac{n}{k^2} \int_0^1 f(y)dy.$$

Passing to the limit in the preceding inequalities and using the limit

$$\lim_{n \to \infty} n \sum_{k=n}^\infty \frac{1}{(k+p)^2} = 1, \quad p > 0,$$

we obtain $\lim_{n \to \infty} \int_0^1 f\left(\left\{\frac{n}{x}\right\}\right) dx = \int_0^1 f(x)dx$.

A more general problem can be found in [26, Problem 1.70, p. 13].

1.54.

(a) The limit equals 1. Using the substitution $\frac{1}{x} = t$, we have

$$\int_0^1 \left\{\frac{1}{x}\right\}^n dx = \int_1^\infty \frac{\{t\}^n}{t^2} dt > \int_1^2 \frac{\{t\}^n}{t^2} dt = \int_1^2 \frac{(t-1)^n}{t^2} dt > \frac{1}{4(n+1)}$$

and this implies that

$$\frac{1}{\sqrt[n]{4(n+1)}} < \sqrt[n]{\int_0^1 \left\{\frac{1}{x}\right\}^n dx} < 1.$$

(b) The limit equals 1. Since f is continuous on $[0, 1]$, there exist $m, M > 0$ such that $0 \neq m \leq f(x) \leq M, \forall x \in [0, 1]$. It follows that

$$\sqrt[n]{m \int_0^1 \left\{\frac{1}{x}\right\}^n dx} \leq \sqrt[n]{\int_0^1 \left\{\frac{1}{x}\right\}^n f(x)dx} \leq \sqrt[n]{M \int_0^1 \left\{\frac{1}{x}\right\}^n dx},$$

and the desired limit holds, based on part (a) of the problem.

1.55.

(a) Using the substitution $\frac{1}{x} = t$, we have

$$\int_0^1 \frac{1}{\left\lfloor\frac{1}{x}\right\rfloor^n} dx = \int_1^\infty \frac{1}{t^2 \lfloor t \rfloor^n} dt = \sum_{i=1}^\infty \frac{1}{i^n} \int_i^{i+1} \frac{1}{t^2} dt = \sum_{i=1}^\infty \frac{1}{i^n} \left(\frac{1}{i} - \frac{1}{i+1}\right)$$

$$= \zeta(n+1) - \sum_{i=1}^\infty \frac{1}{i^n(i+1)} = \zeta(n+1) - \frac{1}{2} - \sum_{i=2}^\infty \frac{1}{i^n(i+1)}.$$

It follows

$$\lim_{n\to\infty} \int_0^1 \frac{1}{\left\lfloor\frac{1}{x}\right\rfloor^n} dx = \lim_{n\to\infty} \zeta(n+1) - \frac{1}{2} - \lim_{n\to\infty} \sum_{i=2}^\infty \frac{1}{i^n(i+1)} = \frac{1}{2}.$$

We used that $\lim_{n\to\infty} \zeta(n+1) = 1$ (see part (a) of Problem 1.41) and

$$\sum_{i=2}^\infty \frac{1}{i^n(i+1)} < \sum_{i=2}^\infty \frac{1}{i^n} = \zeta(n) - 1,$$

which shows $\lim_{n\to\infty} \sum_{i=2}^\infty \frac{1}{i^n(i+1)} = 0$.

(b) Using the substitution $\frac{1}{x} = t$, we have

$$\int_0^1 f\left(\frac{1}{\left\lfloor\frac{1}{x}\right\rfloor^n}\right)dx = \int_1^\infty \frac{f\left(\frac{1}{\lfloor t\rfloor^n}\right)}{t^2}dt = \sum_{i=1}^\infty \int_i^{i+1} \frac{f\left(\frac{1}{i^n}\right)}{t^2}dt$$

$$= \sum_{i=1}^\infty f\left(\frac{1}{i^n}\right)\left(\frac{1}{i} - \frac{1}{i+1}\right)$$

$$= \frac{f(1)}{2} + \sum_{i=2}^\infty f\left(\frac{1}{i^n}\right)\left(\frac{1}{i} - \frac{1}{i+1}\right).$$

Let $\epsilon > 0$. Since f is continuous at 0, there exists $\delta > 0$ such that $|f(x) - f(0)| < \epsilon$, for $|x| < \delta$. Let $n_0 = \left\lfloor \frac{\ln\frac{1}{\delta}}{\ln 2} \right\rfloor + 1$.

We write

$$\sum_{i=2}^\infty f\left(\frac{1}{i^n}\right)\left(\frac{1}{i} - \frac{1}{i+1}\right) = \sum_{i=2}^{n_0}\left[f\left(\frac{1}{i^n}\right) - f(0)\right]\left(\frac{1}{i} - \frac{1}{i+1}\right)$$

$$+ \sum_{i=n_0+1}^\infty \left[f\left(\frac{1}{i^n}\right) - f(0)\right]\left(\frac{1}{i} - \frac{1}{i+1}\right) + \frac{f(0)}{2}.$$

For $n \geq n_0+1$ and $i \geq 2$, we have $\frac{1}{i^n} < \frac{1}{2^n} < \frac{1}{2^{n_0}} < \delta \Rightarrow \left|f\left(\frac{1}{i^n}\right) - f(0)\right| < \epsilon$. It follows that

$$\left|\sum_{i=n_0+1}^\infty \left[f\left(\frac{1}{i^n}\right) - f(0)\right]\left(\frac{1}{i} - \frac{1}{i+1}\right)\right| \leq \epsilon \sum_{i=n_0+1}^\infty \left(\frac{1}{i} - \frac{1}{i+1}\right) < \epsilon,$$

which implies $\lim_{n\to\infty} \sum_{i=n_0+1}^\infty \left[f\left(\frac{1}{i^n}\right) - f(0)\right]\left(\frac{1}{i} - \frac{1}{i+1}\right) = 0$.

The continuity of f at 0 shows that

$$\lim_{n\to\infty} \sum_{i=2}^{n_0}\left[f\left(\frac{1}{i^n}\right) - f(0)\right]\left(\frac{1}{i} - \frac{1}{i+1}\right) = 0.$$

Putting all these together the desired limit holds and the problem is solved.

1.56 and **1.57**. See the solution of Problem 1.55.

1.58. Let $\beta \in (-1, 0)$ be the unique solution of the equation $x + e^{\alpha x} = 0$, $x \in [-1, 0]$.

We have

$$n \int_{-1}^{0} \left(x + e^{\alpha x}\right)^n f(x)dx = n \int_{-1}^{\beta} \left(x + e^{\alpha x}\right)^n f(x)dx + n \int_{\beta}^{0} \left(x + e^{\alpha x}\right)^n f(x)dx.$$

A calculation shows

$$\left| n \int_{-1}^{\beta} \left(x + e^{\alpha x}\right)^n f(x)dx \right| \le n\|f\| \int_{-1}^{\beta} |x + e^{\alpha x}|^n dx \le n\|f\|(1 - e^{-\alpha})^n(1 + \beta),$$

and it follows, by passing to the limit as $n \to \infty$ in the previous inequalities, that

$$\lim_{n \to \infty} n \int_{-1}^{\beta} \left(x + e^{\alpha x}\right)^n f(x)dx = 0.$$

The substitution $h(x) = x + e^{\alpha x} = y$ shows that $x = h^{-1}(y) \Rightarrow dx = \frac{1}{h'(h^{-1}(y))}dy = \frac{1}{1 + \alpha e^{\alpha h^{-1}(y)}}dy$ and we have

$$n \int_{\beta}^{0} \left(x + e^{\alpha x}\right)^n f(x)dx = n \int_{0}^{1} y^n \frac{f(h^{-1}(y))}{1 + \alpha e^{\alpha h^{-1}(y)}}dy.$$

It follows, based on Problem 7.28, that

$$\lim_{n \to \infty} n \int_{\beta}^{0} \left(x + e^{\alpha x}\right)^n f(x)dx = \lim_{n \to \infty} n \int_{0}^{1} y^n \frac{f(h^{-1}(y))}{1 + \alpha e^{\alpha h^{-1}(y)}}dy$$

$$= \frac{f(h^{-1}(1))}{1 + \alpha e^{\alpha h^{-1}(1)}}$$

$$= \frac{f(0)}{1 + \alpha}.$$

1.59. The limit equals $\frac{1}{1+\alpha}$. See the technique used in the solution of Problem 1.58.

1.60. The limit equals α. Since $\sqrt[n]{\alpha} \le \sqrt[n]{x^n + \alpha}$, $x \in [0, 1]$, we have

$$\alpha \le \left(\int_{0}^{1} \sqrt[n]{x^n + \alpha}\,dx \right)^n.$$

We apply Hölder's inequality for the integral[1] and we obtain that

---

[1] *Hölder's inequality for the integral.* If $p > 1$ and $f : [0, 1] \to \mathbb{R}$ is a Riemann integrable function, then $\left(\int_{0}^{1} |f(x)|\,dx \right)^p \le \int_{0}^{1} |f(x)|^p dx.$

$$\left(\int_0^1 \sqrt[n]{x^n + \alpha}\,dx\right)^n \le \int_0^1 (x^n + \alpha)\,dx = \frac{1}{n+1} + \alpha.$$

Thus,

$$\alpha \le \left(\int_0^1 \sqrt[n]{x^n + \alpha}\,dx\right)^n \le \frac{1}{n+1} + \alpha.$$

1.61. The limit equals 1. The problem, in a general form, is discussed in [55].

1.62. The limit equals 1. Observe that

$$\int_0^n \frac{dx}{1 + n^2 \sin^2(x + na)} = \int_{na}^{n(a+1)} \frac{dy}{1 + n^2 \sin^2 y}$$

$$= \int_0^{n(a+1)} \frac{dy}{1 + n^2 \sin^2 y} - \int_0^{na} \frac{dy}{1 + n^2 \sin^2 y}.$$

Now prove that if $b > 0$, then

$$\lim_{n \to \infty} \int_0^{nb} \frac{dx}{1 + n^2 \sin^2 x} = b.$$

1.63. The limit equals $\frac{1}{2}$. First we show that if $a \in \mathbb{R}$, then

$$\int_0^\pi \frac{dx}{1 + a^2 \cos^2 x} = \frac{\pi}{\sqrt{1 + a^2}}. \tag{9.9}$$

We have

$$\int_0^\pi \frac{dx}{1 + a^2 \cos^2 x} = 2 \int_0^{\pi/2} \frac{dx}{1 + a^2 \cos^2 x}$$

$$\overset{\tan x = t}{=} 2 \int_0^\infty \frac{dt}{t^2 + 1 + a^2}$$

$$= \frac{2}{\sqrt{1 + a^2}} \arctan \frac{t}{\sqrt{1 + a^2}} \Big|_0^\infty$$

$$= \frac{\pi}{\sqrt{1 + a^2}}.$$

Let $k = \left\lfloor \frac{n}{\pi} \right\rfloor \in \mathbb{N}$ and we observe that $k\pi \le n < (k+1)\pi$. Let

$$I_n = \int_0^n \frac{x}{1 + n^2 \cos^2 x}\,dx.$$

Since $k\pi \le n < (k+1)\pi$, it follows

$$\int_0^{k\pi} \frac{x}{1+n^2 \cos^2 x}\,dx \le I_n < \int_0^{(k+1)\pi} \frac{x}{1+n^2 \cos^2 x}\,dx. \tag{9.10}$$

Let $m \in \mathbb{N}$ and let $J_m = \int_0^{m\pi} \frac{x}{1+n^2 \cos^2 x}\,dx$. We calculate J_m using the substitution $m\pi - x = t$ and we have

$$J_m = \int_0^{m\pi} \frac{m\pi - t}{1+n^2 \cos^2 t}\,dt$$

$$= m\pi \int_0^{m\pi} \frac{1}{1+n^2 \cos^2 t}\,dt - J_m$$

$$= m^2\pi \int_0^{\pi} \frac{1}{1+n^2 \cos^2 t}\,dt - J_m$$

$$\overset{(9.9)}{=} \frac{\pi^2 m^2}{\sqrt{1+n^2}} - J_m.$$

Therefore,

$$J_m = \frac{\pi^2 m^2}{2\sqrt{1+n^2}}. \tag{9.11}$$

Using relations (9.10) and (9.11), one has

$$\frac{\pi^2 k^2}{2n\sqrt{1+n^2}} \le \frac{I_n}{n} < \frac{\pi^2 (k+1)^2}{2n\sqrt{1+n^2}}, \quad \text{where} \quad k = \left\lfloor \frac{n}{\pi} \right\rfloor.$$

Passing to the limit in the preceding inequalities, we obtain that

$$\lim_{n\to\infty} \frac{1}{n} \int_0^n \frac{x}{1+n^2 \cos^2 x}\,dx = \frac{1}{2}.$$

Remark. Using the same method one can also prove that (see [55])

$$\lim_{n\to\infty} \frac{1}{n} \int_0^n \frac{x}{1+n^2 \sin^2 x}\,dx = \frac{1}{2}.$$

1.65.

(a) The limit equals $\frac{1}{2}$. Integrate by parts.

(b) The limit equals $\frac{1}{k+1}$. The substitution $nx = y$ shows that

$$\frac{1}{n^k} \int_0^1 \ln^k\left(1 + e^{nx}\right) dx = \frac{1}{n^{k+1}} \int_0^n \ln^k\left(1 + e^y\right) dy.$$

Using Stolz–Cesàro Theorem, the ∞/∞ case, we have that

$$\lim_{n\to\infty} \frac{1}{n^{k+1}} \int_0^n \ln^k\left(1 + e^y\right) dy = \lim_{n\to\infty} \frac{\int_n^{n+1} \ln^k\left(1 + e^y\right) dy}{(n+1)^{k+1} - n^{k+1}}$$

$$= \lim_{n\to\infty} \frac{\ln^k\left(1 + e^{\theta_n}\right)}{(n+1)^{k+1} - n^{k+1}}$$

$$= \frac{1}{k+1},$$

where $\theta_n \in (n, n+1)$.

We used, in the previous calculations, the limit $\lim_{n\to\infty} \frac{\ln\left(1+e^{\theta_n}\right)}{n} = 1$, which follows from the inequalities

$$\frac{\ln\left(1 + e^n\right)}{n} < \frac{\ln\left(1 + e^{\theta_n}\right)}{n} < \frac{\ln\left(1 + e^{n+1}\right)}{n}.$$

1.66. The limit equals b. We have

$$\sqrt[n]{\frac{b^{n+1} - a^{n+1}}{n+1}} = \frac{1}{n}\sqrt[n]{\int_a^b \ln^n(e^{nx})dx}$$

$$\leq \frac{1}{n}\sqrt[n]{\int_a^b \ln^n(1 + e^{nx})dx}$$

$$\leq \frac{\ln(1 + e^{nb})}{n} \cdot \sqrt[n]{b - a}.$$

Also, we have

$$\lim_{n\to\infty} \frac{\ln(1 + e^{nb})}{n} \cdot \sqrt[n]{b - a} = b,$$

$$\lim_{n\to\infty} \sqrt[n]{\frac{b^{n+1} - a^{n+1}}{n+1}} = \lim_{n\to\infty} b^{\frac{n+1}{n}} \cdot \lim_{n\to\infty} \sqrt[n]{1 - \left(\frac{a}{b}\right)^{n+1}} \cdot \lim_{n\to\infty} \frac{1}{\sqrt[n]{n+1}} = b,$$

and the result follows based on the Squeeze Theorem.

1.67.

(b) The limit equals 2. Since f is continuous, there exist m and M such that $0 < m \le f(x) \le M$, $\forall x \in [0, 1]$. We have

$$\sqrt[n]{\int_0^1 (1 + x^n)^n f(x)\,dx} \le 2\sqrt[n]{M}.$$

The inequality $1 + x^n \ge 2x^{\frac{n}{2}}$ implies that

$$\sqrt[n]{\int_0^1 (1 + x^n)^n f(x)\,dx} \ge 2\sqrt[n]{\int_0^1 x^{\frac{n^2}{2}} f(x)\,dx}.$$

We have, since $f(x) \ge m$, that

$$2\sqrt[n]{\frac{2m}{n^2 + 2}} \le \sqrt[n]{\int_0^1 (1 + x^n)^n f(x)\,dx} \le 2\sqrt[n]{M},$$

and the desired limit follows based on the Squeeze Theorem.

1.68. The solution is given in [29].
1.69. We have

$$
\begin{aligned}
I_n &= \int_0^{\frac{\pi}{2}} f(x) (\cos x - \sin x)^{2n}\,dx \\
&= \int_0^{\frac{\pi}{2}} f(x) (1 - \sin(2x))^n\,dx \\
&\overset{2x=t}{=} \frac{1}{2} \int_0^{\pi} f\left(\frac{t}{2}\right) (1 - \sin t)^n\,dt \\
&= \frac{1}{2} \int_0^{\frac{\pi}{2}} \left[f(t) + f\left(\frac{\pi - t}{2}\right) \right] (1 - \sin t)^n\,dt.
\end{aligned}
$$

We use the substitution $1 - \sin t = y$ and we have that $t = \arcsin(1 - y)$, which implies $dt = -\frac{1}{\sqrt{2y - y^2}}dy$. It follows that

$$I_n = \int_0^1 F(y)\frac{y^{n - \frac{1}{2}}}{\sqrt{2 - y}}\,dy = \int_0^1 g(y)y^{n - \frac{1}{2}}\,dy,$$

where F and g are the continuous functions

$$F(y) = \frac{1}{2}\left[f\left(\frac{\arcsin(1 - y)}{2}\right) + f\left(\frac{\pi - \arcsin(1 - y)}{2}\right) \right] \quad \text{and} \quad g(y) = \frac{F(y)}{\sqrt{2 - y}}.$$

We prove that if $h : [0, 1] \to \mathbb{R}$ is continuous, then

$$\lim_{n\to\infty} n \int_0^1 h(x)x^{n-\frac{1}{2}}\,dx = h(1),$$

see also Problem 7.28, and this implies that

$$\lim_{n\to\infty} nI_n = g(1) = F(1) = \frac{f(0) + f\left(\frac{\pi}{2}\right)}{2}.$$

Let $\epsilon > 0$. Since h is continuous at 1, there exists $\delta > 0$ such that $|h(x) - h(1)| < \epsilon$, for $1 - \delta < x < 1$. Therefore,

$$\int_0^1 h(x)x^{n-\frac{1}{2}}\,dx = \int_0^{1-\delta} h(x)x^{n-\frac{1}{2}}\,dx + \int_{1-\delta}^1 (h(x) - h(1))x^{n-\frac{1}{2}}\,dx + h(1)\int_{1-\delta}^1 x^{n-\frac{1}{2}}\,dx.$$

$$(9.12)$$

On the one hand

$$\left| n \int_0^{1-\delta} h(x)x^{n-\frac{1}{2}}\,dx \right| \le \|h\|\frac{n(1-\delta)^{n+\frac{1}{2}}}{n+\frac{1}{2}},$$

which implies that $\lim_{n\to\infty} n \int_0^{1-\delta} h(x)x^{n-\frac{1}{2}}\,dx = 0$. Also,

$$\lim_{n\to\infty} n \int_{1-\delta}^1 x^{n-\frac{1}{2}}\,dx = \lim_{n\to\infty} \frac{n}{n+\frac{1}{2}}\left(1 - (1-\delta)^{n+\frac{1}{2}}\right) = 1.$$

On the other hand

$$-\epsilon\frac{n}{n+\frac{1}{2}}\left(1 - (1-\delta)^{n+\frac{1}{2}}\right) < n \int_{1-\delta}^1 (h(x) - h(1))x^{n-\frac{1}{2}}\,dx$$

$$< \epsilon\frac{n}{n+\frac{1}{2}}\left(1 - (1-\delta)^{n+\frac{1}{2}}\right),$$

and we have $-\epsilon \le \lim_{n\to\infty} n \int_{1-\delta}^1 (h(x) - h(1))x^{n-\frac{1}{2}}\,dx \le \epsilon$.

Since ϵ was arbitrary taken, we obtain that $\lim_{n\to\infty} n \int_{1-\delta}^1 (h(x) - h(1))x^{n-\frac{1}{2}}\,dx = 0$.

Equality (9.12) implies $\lim_{n\to\infty} n \int_0^1 h(x)x^{n-\frac{1}{2}}\,dx = h(1)$.

1.71. The limit equals 0. We have

$$n\left(\sqrt[n+1]{\int_a^b f^{n+1}(x)dx} - \sqrt[n]{\int_a^b f^n(x)dx}\right) = n\left(e^{\frac{1}{n+1}\ln\int_a^b f^{n+1}(x)dx} - e^{\frac{1}{n}\ln\int_a^b f^n(x)dx}\right)$$

$$= n\left(\frac{1}{n+1}\ln\int_a^b f^{n+1}(x)dx - \frac{1}{n}\ln\int_a^b f^n(x)dx\right)e^{\theta_n}$$

$$= \left(\frac{n}{n+1}\ln\int_a^b f^{n+1}(x)dx - \ln\int_a^b f^n(x)dx\right)e^{\theta_n}$$

$$= \left(\ln\frac{\int_a^b f^{n+1}(x)dx}{\int_a^b f^n(x)dx} - \ln\sqrt[n+1]{\int_a^b f^{n+1}(x)dx}\right)e^{\theta_n},$$

where θ_n is between $\ln\sqrt[n+1]{\int_a^b f^{n+1}(x)dx}$ and $\ln\sqrt[n]{\int_a^b f^n(x)dx}$.

This shows that $\lim_{n\to\infty}\theta_n = \ln\|f\|$. Passing to the limit in the previous equality, the result follows since $\lim_{n\to\infty}\frac{\int_a^b f^{n+1}(x)dx}{\int_a^b f^n(x)dx} = \|f\|$ (see [26, problem 1.47, p. 9]) and $\lim_{n\to\infty}\sqrt[n+1]{\int_a^b f^{n+1}(x)dx} = \|f\|$.

1.72. The limit equals $\|f\|_\infty \ln\|f\|_\infty$, where $\|f\|_\infty = \sup\{f(x) : x \in [0, \infty)\}$. First we need to prove that the following limit holds:

$$\lim_{n\to\infty}\sqrt[n]{\int_0^\infty f^n(x)e^{-x}dx} = \|f\|_\infty. \tag{9.13}$$

It is easy to see that

$$\overline{\lim}\sqrt[n]{\int_0^\infty f^n(x)e^{-x}dx} \leq \|f\|_\infty. \tag{9.14}$$

Let $b > a \geq 0$ be fixed real numbers. Then

$$\int_0^\infty f^n(x)e^{-x}dx \geq \int_a^b f^n(x)e^{-x}dx \geq e^{-b}\int_a^b f^n(x)dx,$$

and it follows that

$$\sqrt[n]{\int_0^\infty f^n(x)e^{-x}dx} \geq e^{-\frac{b}{n}}\sqrt[n]{\int_a^b f^n(x)dx}.$$

Thus,

$$\underline{\lim} \sqrt[n]{\int_0^\infty f^n(x)e^{-x}dx} \geq ||f||_{[a,b]} = \sup\{f(x) : x \in [a,b]\}.$$

Letting $a \to 0$ and $b \to \infty$ in the preceding inequality, we get that

$$\underline{\lim} \sqrt[n]{\int_0^\infty f^n(x)e^{-x}dx} \geq ||f||_\infty. \tag{9.15}$$

Combining (9.14) and (9.15), we have that (9.13) is proved.

Now we prove that

$$\lim_{n\to\infty} \frac{\int_0^\infty f^{n+1}(x)e^{-x}dx}{\int_0^\infty f^n(x)e^{-x}dx} = ||f||_\infty. \tag{9.16}$$

Let $M = ||f||_\infty$. We have $\int_0^\infty f^{n+1}(x)e^{-x}dx \leq M \int_0^\infty f^n(x)e^{-x}dx$, and hence

$$\frac{\int_0^\infty f^{n+1}(x)e^{-x}dx}{\int_0^\infty f^n(x)e^{-x}dx} \leq M.$$

An application of Hölder's inequality, with $p = \frac{n+1}{n}$ and $q = n+1$, shows that

$$\int_0^\infty f^n(x)e^{-x}dx \leq \left(\int_0^\infty f^{n+1}(x)e^{-x}dx\right)^{\frac{n}{n+1}} \left(\int_0^\infty e^{-x}dx\right)^{\frac{1}{n+1}},$$

and it follows

$$\sqrt[n]{\int_0^\infty f^n(x)e^{-x}dx} \leq \frac{\int_0^\infty f^{n+1}(x)e^{-x}dx}{\int_0^\infty f^n(x)e^{-x}dx}. \tag{9.17}$$

Letting $n \to \infty$ in (9.17) and using (9.13), we get that

$$M \leq \underline{\lim} \frac{\int_0^\infty f^{n+1}(x)e^{-x}dx}{\int_0^\infty f^n(x)e^{-x}dx} \quad \text{and this implies that} \quad \lim_{n\to\infty} \frac{\int_0^\infty f^{n+1}(x)e^{-x}dx}{\int_0^\infty f^n(x)e^{-x}dx} = M.$$

Now we are ready to solve the problem. We have, based on Lagrange Mean Value Theorem, that

$$x_n = n\left(\sqrt[n]{\int_0^\infty f^{n+1}(x)e^{-x}dx} - \sqrt[n]{\int_0^\infty f^n(x)e^{-x}dx}\right)$$

$$= n\left(\exp\left(\frac{1}{n}\ln\int_0^\infty f^{n+1}(x)e^{-x}dx\right) - \exp\left(\frac{1}{n}\ln\int_0^\infty f^n(x)e^{-x}dx\right)\right)$$

$$= \ln\left(\frac{\int_0^\infty f^{n+1}(x)e^{-x}dx}{\int_0^\infty f^n(x)e^{-x}dx}\right)\exp(\theta_n),$$

where θ_n is between $\ln\sqrt[n]{\int_0^\infty f^{n+1}(x)e^{-x}dx}$ and $\ln\sqrt[n]{\int_0^\infty f^n(x)e^{-x}dx}$. This implies that $\lim_{n\to\infty}\theta_n = \ln\|f\|_\infty$. Letting $n\to\infty$ in the preceding equality and using (9.16), we get that

$$\lim_{n\to\infty} x_n = \|f\|_\infty \ln\|f\|_\infty.$$

If f is not bounded, then the limit might not be finite, an example being $f(x) = x$, in which case the limit equals ∞.

1.73.

(a) $L = \int_a^b f(x)dx$. Use the technique given in the solution of Problem 7.37 and prove that

$$\left|\int_a^b f(x)dx - \int_a^b \frac{f(x)}{1 + \sin x \sin(x+1)\cdots\sin(x+n)}dx\right|$$

$$\leq M(b-a)\frac{\left(\frac{1}{2} + \frac{1}{2(n+1)\sin 1}\right)^{\frac{n+1}{2}}}{1 - \left(\frac{1}{2} + \frac{1}{2(n+1)\sin 1}\right)^{\frac{n+1}{2}}},$$

where $M = \sup\{|f(x)| : x \in [a,b]\}$.

(b) The limit equals 0. We have

$$q^n\left|\int_a^b f(x)dx - \int_a^b \frac{f(x)}{1 + \sin x \sin(x+1)\cdots\sin(x+n)}dx\right|$$

$$\leq M(b-a)q^n\frac{\left(\frac{1}{2} + \frac{1}{2(n+1)\sin 1}\right)^{\frac{n+1}{2}}}{1 - \left(\frac{1}{2} + \frac{1}{2(n+1)\sin 1}\right)^{\frac{n+1}{2}}},$$

and the result follows since

$$\lim_{n \to \infty} q^n \left(\frac{1}{2} + \frac{1}{2(n+1)\sin 1} \right)^{\frac{n+1}{2}}$$

$$= \lim_{n \to \infty} \left(\frac{q}{\sqrt{2}} \right)^n \left(1 + \frac{1}{(n+1)\sin 1} \right)^{\frac{n}{2}} \cdot \sqrt{\frac{1}{2} + \frac{1}{2(n+1)\sin 1}} = 0.$$

1.74.

(a) The limit equals 0. It suffices to consider only the case when n is even. Also, since the function $x \to |\sin x|$ is a periodic function of period π, it is enough to consider the case when $a = 0$ and $b = \pi$.

  We have

$$\int_0^\pi \sin^{2n} x \, dx = 2 \int_0^{\frac{\pi}{2}} \sin^{2n} x \, dx = \sqrt{\pi} \frac{\Gamma\left(n + \frac{1}{2}\right)}{\Gamma(n+1)}$$

and it follows, based on Stirling's formula, that

$$\lim_{n \to \infty} \int_0^\pi \sin^{2n} x \, dx = 0.$$

(b) The limit equals 0. We have

$$\left| \int_a^b f(x) \sin^n x \, dx \right| \le M \int_a^b |\sin^n x| \, dx,$$

where $M = \{|f(x)| : x \in [a, b]\}$ and the desired limit follows based on part (a) of the problem.

1.77. We have

$$\int_{-x}^x (f(t+1) - f(t)) \, dt = \int_{1-x}^{1+x} f(y) dy - \int_{-x}^x f(t) dt$$

$$= \int_x^{1+x} f(y) dy - \int_{-x}^{1-x} f(t) dt$$

$$= f(\theta_1) - f(\theta_2),$$

where $\theta_1 \in (x, 1+x)$ and $\theta_2 \in (-x, 1-x)$.

  It follows that

$$\lim_{x \to \infty} \int_{-x}^x (f(t+1) - f(t)) \, dt = f(\infty) - f(-\infty),$$

since $f(\theta_1) \to f(\infty)$ and $f(\theta_2) \to f(-\infty)$, when $x \to \infty$.

1.78. The limit equals $-\int_0^1 f(x)dx$. Approximate f by a polynomial function.

1.79. Since $1 - f \le 1 - f^k = (1 - f)(1 + f + \cdots + f^{k-1}) \le k(1 - f)$, it suffices to solve the problem for the case when $k = 1$. Let $A = f^{-1}(1) = \{x \in [0, 1] : f(x) = 1\}$.

We have

$$n \int_0^1 f^n(x)(1-f(x))dx = n \int_A f^n(x)(1-f(x))dx + n \int_{[0,1]\setminus A} f^n(x)(1-f(x))dx$$

$$= n \int_{[0,1]\setminus A} f^n(x)(1-f(x))dx.$$

Let $f_n : [0, 1] \setminus A \to \mathbb{R}$, $f_n(x) = nf^n(x)(1 - f(x))$. Since

$$\lim_{n \to \infty} \frac{f_{n+1}(x)}{f_n(x)} = \lim_{n \to \infty} \frac{n+1}{n} f(x) = f(x) < 1,$$

we get that $\lim_{n \to \infty} f_n(x) = 0$.

A calculation shows

$$nf^n(x) = f^n(x) + \cdots + f^n(x) < 1 + f(x) + f^2(x) + \cdots + f^{n-1}(x) = \frac{1 - f^n(x)}{1 - f(x)}$$

and this implies $nf^n(x)(1 - f(x)) < 1 - f^n(x) < 1, \forall x \in [0, 1] \setminus A$. It follows, based on Lebesgue Dominated Convergence Theorem, that

$$\lim_{n \to \infty} n \int_0^1 f^n(x)(1-f(x))dx = \lim_{n \to \infty} \int_{[0,1]\setminus A} f_n(x)dx = \int_{[0,1]\setminus A} \left(\lim_{n \to \infty} f_n(x) \right) dx = 0.$$

1.80.

(a) This part of the problem is solved in [95].
(b) Solution due to students Kim Il Jin and Mun Chung Jin. We prove that

$$\sqrt[n]{\int_0^1 x^{1+\frac{1}{2}+\cdots+\frac{1}{n}}(1 - x)(1 - \sqrt{x}) \cdots (1 - \sqrt[n]{x})dx} < \frac{1}{(H_n + 1)\sqrt[n]{H_n + 1}}.$$

The preceding inequality is in fact based on the exact calculation of the integral under the nth root. We need the following lemma.

Lemma 9.1. *Let $\{i_1, i_2, \ldots, i_k\}$ be a subset of $\{1, 2, \ldots, n\}$ with distinct integers $i_1, i_2, \ldots, i_k$ and let $F(x) = x^{H_n} \left(1 - x^{\frac{1}{i_1}}\right) \left(1 - x^{\frac{1}{i_2}}\right) \cdots \left(1 - x^{\frac{1}{i_k}}\right)$. Then*

$$\int_0^1 F(x)dx = M \int_0^1 F(x) \left(\frac{\frac{1}{i_1}}{1 - x^{\frac{1}{i_1}}} + \frac{\frac{1}{i_2}}{1 - x^{\frac{1}{i_2}}} + \cdots + \frac{\frac{1}{i_k}}{1 - x^{\frac{1}{i_k}}} \right) dx,$$

where $M = \dfrac{1}{H_n + 1 + \frac{1}{i_1} + \frac{1}{i_2} + \cdots + \frac{1}{i_k}}$.

Proof. We integrate by parts and we have that

$$\int_0^1 F(x)dx = -\int_0^1 x F'(x)dx.$$

A calculation shows that

$$x F'(x) = F(x) \left[H_n + \sum_{j=1}^k \frac{1}{i_j} - \sum_{j=1}^k \frac{\frac{1}{i_j}}{1 - x^{\frac{1}{i_j}}} \right].$$

It follows that

$$\int_0^1 F(x)dx = - \left(H_n + \sum_{j=1}^k \frac{1}{i_j} \right) \int_0^1 F(x)dx$$

$$+ \int_0^1 F(x) \left(\frac{\frac{1}{i_1}}{1 - x^{\frac{1}{i_1}}} + \frac{\frac{1}{i_2}}{1 - x^{\frac{1}{i_2}}} + \cdots + \frac{\frac{1}{i_k}}{1 - x^{\frac{1}{i_k}}} \right) dx,$$

and the lemma is proved. ∎

We apply the lemma to the integral $\int_0^1 x^{H_n}(1 - x)(1 - \sqrt{x}) \cdots (1 - \sqrt[n]{x})dx$ and we note the lemma allows us to remove the factors $1 - x^{\frac{1}{i_j}}$, $j = 1, \ldots, n$, one by one to obtain n integrals. We apply the lemma one more time to each of these n integrals and we obtain $n(n - 1)$ integrals. We continue this process until only the term x^{H_n} remains under the integral sign, and we see that there are $n!$ integrals of the form $a_{i_1, i_2, \ldots, i_n} \int_0^1 x^{H_n} dx$, where the coefficient $a_{i_1, i_2, \ldots, i_n}$ is obtained by eliminating the product

$$\left(1 - x^{\frac{1}{i_1}} \right) \left(1 - x^{\frac{1}{i_2}} \right) \cdots \left(1 - x^{\frac{1}{i_n}} \right),$$

and the distinct integers $i_1, i_2, \ldots, i_n$ are such that $\{i_1, i_2, \ldots, i_n\} = \{1, 2, \ldots, n\}$.
The coefficient, obtained by eliminating the factor

$$\left(1 - x^{\frac{1}{i_1}} \right) \left(1 - x^{\frac{1}{i_2}} \right) \cdots \left(1 - x^{\frac{1}{i_k}} \right),$$

is given by

$$\frac{\frac{1}{i_1}}{H_n+1+\frac{1}{i_1}+\cdots+\frac{1}{i_k}} \cdot \frac{\frac{1}{i_2}}{H_n+1+\frac{1}{i_2}+\cdots+\frac{1}{i_k}} \cdots \frac{\frac{1}{i_k}}{H_n+1+\frac{1}{i_k}}.$$

It follows that

$$a_{i_1,i_2,\ldots,i_n} = \frac{\frac{1}{i_1}}{H_n+1+\frac{1}{i_1}+\cdots+\frac{1}{i_n}} \cdot \frac{\frac{1}{i_2}}{H_n+1+\frac{1}{i_2}+\cdots+\frac{1}{i_n}} \cdots \frac{\frac{1}{i_n}}{H_n+1+\frac{1}{i_n}}$$

$$= \frac{1}{n!} \cdot \frac{1}{H_n+1+\frac{1}{i_1}+\cdots+\frac{1}{i_n}} \cdot \frac{1}{H_n+1+\frac{1}{i_2}+\cdots+\frac{1}{i_n}} \cdots \frac{1}{H_n+1+\frac{1}{i_n}},$$

since $\prod_{i=1}^{n} i_j = n!$. Therefore, we obtain the following formula

$$\int_0^1 x^{H_n}(1-x)(1-\sqrt{x})\cdots(1-\sqrt[n]{x})dx$$

$$= \int_0^1 \frac{1}{n!}\left[\sum \frac{1}{H_n+1+\frac{1}{i_1}+\cdots+\frac{1}{i_n}} \cdot \frac{1}{H_n+1+\frac{1}{i_2}+\cdots+\frac{1}{i_n}} \cdots \frac{1}{H_n+1+\frac{1}{i_n}}\right] x^{H_n}dx$$

$$= \frac{1}{n!(H_n+1)}\sum \frac{1}{H_n+1+\frac{1}{i_1}+\cdots+\frac{1}{i_n}} \cdot \frac{1}{H_n+1+\frac{1}{i_2}+\cdots+\frac{1}{i_n}} \cdots \frac{1}{H_n+1+\frac{1}{i_n}},$$

where the summation is done by the number of all n-tuples $(i_1, i_2, \ldots, i_n)$ of distinct integers $i_1, i_2, \ldots, i_n$, with $\{i_1, i_2, \ldots, i_n\} = \{1, 2, \ldots, n\}$.

It follows, since the previous sum has $n!$ terms and each of the term under the summation sign is strictly less than $\frac{1}{(H_n+1)^n}$, that

$$\int_0^1 x^{H_n}(1-x)(1-\sqrt{x})\cdots(1-\sqrt[n]{x})dx < \frac{1}{H_n+1} \cdot \frac{1}{n!} \cdot \frac{n!}{(H_n+1)^n} = \frac{1}{(H_n+1)^{n+1}}.$$

As an example, we calculate based on the given technique the previous integral for the cases when $n = 2$ and $n = 3$.

The case $n = 2$. We have

$$\int_0^1 x^{H_2}(1-x)\left(1-\sqrt{x}\right)dx = \frac{1}{2!(H_2+1)}\sum \frac{1}{H_2+1+\frac{1}{i}+\frac{1}{j}} \cdot \frac{1}{H_2+1+\frac{1}{j}},$$

where the summation is done by the number of all pairs (i, j) of distinct integers with $\{i, j\} = \{1, 2\}$.

We have $(i, j) \in \{(1, 2), (2, 1)\}$ and it follows that

$$\int_0^1 x^{H_2} (1-x)\left(1-\sqrt{x}\right) dx = \frac{1}{2!(H_2+1)} \cdot \frac{1}{H_2+1+\frac{1}{1}+\frac{1}{2}} \cdot \frac{1}{H_2+1+\frac{1}{2}}$$

$$+ \frac{1}{2!(H_2+1)} \cdot \frac{1}{H_2+1+\frac{1}{2}+\frac{1}{1}} \cdot \frac{1}{H_2+1+\frac{1}{1}}$$

$$= \frac{13}{420}.$$

The case $n = 3$. We have

$$\int_0^1 x^{H_3} (1-x)\left(1-\sqrt{x}\right)\left(1-\sqrt[3]{x}\right) dx$$

$$= \frac{1}{3!(H_3+1)} \sum \frac{1}{H_3+1+\frac{1}{i}+\frac{1}{j}+\frac{1}{k}} \cdot \frac{1}{H_3+1+\frac{1}{j}+\frac{1}{k}} \cdot \frac{1}{H_3+1+\frac{1}{k}},$$

where the summation is done by the number of all 3-tuples (i, j, k) of distinct integers with $\{i, j, k\} = \{1, 2, 3\}$.

We have $(i, j, k) \in \{(1, 2, 3), (1, 3, 2), (2, 3, 1), (2, 1, 3), (3, 1, 2), (3, 2, 1)\}$ and it follows that

$$\int_0^1 x^{H_3} (1-x)\left(1-\sqrt{x}\right)\left(1-\sqrt[3]{x}\right) dx$$

$$= \frac{1}{3!(H_3+1)} \cdot \frac{1}{H_3+1+\frac{1}{1}+\frac{1}{2}+\frac{1}{3}} \cdot \frac{1}{H_3+1+\frac{1}{2}+\frac{1}{3}} \cdot \frac{1}{H_3+1+\frac{1}{3}}$$

$$+ \frac{1}{3!(H_3+1)} \cdot \frac{1}{H_3+1+\frac{1}{1}+\frac{1}{3}+\frac{1}{2}} \cdot \frac{1}{H_3+1+\frac{1}{3}+\frac{1}{2}} \cdot \frac{1}{H_3+1+\frac{1}{2}}$$

$$+ \frac{1}{3!(H_3+1)} \cdot \frac{1}{H_3+1+\frac{1}{2}+\frac{1}{3}+\frac{1}{1}} \cdot \frac{1}{H_3+1+\frac{1}{3}+\frac{1}{1}} \cdot \frac{1}{H_3+1+\frac{1}{1}}$$

$$+ \frac{1}{3!(H_3+1)} \cdot \frac{1}{H_3+1+\frac{1}{2}+\frac{1}{1}+\frac{1}{3}} \cdot \frac{1}{H_3+1+\frac{1}{1}+\frac{1}{3}} \cdot \frac{1}{H_3+1+\frac{1}{3}}$$

$$+ \frac{1}{3!(H_3+1)} \cdot \frac{1}{H_3+1+\frac{1}{3}+\frac{1}{1}+\frac{1}{2}} \cdot \frac{1}{H_3+1+\frac{1}{1}+\frac{1}{2}} \cdot \frac{1}{H_3+1+\frac{1}{2}}$$

$$+ \frac{1}{3!(H_3+1)} \cdot \frac{1}{H_3+1+\frac{1}{3}+\frac{1}{2}+\frac{1}{1}} \cdot \frac{1}{H_3+1+\frac{1}{2}+\frac{1}{1}} \cdot \frac{1}{H_3+1+\frac{1}{1}}$$

$$= \frac{1021194}{185910725}.$$

We mention that another solution of part (b) of the problem, based on an $\epsilon - \delta$ argument, was given by I.J. Kim.

Series of Real Numbers   **10**

10.1   Miscellaneous Series

2.1.

(a) $\frac{1}{8}$;   (b) $\frac{1}{2}$;   (c) $\frac{11}{18}$;   (d) $\frac{1}{12}$;   (e) using the identity

$$\ln\left(1 - \frac{1}{n^2}\right) = \ln(n+1) + \ln(n-1) - 2\ln n$$

it follows that the series equals $-\ln 2$;   (f) $\frac{a}{(a-1)^2}$;   (g) using the identity

$$\frac{n}{n^4 + n^2 + 1} = \frac{1}{2}\left(\frac{1}{n^2 - n + 1} - \frac{1}{n^2 + n + 1}\right)$$

we have that the series equals $\frac{1}{2}$;

(h) $\frac{n}{(n-2)!+(n-1)!+n!} = \frac{1}{(n-1)!} - \frac{1}{n!}$ and the series equals 1;

(i) e;   (j) 20 e.

2.2.

(a) We have, see the solution of Problem 7.47, that the following equality holds $\left\{(1+\sqrt{2})^{2n-1}\right\} = (\sqrt{2}-1)^{2n-1}$, $\forall n \geq 1$. It follows that

$$\sum_{n=1}^{\infty}\left\{(1+\sqrt{2})^{2n-1}\right\} = \sum_{n=1}^{\infty}\left(\sqrt{2}-1\right)^{2n-1} = \frac{1}{2}.$$

The other parts of the problem can be solved similarly.

© The Author(s), under exclusive license to Springer Nature Switzerland AG 2021
A. Sîntămărian, O. Furdui, *Sharpening Mathematical Analysis Skills*, Problem Books
in Mathematics, https://doi.org/10.1007/978-3-030-77139-3_10

2.3. Use that $\frac{1}{(2n+1)^2-1} = \frac{1}{4}\left(\frac{1}{n} - \frac{1}{n+1}\right)$, $\forall n \geq 1$.

2.4. (a) $\frac{\pi^2}{3} - 3$;   (b) $10 - \pi^2$.

2.5. (a) $\ln 2$;   (b) $\frac{\pi\sqrt{3}}{12} - \frac{1}{4}\ln 3$;   (c) $\frac{1}{4}\ln 2 - \frac{\pi}{24}$.

2.6. Let $N = \max\{m, n, p : 2^m 3^n 5^p \in A\}$. We have

$$\sum_{x \in A} \frac{1}{x} \leq \sum_{m=0}^{N}\sum_{n=0}^{N}\sum_{p=0}^{N} \frac{1}{2^m 3^n 5^p}$$

$$= \sum_{m=0}^{N} \frac{1}{2^m} \sum_{n=0}^{N} \frac{1}{3^n} \sum_{p=0}^{N} \frac{1}{5^p}$$

$$= 2\left(1 - \frac{1}{2^{N+1}}\right)\frac{3}{2}\left(1 - \frac{1}{3^{N+1}}\right)\frac{5}{4}\left(1 - \frac{1}{5^{N+1}}\right)$$

$$< \frac{15}{4}.$$

2.7.

(a) We calculate the $4n$th partial sum of the series and we have that

$$S_{4n} = 1 + \frac{1}{2} - \frac{1}{3} - \frac{1}{4} + \cdots + \frac{1}{4n-3} + \frac{1}{4n-2} - \frac{1}{4n-1} - \frac{1}{4n}$$

$$= H_{4n} - 2\sum_{i=1}^{n} \frac{1}{4i-1} - 2\sum_{i=1}^{n} \frac{1}{4i}$$

$$= \sum_{i=1}^{4n} \int_0^1 x^{i-1}dx - 2\sum_{i=1}^{n} \int_0^1 x^{4i-2}dx - 2\sum_{i=1}^{n} \int_0^1 x^{4i-1}dx$$

$$= \int_0^1 \frac{1-x^{4n}}{1-x}dx - 2\int_0^1 x^2\frac{1-x^{4n}}{1-x^4}dx - 2\int_0^1 x^3\frac{1-x^{4n}}{1-x^4}dx$$

$$= \int_0^1 \left[\frac{1-x^{4n}}{1-x} - 2x^2(1+x)\frac{1-x^{4n}}{1-x^4}\right]dx$$

$$= \int_0^1 \frac{1-x^{4n}}{1-x}\left(1 - \frac{2x^2}{1+x^2}\right)dx$$

$$= \int_0^1 \frac{(1-x^{4n})(1+x)}{1+x^2}dx.$$

It follows that

$$\lim_{n\to\infty} S_{4n} = \int_0^1 \frac{1+x}{1+x^2}dx = \frac{\pi}{4} + \frac{\ln 2}{2}.$$

(b) We have

$$1+\frac{1}{2}+\frac{1}{3}-\frac{1}{4}-\frac{1}{5}-\frac{1}{6}+\frac{1}{7}+\frac{1}{8}+\frac{1}{9}-\frac{1}{10}-\frac{1}{11}-\frac{1}{12}+\cdots$$

$$=\int_0^1 (1+x+x^2-x^3-x^4-x^5+x^6+x^7+x^8-x^9-x^{10}-x^{11}+\cdots)dx$$

$$=\int_0^1 (1+x+x^2)(1-x^3+x^6-x^9+\cdots)dx$$

$$=\int_0^1 \frac{1+x+x^2}{1+x^3}dx$$

$$=\frac{2\sqrt{3}}{9}\pi+\frac{\ln 2}{3}.$$

2.8. Use the identity

$$\frac{1}{n!(n^4+n^2+1)}=\frac{1}{2}\left(\frac{n}{(n+1)!(n^2+n+1)}-\frac{n-1}{n!(n^2-n+1)}+\frac{1}{(n+1)!}\right), \quad n \geq 1.$$

2.9. (b) $(a, b, c) \in \{(1, 2, 3), (4, 8, 6)\}$.
2.10.

(a) We have

$$0.9999999\ldots=\sum_{n=1}^{\infty}\frac{9}{10^n}=\frac{9}{10}\sum_{n=1}^{\infty}\frac{1}{10^{n-1}}=1.$$

The other parts of the problem can be solved similarly.

2.11. We have

$$\sum_{n=1}^{\infty}\frac{1}{(3+(-1)^n)^n}=\frac{11}{15} \quad \text{and} \quad \sum_{n=1}^{\infty}\frac{(-1)^n}{(3+(-1)^n)^n}=-\frac{3}{5}.$$

Calculate the $2n$th partial sum of the series.
2.12. Both series are equal to $\frac{\pi}{4}$.
2.13.

(a) Let $\alpha=\frac{1+\sqrt{5}}{2}$ and $\beta=\frac{1-\sqrt{5}}{2}$. We have

$$F_{n+1}^2-F_{n-1}^2=\left(\frac{\alpha^{n+1}-\beta^{n+1}}{\sqrt{5}}\right)^2-\left(\frac{\alpha^{n-1}-\beta^{n-1}}{\sqrt{5}}\right)^2$$

$$= \frac{1}{\sqrt{5}} \left(\alpha^{2n} \frac{\alpha^2 - \alpha^{-2}}{\sqrt{5}} - \beta^{2n} \frac{\beta^{-2} - \beta^2}{\sqrt{5}} \right)$$

$$= \frac{\alpha^{2n} - \beta^{2n}}{\sqrt{5}}$$

$$= F_{2n}.$$

(b) The series equals 2. Use that

$$\frac{F_{2n}}{F_{n-1}^2 F_{n+1}^2} = \frac{1}{F_{n-1}^2} - \frac{1}{F_{n+1}^2}, \quad n \geq 2.$$

2.14. The series equals γ.

2.15. The series equals 0. We have $S_{2n} = \sum_{k=2}^{2n} \ln\left(1 + \frac{(-1)^n}{n}\right) = \ln \frac{2n+1}{2n}, n \geq 1.$

2.17. (a) $P(x) = -2x^2 - 4x - 6$; (b) $Q(x) = -2x^3 - 6x^2 - 18x - 26.$

2.18. Use that $\frac{1}{n(n+k)(n+2k)} = \frac{1}{2k^2}\left(\frac{1}{n} - \frac{2}{n+k} + \frac{1}{n+2k}\right)$ and calculate the nth partial sum of the series.

2.19.

(a) Let

$$x_n = \sum_{k=1}^{\infty} \frac{1}{k(k+1)(k+2)\cdots(k+n)}.$$

We have

$$\frac{1}{k(k+1)(k+2)\cdots(k+n)}$$

$$= \frac{1}{n}\left(\frac{1}{k(k+1)\cdots(k+n-1)} - \frac{1}{(k+1)(k+2)\cdots(k+n)}\right)$$

and it follows that

$$x_n = \frac{1}{n}\left(x_{n-1} - \sum_{k=1}^{\infty} \frac{1}{(k+1)(k+2)\cdots(k+n)}\right)$$

$$\overset{k+1=m}{=} \frac{1}{n}\left[x_{n-1} - \left(\sum_{m=1}^{\infty} \frac{1}{m(m+1)\cdots(m+n-1)} - \frac{1}{n!}\right)\right]$$

$$= \frac{1}{n}\left(x_{n-1} - x_{n-1} + \frac{1}{n!}\right)$$

$$= \frac{1}{n \cdot n!}.$$

2.21. We have

$$\int_0^1 x \left\{\frac{1}{x}\right\} dx = \int_1^\infty \frac{\{y\}}{y^3} dy$$

$$= \sum_{k=1}^\infty \int_k^{k+1} \frac{y-k}{y^3} dy$$

$$= \sum_{k=1}^\infty \left(-\frac{1}{y} + \frac{k}{2y^2}\right) \Big|_k^{k+1}$$

$$= \frac{1}{2} \sum_{k=1}^\infty \left(\frac{1}{k} - \frac{1}{k+1} - \frac{1}{(k+1)^2}\right)$$

$$= 1 - \frac{\zeta(2)}{2}.$$

2.24. Use that $\frac{1}{\sqrt{n(n+1)}} < \frac{2}{\sqrt{n}} - \frac{2}{\sqrt{n+1}}$, $\forall n \geq 1$.

2.26. We have

$$\sum_{n=1}^\infty \frac{1}{n(n+1)} \left(1 - \frac{1}{2} + \cdots + \frac{(-1)^{n-1}}{n}\right)$$

$$= \sum_{n=1}^\infty \left(\frac{1 - \frac{1}{2} + \cdots + \frac{(-1)^{n-1}}{n}}{n} - \frac{1 - \frac{1}{2} + \cdots + \frac{(-1)^n}{n+1}}{n+1} + \frac{(-1)^n}{(n+1)^2}\right)$$

$$= \sum_{n=1}^\infty \left(\frac{1 - \frac{1}{2} + \cdots + \frac{(-1)^{n-1}}{n}}{n} - \frac{1 - \frac{1}{2} + \cdots + \frac{(-1)^n}{n+1}}{n+1}\right) + \sum_{n=1}^\infty \frac{(-1)^n}{(n+1)^2}$$

$$= 1 + \sum_{n=1}^\infty \frac{(-1)^n}{(n+1)^2}$$

$$= \frac{\pi^2}{12}.$$

2.27. Let $l = \lim_{n \to \infty} \left(\frac{x_{n+1}}{x_n}\right)^n < 1$. For $\epsilon \in (0, 1-l)$, $\exists n_0(\epsilon) \in \mathbb{N}$ such that $0 < \frac{x_{n+1}}{x_n} < \sqrt[n]{l + \epsilon}$, $\forall n \geq n_0$. It follows that $0 < x_n < x_{n_0}(l+\epsilon)^{\frac{1}{n_0} + \frac{1}{n_0+1} + \cdots + \frac{1}{n-1}}$, $\forall n \geq n_0$. Passing to the limit, as $n \to \infty$, in the previous inequalities, we get that $\lim_{n \to \infty} x_n = 0$.

2.28. Let $a_n = \frac{(2n-1)!!}{(2n)!!(n+1)}$. Observe that

$$a_n = 2(n+1)a_n - 2(n+2)a_{n+1}, \quad \forall n \geq 1.$$

2.29.

(a) Use Problem 2.27.

(b) Let $a_n = \frac{q(q+p)(q+2p)\cdots(q+np-p)}{p(2p)\cdots(np)} \cdot \frac{1}{n+1}$. Calculate the nth partial sum of the series and use that $a_n = \frac{p}{p-q}[(n+1)a_n - (n+2)a_{n+1}]$, $n \geq 1$.

2.31. We have

$$\sum_{n=2}^{\infty} (\zeta(n) - 1) = \sum_{n=2}^{\infty}\sum_{k=2}^{\infty} \frac{1}{k^n} = \sum_{k=2}^{\infty}\sum_{n=2}^{\infty} \frac{1}{k^n} = \sum_{k=2}^{\infty} \frac{1}{k^2 - k} = 1.$$

2.32. We have

$$\sum_{n=2}^{\infty}(-1)^n (\zeta(n) - 1) = \sum_{n=2}^{\infty}(-1)^n \sum_{k=2}^{\infty} \frac{1}{k^n} = \sum_{k=2}^{\infty}\sum_{n=2}^{\infty} \left(-\frac{1}{k}\right)^n = \sum_{k=2}^{\infty} \frac{1}{k^2 + k} = \frac{1}{2}$$

and

$$\sum_{n=1}^{\infty}(\zeta(2n) - 1) = \sum_{n=1}^{\infty}\sum_{k=2}^{\infty} \frac{1}{k^{2n}} = \sum_{k=2}^{\infty}\sum_{n=1}^{\infty} \left(\frac{1}{k^2}\right)^n = \sum_{k=2}^{\infty} \frac{1}{k^2 - 1} = \frac{3}{4}.$$

2.33. We have

$$\sum_{n=2}^{\infty}(-1)^n(\zeta(n+1) - \zeta(n)) = \sum_{n=2}^{\infty}(-1)^n \sum_{i=2}^{\infty} \frac{1}{i^n} \left(\frac{1}{i} - 1\right)$$

$$= \sum_{i=2}^{\infty} \left(\frac{1}{i} - 1\right) \sum_{n=2}^{\infty} \left(-\frac{1}{i}\right)^n$$

$$= \sum_{i=2}^{\infty} \left(\frac{1}{i} - 1\right) \cdot \frac{1}{i^2 + i}$$

$$= \sum_{i=2}^{\infty} \left[\frac{1}{i^2} - 2\left(\frac{1}{i} - \frac{1}{i+1}\right)\right]$$

$$= \zeta(2) - 2.$$

2.34.

(a) We have

$$(n, p) = \begin{cases} p & \text{if} \quad n = pi \\ 1 & \text{if} \quad n = pi + j, \ j = 1, 2, \ldots, p - 1 \end{cases}$$

and it follows that

$$\sum_{n=1}^{\infty} (-1)^{n-1} \frac{(n, p)}{n} = \sum_{n=pi} (-1)^{pi-1} \frac{p}{pi} + \sum_{\substack{n=pi+j \\ j=1,\ldots,p-1}} (-1)^{pi+j-1} \frac{1}{pi+j}$$

$$= \sum_{i=1}^{\infty} (-1)^{i-1} \frac{1}{i} + \sum_{\substack{n=pi+j \\ j=0,\ldots,p-1}} (-1)^{pi+j-1} \frac{1}{pi+j} - \sum_{n=pi} (-1)^{pi-1} \frac{1}{pi}$$

$$= \ln 2 + \sum_{n=1}^{\infty} (-1)^{n-1} \frac{1}{n} - \frac{1}{p} \sum_{i=1}^{\infty} (-1)^{i-1} \frac{1}{i}$$

$$= \ln 2 + \ln 2 - \frac{1}{p} \ln 2$$

$$= \frac{2p - 1}{p} \ln 2.$$

(b) We have

$$(n, 4) = \begin{cases} 1 & \text{if} \quad n = 4k + 1, \ n = 4k + 3 \\ 2 & \text{if} \quad n = 4k + 2 \\ 4 & \text{if} \quad n = 4k \end{cases}$$

and it follows that

$$\sum_{n=1}^{\infty} \frac{(n, 4)}{n^2} = \sum_{k=0}^{\infty} \frac{1}{(4k+1)^2} + \sum_{k=0}^{\infty} \frac{1}{(4k+3)^2} + 2 \sum_{k=0}^{\infty} \frac{1}{(4k+2)^2} + 4 \sum_{k=1}^{\infty} \frac{1}{(4k)^2}$$

$$= \sum_{k=0}^{\infty} \left(\frac{1}{(4k+1)^2} + \frac{1}{(4k+2)^2} + \frac{1}{(4k+3)^2} + \frac{1}{(4k+4)^2} \right)$$

$$+ \sum_{k=0}^{\infty} \frac{1}{(4k+2)^2} + \frac{3}{16} \sum_{k=1}^{\infty} \frac{1}{k^2}$$

$$= \zeta(2) + \frac{1}{4} \sum_{k=0}^{\infty} \frac{1}{(2k+1)^2} + \frac{3}{16} \zeta(2)$$

$$= \frac{11}{8} \zeta(2).$$

The second series of this part of the problem can be calculated similarly.

2.35. Let S_n be the nth partial sum of the series. A calculation shows that

$$S_n = \sum_{k=1}^{n} \left(k \ln \left(1 + \frac{1}{k} \right) - 1 + \frac{1}{2k} \right) = n \ln(n+1) - \ln n! - n + \frac{1}{2} H_n,$$

where $H_n = 1 + \frac{1}{2} + \cdots + \frac{1}{n}$ denotes the nth harmonic number.
Using Stirling's formula

$$\ln n! \sim \frac{1}{2} \ln(2\pi) + \frac{1}{2} \ln n + n \ln n - n,$$

we have that

$$S_n \sim n \ln \frac{n+1}{n} - \frac{1}{2} \ln(2\pi) + \frac{1}{2} \left(H_n - \ln n \right),$$

which implies $\lim\limits_{n \to \infty} S_n = 1 + \frac{\gamma}{2} - \frac{1}{2} \ln(2\pi)$.
The second series can be calculated similarly.

2.36. Add and subtract the series from Problem 2.35. Another method would be to calculate the nth partial sum of the series.

2.37. We calculate S_{2n}, the $2n$th partial sum of the series, and we have

$$S_{2n} = \ln \left(\frac{2n+1}{n} \cdot \frac{(1 \cdot 3 \cdot 5 \cdots (2n-1))^4}{(2 \cdot 4 \cdots 2n)^2 \cdot (2 \cdot 4 \cdots (2n-2))^2} \right)$$

$$= \ln \frac{2n+1}{n} + 2 \ln \frac{((2n-1)!)^2}{2^{4n-3} \cdot n! \cdot ((n-1)!)^3}.$$

It follows, based on Stirling's formula, that

$$S_{2n} \sim \ln \frac{2n+1}{n} + 2 \ln \frac{2}{\pi e} + 2 \ln \left(\frac{2n-1}{2n-2} \right)^{4n-1} + \ln \frac{n-1}{n} + 2 \ln \left(\frac{n-1}{n} \right)^n$$

and $\lim\limits_{n \to \infty} S_{2n} = \ln \frac{8}{\pi^2}$.

2.38. Use the formula $\ln \left(1 + \frac{1}{2n} \right) + \ln \left(1 + \frac{1}{2n+1} \right) = \ln \left(1 + \frac{1}{n} \right)$, $\forall n \geq 1$. It follows that

$$\ln \left(1 + \frac{1}{2n} \right) \ln \left(1 + \frac{1}{2n+1} \right)$$

$$= \frac{1}{2} \left[\ln^2 \left(1 + \frac{1}{n} \right) - \ln^2 \left(1 + \frac{1}{2n} \right) - \ln^2 \left(1 + \frac{1}{2n+1} \right) \right],$$

and calculate, based on the preceding formula, the nth partial sum of the series.

2.39. Using formula $\ln\left(1+\frac{1}{2n}\right)+\ln\left(1+\frac{1}{2n+1}\right)=\ln\left(1+\frac{1}{n}\right)$, $\forall n \geq 1$, we have that

$$\ln\left(1+\frac{1}{n}\right)\ln\left(1+\frac{1}{2n}\right)\ln\left(1+\frac{1}{2n+1}\right)$$
$$=\frac{1}{3}\left[\ln^3\left(1+\frac{1}{n}\right)-\ln^3\left(1+\frac{1}{2n}\right)-\ln^3\left(1+\frac{1}{2n+1}\right)\right].$$

We calculate, based on the preceding formula, the nth partial sum of the series and we have

$$S_n=\frac{\ln^3 2}{3}-\frac{1}{3}\sum_{k=n+1}^{2n+1}\ln^3\left(1+\frac{1}{k}\right).$$

Since

$$(n+1)\ln^3\left(1+\frac{1}{2n+1}\right)<\sum_{k=n+1}^{2n+1}\ln^3\left(1+\frac{1}{k}\right)<(n+1)\ln^3\left(1+\frac{1}{n+1}\right),$$

we obtain that $\lim\limits_{n\to\infty} S_n = \frac{\ln^3 2}{3}$.

2.40.

(a) Since $\frac{H_n}{n(n+1)}=\frac{H_n}{n}-\frac{H_{n+1}}{n+1}+\frac{1}{(n+1)^2}$, $\forall n \geq 1$, we have that

$$\sum_{n=1}^{\infty}\frac{H_n}{n(n+1)}=\sum_{n=1}^{\infty}\left(\frac{H_n}{n}-\frac{H_{n+1}}{n+1}\right)+\sum_{n=1}^{\infty}\frac{1}{(n+1)^2}=1+\zeta(2)-1=\zeta(2).$$

The other series can be calculated similarly.

2.41.

(a) Let $I_{n,n-1}=\int_0^1 x^n(1-x)^{n-1}dx$. We integrate by parts and we obtain the recurrence formula $I_{n,n-1}=\frac{n-1}{n+1}I_{n+1,n-2}$.

(b) We have, based on part (a), that

$$\sum_{n=0}^{\infty} \frac{1}{C_n} = \sum_{n=0}^{\infty} \frac{(n+1)(n!)^2}{(2n)!}$$

$$= 1 + \sum_{n=1}^{\infty} (n+1)n \int_0^1 x^n (1-x)^{n-1} dx$$

$$= 1 + \int_0^1 x \sum_{n=1}^{\infty} n(n+1)(x(1-x))^{n-1} dx$$

$$= 1 + \int_0^1 \frac{2x}{(1-x+x^2)^3} dx$$

$$= 1 + \int_0^1 \frac{2x-1}{(1-x+x^2)^3} dx + \int_0^1 \frac{1}{(1-x+x^2)^3} dx$$

$$= 1 - \frac{1}{2(x^2-x+1)^2} \Big|_0^1 + \int_0^1 \frac{1}{((x-1/2)^2 + 3/4)^3} dx$$

$$\stackrel{x-1/2=u}{=\!=\!=} 1 + 2 \int_0^{1/2} \frac{1}{(u^2+3/4)^3} du$$

$$= 1 + 2 \left(\frac{16u}{3(4u^2+3)^2} + \frac{8u}{3(4u^2+3)} + \frac{4\sqrt{3}}{9} \arctan \frac{2\sqrt{3}u}{3} \right) \Big|_0^{1/2}$$

$$= 2 + \frac{4\sqrt{3}}{27} \pi.$$

(c) This part of the problem can be solved in the same way as part (b).

2.42. We have

$$\sum_{n=1}^{\infty} \frac{n^k}{(n+1)^p} = \sum_{n=1}^{\infty} \frac{(n+1-1)^k}{(n+1)^p} = \sum_{n=1}^{\infty} \frac{1}{(n+1)^p} \sum_{i=0}^{k} C_k^i (n+1)^i (-1)^{k-i}$$

$$= \sum_{i=0}^{k} C_k^i (-1)^{k-i} \sum_{n=1}^{\infty} \frac{1}{(n+1)^{p-i}} = \sum_{i=0}^{k} (-1)^{k-i} C_k^i \zeta(p-i).$$

10.2   Applications of Abel's Summation Formula

2.43.

(a) We use Abel's summation formula with $a_k = \cos k$ and $b_k = \frac{1}{n+k}$, and we have that

$$\frac{\cos 1}{n+1} + \frac{\cos 2}{n+2} + \cdots + \frac{\cos n}{2n} = \frac{\cos 1 + \cos 2 + \cdots + \cos n}{2n+1}$$

$$+ \sum_{k=1}^{n} \frac{\cos 1 + \cos 2 + \cdots + \cos k}{(n+k)(n+k+1)}$$

$$= \frac{\sin \frac{n}{2} \cos \frac{n+1}{2}}{(2n+1) \sin \frac{1}{2}} + \sum_{k=1}^{n} \frac{\sin \frac{k}{2} \cos \frac{k+1}{2}}{(n+k)(n+k+1) \sin \frac{1}{2}}.$$

We used that $\cos 1 + \cos 2 + \cdots + \cos n = \frac{\sin \frac{n}{2} \cos \frac{n+1}{2}}{\sin \frac{1}{2}}$, $\forall n \geq 1$.

It follows that

$$\lim_{n \to \infty} \left(\frac{\cos 1}{n+1} + \frac{\cos 2}{n+2} + \cdots + \frac{\cos n}{2n} \right) = 0,$$

since

$$\left| \sum_{k=1}^{n} \frac{\sin \frac{k}{2} \cos \frac{k+1}{2}}{(n+k)(n+k+1) \sin \frac{1}{2}} \right| \leq \frac{1}{\sin \frac{1}{2}} \sum_{k=1}^{n} \frac{1}{n^2} = \frac{1}{n \sin \frac{1}{2}}.$$

(b) We have

$$\frac{\cos^2 1}{n+1} + \frac{\cos^2 2}{n+2} + \cdots + \frac{\cos^2 n}{2n} = \frac{1}{2} \left(\frac{1}{n+1} + \frac{1}{n+2} + \cdots + \frac{1}{2n} \right)$$

$$+ \frac{1}{2} \left(\frac{\cos 2}{n+1} + \frac{\cos 4}{n+2} + \cdots + \frac{\cos(2n)}{2n} \right).$$

Exactly as in the solution of part (a), it can be proved that

$$\lim_{n \to \infty} \left(\frac{\cos 2}{n+1} + \frac{\cos 4}{n+2} + \cdots + \frac{\cos(2n)}{2n} \right) = 0,$$

and the result follows since $\lim_{n \to \infty} \left(\frac{1}{n+1} + \frac{1}{n+2} + \cdots + \frac{1}{2n} \right) = \ln 2$.

The limits in Remark 2.3 can be proved by using the formulae [100, p. 130]

• $\cos^{2k-1} x = \frac{1}{2^{2k-2}} \sum_{i=0}^{k-1} C_{2k-1}^i \cos(2k - 2i - 1)x$;

• $\cos^{2k} x = \frac{1}{2^{2k}} \left(\sum_{i=0}^{k-1} 2C_{2k}^i \cos 2(k - i)x + C_{2k}^k \right).$

2.44. (a) $x_{n+1} - x_n = C_{n+1}^k \left(e - 1 - \frac{1}{1!} - \frac{1}{2!} - \cdots - \frac{1}{(n+1)!} \right) > 0, \quad \forall n \geq k$.

(b) and (c) We have

$$x_n < \sum_{i=k}^{\infty} C_i^k \left(e - 1 - \frac{1}{1!} - \frac{1}{2!} - \cdots - \frac{1}{i!} \right),$$

and we prove that the series converges by calculating its value.

One can check, using integration by parts, that the following formula holds:

$$e - 1 - \frac{1}{1!} - \frac{1}{2!} - \cdots - \frac{1}{i!} = \frac{1}{i!} \int_0^1 (1-t)^i e^t \, dt, \quad i \geq 0.$$

It follows that

$$\lim_{n \to \infty} x_n = \sum_{i=k}^{\infty} C_i^k \left(e - 1 - \frac{1}{1!} - \frac{1}{2!} - \cdots - \frac{1}{i!} \right)$$

$$= \sum_{i=k}^{\infty} \frac{C_i^k}{i!} \int_0^1 (1-t)^i e^t \, dt$$

$$= \frac{1}{k!} \int_0^1 (1-t)^k e^t \left(\sum_{i=k}^{\infty} \frac{(1-t)^{i-k}}{(i-k)!} \right) dt$$

$$= \frac{e}{k!} \int_0^1 (1-t)^k \, dt$$

$$= \frac{e}{(k+1)!}.$$

2.45. *Solution I.* We use Abel's summation formula, with

$$a_n = 1 \quad \text{and} \quad b_n = e^x - 1 - \frac{x}{1!} - \frac{x^2}{2!} - \cdots - \frac{x^n}{n!}$$

for the first series, and

$$a_n = n \quad \text{and} \quad b_n = e^x - 1 - \frac{x}{1!} - \frac{x^2}{2!} - \cdots - \frac{x^n}{n!}$$

for the second series, and

$$a_n = n^2 \quad \text{and} \quad b_n = e^x - 1 - \frac{x}{1!} - \frac{x^2}{2!} - \cdots - \frac{x^n}{n!}$$

for the third series.

Solution II. We calculate the first series. Let $y(x)$ be the sum of the series. We have

$$y(x) = \sum_{n=0}^{\infty} \left(e^x - 1 - \frac{x}{1!} - \frac{x^2}{2!} - \cdots - \frac{x^n}{n!} \right)$$

$$= e^x - 1 + \sum_{n=1}^{\infty} \left(e^x - 1 - \frac{x}{1!} - \frac{x^2}{2!} - \cdots - \frac{x^n}{n!} \right).$$

It follows that

$$y'(x) = e^x + \sum_{n=1}^{\infty} \left(e^x - 1 - \frac{x}{1!} - \frac{x^2}{2!} - \cdots - \frac{x^{n-1}}{(n-1)!} \right)$$

$$= e^x + \sum_{n=1}^{\infty} \left(e^x - 1 - \frac{x}{1!} - \frac{x^2}{2!} - \cdots - \frac{x^n}{n!} \right) + \sum_{n=1}^{\infty} \frac{x^n}{n!}$$

$$= e^x + \left(y(x) - e^x + 1 \right) + e^x - 1$$

$$= e^x + y(x).$$

Therefore, $y'(x) - y(x) = e^x \Leftrightarrow y'(x)e^{-x} - y(x)e^{-x} = 1 \Leftrightarrow \left(y(x)e^{-x} \right)' = 1 \Rightarrow y(x)e^{-x} = x + \mathscr{C}, \mathscr{C} \in \mathbb{R}$. We obtain that $y(x) = (x + \mathscr{C})e^x$. Since $y(0) = 0$ we have that $y(x) = xe^x$.

The other two series can be calculated similarly.

2.46. Observe that the $2n$th partial sum of the series is $S_{2n} = \frac{1}{1^2} + \frac{1}{3^2} + \cdots + \frac{1}{(2n-1)^2}$, $\forall n \geq 1$. An alternative solution is based on the use of Abel's summation formula.
2.47. Use Abel's summation formula with:

(a) $a_n = 1$ and $b_n = \zeta(2) - 1 - \frac{1}{2^2} - \cdots - \frac{1}{n^2} - \frac{1}{n+k}$;
(b) $a_n = 1$ and $b_n = \left(\frac{1}{n^2} + \frac{1}{(n+1)^2} + \cdots \right) - \frac{1}{n+k}$.

Alternatively, the series in part (b) follows directly from the series in part (a).
2.48. Parts (a) and (b) follow by direct computations (see the solution of parts (a) and (b) of Problem 2.50).

(c) *Solution I.* Use part (a) of the problem.

Solution II. Let

$$A = \sum_{n=1}^{\infty} \left[\left(\frac{1}{n^2} + \frac{1}{(n+2)^2} + \frac{1}{(n+4)^2} + \cdots \right) - \frac{1}{2n} \right]$$

$$B = \sum_{n=1}^{\infty} \left[\left(\frac{1}{(n+1)^2} + \frac{1}{(n+3)^2} + \frac{1}{(n+5)^2} + \cdots \right) - \frac{1}{2n} \right].$$

We have, based on Problem 2.47, that

$$A + B = \sum_{n=1}^{\infty} \left[\left(\frac{1}{n^2} + \frac{1}{(n+1)^2} + \cdots \right) - \frac{1}{n} \right] = 1.$$

On the other hand,

$$B = \sum_{n=1}^{\infty} \left[\left(\frac{1}{(n+1)^2} + \frac{1}{(n+3)^2} + \frac{1}{(n+5)^2} + \cdots \right) - \frac{1}{2n} \right]$$

$$\overset{n+1=i}{=} \sum_{i=2}^{\infty} \left[\left(\frac{1}{i^2} + \frac{1}{(i+2)^2} + \frac{1}{(i+4)^2} + \cdots \right) - \frac{1}{2(i-1)} \right]$$

$$= \sum_{i=2}^{\infty} \left[\left(\frac{1}{i^2} + \frac{1}{(i+2)^2} + \frac{1}{(i+4)^2} + \cdots \right) - \frac{1}{2i} \right] + \sum_{i=2}^{\infty} \left(\frac{1}{2i} - \frac{1}{2(i-1)} \right)$$

$$= \sum_{i=2}^{\infty} \left[\left(\frac{1}{i^2} + \frac{1}{(i+2)^2} + \frac{1}{(i+4)^2} + \cdots \right) - \frac{1}{2i} \right] - \frac{1}{2}$$

$$= \sum_{i=1}^{\infty} \left[\left(\frac{1}{i^2} + \frac{1}{(i+2)^2} + \frac{1}{(i+4)^2} + \cdots \right) - \frac{1}{2i} \right]$$

$$\quad - \left[\left(\frac{1}{1^2} + \frac{1}{3^2} + \frac{1}{5^2} + \cdots \right) - \frac{1}{2} \right] - \frac{1}{2}$$

$$= A - \frac{\pi^2}{8}.$$

It follows that

$$A = \sum_{n=1}^{\infty} \left[\left(\frac{1}{n^2} + \frac{1}{(n+2)^2} + \frac{1}{(n+4)^2} + \cdots \right) - \frac{1}{2n} \right] = \frac{1}{2} + \frac{\pi^2}{16}$$

$$B = \sum_{n=1}^{\infty} \left[\left(\frac{1}{(n+1)^2} + \frac{1}{(n+3)^2} + \frac{1}{(n+5)^2} + \cdots \right) - \frac{1}{2n} \right] = \frac{1}{2} - \frac{\pi^2}{16}.$$

2.50.

(a) We have

$$\frac{1}{n^3} + \frac{1}{(n+k)^3} + \frac{1}{(n+2k)^3} + \cdots = \sum_{i=0}^{\infty} \frac{1}{(n+ki)^3}$$

$$= \sum_{i=0}^{\infty} \frac{1}{2} \int_0^1 x^{n+ki-1} \ln^2 x \, dx$$

$$\overset{(*)}{=} \frac{1}{2} \int_0^1 \sum_{i=0}^{\infty} x^{n+ki-1} \ln^2 x \, dx$$

$$= \frac{1}{2} \int_0^1 \frac{x^{n-1}}{1-x^k} \ln^2 x \, dx.$$

We used at step (*) Tonelli's theorem for nonnegative functions.

(b) We have, based on part (a), that

$$n^2 \left(\frac{1}{n^3} + \frac{1}{(n+k)^3} + \frac{1}{(n+2k)^3} + \cdots \right) = \frac{n^2}{2} \int_0^1 \frac{x^{n-1}}{1-x^k} \ln^2 x \, dx$$

$$\overset{x^n=y}{=} \frac{1}{2n} \int_0^1 \frac{\ln^2 y}{1 - y^{\frac{k}{n}}} dy$$

$$= \frac{1}{2k} \int_0^1 \frac{\frac{k}{n}}{1 - y^{\frac{k}{n}}} \ln^2 y \, dy.$$

Let $f_n(y) = \dfrac{\frac{k}{n}}{1-y^{\frac{k}{n}}} \ln^2 y$, $y \in (0,1)$. We have $\lim\limits_{n\to\infty} f_n(y) = -\ln y$.

On the other hand, for $n > k$, we have $f_n(y) = \dfrac{\frac{k}{n}}{1-y^{\frac{k}{n}}} \ln^2 y < \dfrac{\ln^2 y}{1-y}$, since

$\dfrac{\frac{k}{n}}{1-y^{\frac{k}{n}}} \le \dfrac{1}{1-y}, \forall n \ge k$. The last inequality follows from the fact that the function

$g : [0,1] \to \mathbb{R}$, $g(x) = \dfrac{x}{1-y^x}$ is an increasing function when $y \in (0,1)$. The

function $y \to \dfrac{\ln^2 y}{1-y}$ is integrable over the interval $[0,1]$ and we have, based on

Lebesgue Dominated Convergence Theorem, that

$$\lim_{n\to\infty} n^2 \left(\frac{1}{n^3} + \frac{1}{(n+k)^3} + \frac{1}{(n+2k)^3} + \cdots \right) = \lim_{n\to\infty} \frac{1}{2k} \int_0^1 \frac{\frac{k}{n}}{1-y^{\frac{k}{n}}} \ln^2 y \, dy$$

$$= \frac{1}{2k} \int_0^1 (-\ln x) dx$$

$$= \frac{1}{2k}.$$

(c) *Solution I.* Use part (a) of the problem.

Solution II. Let

$$A = \sum_{n=1}^{\infty} \left(\frac{1}{n^3} + \frac{1}{(n+2)^3} + \frac{1}{(n+4)^3} + \cdots \right)$$

$$B = \sum_{n=1}^{\infty} \left(\frac{1}{(n+1)^3} + \frac{1}{(n+3)^3} + \frac{1}{(n+5)^3} + \cdots \right).$$

We have, based on Abel's summation formula, that

$$A + B = \sum_{n=1}^{\infty} \left(\frac{1}{n^3} + \frac{1}{(n+1)^3} + \frac{1}{(n+2)^3} + \cdots \right) = \sum_{n=1}^{\infty} \frac{1}{n^2} = \zeta(2).$$

On the other hand,

$$B = \sum_{n=1}^{\infty} \left(\frac{1}{(n+1)^3} + \frac{1}{(n+3)^3} + \frac{1}{(n+5)^3} + \cdots \right)$$

$$\overset{n+1=i}{=} \sum_{i=2}^{\infty} \left(\frac{1}{i^3} + \frac{1}{(i+2)^3} + \frac{1}{(i+4)^3} + \cdots \right)$$

$$= \sum_{i=1}^{\infty} \left(\frac{1}{i^3} + \frac{1}{(i+2)^3} + \frac{1}{(i+4)^3} + \cdots \right)$$

$$\quad - \left(\frac{1}{1^3} + \frac{1}{3^3} + \frac{1}{5^3} + \cdots \right)$$

$$= A - \frac{7}{8}\zeta(3).$$

It follows that

$$A = \sum_{n=1}^{\infty} \left(\frac{1}{n^3} + \frac{1}{(n+2)^3} + \frac{1}{(n+4)^3} + \cdots \right) = \frac{1}{2}\zeta(2) + \frac{7}{16}\zeta(3)$$

$$B = \sum_{n=1}^{\infty} \left(\frac{1}{(n+1)^3} + \frac{1}{(n+3)^3} + \frac{1}{(n+5)^3} + \cdots \right) = \frac{1}{2}\zeta(2) - \frac{7}{16}\zeta(3).$$

2.52. Use Abel's summation formula with

$$a_n = (-1)^n \quad \text{and} \quad b_n = 1 + \frac{1}{2} + \frac{1}{3} + \cdots + \frac{1}{n} - \ln n - \gamma.$$

Alternatively, calculate the $2n$th partial sum of the series and use Wallis formula.

2.53. Use Abel's summation formula with:

(a) $a_n = (-1)^n$ and $b_n = 1 + \frac{1}{2} + \frac{1}{3} + \cdots + \frac{1}{n} - \ln \sqrt{n(n+1)} - \gamma$;

(b) $a_n = (-1)^n$ and $b_n = 1 + \frac{1}{2} + \frac{1}{3} + \cdots + \frac{1}{n} - \ln \sqrt[3]{n(n+1)(n+2)} - \gamma$;

(c) $a_n = 1$ and $b_n = 1 + \frac{1}{2} + \frac{1}{3} + \cdots + \frac{1}{n} - \ln \sqrt{n(n+1)} - \gamma$.

2.54.

(a) We calculate the nth partial sum of the series, using Abel's summation formula with $a_k = 1$ and $b_k = 1 + \frac{1}{2} + \cdots + \frac{1}{k} - \ln\left(k + \frac{1}{2}\right) - \gamma$, and we have that

$$S_n = \sum_{k=1}^{n} \left(1 + \frac{1}{2} + \cdots + \frac{1}{k} - \ln\left(k + \frac{1}{2}\right) - \gamma\right)$$

$$= n\left(1 + \frac{1}{2} + \cdots + \frac{1}{n+1} - \ln\left(n + \frac{3}{2}\right) - \gamma\right)$$

$$+ \sum_{k=1}^{n} k\left(-\frac{1}{k+1} + \ln(2k+3) - \ln(2k+1)\right)$$

$$= n\left(H_{n+1} - \ln\left(n + \frac{3}{2}\right) - \gamma\right)$$

$$- (n+1) + H_{n+1} + n\ln(2n+3) - \ln(2n+1)! + n\ln 2 + \ln n!,$$

where $H_k = 1 + \frac{1}{2} + \cdots + \frac{1}{k}$ denotes the kth harmonic number.

Using Stirling's formula,

$$\ln n! \sim \frac{1}{2}\ln(2\pi) + \frac{2n+1}{2}\ln n - n,$$

we obtain

$$S_n \sim n\left(H_{n+1} - \ln\left(n + \frac{3}{2}\right) - \gamma\right) + H_{n+1} - \ln(n+1)$$

$$+ n\ln\frac{4n^2 + 6n}{4n^2 + 4n + 1} + \frac{1}{2}\ln\frac{n}{2n+1} + \ln\frac{n+1}{2n+1},$$

and it follows that

$$\lim_{n\to\infty} S_n = \gamma + \frac{1}{2} + \frac{1}{2}\ln\frac{1}{2} + \ln\frac{1}{2} = \gamma + \frac{1}{2} - \frac{3}{2}\ln 2.$$

(b) First we note that, if $p \in \mathbb{N}$, then one has

$$\sum_{k=1}^{\infty}\left(\ln\frac{kp}{kp+1} + \frac{1}{kp}\right) = \ln\Gamma\left(1 + \frac{1}{p}\right) + \frac{\gamma}{p}. \tag{10.1}$$

Formula (10.1) follows from the Weierstrass product formula for the Gamma function [150, p. 236]

$$\frac{1}{\Gamma(z)} = ze^{\gamma z}\prod_{n=1}^{\infty}\left\{\left(1 + \frac{z}{n}\right)e^{-\frac{z}{n}}\right\}, \quad z \neq 0, -1, -2, \ldots,$$

by setting $z = \frac{1}{p}$ and by taking logarithms of both sides.

Now we are ready to calculate the series for any $p \geq 1$. We have

$$1 + \frac{1}{2} + \cdots + \frac{1}{k} - \ln\left(k + \frac{1}{p}\right) - \gamma + \frac{2-p}{2pk}$$

$$= 1 + \frac{1}{2} + \cdots + \frac{1}{k} - \ln\left(k + \frac{1}{2}\right) - \gamma + \left(\ln\frac{kp}{kp+1} + \frac{1}{kp}\right) - \left(\ln\frac{2k}{2k+1} + \frac{1}{2k}\right),$$

and this implies, based on part (a) and formula (10.1), that

$$\sum_{k=1}^{\infty}\left(1 + \frac{1}{2} + \cdots + \frac{1}{k} - \ln\left(k + \frac{1}{p}\right) - \gamma + \frac{2-p}{2pk}\right)$$

$$= \sum_{k=1}^{\infty}\left(1 + \frac{1}{2} + \cdots + \frac{1}{k} - \ln\left(k + \frac{1}{2}\right) - \gamma\right)$$

$$+ \sum_{k=1}^{\infty}\left(\ln\frac{kp}{kp+1} + \frac{1}{kp}\right) - \sum_{k=1}^{\infty}\left(\ln\frac{2k}{2k+1} + \frac{1}{2k}\right)$$

$$= \gamma + \frac{1}{2} - \frac{3}{2}\ln 2 + \ln\Gamma\left(1 + \frac{1}{p}\right) + \frac{\gamma}{p} - \ln\Gamma\left(1 + \frac{1}{2}\right) - \frac{\gamma}{2}$$

$$= \gamma\left(\frac{1}{2} + \frac{1}{p}\right) + \frac{1}{2} - \ln\sqrt{2\pi} + \ln\Gamma\left(1 + \frac{1}{p}\right).$$

It is worth mentioning that, by letting $p \to \infty$ in the previous series formula, we obtain

$$\sum_{k=1}^{\infty}\left(1+\frac{1}{2}+\cdots+\frac{1}{k}-\ln k-\gamma-\frac{1}{2k}\right)=\frac{\gamma}{2}+\frac{1}{2}-\ln\sqrt{2\pi},$$

which is problem 3.42 in [26].

2.55. Use Abel's summation formula with:

(a) $a_n = 1$ and $b_n = O_n - \frac{\gamma}{2} - \ln 2 - \frac{\ln n}{2}$;
(b) $a_n = (-1)^n$ and $b_n = O_n - \frac{\gamma}{2} - \ln 2 - \frac{\ln n}{2}$.

2.56.

(a) Use mathematical induction.
(b) Use Abel's summation formula with

$$a_n = H_n \quad \text{and} \quad b_n = \zeta(3) - 1 - \frac{1}{2^3} - \cdots - \frac{1}{n^3}.$$

10.3   Series with Positive Terms

2.57. (a) Converges; (b) diverges; (c) diverges; (d) converges; (e) diverges; (f) converges; (g) diverges; (h) converges; (i) converges.

2.60. (a) Converges for $a \in (0, 1)$ and diverges for $a > 1$. When $a = 1$ we obtain the generalized harmonic series, which converges for $p > 1$ and diverges for $p \le 1$; (b) converges for $a \in (0, 7)$ and diverges for $a \ge 7$; (c) converges; (d) converges.

2.61. (a) Converges; (b) converges; (c) diverges. Parts (b) and (c) can also be solved using the comparison criterion. For n large enough, we have $\frac{1}{(\ln n)^{\ln n}} = \frac{1}{n^{\ln \ln n}} < \frac{1}{n^2}$ and $\frac{1}{(\ln n)^{\ln \ln n}} = \frac{1}{e^{(\ln \ln n)^2}} > \frac{1}{e^{\ln n}} = \frac{1}{n}$.

2.62. (a) Converges for $a \in (0, 1)$ and diverges for $a \ge 1$; (b) converges; (c) converges; (d) converges.

2.63. (a) Converges; (b) converges; (c) converges for $a \in \left(0, \frac{1}{e}\right)$ and diverges for $a \ge \frac{1}{e}$; (d) diverges for $a \in \left(0, \frac{5}{4}\right]$ and converges for $a > \frac{5}{4}$; (e) converges for $x \in \left(0, \frac{1}{e}\right)$ and diverges for $x \ge \frac{1}{e}$; (f) converges for $x \in \left(0, \frac{1}{e}\right)$ and diverges for $x \ge \frac{1}{e}$.

2.64. (a) Let $t = \max\{a, b\}$ and let $x_n = \frac{c(a+c)\cdots(an+c-a)}{b(a+b)\cdots(an+b-a)}$.
   We have

$$\ln x_n = \sum_{k=1}^{n} \ln \frac{ak+c-a}{ak+b-a} = \sum_{k=1}^{n} \ln\left(1 - \frac{b-c}{ak+b-a}\right) \le -(b-c)\sum_{k=1}^{n} \frac{1}{ak+b-a},$$

where the last inequality follows since $\ln(1-x) \le -x, \forall x \in [0, 1)$.

Thus,

$$\ln x_n \le -(b-c)\left(\frac{1}{b} + \frac{1}{a+b} + \cdots + \frac{1}{(n-1)a+b}\right)$$

$$\le -(b-c)\left(\frac{1}{t} + \frac{1}{2t} + \cdots + \frac{1}{nt}\right)$$

$$= -\frac{b-c}{t}H_n$$

$$= -\frac{b-c}{t}(H_n - \ln n) - \frac{b-c}{t}\ln n.$$

This implies that

$$x_n \le \frac{e^{-\frac{b-c}{\max\{a,b\}}(H_n - \ln n)}}{n^{\frac{b-c}{\max\{a,b\}}}}, \qquad \forall n \ge 1.$$

Part (a) of the problem is used for proving that the series in part (c) converges.
2.66. When $p \le 0$ the series $\sum_{n=1}^{\infty}\frac{1}{n^p}$ diverges since:

$$p < 0 \qquad u_n = \frac{1}{n^p} = n^{-p} \longrightarrow \infty \implies \sum_{n=1}^{\infty} u_n \text{ diverges};$$

$$p = 0 \qquad u_n = 1 \nrightarrow 0 \implies \sum_{n=1}^{\infty} u_n \text{ diverges}.$$

When $p > 0$ the function $f(x) = \frac{1}{x^p}$ is decreasing on $[1, \infty)$ and we have

$$p \ne 1 \qquad v_n = \int_1^n \frac{1}{x^p}\,dx = \frac{n^{-p+1}}{-p+1} - \frac{1}{-p+1};$$

$$p = 1 \qquad v_n = \int_1^n \frac{1}{x}\,dx = \ln n.$$

It follows, based on Cauchy's integral criterion, that

$$0 < p < 1 \lim_{n\to\infty} v_n = \lim_{n\to\infty}\left(\frac{n^{-p+1}}{-p+1} - \frac{1}{-p+1}\right) = \infty \implies \sum_{n=1}^{\infty}\frac{1}{n^p} \text{ diverges};$$

$$p = 1 \lim_{n\to\infty} v_n = \lim_{n\to\infty} \ln n = \infty \implies \sum_{n=1}^{\infty}\frac{1}{n} \text{ diverges};$$

$$p > 1 \lim_{n \to \infty} v_n = \lim_{n \to \infty} \left(\frac{n^{-p+1}}{-p+1} - \frac{1}{-p+1} \right) = \frac{1}{p-1} \Rightarrow \sum_{n=1}^{\infty} \frac{1}{n^p} \text{ converges.}$$

2.68. (a) Diverges; (b) converges for $p > 1$ and diverges for $p \leq 1$; (c) diverges; (d) converges for $p > 1$ and diverges for $p \leq 1$.

2.69. The series diverges. Compare the series with $\sum_{n=2}^{\infty} \frac{\ln n}{n}$.

10.4 Alternating Series

2.70. (a) Semiconvergent; (b) absolutely convergent; (c) semiconvergent; (d) semiconvergent; (e) semiconvergent; (f) absolutely convergent.

2.71. (a) The series converges. The series alternates since $n + 1 + \sin(n+1) > n + \sin n, \forall n \geq 1$; (b) The series converges. We have an alternating series since $n + 1 + \cos(n+1) > n + \cos n, \forall n \geq 1$; (c) The series converges. Observe that $\left| \sum_{n=2}^{\infty} \frac{(-1)^n}{n+(-1)^n \sin n} - \sum_{n=2}^{\infty} \frac{(-1)^n}{n} \right| \leq \sum_{n=2}^{\infty} \frac{1}{(n-1)n} = 1$; (d) The series converges. Observe that $\left| \sum_{n=2}^{\infty} \frac{(-1)^n}{n+(-1)^n \cos n} - \sum_{n=2}^{\infty} \frac{(-1)^n}{n} \right| \leq \sum_{n=2}^{\infty} \frac{1}{(n-1)n} = 1$.

2.74.

(a) We have

$$1 - \frac{1}{2} + \frac{1}{3} - \frac{1}{4} + \cdots + \frac{1}{2n-1} - \frac{1}{2n} = 1 + \frac{1}{2} + \frac{1}{3} + \frac{1}{4} + \cdots + \frac{1}{2n-1} + \frac{1}{2n}$$

$$- 2 \left(\frac{1}{2} + \frac{1}{4} + \cdots + \frac{1}{2n} \right)$$

$$= 1 + \frac{1}{2} + \frac{1}{3} + \frac{1}{4} + \cdots + \frac{1}{2n-1} + \frac{1}{2n}$$

$$- \left(1 + \frac{1}{2} + \frac{1}{3} + \cdots + \frac{1}{n} \right)$$

$$= \frac{1}{n+1} + \frac{1}{n+2} + \cdots + \frac{1}{2n}.$$

(b) Let S_{2n} be the $2n$th partial sum of the series. We have, based on part (a), that

$$S_{2n} = 1 - \frac{1}{2} + \frac{1}{3} - \frac{1}{4} + \cdots + \frac{1}{2n-1} - \frac{1}{2n}$$

$$= \left[1 + \frac{1}{2} + \frac{1}{3} + \frac{1}{4} + \cdots + \frac{1}{2n-1} + \frac{1}{2n} - \ln(2n) \right]$$

$$- \left(1 + \frac{1}{2} + \frac{1}{3} + \cdots + \frac{1}{n} - \ln n \right) + \ln 2,$$

and it follows that $\lim\limits_{n \to \infty} S_{2n} = \ln 2$.

Remark. From the solution of part (b) of Problem 2.74 it follows that

$$\lim_{n \to \infty} \left(\frac{1}{n+1} + \frac{1}{n+2} + \cdots + \frac{1}{2n} \right) = \ln 2.$$

More generally, when $k \geq 2$ is an integer, it can be proved that

$$\lim_{n \to \infty} \left(\frac{1}{n+1} + \frac{1}{n+2} + \cdots + \frac{1}{kn} \right) = \ln k.$$

2.75. By calculating the $2n$th partial sum of the series, we have that $\sum\limits_{n=1}^{\infty} \frac{(-1)^{n-1}}{n+x} =$
$\sum\limits_{n=1}^{\infty} \frac{1}{(x+2n)(x+2n-1)}$. Since $\frac{1}{(x+2n)^2} < \frac{1}{(x+2n)(x+2n-1)} < \frac{1}{(x+2n-1)^2}$, we get that

$$x \sum_{n=1}^{\infty} \frac{1}{(x+2n)^2} < x \sum_{n=1}^{\infty} \frac{1}{(x+2n)(x+2n-1)} < x \sum_{n=1}^{\infty} \frac{1}{(x+2n-1)^2}.$$

Now the problem is solved if we prove that $\lim\limits_{x \to \infty} x \sum\limits_{n=1}^{\infty} \frac{1}{(x+n)^2} = 1$.

We have

$$\frac{1}{x+1} = \int_1^{\infty} \frac{1}{(x+t)^2} dt = \sum_{n=1}^{\infty} \int_n^{n+1} \frac{1}{(x+t)^2} dt < \sum_{n=1}^{\infty} \frac{1}{(x+n)^2}.$$

Similarly, one can prove that $\sum_{n=1}^{\infty} \frac{1}{(x+n)^2} < \frac{1}{x+1} + \frac{1}{(x+1)^2}$. It follows that

$$\frac{x}{x+1} < x\sum_{n=1}^{\infty} \frac{1}{(x+n)^2} < \frac{x}{x+1} + \frac{x}{(x+1)^2}, \quad x > 0.$$

2.76. We prove the first limit in Remark 2.8. Let $h > 0$.
Since f decreases, we have that

$$\sum_{k=0}^{n-1} f((k+1)h) \le \int_0^n f(xh)dx = \sum_{k=0}^{n-1} \int_k^{k+1} f(xh)dx \le \sum_{k=0}^{n-1} f(kh)$$

and it follows that

$$f(h) + f(2h) + \cdots + f(nh) \le \int_0^n f(xh)dx \le f(0) + f(h) + \cdots + f((n-1)h).$$

This implies, since $\int_0^n f(xh)dx = \frac{1}{h}\int_0^{nh} f(t)dt$, that

$$h\left(f(h)+f(2h)+\cdots+f(nh)\right) \le \int_0^{nh} f(t)dt \le h\left(f(0)+f(h)+\cdots+f((n-1)h)\right).$$

Passing to the limit, as $n \to \infty$, in the previous inequalities we get that

$$h\sum_{n=1}^{\infty} f(nf) \le \int_0^{\infty} f(t)dt \le hf(0) + h\sum_{n=1}^{\infty} f(nh).$$

This implies that $\lim_{h\to 0^+} h\sum_{n=1}^{\infty} f(nh) = \int_0^{\infty} f(t)dt$. The second limit can be proved similarly.

10.5 Series with Harmonic Numbers and Factorials

2.77.

(b) We have, based on part (a), that

$$\sum_{i=1}^{\infty}\sum_{j=1}^{\infty}\frac{(i-1)!(j-1)!}{(i+j)!}H_{i+j}=\sum_{i=1}^{\infty}(i-1)!\sum_{j=1}^{\infty}\frac{(j-1)!}{(i+j)!}H_{i+j}$$

$$=\sum_{i=1}^{\infty}(i-1)!\left(\frac{H_i}{i\cdot i!}+\frac{1}{i^2\cdot i!}\right)$$

$$=\sum_{i=1}^{\infty}\frac{H_i}{i^2}+\sum_{i=1}^{\infty}\frac{1}{i^3}$$

$$=3\zeta(3).$$

We used that $\sum_{i=1}^{\infty}\frac{H_i}{i^2}=2\zeta(3)$ (see Problems 7.60 and 7.61).

(c) We have, based on part (a), that

$$\sum_{i=1}^{\infty}\sum_{j=1}^{\infty}(-1)^{i-1}\frac{(i-1)!(j-1)!}{(i+j)!}H_{i+j}=\sum_{i=1}^{\infty}(-1)^{i-1}(i-1)!\sum_{j=1}^{\infty}\frac{(j-1)!}{(i+j)!}H_{i+j}$$

$$=\sum_{i=1}^{\infty}(-1)^{i-1}(i-1)!\left(\frac{H_i}{i\cdot i!}+\frac{1}{i^2\cdot i!}\right)$$

$$=\sum_{i=1}^{\infty}(-1)^{i-1}\frac{H_i}{i^2}+\sum_{i=1}^{\infty}(-1)^{i-1}\frac{1}{i^3}$$

$$=\frac{11}{8}\zeta(3).$$

We used that $\sum_{i=1}^{\infty}(-1)^{i-1}\frac{H_i}{i^2}=\frac{5}{8}\zeta(3)$. This series can be calculated by using the formula $\int_0^1 x^n\ln(1-x)dx=-\frac{H_{n+1}}{n+1}$, $n\geq 0$, and by observing that

$$\sum_{i=1}^{\infty}(-1)^{i-1}\frac{H_i}{i^2}=-\int_0^1\frac{\ln(1-x)\ln(1+x)}{x}dx,$$

where the integral is calculated in the solution of Problem 2.81.

(d) Use symmetry combined to part (c) of the problem.

2.78. We have, based on part (a) of Problem 2.77, with i replaced by $i+1$, that

$$\sum_{i=1}^{\infty}\sum_{j=1}^{\infty}\frac{(i-1)!(j-1)!}{(i+j+1)!}H_{i+j+1}=\sum_{i=1}^{\infty}(i-1)!\sum_{j=1}^{\infty}\frac{(j-1)!}{(i+j+1)!}H_{i+j+1}$$

$$=\sum_{i=1}^{\infty}(i-1)!\left(\frac{H_{i+1}}{(i+1)\cdot(i+1)!}+\frac{1}{(i+1)^2\cdot(i+1)!}\right)$$

$$=\sum_{i=1}^{\infty}\frac{H_{i+1}}{i(i+1)^2}+\sum_{i=1}^{\infty}\frac{1}{i(i+1)^3}$$

$$=\sum_{i=1}^{\infty}\left(\frac{H_i}{i}-\frac{H_{i+1}}{i+1}+\frac{1}{i(i+1)}-\frac{H_{i+1}}{(i+1)^2}\right)+\sum_{i=1}^{\infty}\frac{1}{i(i+1)^3}$$

$$=1+1-2\zeta(3)+1+3-\zeta(2)-\zeta(3)$$

$$=6-\zeta(2)-3\zeta(3).$$

10.6   A Mosaic of Series

2.79. *Solution I (A jewel in the theory of series).*
We have

$$S=\sum_{n=1}^{\infty}(-1)^{n-1}\left[n\left(\zeta(2)-1-\frac{1}{2^2}-\cdots-\frac{1}{n^2}\right)-1\right]$$

$$=\zeta(2)-2+\sum_{n=2}^{\infty}(-1)^{n-1}\left[n\left(\zeta(2)-1-\frac{1}{2^2}-\cdots-\frac{1}{n^2}\right)-1\right]$$

$$\overset{n-1=m}{=}\zeta(2)-2+\sum_{m=1}^{\infty}(-1)^m\left[(m+1)\left(\zeta(2)-1-\frac{1}{2^2}-\cdots-\frac{1}{(m+1)^2}\right)-1\right]$$

$$=\zeta(2)-2-\sum_{m=1}^{\infty}(-1)^{m-1}\left[m\left(\zeta(2)-1-\frac{1}{2^2}-\cdots-\frac{1}{m^2}\right)-1\right]$$

$$-\sum_{m=1}^{\infty}(-1)^{m-1}\left(\zeta(2)-1-\frac{1}{2^2}-\cdots-\frac{1}{m^2}\right)+\sum_{m=1}^{\infty}\frac{(-1)^{m-1}}{m+1}$$

$$=\zeta(2)-2-S-\sum_{m=1}^{\infty}(-1)^{m-1}\left(\frac{1}{(m+1)^2}+\frac{1}{(m+2)^2}+\cdots\right)+\sum_{j=2}^{\infty}\frac{(-1)^{j-2}}{j}$$

(continued)

$$= \zeta(2) - 2 - S + \sum_{k=2}^{\infty} (-1)^{k-1} \left(\frac{1}{k^2} + \frac{1}{(k+1)^2} + \cdots \right) - \ln 2 + 1$$

$$\overset{(*)}{=} \zeta(2) - 1 - S + \frac{\pi^2}{8} - \zeta(2)$$

$$= \frac{\pi^2}{8} - 1 - \ln 2 - S.$$

We used at step (*) Problem 2.46.

Solution II. First we observe that

$$\zeta(2) - 1 - \frac{1}{2^2} - \cdots - \frac{1}{n^2} = \sum_{k=1}^{\infty} \frac{1}{(n+k)^2} > \sum_{k=1}^{\infty} \frac{1}{(n+k)(n+k+1)} = \frac{1}{n+1}$$

and

$$\zeta(2) - 1 - \frac{1}{2^2} - \cdots - \frac{1}{n^2} = \sum_{k=1}^{\infty} \frac{1}{(n+k)^2} < \sum_{k=1}^{\infty} \frac{1}{(n+k-1)(n+k)} = \frac{1}{n}.$$

These imply that $\frac{1}{n+1} < \zeta(2) - 1 - \frac{1}{2^2} - \cdots - \frac{1}{n^2} < \frac{1}{n}, \forall n \geq 1$.
Let $x_n = n \left(\zeta(2) - 1 - \frac{1}{2^2} - \cdots - \frac{1}{n^2} \right) - 1$. A calculation shows that

$$x_{n+1} - x_n = \zeta(2) - 1 - \frac{1}{2^2} - \cdots - \frac{1}{n^2} - \frac{1}{n+1} > 0.$$

Also, from the previous inequalities, we have that

$$\lim_{n \to \infty} n \left(\zeta(2) - 1 - \frac{1}{2^2} - \cdots - \frac{1}{n^2} \right) = 1,$$

which implies that $\lim_{n \to \infty} x_n = 0$. Thus, the series $\sum_{n=1}^{\infty} (-1)^{n-1}(-x_n)$ is a Leibniz series; hence, it converges.

Now we calculate the series. We apply Abel's summation formula with

$$a_n = (-1)^{n-1} \quad \text{and} \quad b_n = n \left(\zeta(2) - 1 - \frac{1}{2^2} - \cdots - \frac{1}{n^2} \right) - 1$$

and we have, since $\lim_{n \to \infty} (a_1 + a_2 + \cdots + a_n) b_{n+1} = 0$, that

$$\sum_{n=1}^{\infty} (-1)^{n-1} \left[n \left(\zeta(2) - 1 - \frac{1}{2^2} - \cdots - \frac{1}{n^2} \right) - 1 \right]$$

$$= \sum_{n=1}^{\infty} \left((-1)^{1-1} + \cdots + (-1)^{n-1} \right) \left[\frac{1}{n+1} - \left(\zeta(2) - 1 - \frac{1}{2^2} - \cdots - \frac{1}{n^2} \right) \right]$$

$$= \sum_{k=1}^{\infty} \left[\frac{1}{2k} - \left(\zeta(2) - 1 - \frac{1}{2^2} - \cdots - \frac{1}{(2k-1)^2} \right) \right].$$

We apply one more time Abel's summation formula, with $a_k = 1$ and $b_k = \frac{1}{2k} - \left(\zeta(2) - 1 - \frac{1}{2^2} - \cdots - \frac{1}{(2k-1)^2} \right)$, and we have that

$$\sum_{k=1}^{\infty} \left[\frac{1}{2k} - \left(\zeta(2) - 1 - \frac{1}{2^2} - \cdots - \frac{1}{(2k-1)^2} \right) \right]$$

$$= \lim_{k \to \infty} k \left[\frac{1}{2k+2} - \left(\zeta(2) - 1 - \frac{1}{2^2} - \cdots - \frac{1}{(2k+1)^2} \right) \right]$$

$$+ \sum_{k=1}^{\infty} k \left(\frac{1}{2k} - \frac{1}{2k+2} - \frac{1}{(2k)^2} - \frac{1}{(2k+1)^2} \right)$$

$$= \sum_{k=1}^{\infty} \left(\frac{1}{2(k+1)} - \frac{1}{4k} - \frac{1}{2(2k+1)} + \frac{1}{2(2k+1)^2} \right)$$

$$= \sum_{k=1}^{\infty} \left(\frac{1}{2(k+1)} - \frac{1}{4k} - \frac{1}{2(2k+1)} \right) + \frac{1}{2} \sum_{k=1}^{\infty} \frac{1}{(2k+1)^2}$$

$$= -\frac{\ln 2}{2} + \frac{\pi^2}{16} - \frac{1}{2}.$$

2.80. Part (a) follows by direct computations and the series in part (b) telescopes.
2.81.

(a) Let $k \geq 1$ be an integer. We have

$$\frac{1}{n+1} - \frac{1}{n+2} + \frac{1}{n+3} + \cdots + \frac{(-1)^{k-1}}{n+k}$$

$$= \int_0^1 \left(x^n - x^{n+1} + \cdots + (-1)^{k-1} x^{n+k-1} \right) dx$$

$$= \int_0^1 x^n \frac{1 - (-x)^k}{1+x} dx$$

and it follows that

$$\lim_{k \to \infty} \left(\frac{1}{n+1} - \frac{1}{n+2} + \frac{1}{n+3} + \cdots + \frac{(-1)^{k-1}}{n+k} \right) = \int_0^1 \frac{x^n}{1+x} dx.$$

(b) First we prove that the following logarithmic integral formula holds

$$\int_0^1 \frac{\ln(1-x) \ln(1+x)}{x} dx = -\frac{5}{8} \zeta(3).$$ (10.2)

We have, since $ab = \frac{1}{2}(a^2 + b^2 - (a-b)^2)$, that

$$\int_0^1 \frac{\ln(1-x) \ln(1+x)}{x} dx = \frac{1}{2} \int_0^1 \frac{\ln^2(1-x)}{x} dx + \frac{1}{2} \int_0^1 \frac{\ln^2(1+x)}{x} dx$$

$$- \frac{1}{2} \int_0^1 \frac{\ln^2 \left(\frac{1-x}{1+x} \right)}{x} dx$$

$$= \zeta(3) + \frac{\zeta(3)}{8} - \frac{1}{2} \int_0^1 \frac{\ln^2 \left(\frac{1-x}{1+x} \right)}{x} dx$$

$$\overset{\frac{1-x}{1+x}=t}{=\!=\!=} \frac{9}{8} \zeta(3) - \int_0^1 \frac{\ln^2 t}{1-t^2} dt$$

$$= \frac{9}{8} \zeta(3) - \frac{1}{2} \int_0^1 \frac{\ln^2 t}{1-t} dt - \frac{1}{2} \int_0^1 \frac{\ln^2 t}{1+t} dt$$

$$= -\frac{5}{8} \zeta(3),$$

since $\int_0^1 \frac{\ln^2(1+x)}{x} dx = \frac{\zeta(3)}{4}$ (see [6, pp. 291–292]).

Now we are ready to solve part (b) of the problem. We have, based on part (a), that

$$\sum_{n=1}^{\infty} \frac{H_n}{n} \left(\frac{1}{n+1} - \frac{1}{n+2} + \frac{1}{n+3} - \cdots \right) = \sum_{n=1}^{\infty} \frac{H_n}{n} \int_0^1 \frac{x^n}{1+x} dx$$

$$= \int_0^1 \frac{1}{1+x} \left(\sum_{n=1}^{\infty} \frac{H_n}{n} x^n \right) dx$$

$$\overset{7.86}{=\!=\!=} \int_0^1 \frac{1}{1+x} \left(\text{Li}_2(x) + \frac{1}{2} \ln^2(1-x) \right) dx$$

$$= \int_0^1 \frac{\text{Li}_2(x)}{1+x} dx + \frac{1}{2} \int_0^1 \frac{\ln^2(1-x)}{1+x} dx.$$

We calculate the first integral by parts and we have, based on (10.2), that

$$\int_0^1 \frac{\text{Li}_2(x)}{1+x} dx = \text{Li}_2(x) \ln(1+x) \Big|_0^1 + \int_0^1 \frac{\ln(1-x)\ln(1+x)}{x} dx$$

$$= \zeta(2) \ln 2 - \frac{5}{8} \zeta(3).$$

We calculate the second integral and we have

$$\int_0^1 \frac{\ln^2(1-x)}{1+x} dx = \int_0^1 \frac{\ln^2 t}{2-t} dt$$

$$= \frac{1}{2} \int_0^1 \frac{\ln^2 t}{1 - \frac{t}{2}} dt$$

$$= \frac{1}{2} \int_0^1 \ln^2 t \sum_{n=0}^{\infty} \left(\frac{t}{2}\right)^n dt$$

$$= \frac{1}{2} \sum_{n=0}^{\infty} \frac{1}{2^n} \int_0^1 t^n \ln^2 t \, dt$$

$$= 2 \sum_{n=0}^{\infty} \frac{1}{2^{n+1}(n+1)^3}$$

$$= 2\text{Li}_3 \left(\frac{1}{2}\right).$$

Putting all these together and using that (see [21, p. 44])

$$\text{Li}_3 \left(\frac{1}{2}\right) = \frac{1}{24} \left(-2\pi^2 \ln 2 + 4 \ln^3 2 + 21\zeta(3)\right),$$

part (b) of the problem is solved.

(c) We have, based on part (a), that

$$\sum_{n=1}^{\infty} (-1)^n \frac{H_n}{n} \left(\frac{1}{n+1} - \frac{1}{n+2} + \frac{1}{n+3} - \cdots \right)$$

$$= \sum_{n=1}^{\infty} (-1)^n \frac{H_n}{n} \int_0^1 \frac{x^n}{1+x} \, dx$$

$$= \int_0^1 \frac{1}{1+x} \left(\sum_{n=1}^{\infty} \frac{H_n}{n} (-x)^n \right) dx$$

$$\overset{7.86}{=} \int_0^1 \frac{1}{1+x} \left(\text{Li}_2(-x) + \frac{1}{2} \ln^2(1+x) \right) dx$$

$$= \int_0^1 \frac{\text{Li}_2(-x)}{1+x} \, dx + \frac{1}{2} \int_0^1 \frac{\ln^2(1+x)}{1+x} \, dx$$

$$= \int_0^1 \frac{\text{Li}_2(-x)}{1+x} \, dx + \frac{1}{6} \ln^3 2.$$

We calculate the preceding integral by parts and we have that

$$\int_0^1 \frac{\text{Li}_2(-x)}{1+x} \, dx = \text{Li}_2(-x) \ln(1+x) \Big|_0^1 + \int_0^1 \frac{\ln^2(1+x)}{x} \, dx = -\frac{\pi^2}{12} \ln 2 + \frac{\zeta(3)}{4},$$

and part (c) of the problem is solved.

(d) We have, based on part (a), that

$$\sum_{n=1}^{\infty} H_n \left(\frac{1}{n+1} - \frac{1}{n+2} + \frac{1}{n+3} - \cdots \right)^2 = \sum_{n=1}^{\infty} \int_0^1 \int_0^1 \frac{x^n y^n}{(1+x)(1+y)} \, dx \, dy$$

$$= \int_0^1 \int_0^1 \frac{1}{(1+x)(1+y)} \sum_{n=1}^{\infty} H_n (xy)^n \, dx \, dy$$

$$\overset{3.63 \,(a)}{=} -\int_0^1 \int_0^1 \frac{\ln(1-xy)}{(1+x)(1+y)(1-xy)} \, dx \, dy$$

$$= -\int_0^1 \frac{1}{1+x} \left(\int_0^1 \frac{\ln(1-xy)}{(1+y)(1-xy)} \, dy \right) dx.$$

We calculate the inner integral and we have that

$$\int_0^1 \frac{\ln(1-xy)}{(1+y)(1-xy)}dy = \frac{1}{1+x}\int_0^1 \frac{\ln(1-xy)}{1+y}dy + \frac{1}{1+x}\int_0^1 \frac{x\ln(1-xy)}{1-xy}dy$$

$$= \frac{1}{1+x}\int_0^1 \frac{\ln(1-xy)}{1+y}dy - \frac{\ln^2(1-x)}{2(1+x)}.$$

We calculate the preceding logarithmic integral and we have

$$\int_0^1 \frac{\ln(1-xy)}{1+y}dy \overset{xy=t}{=} \int_0^x \frac{\ln(1-t)}{x+t}dt$$

$$\overset{x+t=u}{=} \int_x^{2x} \frac{\ln(1+x-u)}{u}du$$

$$\overset{u=(1+x)v}{=} \ln 2\ln(1+x) + \int_{\frac{x}{1+x}}^{\frac{2x}{1+x}} \frac{\ln(1-v)}{v}dv.$$

Putting all these together, we have that

$$\int_0^1 \frac{1}{1+x}\left(\int_0^1 \frac{\ln(1-xy)}{(1+y)(1-xy)}dy\right)dx$$

$$= \int_0^1 \frac{1}{(1+x)^2}\left(\ln 2\ln(1+x) + \int_{\frac{x}{1+x}}^{\frac{2x}{1+x}} \frac{\ln(1-v)}{v}dv - \frac{\ln^2(1-x)}{2}\right)dx$$

$$= \ln 2\int_0^1 \frac{\ln(1+x)}{(1+x)^2}dx + \int_0^1 \frac{1}{(1+x)^2}\left(\int_{\frac{x}{1+x}}^{\frac{2x}{1+x}} \frac{\ln(1-v)}{v}dv\right)dx - \frac{1}{2}\int_0^1 \frac{\ln^2(1-x)}{(1+x)^2}dx$$

$$= \ln 2\frac{1-\ln 2}{2} + \int_0^1 \frac{1}{(1+x)^2}\left(\int_{\frac{x}{1+x}}^{\frac{2x}{1+x}} \frac{\ln(1-v)}{v}dv\right)dx - \frac{1}{2}\text{Li}_2\left(\frac{1}{2}\right),$$

where the last equality follows based on part (ii) of Remark 7.3 after Problem 7.66.
We calculate the logarithmic integral by parts, with

$$f(x) = \int_{\frac{x}{1+x}}^{\frac{2x}{1+x}} \frac{\ln(1-v)}{v}dv \quad \text{and} \quad f'(x) = \frac{\ln(1-x)}{x(1+x)},$$

$g'(x) = \frac{1}{(1+x)^2}$ and $g(x) = -\frac{1}{1+x}$, and we have that

$$\int_0^1 \frac{1}{(1+x)^2} \left(\int_{\frac{x}{1+x}}^{\frac{2x}{1+x}} \frac{\ln(1-v)}{v} dv \right) dx$$

$$= -\frac{1}{1+x} \int_{\frac{x}{1+x}}^{\frac{2x}{1+x}} \frac{\ln(1-v)}{v} dv \Big|_0^1 + \int_0^1 \frac{\ln(1-x)}{x(1+x)^2} dx$$

$$= -\frac{1}{2} \int_{\frac{1}{2}}^1 \frac{\ln(1-v)}{v} dv + \int_0^1 \ln(1-x) \left(\frac{1}{x} - \frac{1}{1+x} - \frac{1}{(1+x)^2} \right) dx$$

$$= -\frac{1}{2} \int_{\frac{1}{2}}^1 \frac{\ln(1-v)}{v} dv + \int_0^1 \frac{\ln(1-x)}{x} dx - \int_0^1 \frac{\ln(1-x)}{1+x} dx - \int_0^1 \frac{\ln(1-x)}{(1+x)^2} dx$$

$$= \frac{1}{2} \left[\text{Li}_2(1) - \text{Li}_2 \left(\frac{1}{2} \right) \right] - \text{Li}_2(1) + \text{Li}_2 \left(\frac{1}{2} \right) + \frac{\ln 2}{2}$$

$$= -\frac{1}{2} \text{Li}_2(1) + \frac{1}{2} \text{Li}_2 \left(\frac{1}{2} \right) + \frac{\ln 2}{2}.$$

We used in the preceding calculations the integrals $\int_0^1 \frac{\ln(1-x)}{1+x} dx = -\text{Li}_2 \left(\frac{1}{2} \right)$ (see Problem 7.66) and $\int_0^1 \frac{\ln(1-x)}{(1+x)^2} dx = -\frac{\ln 2}{2}$, which can be calculated by the method given in part (ii) of Remark 7.3 after Problem 7.66.

It follows that

$$\int_0^1 \frac{1}{1+x} \left(\int_0^1 \frac{\ln(1-xy)}{(1+y)(1-xy)} dy \right) dx = \ln 2 - \frac{1}{2} \ln^2 2 - \frac{\pi^2}{12},$$

and the problem is solved.

2.82.

(a) Prove the formula by mathematical induction.
(b) Use Abel's summation formula with
$$a_n = (-1)^n n \text{ and } b_n = \zeta(2) - 1 - \frac{1}{2^2} - \cdots - \frac{1}{n^2} - \frac{1}{n}.$$
(c) Prove the formula by mathematical induction.
(d) Use Abel's summation formula with
$$a_n = (-1)^n n^2 \text{ and } b_n = \zeta(3) - 1 - \frac{1}{2^3} - \cdots - \frac{1}{n^3} - \frac{1}{2n^2}.$$

2.83. Use Abel's summation formula with:

(a) $a_n = (-1)^{n-1}$ and $b_n = \zeta(k) - 1 - \frac{1}{2^k} - \cdots - \frac{1}{n^k}$;
(b) $a_n = (-1)^{n-1} n$ and $b_n = \zeta(k) - 1 - \frac{1}{2^k} - \cdots - \frac{1}{n^k}$.

2.84.

(a) We use Abel's summation formula with $a_n = 1$ and $b_n = \frac{1}{n!} + \frac{2}{(n+1)!} + \frac{3}{(n+2)!} + \cdots$ and we have, since $b_n - b_{n+1} = \frac{1}{n!} + \frac{1}{(n+1)!} + \cdots$, that

$$\sum_{n=1}^{\infty} \left(\frac{1}{n!} + \frac{2}{(n+1)!} + \frac{3}{(n+2)!} + \cdots \right)$$

$$= \lim_{n \to \infty} n \left(\frac{1}{(n+1)!} + \frac{2}{(n+2)!} + \frac{3}{(n+3)!} + \cdots \right)$$

$$+ \sum_{n=1}^{\infty} n \left(\frac{1}{n!} + \frac{1}{(n+1)!} + \cdots \right)$$

$$= \sum_{n=1}^{\infty} n \left(\frac{1}{n!} + \frac{1}{(n+1)!} + \cdots \right)$$

$$\stackrel{(*)}{=} \sum_{n=1}^{\infty} \frac{n(n+1)}{2} \cdot \frac{1}{n!}$$

$$= \frac{1}{2} \sum_{n=1}^{\infty} \frac{n+1}{(n-1)!}$$

$$= \frac{3}{2}e.$$

We used at step (*) Abel's summation formula with $a_n = n$ and $b_n = \frac{1}{n!} + \frac{1}{(n+1)!} + \cdots$.

Parts (b) and (c) of the problem can be solved similarly.

2.85.

(a) We use Abel's summation formula with $a_n = 1$ and $b_n = \frac{1}{n^k} + \frac{2}{(n+1)^k} + \frac{3}{(n+2)^k} + \cdots$, and we have that

$$\sum_{n=1}^{\infty} \left(\frac{1}{n^k} + \frac{2}{(n+1)^k} + \frac{3}{(n+2)^k} + \cdots \right) = \sum_{n=1}^{\infty} n \left(\frac{1}{n^k} + \frac{1}{(n+1)^k} + \frac{1}{(n+2)^k} + \cdots \right)$$

$$\stackrel{(*)}{=} \sum_{n=1}^{\infty} \frac{n(n+1)}{2} \cdot \frac{1}{n^k}$$

$$= \frac{1}{2}(\zeta(k-2) + \zeta(k-1)).$$

We used at step (*) Abel's summation formula with
$a_n = n$ and $b_n = \frac{1}{n^k} + \frac{1}{(n+1)^k} + \cdots$.
Part (b) of the problem can be solved similarly.

2.86.

(a) We use Abel's summation formula with

$$a_n = n \quad \text{and} \quad b_n = \left(\zeta(2) - 1 - \frac{1}{2^2} - \cdots - \frac{1}{n^2} - \frac{1}{n} \right)^2,$$

and we have, since

$$b_n - b_{n+1}$$
$$= -\frac{1}{n(n+1)^2} \left[2 \left(\zeta(2) - 1 - \frac{1}{2^2} - \cdots - \frac{1}{(n+1)^2} \right) + \frac{1}{(n+1)^2} - \frac{1}{n+1} - \frac{1}{n} \right],$$

that

$$\sum_{n=1}^{\infty} n \left(\zeta(2) - 1 - \frac{1}{2^2} - \cdots - \frac{1}{n^2} - \frac{1}{n} \right)^2$$

$$= -\frac{1}{2} \sum_{n=1}^{\infty} \frac{1}{n+1} \left[2 \left(\zeta(2) - 1 - \frac{1}{2^2} - \cdots - \frac{1}{(n+1)^2} \right) + \frac{1}{(n+1)^2} - \frac{1}{n+1} - \frac{1}{n} \right]$$

$$= -\sum_{n=1}^{\infty} \frac{1}{n+1} \left(\zeta(2) - 1 - \frac{1}{2^2} - \cdots - \frac{1}{(n+1)^2} \right) - \frac{1}{2} \sum_{n=1}^{\infty} \frac{1}{(n+1)^3}$$

$$+ \frac{1}{2} \sum_{n=1}^{\infty} \frac{1}{(n+1)^2} + \frac{1}{2} \sum_{n=1}^{\infty} \frac{1}{n(n+1)}$$

$$= -\sum_{m=1}^{\infty} \frac{1}{m} \left(\zeta(2) - 1 - \frac{1}{2^2} - \cdots - \frac{1}{m^2} \right) + \zeta(2) - 1 - \frac{1}{2}(\zeta(3) - 1)$$

$$+ \frac{1}{2}(\zeta(2) - 1) + \frac{1}{2}$$

$$= \frac{3}{2} \zeta(2) - \frac{3}{2} \zeta(3) - \frac{1}{2}.$$

We used the formula $\sum_{m=1}^{\infty} \frac{1}{m} \left(\zeta(2) - 1 - \frac{1}{2^2} - \cdots - \frac{1}{m^2} \right) = \zeta(3)$, which can be proved using Abel's summation formula with
$a_m = \frac{1}{m}$ and $b_m = \zeta(2) - 1 - \frac{1}{2^2} - \cdots - \frac{1}{m^2}$.

(b) We use Abel's summation formula with

$$a_n = n^3 \quad \text{and} \quad b_n = \left(\zeta(3) - 1 - \frac{1}{2^3} - \cdots - \frac{1}{n^3} - \frac{1}{2n^2} \right)^2,$$

and we have, since

$$b_n - b_{n+1} = -\frac{3n+1}{2n^2(n+1)^3} \left[2 \left(\zeta(3) - 1 - \frac{1}{2^3} - \cdots - \frac{1}{(n+1)^3} \right) \right.$$
$$\left. + \frac{1}{(n+1)^3} - \frac{1}{2(n+1)^2} - \frac{1}{2n^2} \right],$$

that

$$\sum_{n=1}^{\infty} n^3 \left(\zeta(3) - 1 - \frac{1}{2^3} - \cdots - \frac{1}{n^3} - \frac{1}{2n^2} \right)^2 = \sum_{n=1}^{\infty} A_n (b_n - b_{n+1}),$$

where

$$A_n (b_n - b_{n+1}) = -\frac{3n+1}{8(n+1)} \left[2 \left(\zeta(3) - 1 - \frac{1}{2^3} - \cdots - \frac{1}{(n+1)^3} \right) \right.$$
$$\left. + \frac{1}{(n+1)^3} - \frac{1}{2(n+1)^2} - \frac{1}{2n^2} \right]$$
$$= -\frac{1}{8} \left(3 - \frac{2}{n+1} \right) \left[2 \left(\zeta(3) - 1 - \frac{1}{2^3} - \cdots - \frac{1}{(n+1)^3} \right) \right.$$
$$\left. + \frac{1}{(n+1)^3} - \frac{1}{2(n+1)^2} - \frac{1}{2n^2} \right],$$

and it follows, after some calculations, that

$$\sum_{n=1}^{\infty} n^3 \left(\zeta(3) - 1 - \frac{1}{2^3} - \cdots - \frac{1}{n^3} - \frac{1}{2n^2} \right)^2$$

$$= -\frac{3}{4} \sum_{n=1}^{\infty} \left(\zeta(3) - 1 - \frac{1}{2^3} - \cdots - \frac{1}{(n+1)^3} \right) - \frac{3}{8} \sum_{n=1}^{\infty} \frac{1}{(n+1)^3} + \frac{3}{16} \sum_{n=1}^{\infty} \frac{1}{(n+1)^2}$$

$$+ \frac{3}{16} \sum_{n=1}^{\infty} \frac{1}{n^2} + \frac{1}{2} \sum_{n=1}^{\infty} \frac{1}{n+1} \left(\zeta(3) - 1 - \frac{1}{2^3} - \cdots - \frac{1}{(n+1)^3} \right)$$

$$+ \frac{1}{4} \sum_{n=1}^{\infty} \frac{1}{(n+1)^4} - \frac{1}{8} \sum_{n=1}^{\infty} \frac{1}{(n+1)^3} - \frac{1}{8} \sum_{n=1}^{\infty} \frac{1}{n^2(n+1)}$$

$$= -\frac{1}{2}\zeta(2) + \frac{1}{2}\zeta(3) + \frac{3}{8}\zeta(4) - \frac{1}{16}.$$

We used in the preceding computations the formulae:

- $$\sum_{n=1}^{\infty} \left(\zeta(3) - 1 - \frac{1}{2^3} - \cdots - \frac{1}{(n+1)^3} \right) = \zeta(2) - 2\zeta(3) + 1;$$

- $$\sum_{n=1}^{\infty} \frac{1}{n+1} \left(\zeta(3) - 1 - \frac{1}{2^3} - \cdots - \frac{1}{(n+1)^3} \right) = \frac{\zeta(4)}{4} - \zeta(3) + 1.$$

Power Series

11

11.1 Convergence and Sum of Power Series

3.1. (a) $(-5, 5)$; (b) $[-3, 3)$; (c) $(-2, 0]$; (d) $[-3, 7)$; (e) $(-1, 3)$; (f) $\mathbb{R}$; (g) $(-e, e)$; (h) $\mathbb{R}$; (i) $(-1, 1)$; (j) $\{0\}$; (k) $[-1, 1]$; (l) $[-1, 1]$.

3.2. $(-1, 1)$. Observe that $a_k = \begin{cases} 1 & \text{if } k = n! \\ 0 & \text{if } k \neq n! \end{cases} \rightarrow R = 1.$

3.3. (a) We have

$$\sum_{n=0}^{\infty}(n + 1)x^{n+1} = \sum_{m=1}^{\infty} mx^m = \frac{x}{(1 - x)^2}, \quad x \in (-1, 1).$$

(b) We have

$$\sum_{n=0}^{\infty}(n + 1)(n + 2)x^n = \sum_{m=1}^{\infty} m(m + 1)x^{m-1}, \quad x \in (-1, 1).$$

For the calculation of the preceding series, we use the power series

$$\sum_{m=1}^{\infty} mx^{m+1} = \frac{x^2}{(1 - x)^2}, \quad x \in (-1, 1),$$

which can be obtained by multiplying the derivative of the geometric series by x^2.

© The Author(s), under exclusive license to Springer Nature Switzerland AG 2021
A. Sîntămărian, O. Furdui, *Sharpening Mathematical Analysis Skills*, Problem Books in Mathematics, https://doi.org/10.1007/978-3-030-77139-3_11

We differentiate the preceding series, and we divide it by $x \neq 0$ to obtain

$$\sum_{m=1}^{\infty} m(m+1)x^{m-1} = \frac{2}{(1-x)^3}, \quad x \in (-1, 1).$$

(c) Let

$$f(x) = \sum_{n=1}^{\infty} (-1)^{n-1} \frac{x^{3n-1}}{3n-1}, \quad x \in (-1, 1].$$

We have

$$f'(x) = x \sum_{n=1}^{\infty} (-x^3)^{n-1} = \frac{x}{1+x^3}, \quad x \in (-1, 1).$$

For $x \in (-1, 1)$, it follows that

$$f(x)-f(0)=f(x)=\int_0^x \frac{t}{1+t^3}dt=\frac{1}{6}\ln\frac{x^2-x+1}{(x+1)^2}+\frac{1}{\sqrt{3}}\arctan\frac{2x-1}{\sqrt{3}}+\frac{\pi}{6\sqrt{3}}.$$

When $x = 1$ we have

$$f(1) = \lim_{\substack{x \to 1 \\ x < 1}} f(x) = \frac{\pi}{3\sqrt{3}} - \frac{\ln 2}{3}.$$

(d) Let

$$f(x) = \sum_{n=1}^{\infty} (-1)^{n-1}(5n-1)x^{5n-2}, \quad x \in (-1, 1).$$

We have

$$\int_0^x f(t)dt = \int_0^x \left(\sum_{n=1}^{\infty} (-1)^{n-1}(5n-1)t^{5n-2} \right) dt$$

$$= \sum_{n=1}^{\infty} (-1)^{n-1}(5n-1) \int_0^x t^{5n-2}dt$$

$$= x^4 \sum_{n=1}^{\infty} (-x^5)^{n-1}$$

$$= \frac{x^4}{1+x^5}.$$

It follows that

$$f(x) = \left(\frac{x^4}{1+x^5}\right)' = \frac{4x^3 - x^8}{(1+x^5)^2}, \quad x \in (-1, 1).$$

(e) We observe that

$$\sum_{n=0}^{\infty} (-1)^n (n+1)^3 x^n = \sum_{m=1}^{\infty} (-1)^{m-1} m^3 x^{m-1}.$$

We differentiate the geometric series and we multiply it by x to obtain

$$\sum_{n=1}^{\infty} n x^n = \frac{x}{(1-x)^2}, \quad x \in (-1, 1).$$

We differentiate the preceding power series, we multiply it by x, and we have

$$\sum_{n=1}^{\infty} n^2 x^n = \frac{x + x^2}{(1-x)^3}, \quad x \in (-1, 1)$$

We differentiate the above series and replace x by $-x$ to obtain

$$\sum_{n=1}^{\infty} (-1)^{n-1} n^3 x^{n-1} = \frac{x^2 - 4x + 1}{(1+x)^4}, \quad x \in (-1, 1).$$

(f) We have, for $x \in (-1, 1)$, that

$$\sum_{n=1}^{\infty} \frac{x^{4n-1}}{4n - 3} = x^2 \sum_{n=1}^{\infty} \int_0^x t^{4n-4} dt = x^2 \int_0^x \left(\sum_{n=1}^{\infty} t^{4n-4}\right) dt$$

$$= x^2 \int_0^x \frac{1}{1-t^4} dt = \frac{x^2}{2} \arctan x + \frac{x^2}{4} \ln \frac{1+x}{1-x}.$$

3.4. A. (a) $[-1, 1)$; (b) $\mathbb{R}$; (c) $[-1, 1)$, for $k = 2$ and $[-1, 1]$, for $k \geq 3$.
B. Let $x \neq 1$. We have

$$S = \sum_{n=1}^{\infty} \left(e - 1 - \frac{1}{1!} - \frac{1}{2!} - \cdots - \frac{1}{n!} \right) x^n$$

$$= (e - 2)x + \sum_{n=2}^{\infty} \left(e - 1 - \frac{1}{1!} - \frac{1}{2!} - \cdots - \frac{1}{n!} \right) x^n$$

$$\overset{n-1=m}{=} (e - 2)x + \sum_{m=1}^{\infty} \left(e - 1 - \frac{1}{1!} - \frac{1}{2!} - \cdots - \frac{1}{(m + 1)!} \right) x^{m+1}$$

$$= (e - 2)x + \sum_{m=1}^{\infty} \left(e - 1 - \frac{1}{1!} - \frac{1}{2!} - \cdots - \frac{1}{m!} \right) x^{m+1} - \sum_{m=1}^{\infty} \frac{x^{m+1}}{(m + 1)!}$$

$$= (e - 2)x + xS - \left(e^x - 1 - x \right),$$

and it follows that $S = \frac{e^x - ex}{x-1} + 1$.

When $x = 1$, the series can be calculated by Abel's summation formula. Part **C** can be solved similarly.

11.2   Maclaurin Series of Elementary Functions

3.5. Solve the problem by mathematical induction.

3.6. (a) The series equals $\frac{1}{e} - \sin 1$. We have

$$u_n = \frac{16n^2 + 4n - 1}{(4n + 2)!} = \frac{(4n + 1)(4n + 2) - 8n - 3}{(4n + 2)!}$$

$$= \frac{1}{(4n)!} - \frac{8n + 3}{(4n + 2)!} = \frac{1}{(4n)!} - \frac{2(4n + 2) - 1}{(4n + 2)!}$$

$$= \frac{1}{(4n)!} - \frac{2}{(4n + 1)!} + \frac{1}{(4n + 2)!}$$

$$= \frac{1}{(4n)!} - \frac{1}{(4n + 1)!} + \frac{1}{(4n + 2)!} - \frac{1}{(4n + 3)!} - \left[\frac{1}{(4n + 1)!} - \frac{1}{(4n + 3)!} \right].$$

Therefore,

$$\sum_{k=0}^{n} u_k = \sum_{k=0}^{n} \left[\frac{1}{(4k)!} - \frac{1}{(4k + 1)!} + \frac{1}{(4k + 2)!} - \frac{1}{(4k + 3)!} \right]$$

$$- \sum_{k=0}^{n} \left[\frac{1}{(4k + 1)!} - \frac{1}{(4k + 3)!} \right]$$

$$= \sum_{k=0}^{4n+3}(-1)^k\frac{1}{k!} - \sum_{k=0}^{2n+1}(-1)^k\frac{1}{(2k+1)!}.$$

It follows that

$$\sum_{n=0}^{\infty}\frac{16n^2+4n-1}{(4n+2)!} = \lim_{n\to\infty}\sum_{k=0}^{n}u_k = \frac{1}{e} - \sin 1.$$

(b) The series equals $\cos 1 - \frac{1}{e}$. We have

$$v_n = \frac{16n^2+12n+1}{(4n+3)!} = \frac{(4n+2)(4n+3)-8n-5}{(4n+3)!}$$

$$= \frac{1}{(4n+1)!} - \frac{8n+5}{(4n+3)!} = \frac{1}{(4n+1)!} - \frac{2(4n+3)-1}{(4n+3)!}$$

$$= \frac{1}{(4n+1)!} - \frac{2}{(4n+2)!} + \frac{1}{(4n+3)!}$$

$$= \frac{1}{(4n)!} - \frac{1}{(4n+2)!} - \left[\frac{1}{(4n)!} - \frac{1}{(4n+1)!} + \frac{1}{(4n+2)!} - \frac{1}{(4n+3)!}\right].$$

Therefore,

$$\sum_{k=0}^{n}v_k = \sum_{k=0}^{n}\left[\frac{1}{(4k)!} - \frac{1}{(4k+2)!}\right]$$

$$- \sum_{k=0}^{n}\left[\frac{1}{(4k)!} - \frac{1}{(4k+1)!} + \frac{1}{(4k+2)!} - \frac{1}{(4k+3)!}\right]$$

$$= \sum_{k=0}^{2n+1}(-1)^k\frac{1}{(2k)!} - \sum_{k=0}^{4n+3}(-1)^k\frac{1}{k!}.$$

It follows that

$$\sum_{n=0}^{\infty}\frac{16n^2+12n+1}{(4n+3)!} = \lim_{n\to\infty}\sum_{k=0}^{n}v_k = \cos 1 - \frac{1}{e}.$$

3.7. We have

$$u_n = \frac{1}{n(2n+1)(4n+1)} = \frac{1}{n} + \frac{2}{2n+1} - \frac{8}{4n+1}$$

$$= 4\left(\frac{1}{4n} - \frac{1}{4n+1} + \frac{1}{4n+2} - \frac{1}{4n+3}\right) - 4\left(\frac{1}{4n+1} - \frac{1}{4n+3}\right).$$

Therefore,

$$\sum_{k=1}^{n} u_k = 4\sum_{k=1}^{n}\left(\frac{1}{4k} - \frac{1}{4k+1} + \frac{1}{4k+2} - \frac{1}{4k+3}\right) - 4\sum_{k=1}^{n}\left(\frac{1}{4k+1} - \frac{1}{4k+3}\right)$$

$$= -4\sum_{k=1}^{4n+3}(-1)^{k-1}\frac{1}{k} + 4\left(1 - \frac{1}{2} + \frac{1}{3}\right) - 4\sum_{k=0}^{2n+1}(-1)^k\frac{1}{2k+1} + 4\left(1 - \frac{1}{3}\right)$$

$$= 6 - 4\sum_{k=1}^{4n+3}(-1)^{k-1}\frac{1}{k} - 4\sum_{k=0}^{2n+1}(-1)^k\frac{1}{2k+1}.$$

It follows that

$$\sum_{n=1}^{\infty}\frac{1}{n(2n+1)(4n+1)} = \lim_{n\to\infty}\sum_{k=1}^{n}u_k = 6 - 4\ln 2 - 4\arctan 1 = 6 - 4\ln 2 - \pi.$$

3.8.

(a) We have

$$\sinh x = \frac{e^x - e^{-x}}{2} = \frac{1}{2}\sum_{n=0}^{\infty}\frac{x^n}{n!} - \frac{1}{2}\sum_{n=0}^{\infty}\frac{(-x)^n}{n!}$$

$$= \frac{1}{2}\sum_{n=0}^{\infty}\frac{1 - (-1)^n}{n!}x^n \overset{n=2m-1}{=} \sum_{m=1}^{\infty}\frac{x^{2m-1}}{(2m-1)!}, \quad x \in \mathbb{R}.$$

(b) We have

$$\cos^2 x = \frac{1 + \cos(2x)}{2} = 1 + \sum_{n=1}^{\infty}(-1)^n\frac{2^{2n-1}}{(2n)!}x^{2n}, \quad x \in \mathbb{R}.$$

(c) We use the identity $\sin(3x) = 3\sin x - 4\sin^3 x$, $x \in \mathbb{R}$, and we have that

$$\sin^3 x = \frac{3\sin x - \sin(3x)}{4} = \frac{1}{4}\left(3\sum_{n=0}^{\infty}(-1)^n\frac{x^{2n+1}}{(2n+1)!} - \sum_{n=0}^{\infty}(-1)^n\frac{(3x)^{2n+1}}{(2n+1)!}\right)$$

$$= \frac{1}{4}\sum_{n=0}^{\infty}(-1)^n\frac{3 - 3^{2n+1}}{(2n+1)!}x^{2n+1}, \quad x \in \mathbb{R}.$$

(d) We have

$$x^2 e^{-3x} = x^2 \sum_{n=0}^{\infty} \frac{(-3x)^n}{n!} = \sum_{n=0}^{\infty} (-1)^n \frac{3^n}{n!} x^{n+2}, \quad x \in \mathbb{R}.$$

(e) We have

$$\frac{3x-1}{(x-1)^2} = \frac{-3}{1-x} + \frac{2}{(1-x)^2} = -3 \sum_{n=0}^{\infty} x^n + 2 \sum_{n=0}^{\infty} (n+1)x^n = \sum_{n=0}^{\infty} (2n-1)x^n,$$

for $x \in (-1, 1)$.

(f) We use the binomial series with $\alpha = -\frac{1}{2}$ and $x \to -\frac{x^2}{9}$ and we have that

$$\frac{1}{\sqrt{9-x^2}} = \frac{1}{3} \left(1 - \frac{x^2}{9} \right)^{-\frac{1}{2}} = \frac{1}{3} \left[1 + \sum_{n=1}^{\infty} \frac{\left(-\frac{1}{2}\right)\left(-\frac{3}{2}\right) \cdots \left(-n+\frac{1}{2}\right)}{n!} \left(-\frac{x^2}{9}\right)^n \right]$$

$$= \frac{1}{3} + \frac{1}{3} \sum_{n=1}^{\infty} \frac{(2n-1)!!}{(2n)!! \cdot 3^{2n}} x^{2n}, \quad x \in (-3, 3).$$

(g) We have

$$\frac{1}{2} \ln \frac{1+x}{1-x} = \frac{1}{2} \left(\sum_{n=0}^{\infty} (-1)^n \frac{x^{n+1}}{n+1} + \sum_{n=0}^{\infty} \frac{x^{n+1}}{n+1} \right)$$

$$= \frac{1}{2} \sum_{n=0}^{\infty} \frac{(-1)^n + 1}{n+1} x^{n+1}$$

$$= \sum_{n=0}^{\infty} \frac{x^{2n+1}}{2n+1}, \quad x \in (-1, 1).$$

(h) We have

$$f(x) = \ln(1-x) + \ln(1+5x) = -\sum_{n=1}^{\infty} \frac{x^n}{n} + \sum_{n=1}^{\infty} (-1)^{n-1} \frac{(5x)^n}{n}$$

$$= \sum_{n=1}^{\infty} \frac{(-1)^{n-1} 5^n - 1}{n} x^n, \quad x \in \left(-\frac{1}{5}, \frac{1}{5} \right].$$

(i) We use the binomial series and we have

$$f'(x) = \frac{1}{\sqrt{1+x^2}} = (1+x^2)^{-\frac{1}{2}} = 1 + \sum_{n=1}^{\infty}(-1)^n \frac{(2n-1)!!}{(2n)!!}x^{2n}, \quad x \in (-1, 1).$$

We integrate the preceding relation and we obtain

$$f(x) = x + \sum_{n=1}^{\infty}(-1)^n \frac{(2n-1)!!}{(2n)!!(2n+1)}x^{2n+1}, \quad x \in [-1, 1].$$

(j) We have

$$\cos(3x) + x\sin(3x) = \sum_{n=0}^{\infty}(-1)^n \frac{(3x)^{2n}}{(2n)!} + x\sum_{n=1}^{\infty}(-1)^{n-1}\frac{(3x)^{2n-1}}{(2n-1)!}$$

$$= 1 + \sum_{n=1}^{\infty}(-1)^n \frac{(3x)^{2n}}{(2n)!} + x\sum_{n=1}^{\infty}(-1)^{n-1}\frac{(3x)^{2n-1}}{(2n-1)!}$$

$$= 1 + \sum_{n=1}^{\infty}(-1)^{n-1}\frac{3^{2n-1}(2n-3)}{(2n)!}x^{2n}, \quad x \in \mathbb{R}.$$

(k) We have

$$\frac{1+x^2}{1-x} = 1 + x + 2x^2 \cdot \frac{1}{1-x} = 1 + x + 2\sum_{n=2}^{\infty}x^n, \quad x \in (-1, 1).$$

3.9. A method for calculating series of this form is given in [130].
3.10. Use that

$$\frac{1}{x^2+x+1} = \frac{1}{\alpha-\beta}\left(\frac{1}{\beta(1-\beta^2x)} - \frac{1}{\alpha(1-\alpha^2x)}\right),$$

where $\alpha = -\frac{1}{2} + \frac{\sqrt{3}}{2}i$ and $\beta = -\frac{1}{2} - \frac{\sqrt{3}}{2}i$.
3.12. Observe that $(n-2)! + (n+2)! = (n-2)!(n^2+n-1)^2$, $\forall n \geq 2$. The second series of part (a) is due to T. Andreescu [1].
3.13. (a) We have

$$\sum_{n=1}^{\infty} \frac{\lfloor \log_2 n \rfloor}{n(n+1)} = \sum_{k=0}^{\infty} \left(\sum_{2^k \le n < 2^{k+1}} \frac{\lfloor \log_2 n \rfloor}{n(n+1)} \right)$$

$$= \sum_{k=0}^{\infty} k \sum_{2^k \le n < 2^{k+1}} \left(\frac{1}{n} - \frac{1}{n+1} \right)$$

$$= \sum_{k=0}^{\infty} \frac{k}{2^{k+1}}$$

$$= 1.$$

(b) We have

$$\sum_{n=2}^{\infty} \frac{1}{n(n+1) \lfloor \log_2 n \rfloor} = \sum_{k=1}^{\infty} \left(\sum_{2^k \le n < 2^{k+1}} \frac{1}{n(n+1) \lfloor \log_2 n \rfloor} \right)$$

$$= \sum_{k=1}^{\infty} \frac{1}{k} \sum_{2^k \le n < 2^{k+1}} \left(\frac{1}{n} - \frac{1}{n+1} \right)$$

$$= \sum_{k=1}^{\infty} \frac{1}{k 2^{k+1}}$$

$$= \frac{\ln 2}{2}.$$

The other parts of the problem can be solved similarly.

3.14. We use the geometric series and we have

$$\frac{1}{x+5} = \frac{1}{x-2+7} = \frac{1}{7} \cdot \frac{1}{1 + \frac{x-2}{7}} = \frac{1}{7} \sum_{n=0}^{\infty} \left(-\frac{x-2}{7} \right)^n = \sum_{n=0}^{\infty} (-1)^n \frac{(x-2)^n}{7^{n+1}},$$

for $x \in (-5, 9)$.

Remark. Another method for solving the problem is based on the calculation of the derivative of order n of the function $f(x) = \frac{1}{x+5}$ and its evaluation at 2, i.e.

$$\left(\frac{1}{x+5} \right)^{(n)} (2).$$

3.15. We use the logarithmic series and we have

$$\ln(1+x) = \ln 4\left(1 + \frac{x-3}{4}\right) = \ln 4 + \ln\left(1 + \frac{x-3}{4}\right) = \ln 4 + \sum_{n=1}^{\infty}(-1)^{n-1}\frac{(x-3)^n}{4^n\, n},$$

for $x \in (-1, 7]$.

Remark. Another method for solving the problem is based on the calculation of the derivative of order n of the function $f(x) = \ln(1 + x)$ and its evaluation at 3, i.e. $(\ln(1 + x))^{(n)}(3)$.

3.16. We use the power series expansion

$$\frac{1}{(1-x)^2} = \sum_{n=1}^{\infty} nx^{n-1}, \quad x \in (-1, 1),$$

and we have that

$$\frac{1}{(x-1)^2} = \frac{1}{4\left(1 - \frac{x+1}{2}\right)^2} = \frac{1}{4}\sum_{n=1}^{\infty} n\left(\frac{x+1}{2}\right)^{n-1} = \sum_{n=1}^{\infty}\frac{n}{2^{n+1}}(x+1)^{n-1},$$

for $x \in (-3, 1)$.

Remark. Another method for solving the problem is based on the calculation of the derivative of order n of the function $f(x) = \frac{1}{(x-1)^2}$ and its evaluation at -1, i.e.

$$\left(\frac{1}{(x-1)^2}\right)^{(n)}(-1).$$

3.17. We have

$$\frac{1}{x^2 + 4x + 3} = \frac{1}{2}\left(\frac{1}{x+1} - \frac{1}{x+3}\right) = \frac{1}{4\left(1 - \frac{x+5}{2}\right)} - \frac{1}{8\left(1 - \frac{x+5}{4}\right)}$$

$$= \frac{1}{4}\sum_{n=0}^{\infty}\frac{(x+5)^n}{2^n} - \frac{1}{8}\sum_{n=0}^{\infty}\frac{(x+5)^n}{4^n}$$

$$= \sum_{n=0}^{\infty}\left(\frac{1}{2^{n+2}} - \frac{1}{2^{2n+3}}\right)(x+5)^n,$$

for $x \in (-7, -3)$.

3.18. We use the geometric series and we have

$$\frac{1}{x^2 - 4x + 8} = \frac{1}{(x-2)^2 + 4} = \frac{1}{4} \cdot \frac{1}{1 + \left(\frac{x-2}{2}\right)^2}$$

$$= \frac{1}{4} \sum_{n=0}^{\infty} (-1)^n \left(\frac{x-2}{2}\right)^{2n} = \sum_{n=0}^{\infty} (-1)^n \frac{(x-2)^{2n}}{2^{2n+2}},$$

for $x \in (0, 4)$.

3.19. We use the power series for the exponential function and we have

$$e^{2x-1} = e^{2(x+1)-3} = \frac{1}{e^3} \cdot e^{2(x+1)} = \frac{1}{e^3} \sum_{n=0}^{\infty} \frac{(2(x+1))^n}{n!} = \sum_{n=0}^{\infty} \frac{2^n}{e^3 n!} (x+1)^n,$$

for $x \in \mathbb{R}$.

Remark. Another method for solving the problem is based on the calculation of the derivative of order n of the function $f(x) = e^{2x-1}$ and its evaluation at -1, i.e.

$$(e^{2x-1})^{(n)}(-1).$$

3.20. We use the power series expansion for the sine function and we have

$$\sin(3x+\pi) = \sin(3(x-\pi)+4\pi) = \sin(3(x-\pi)) = \sum_{n=0}^{\infty} (-1)^n \frac{3^{2n+1}}{(2n+1)!} (x-\pi)^{2n+1},$$

for $x \in \mathbb{R}$.

Remark. Another method for solving the problem is based on the calculation of the derivative of order n of the function $f(x) = \sin(3x + \pi)$ and its evaluation at π, i.e.

$$(\sin(3x + \pi))^{(n)}(\pi).$$

3.21. We use the trigonometric formula

$$\arctan a + \arctan b = \arctan \frac{a+b}{1-ab}, \quad ab < 1,$$

and we have that

$$\arctan x - \arctan \frac{1}{x^2 - x + 1} = -\left[\arctan(-x) + \arctan \frac{1}{x^2 - x + 1}\right]$$

$$= -\arctan \frac{-x + \frac{1}{x^2 - x + 1}}{1 + \frac{x}{x^2 - x + 1}}$$

$$= -\arctan(-(x - 1))$$

$$= \arctan(x - 1), \quad \forall x \in \mathbb{R}.$$

Therefore, $f(x) = \arctan(x - 1)$, for $x \in \mathbb{R}$. Now, using the power series expansion of the arctan function we write

$$\arctan x - \arctan \frac{1}{x^2 - x + 1} = \arctan(x - 1) = \sum_{n=0}^{\infty} (-1)^n \frac{(x - 1)^{2n+1}}{2n + 1},$$

for $x \in [0, 2]$.

11.3 Gems with Numerical and Power Series

3.22. (c) We have

$$\sum_{n=2}^{\infty} \frac{n}{(n - 2)! + (n - 1)! + n!} x^n = \sum_{n=2}^{\infty} \frac{x^n}{(n - 2)! n}$$

$$= \sum_{n=2}^{\infty} \frac{n - 1}{n!} x^n$$

$$= \sum_{n=2}^{\infty} \frac{x^n}{(n - 1)!} - \sum_{n=2}^{\infty} \frac{x^n}{n!}$$

$$= x\left(e^x - 1\right) - \left(e^x - 1 - x\right)$$

$$= (x - 1)e^x + 1.$$

The other parts of the problem can be solved similarly.
3.23. (a) We have, based on part (a) of Problem 2.81, that

$$\sum_{n=0}^{\infty}(-1)^n\left(\frac{1}{n+1}-\frac{1}{n+2}+\frac{1}{n+3}-\cdots\right)=\sum_{n=0}^{\infty}(-1)^n\int_0^1\frac{x^n}{1+x}dx$$

$$=\int_0^1\frac{1}{1+x}\sum_{n=0}^{\infty}(-x)^n dx$$

$$=\int_0^1\frac{1}{(1+x)^2}dx$$

$$=\frac{1}{2}.$$

(c) Observe that $S_{4n-1}=1-\frac{1}{3}+\frac{1}{5}-\cdots-\frac{1}{4n-1}$, $n\geq 1$.

(d) We have

$$\sum_{n=0}^{\infty}(-1)^{\lfloor\frac{n}{k}\rfloor}x^n=1+x+x^2+\cdots+x^{k-1}-(x^k+x^{k+1}+\cdots+x^{2k-1})$$

$$+(x^{2k}+x^{2k+1}+\cdots+x^{3k-1})-\cdots$$

$$=(1+x+x^2+\cdots+x^{k-1})(1-x^k+x^{2k}-x^{3k}+\cdots)$$

$$=\frac{1+x+x^2+\cdots+x^{k-1}}{1+x^k}.$$

(e) We have

$$\sum_{n=0}^{\infty}(-1)^{\lfloor\frac{n}{k}\rfloor}\left(\frac{1}{n+1}-\frac{1}{n+2}+\frac{1}{n+3}-\cdots\right)\overset{2.81\,(a)}{=}\sum_{n=0}^{\infty}(-1)^{\lfloor\frac{n}{k}\rfloor}\int_0^1\frac{x^n}{1+x}dx$$

$$=\int_0^1\frac{1}{1+x}\left(\sum_{n=0}^{\infty}(-1)^{\lfloor\frac{n}{k}\rfloor}x^n\right)dx$$

$$\overset{(d)}{=}\int_0^1\frac{1+x+\cdots+x^{k-1}}{(1+x)(1+x^k)}dx.$$

(f) We have

$$\sum_{n=0}^{\infty}(-1)^{\lfloor\frac{n}{2}\rfloor}\frac{x^{n+1}}{n+1}=\sum_{n=0}^{\infty}(-1)^{\lfloor\frac{n}{2}\rfloor}\int_0^x t^n dt$$

$$=\int_0^x\left(\sum_{n=0}^{\infty}(-1)^{\lfloor\frac{n}{2}\rfloor}t^n\right)dt$$

$$\overset{(b)}{=}\int_0^x\frac{1+t}{1+t^2}dt$$

$$=\arctan x+\frac{1}{2}\ln(1+x^2).$$

3.24. We have, based on Problem 2.81, that

$$\sum_{n=1}^{\infty} \left(\frac{1}{n+1} - \frac{1}{n+2} + \cdots \right) \left(\zeta(2) - 1 - \frac{1}{2^2} - \cdots - \frac{1}{n^2} \right)$$

$$= \sum_{n=1}^{\infty} \int_0^1 \frac{x^n}{1+x} dx \left(\zeta(2) - 1 - \frac{1}{2^2} - \cdots - \frac{1}{n^2} \right)$$

$$= \int_0^1 \frac{1}{1+x} \sum_{n=1}^{\infty} \left(\zeta(2) - 1 - \frac{1}{2^2} - \cdots - \frac{1}{n^2} \right) x^n dx$$

$$\overset{3.4\ C}{=} \int_0^1 \frac{x \zeta(2) - \mathrm{Li}_2(x)}{1 - x^2} dx$$

$$= \frac{\zeta(2)}{2} \int_0^1 \frac{x}{1+x} - \frac{1}{2} \int_0^1 \frac{\mathrm{Li}_2(x)}{1+x} dx + \frac{1}{2} \int_0^1 \frac{x \zeta(2) - \mathrm{Li}_2(x)}{1 - x} dx.$$

We calculate the second integral. We have

$$\int_0^1 \frac{\mathrm{Li}_2(x)}{1+x} dx = \mathrm{Li}_2(x) \ln(1+x) \Big|_0^1 + \int_0^1 \frac{\ln(1-x)\ln(1+x)}{x} dx$$

$$= \zeta(2) \ln 2 - \frac{5}{8} \zeta(3).$$

The last equality holds based on formula (10.2).
    On the other hand,

$$\int_0^1 \frac{x \zeta(2) - \mathrm{Li}_2(x)}{1 - x} dx = - (x\zeta(2) - \mathrm{Li}_2(x)) \ln(1 - x) \Big|_0^1$$

$$+ \zeta(2) \int_0^1 \ln(1 - x) dx + \int_0^1 \frac{\ln^2(1-x)}{x} dx$$

$$= \zeta(2) \int_0^1 \ln(1 - x) dx + \int_0^1 \frac{\ln^2(1-x)}{x} dx$$

$$= -\zeta(2) + 2\zeta(3).$$

Putting all these together, we have that

$$\sum_{n=1}^{\infty} \left(\frac{1}{n+1} - \frac{1}{n+2} + \cdots \right) \left(\zeta(2) - 1 - \frac{1}{2^2} - \cdots - \frac{1}{n^2} \right)$$

$$= \int_0^1 \frac{x \zeta(2) - \mathrm{Li}_2(x)}{1 - x^2} dx = \frac{21}{16} \zeta(3) - \zeta(2) \ln 2.$$

3.25. First we prove the following formula:

$$\frac{1}{n+1}+\frac{1}{n+2}-\frac{1}{n+3}-\frac{1}{n+4}+\cdots = \int_0^1 x^n \frac{1+x}{1+x^2}dx. \qquad (11.1)$$

We have

$$\frac{1}{n+1}+\frac{1}{n+2}-\frac{1}{n+3}-\frac{1}{n+4}+\cdots = \int_0^1 \left(x^n+x^{n+1}-x^{n+2}-x^{n+3}+\cdots\right)dx$$

$$= \int_0^1 x^n(1+x-x^2-x^3+x^4+x^5-\cdots)dx$$

$$= \int_0^1 x^n(1+x)(1-x^2+x^4-x^6+\cdots)dx$$

$$= \int_0^1 x^n \frac{1+x}{1+x^2}dx.$$

(a) We have, based on formula (11.1), that

$$\sum_{n=1}^{\infty}(-1)^{n-1}\left(\frac{1}{n+1}+\frac{1}{n+2}-\frac{1}{n+3}-\frac{1}{n+4}+\cdots\right) = \sum_{n=1}^{\infty}(-1)^{n-1}\int_0^1 x^n \frac{1+x}{1+x^2}dx$$

$$= -\int_0^1 \frac{1+x}{1+x^2}\sum_{n=1}^{\infty}(-x)^n dx$$

$$= \int_0^1 \frac{x}{1+x^2}dx$$

$$= \frac{\ln 2}{2}.$$

(b) We have, based on formula (11.1), that

$$\sum_{n=1}^{\infty}(-1)^{n-1}H_n\left(\frac{1}{n+1}+\frac{1}{n+2}-\frac{1}{n+3}-\frac{1}{n+4}+\cdots\right)$$

$$=\sum_{n=1}^{\infty}(-1)^{n-1}H_n\int_0^1 x^n\frac{1+x}{1+x^2}dx$$

$$=-\int_0^1\frac{1+x}{1+x^2}\sum_{n=1}^{\infty}H_n(-x)^n dx$$

$$\underset{=}{\overset{3.63\,(a)}{=}}\int_0^1\frac{\ln(1+x)}{1+x^2}dx.$$

We calculate the preceding integral. Using the substitution $x=\frac{1-t}{1+t}$, we have that

$$I=\int_0^1\frac{\ln(1+x)}{1+x^2}dx=\ln 2\int_0^1\frac{1}{1+t^2}dt-I,$$

and it follows that $I=\frac{\pi\ln 2}{8}$.

(c) We have, based on formula (11.1), that

$$S=\sum_{n=1}^{\infty}\left(\frac{1}{n}+\frac{1}{n+1}-\frac{1}{n+2}-\frac{1}{n+3}+\cdots\right)^2$$

$$=\sum_{n=1}^{\infty}\int_0^1 x^{n-1}\frac{1+x}{1+x^2}dx\int_0^1 y^{n-1}\frac{1+y}{1+y^2}dy$$

$$=\int_0^1\int_0^1\frac{(1+x)(1+y)}{(1+x^2)(1+y^2)}\sum_{n=1}^{\infty}(xy)^{n-1}dxdy$$

$$=\int_0^1\int_0^1\frac{(1+x)(1+y)}{(1+x^2)(1+y^2)(1-xy)}dxdy$$

$$=\int_0^1\frac{1+x}{1+x^2}\left(\int_0^1\frac{1+y}{(1+y^2)(1-xy)}dy\right)dx.$$

We calculate the inner integral and we have that

$$\int_0^1\frac{1+y}{(1+y^2)(1-xy)}dy=\int_0^1\left(\frac{\frac{1-x}{1+x^2}+\frac{1+x}{1+x^2}y}{1+y^2}+\frac{\frac{x^2+x}{1+x^2}}{1-xy}\right)dy$$

$$=\frac{\pi}{4}\cdot\frac{1-x}{1+x^2}+\frac{\ln 2}{2}\cdot\frac{1+x}{1+x^2}-\frac{1+x}{1+x^2}\ln(1-x).$$

It follows that

$$S = \frac{\pi}{4} \int_0^1 \frac{1-x^2}{(1+x^2)^2}dx + \frac{\ln 2}{2} \int_0^1 \left(\frac{1+x}{1+x^2}\right)^2 dx - \int_0^1 \left(\frac{1+x}{1+x^2}\right)^2 \ln(1-x)dx$$

$$= \frac{\pi}{8} + \frac{\ln 2}{2}\left(\frac{\pi}{4}+\frac{1}{2}\right) - \int_0^1 \left(\frac{1+x}{1+x^2}\right)^2 \ln(1-x)dx$$

$$= \frac{\pi}{4} + G + \frac{\ln 2}{2},$$

since

$$\int_0^1 \left(\frac{1+x}{1+x^2}\right)^2 \ln(1-x)dx = \frac{\pi \ln 2}{8} - G - \frac{\pi}{8} - \frac{\ln 2}{4}.$$

To calculate the previous integral we integrate by parts, with $f(x) = \ln(1-x)$, $f'(x) = \frac{1}{1-x}$, $g'(x) = \left(\frac{1+x}{1+x^2}\right)^2$, $g(x) = \arctan x - \frac{1}{1+x^2} + \frac{1}{2} - \frac{\pi}{4}$, and we get that

$$\int_0^1 \left(\frac{1+x}{1+x^2}\right)^2 \ln(1-x)dx = \ln(1-x)\left(\arctan x - \frac{1}{1+x^2} + \frac{1}{2} - \frac{\pi}{4}\right)\Big|_0^1$$

$$+ \int_0^1 \frac{\arctan x - \arctan 1 + \frac{1}{2} - \frac{1}{1+x^2}}{1-x}dx$$

$$= \int_0^1 \frac{\arctan x - \arctan 1}{1-x}dx - \frac{1}{2}\int_0^1 \frac{1+x}{1+x^2}dx$$

$$= \int_0^1 \frac{\arctan x - \arctan 1}{1-x}dx - \frac{1}{2}\left(\frac{\pi}{4}+\frac{\ln 2}{2}\right)$$

$$= \frac{\pi \ln 2}{8} - G - \frac{\pi}{8} - \frac{\ln 2}{4},$$

since

A Catalan integral $$\int_0^1 \frac{\arctan x - \arctan 1}{1-x}dx = \frac{\pi \ln 2}{8} - G.$$

We prove the previous equality using integration by parts and we have that

$$\int_0^1 \frac{\arctan x - \arctan 1}{1-x} dx = -\ln(1-x)(\arctan x - \arctan 1)\Big|_0^1$$

$$+ \int_0^1 \frac{\ln(1-x)}{1+x^2} dx$$

$$= \int_0^1 \frac{\ln(1-x)}{1+x^2} dx$$

$$\overset{x=\frac{1-t}{1+t}}{=} \int_0^1 \frac{\ln 2 + \ln t - \ln(1+t)}{1+t^2} dt$$

$$= \frac{\pi \ln 2}{4} + \int_0^1 \frac{\ln t}{1+t^2} dt - \int_0^1 \frac{\ln(1+t)}{1+t^2} dt$$

$$= \frac{\pi \ln 2}{8} - G,$$

(11.2)

since

$$\int_0^1 \frac{\ln x}{1+x^2} dx = -G \quad \text{and} \quad \int_0^1 \frac{\ln(1+x)}{1+x^2} dx = \frac{\pi \ln 2}{8}.$$

The last integral was calculated in the solution of part (b).

Part (d) can be solved similarly to part (c) of the problem.

3.26. We have, based on part (a) of Problem 2.81 and formula (11.1), that

$$\sum_{n=1}^\infty \left(\frac{1}{n} + \frac{1}{n+1} - \frac{1}{n+2} - \frac{1}{n+3} + \cdots \right) \left(\frac{1}{n} - \frac{1}{n+1} + \frac{1}{n+2} - \cdots \right)$$

$$= \sum_{n=1}^\infty \int_0^1 x^{n-1} \frac{1+x}{1+x^2} dx \int_0^1 \frac{y^{n-1}}{1+y} dy$$

$$= \int_0^1 \int_0^1 \frac{1+x}{(1+x^2)(1+y)} \sum_{n=1}^\infty (xy)^{n-1} dx dy$$

$$= \int_0^1 \int_0^1 \frac{1+x}{(1+x^2)(1+y)(1-xy)} dx dy$$

$$= \int_0^1 \frac{1+x}{1+x^2} \left(\int_0^1 \frac{1}{(1+y)(1-xy)} dy \right) dx$$

$$= \int_0^1 \frac{\ln 2 - \ln(1-x)}{1+x^2} dx$$

$$= \frac{\pi \ln 2}{4} - \int_0^1 \frac{\ln(1-x)}{1+x^2} dx$$

$$= \frac{\pi \ln 2}{8} + G.$$

The last integral follows based on formula (11.2).

11.4 Single Zeta Series

3.27. We have

$$\sum_{n=1}^{\infty} n\left(\zeta(n+1)-1\right) = \sum_{n=1}^{\infty} n \sum_{i=2}^{\infty} \frac{1}{i^{n+1}} = \sum_{i=2}^{\infty} \frac{1}{i} \sum_{n=1}^{\infty} \frac{n}{i^{n}} = \sum_{i=2}^{\infty} \frac{1}{(i-1)^2} = \zeta(2).$$

3.28. We have

$$\sum_{n=2}^{\infty} \frac{\zeta(n)-1}{n} = \sum_{n=2}^{\infty} \frac{1}{n} \sum_{k=2}^{\infty} \frac{1}{k^n} = \sum_{k=2}^{\infty} \sum_{n=2}^{\infty} \frac{\left(\frac{1}{k}\right)^n}{n} = -\sum_{k=2}^{\infty} \left(\ln\left(1-\frac{1}{k}\right)+\frac{1}{k}\right) = 1-\gamma$$

and

$$\sum_{n=1}^{\infty} \frac{\zeta(?n)-1}{n} = \sum_{n=1}^{\infty} \frac{1}{n} \sum_{k=2}^{\infty} \frac{1}{k^{2n}} = \sum_{k=2}^{\infty} \sum_{n=1}^{\infty} \frac{\left(\frac{1}{k^2}\right)^n}{n} = -\sum_{k=2}^{\infty} \ln\left(1-\frac{1}{k^2}\right) = \ln 2.$$

3.29. (e) We have

$$\sum_{n=2}^{\infty} \left[(k+1)^n \left(\zeta(n)-1-\frac{1}{2^n}-\cdots-\frac{1}{k^n}\right)-1\right] = \sum_{n=2}^{\infty} \left((k+1)^n \sum_{i=k+1}^{\infty} \frac{1}{i^n}-1\right)$$

$$= \sum_{n=2}^{\infty} \sum_{i=k+2}^{\infty} \left(\frac{k+1}{i}\right)^n$$

$$= \sum_{i=k+2}^{\infty} \sum_{n=2}^{\infty} \left(\frac{k+1}{i}\right)^n$$

$$= \sum_{i=k+2}^{\infty} \frac{(k+1)^2}{i(i-k-1)}$$

$$= (k+1)H_{k+1}.$$

The other parts of the problem are solved similarly.

3.30. (a) We have

$$\sum_{n=2}^{\infty} \left[2^{2n-1} \left(\zeta(2n-1) - 1 \right) - 1 \right] = \sum_{n=2}^{\infty} \left(2^{2n-1} \sum_{i=2}^{\infty} \frac{1}{i^{2n-1}} - 1 \right)$$

$$= \sum_{n=2}^{\infty} \sum_{i=3}^{\infty} \left(\frac{2}{i} \right)^{2n-1}$$

$$= \sum_{i=3}^{\infty} \frac{i}{2} \sum_{n=2}^{\infty} \left(\frac{2}{i} \right)^{2n}$$

$$= 8 \sum_{i=3}^{\infty} \frac{1}{i(i^2 - 4)}$$

$$= \frac{11}{12}.$$

Part (b) is solved similarly.

3.31. (b) We have

$$\sum_{n=2}^{\infty} (-1)^n \left[(k+1)^n \left(\zeta(n) - 1 - \frac{1}{2^n} - \cdots - \frac{1}{k^n} \right) - 1 \right]$$

$$= \sum_{n=2}^{\infty} (-1)^n \left((k+1)^n \sum_{i=k+1}^{\infty} \frac{1}{i^n} - 1 \right)$$

$$= \sum_{n=2}^{\infty} (-1)^n \sum_{i=k+2}^{\infty} \left(\frac{k+1}{i} \right)^n$$

$$= \sum_{i=k+2}^{\infty} \sum_{n=2}^{\infty} \left(-\frac{k+1}{i} \right)^n$$

$$= \sum_{i=k+2}^{\infty} \frac{(k+1)^2}{i(i+k+1)}$$

$$= (k+1)(H_{2k+2} - H_{k+1}).$$

Part (a) follows from part (b) with $k = 1$.

3.32. (b) We have

$$\sum_{n=2}^{\infty}(n-1)\left[(k+1)^n\left(\zeta(n)-1-\frac{1}{2^n}-\cdots-\frac{1}{k^n}\right)-1\right]$$

$$=\sum_{n=2}^{\infty}(n-1)\left((k+1)^n\sum_{i=k+1}^{\infty}\frac{1}{i^n}-1\right)$$

$$=\sum_{n=2}^{\infty}(n-1)\sum_{i=k+2}^{\infty}\left(\frac{k+1}{i}\right)^n$$

$$=\sum_{i=k+2}^{\infty}\sum_{n=2}^{\infty}(n-1)\left(\frac{k+1}{i}\right)^n$$

$$=\sum_{i=k+2}^{\infty}\frac{(k+1)^2}{(i-k-1)^2}$$

$$=(k+1)^2\zeta(2).$$

3.33. We have

$$\sum_{n=1}^{\infty}\frac{1}{n}\left[4^n\left(\zeta(2n)-1\right)-1\right]=\sum_{n=1}^{\infty}\frac{1}{n}\left(4^n\sum_{i=2}^{\infty}\frac{1}{i^{2n}}-1\right)$$

$$=\sum_{n=1}^{\infty}\frac{1}{n}\sum_{i=3}^{\infty}\left(\frac{4}{i^2}\right)^n$$

$$=\sum_{i=3}^{\infty}\sum_{n=1}^{\infty}\frac{1}{n}\left(\frac{4}{i^2}\right)^n$$

$$=-\sum_{i=3}^{\infty}\ln\left(1-\frac{4}{i^2}\right)$$

$$=\ln 6.$$

The first series is calculated similarly.

3.34. Use the identity

$$H_n\left(\zeta(n)-\zeta(n+1)\right)=H_n(\zeta(n)-1)-H_{n+1}(\zeta(n+1)-1)+\frac{\zeta(n+1)-1}{n+1},\quad n\geq 2,$$

and the first series of Problem 3.28.

3.35. (a) See [16]. A different solution, for both parts (a) and (b), is based on Abel's summation formula. Part (c) follows from parts (a) and (b) by subtraction.

3.36. To prove the first equality use the definition of the zeta function and change the order of summation of terms, which is possible since the terms of the series are positive, and to prove the second equality calculate the nth partial sum of the series.

3.37. We have

$$\sum_{n=1}^{\infty} \frac{n\zeta(2n)}{4^{n-2}} = \sum_{n=1}^{\infty} \frac{n}{4^{n-2}} \sum_{i=1}^{\infty} \frac{1}{i^{2n}} = 16 \sum_{i=1}^{\infty} \sum_{n=1}^{\infty} \frac{n}{(4i^2)^n} = 64 \sum_{i=1}^{\infty} \frac{i^2}{(4i^2-1)^2} = \pi^2.$$

3.38. (a) This is problem 3.8 in [26, p. 140].

(e) We have, based on part (a), that

$$\sum_{n=2}^{\infty} \left[2^n \left(n - \zeta(2) - \zeta(3) - \cdots - \zeta(n) \right) - 1 \right] = \sum_{n=2}^{\infty} \left(2^n \sum_{k=1}^{\infty} \frac{1}{k(k+1)^n} - 1 \right)$$

$$= \sum_{n=2}^{\infty} \sum_{k=2}^{\infty} \frac{2^n}{k(k+1)^n}$$

$$= \sum_{k=2}^{\infty} \frac{1}{k} \sum_{n=2}^{\infty} \left(\frac{2}{k+1} \right)^n$$

$$= \sum_{k=2}^{\infty} \frac{4}{(k-1)k(k+1)}$$

$$= 1.$$

The other parts of the problem are solved similarly.

3.39. (a) We have

$$\sum_{n=1}^{\infty} (\beta(n) - 1) = \beta(1) - 1 + \sum_{n=2}^{\infty} \sum_{i=1}^{\infty} \frac{(-1)^i}{(2i+1)^n} = \frac{\pi}{4} - 1 + \sum_{i=1}^{\infty} (-1)^i \sum_{n=2}^{\infty} \frac{1}{(2i+1)^n}$$

$$= \frac{\pi}{4} - 1 + \sum_{i=1}^{\infty} \frac{(-1)^i}{2i(2i+1)} = -\frac{\ln 2}{2}.$$

The other parts of the problem are solved similarly.

3.40. We have, based on Problem 2.31 and part (a) of Problem 3.39, that

$$\sum_{n=2}^{\infty} (\zeta(n) - \beta(n)) = \sum_{n=2}^{\infty} (\zeta(n) - 1) - \sum_{n=2}^{\infty} (\beta(n) - 1) = \beta(1) + \frac{\ln 2}{2} = \frac{\pi}{4} + \frac{\ln 2}{2}.$$

11.5   Polylogarithm Series

3.41. (m) We have

$$\sum_{n=2}^{\infty} (n-1)\left[(k+1)^n\left(\mathrm{Li}_n(x) - x - \frac{x^2}{2^n} - \cdots - \frac{x^k}{k^n}\right) - x^{k+1}\right]$$

$$= \sum_{n=2}^{\infty} (n-1)\left[(k+1)^n \sum_{i=k+1}^{\infty} \frac{x^i}{i^n} - x^{k+1}\right]$$

$$= \sum_{n=2}^{\infty} (n-1) \sum_{i=k+2}^{\infty} x^i \left(\frac{k+1}{i}\right)^n$$

$$= \sum_{i=k+2}^{\infty} x^i \left(\frac{k+1}{i}\right)^2 \sum_{n=2}^{\infty} (n-1)\left(\frac{k+1}{i}\right)^{n-2}$$

$$= (k+1)^2 \sum_{i=k+2}^{\infty} \frac{x^i}{(i-k-1)^2}$$

$$= (k+1)^2 \sum_{j=1}^{\infty} \frac{x^{k+1+j}}{j^2}$$

$$= (k+1)^2 x^{k+1} \mathrm{Li}_2(x).$$

The other parts of the problem are solved similarly.
3.42. (a) We have

$$S_n = \sum_{i=2}^{\infty} \frac{x^i}{i^n(i-1)} = \sum_{i=2}^{\infty} \frac{x^i}{i^{n-1}}\left(\frac{1}{i-1} - \frac{1}{i}\right) = S_{n-1} - (\mathrm{Li}_n(x) - x).$$

It follows that

$$S_n = S_1 - \mathrm{Li}_2(x) - \mathrm{Li}_3(x) - \cdots - \mathrm{Li}_n(x) + (n-1)x$$

$$= nx + (1-x)\ln(1-x) - \mathrm{Li}_2(x) - \mathrm{Li}_3(x) - \cdots - \mathrm{Li}_n(x),$$

since $S_1 = (1-x)\ln(1-x) + x$.
  (b) We have, based on part (a), that

$$\sum_{n=1}^{\infty} (\mathrm{Li}_2(x) + \mathrm{Li}_3(x) + \cdots + \mathrm{Li}_{n+1}(x) - (n+1)x - (1-x)\ln(1-x))$$

$$= -\sum_{n=1}^{\infty} \sum_{i=2}^{\infty} \frac{x^i}{i^{n+1}(i-1)} = -\sum_{i=2}^{\infty} \frac{x^i}{i-1} \sum_{n=1}^{\infty} \frac{1}{i^{n+1}} = -\sum_{i=2}^{\infty} \frac{x^i}{(i-1)(i^2-i)}$$

$$= -\sum_{i=2}^{\infty} \left(\frac{x^i}{(i-1)^2} - \frac{x^i}{i-1} + \frac{x^i}{i} \right) = (1-x)\ln(1-x) - x\mathrm{Li}_2(x) + x.$$

The other parts of the problem are solved similarly.

3.43. (a) We have

$$S_n = \sum_{i=2}^{\infty} \frac{x^i}{i^n(i+1)} = \sum_{i=2}^{\infty} \frac{x^i}{i^{n-1}} \left(\frac{1}{i} - \frac{1}{i+1} \right) = (\mathrm{Li}_n(x) - x) - S_{n-1}.$$

It follows, since $S_1 = \frac{1-x}{x}\ln(1-x) - \frac{x}{2} + 1$, that

$$(-1)^n S_n = \sum_{k=2}^{n} (-1)^k \left(\mathrm{Li}_k(x) - x \right) - \left(\frac{1-x}{x}\ln(1-x) - \frac{x}{2} + 1 \right)$$

$$= \sum_{k=2}^{n} (-1)^k \mathrm{Li}_k(x) - \frac{x}{2}(1 - (-1)^{n-1}) - \frac{1-x}{x}\ln(1-x) + \frac{x}{2} - 1$$

$$= \sum_{k=2}^{n} (-1)^k \mathrm{Li}_k(x) - \frac{1-x}{x}\ln(1-x) - 1 + \frac{x}{2}(-1)^{n-1}.$$

(c) We have, based on part (a), that

$$\sum_{n=2}^{\infty} \left[(-1)^n \left(\mathrm{Li}_2(x) + \cdots + (-1)^n \mathrm{Li}_n(x) - \frac{1-x}{x}\ln(1-x) - 1 \right) - \frac{x}{2} \right]$$

$$= \sum_{n=2}^{\infty} \sum_{i=2}^{\infty} \frac{x^i}{i^n(i+1)} = \sum_{i=2}^{\infty} \frac{x^i}{i+1} \sum_{n=2}^{\infty} \frac{1}{i^n} = \sum_{i=2}^{\infty} x^i \left(\frac{1}{2(i-1)} - \frac{1}{i} + \frac{1}{2(i+1)} \right)$$

$$= -\frac{(1-x)^2}{2x}\ln(1-x) + \frac{3x}{4} - \frac{1}{2}.$$

Part (d) is solved similarly.

11.6   Inequalities and Integrals

3.44. It suffices to prove that $x \ln x \le \ln(x^2 - x + 1)$, for $x \in (0, 1]$. We change variables $x = 1 - y$ and we obtain the inequality $(1 - y) \ln(1 - y) \le \ln(1 - y + y^2)$, for $y \in [0, 1)$. Now we use the power series expansion of the logarithmic function and we have that

$$\ln(1 - (y - y^2)) - (1 - y) \ln(1 - y) = -\sum_{n=1}^{\infty} \frac{(y - y^2)^n}{n} + (1 - y) \sum_{n=1}^{\infty} \frac{y^n}{n}$$

$$= (1 - y) \sum_{n=1}^{\infty} \frac{y^n}{n} \left(1 - (1 - y)^{n-1}\right)$$

$$\ge 0.$$

Remark. From the solution of Problem 3.44 we also have that the following inequality holds true $(1 - x)^{1-x} \le x^2 - x + 1$, for $x \in [0, 1)$. For another solution of Problem 3.44, which uses concavity, see the remark after the solution of Problem 4.6.

3.45. Since $(2n)! \ge 2^n n!$, for all $n \ge 0$, we have that

$$\frac{e^x + e^{-x}}{2} = \sum_{n=0}^{\infty} \frac{x^{2n}}{(2n)!} \le \sum_{n=0}^{\infty} \frac{x^{2n}}{2^n n!} = e^{\frac{x^2}{2}}.$$

3.46. (a) Let $f : [0, \infty) \to \mathbb{R}$, $f(x) = qx^p + py^q - pqxy$. A calculation shows that $f'(x) = pq(x^{p-1} - y)$ and $f'(x) = 0 \Rightarrow x = y^{\frac{1}{p-1}}$. We have that f decreases on $[0, y^{\frac{1}{p-1}})$ and increases on $(y^{\frac{1}{p-1}}, \infty)$, and it follows, since $f(y^{\frac{1}{p-1}}) = 0$, that $f(x) \ge 0$, for all $x \ge 0$.

(b) Solution due to C. Mortici [111]. The inequality is equivalent to

$$\frac{1}{p(1 - x^p)} + \frac{1}{q(1 - y^q)} \ge \frac{1}{1 - xy}, \quad x, y \in (0, 1).$$

We have

$$\frac{1}{p(1 - x^p)} + \frac{1}{q(1 - y^q)} = \sum_{n=0}^{\infty} \left[\frac{(x^n)^p}{p} + \frac{(y^n)^q}{q} \right] \overset{(a)}{\ge} \sum_{n=0}^{\infty} x^n y^n = \frac{1}{1 - xy}.$$

3.47. (a) Solution due to C. Mortici [111]. Let $S = a + b + c$ and $x = \frac{a}{S}$, $y = \frac{b}{S}$ and $z = \frac{c}{S}$. The inequality is equivalent to

$$\frac{x}{1-x} + \frac{y}{1-y} + \frac{z}{1-z} \geq \frac{3}{2}.$$

Jensen's inequality for the convex function $f(x) = x^n$, $n \geq 2$, states that

$$\frac{x^n + y^n + z^n}{3} \geq \left(\frac{x+y+z}{3}\right)^n, \quad x, y, z \geq 0.$$

We have, since $x + y + z = 1$, that

$$\frac{x}{1-x} + \frac{y}{1-y} + \frac{z}{1-z} = \sum_{n=0}^{\infty} \left(x^{n+1} + y^{n+1} + z^{n+1}\right)$$

$$\geq \sum_{n=0}^{\infty} 3\left(\frac{x+y+z}{3}\right)^{n+1}$$

$$= \sum_{n=0}^{\infty} \frac{1}{3^n}$$

$$= \frac{3}{2}.$$

Two new proofs of Nesbitt's inequality are given in Chap. 7, after Problem 7.132.

3.48. (a) Solution due to C. Mortici [111]. First we note that, if x, y, and z are nonnegative real numbers, then $x^2 + y^2 + z^2 \geq xy + yz + xz$. We have

$$\frac{1}{1-a^2} + \frac{1}{1-b^2} + \frac{1}{1-c^2} = \sum_{n=0}^{\infty} \left(a^{2n} + b^{2n} + c^{2n}\right)$$

$$\geq \sum_{n=0}^{\infty} \left(a^n b^n + b^n c^n + a^n c^n\right)$$

$$= \frac{1}{1-ab} + \frac{1}{1-bc} + \frac{1}{1-ac}.$$

(b) We observe that if a, b, c, and d are nonnegative real numbers, then

(continued)

$$a^3 + b^3 + c^3 + d^3 \geq abc + bcd + cda + dab.$$

We prove the previous inequality. We have, based on the $AM - GM$ inequality, that

$$3(a^3 + b^3 + c^3 + d^3) = \left(a^3 + b^3 + c^3\right) + \left(b^3 + c^3 + d^3\right)$$
$$+ \left(c^3 + d^3 + a^3\right) + \left(d^3 + a^3 + b^3\right)$$
$$\geq 3abc + 3bcd + 3cda + 3dab.$$

Now, based on the previous inequality, we have that

$$\frac{1}{1-a^3} + \frac{1}{1-b^3} + \frac{1}{1-c^3} + \frac{1}{1-d^3} = \sum_{n=0}^{\infty} \left(a^{3n} + b^{3n} + c^{3n} + d^{3n}\right)$$

$$\geq \sum_{n=0}^{\infty} \left[(abc)^n + (bcd)^n + (cda)^n + (dab)^n\right]$$

$$= \frac{1}{1-abc} + \frac{1}{1-bcd} + \frac{1}{1-cda} + \frac{1}{1-dab}.$$

3.49. (a) Use the recurrence relation $\frac{I_n}{n} = 2(I_n - I_{n+1})$, $\forall n \geq 1$.

(b) A direct calculation shows that

$$I_n = \frac{(2n-3)!!}{(2n-2)!!} \cdot \frac{\pi}{2}, \quad \forall n \geq 2,$$

and use part (a) of the problem.

3.50. (a) We prove that $I_n = -\frac{1}{2n} + 2n\left(I_n - I_{n+1}\right)$, $n \geq 1$. We calculate I_n using integration by parts, with

$$f(x) = \frac{\arctan x}{(1+x^2)^n}, \quad f'(x) = \frac{1}{(1+x^2)^{n+1}} - \frac{2nx}{(1+x^2)^{n+1}} \arctan x,$$

$g'(x) = 1, g(x) = x$, and we have that

$$I_n = \frac{x \arctan x}{(1+x^2)^n} \Big|_0^{\infty} - \int_0^{\infty} \left(\frac{x}{(1+x^2)^{n+1}} - 2n\frac{x^2}{(1+x^2)^{n+1}} \arctan x\right) dx$$

$$= -\int_0^{\infty} \frac{x}{(1+x^2)^{n+1}} dx + 2n \int_0^{\infty} \frac{x^2}{(1+x^2)^{n+1}} \arctan x \, dx$$

$$= \frac{1}{2n(1+x^2)^n} \Big|_0^\infty + 2n \int_0^\infty \left(\frac{\arctan x}{(1+x^2)^n} - \frac{\arctan x}{(1+x^2)^{n+1}} \right) dx$$

$$= -\frac{1}{2n} + 2n(I_n - I_{n+1}).$$

It follows that $\frac{I_n}{n} = -\frac{1}{2n^2} + 2(I_n - I_{n+1})$ and this implies

$$\sum_{n=1}^\infty \frac{I_n}{n} = -\frac{1}{2} \sum_{n=1}^\infty \frac{1}{n^2} + 2 \sum_{n=1}^\infty (I_n - I_{n+1}) = -\frac{1}{2}\zeta(2) + 2I_1 = \frac{\pi^2}{6},$$

since $I_1 = \int_0^\infty \frac{\arctan x}{1+x^2} dx = \frac{\arctan^2 x}{2} \Big|_0^\infty = \frac{\pi^2}{8}$.
(b) We have

$$\int_0^\infty \arctan x \ln \left(1 + \frac{1}{x^2} \right) dx = -\int_0^\infty \arctan x \ln \left(1 - \frac{1}{1+x^2} \right) dx$$

$$= \int_0^\infty \arctan x \left(\sum_{n=1}^\infty \frac{\left(\frac{1}{1+x^2} \right)^n}{n} \right) dx$$

$$= \sum_{n=1}^\infty \frac{I_n}{n}$$

$$= \zeta(2).$$

(c) Use the identity $\arctan x + \text{arccot} \, x = \frac{\pi}{2}$, $\forall x > 0$, and part (b) of the problem.
3.52. Use that $\frac{1}{x^x} = e^{-x \ln x}$ and $x^x = e^{x \ln x}$.
3.53. (a) The integral equals $1 + \frac{1-a}{a} \ln(1-a)$. We have

$$\int_0^1 a^{\lfloor \frac{1}{x} \rfloor} dx \overset{x=\frac{1}{y}}{=} \int_1^\infty \frac{a^{\lfloor y \rfloor}}{y^2} dy = \sum_{k=1}^\infty \int_k^{k+1} \frac{a^k}{y^2} dy$$

$$= \sum_{k=1}^\infty a^k \left(\frac{1}{k} - \frac{1}{k+1} \right)$$

$$= \sum_{k=1}^\infty \frac{a^k}{k} - \sum_{k=1}^\infty \frac{a^k}{k+1}$$

$$= -\ln(1-a) - \frac{1}{a} (-\ln(1-a) - a)$$

$$= 1 + \frac{1-a}{a} \ln(1-a).$$

(b) This part follows from part (a) with $a = \frac{1}{2}$.

3.54. We have

$$\int_0^\infty 2^{-\lfloor x \rfloor} dx = \sum_{k=0}^\infty \int_k^{k+1} 2^{-\lfloor x \rfloor} dx = \sum_{k=0}^\infty \frac{1}{2^k} = 2.$$

The second integral can be calculated similarly.

3.55. We have

$$\int_0^\infty \lfloor x \rfloor e^{-x} dx = \sum_{k=0}^\infty \int_k^{k+1} \lfloor x \rfloor e^{-x} dx = \left(1 - e^{-1}\right) \sum_{k=0}^\infty \frac{k}{e^k} = \frac{1}{e-1}.$$

3.56. We have

$$\int_0^1 \frac{\ln(1-x)}{x} dx = -\int_0^1 \left(\sum_{n=1}^\infty \frac{x^{n-1}}{n}\right) dx = -\sum_{n=1}^\infty \frac{1}{n} \int_0^1 x^{n-1} dx = -\sum_{n=1}^\infty \frac{1}{n^2} = -\zeta(2)$$

and

$$\int_0^1 \left(\frac{\ln(1-x)}{x}\right)^2 dx = \int_0^1 \frac{\ln^2 x}{(1-x)^2} dx$$

$$= \int_0^1 \ln^2 x \left(\sum_{n=0}^\infty (n+1)x^n\right) dx$$

$$= \sum_{n=0}^\infty (n+1) \int_0^1 x^n \ln^2 x \, dx$$

$$= 2 \sum_{n=0}^\infty \frac{1}{(n+1)^2}$$

$$= 2\zeta(2).$$

Another method for calculating the second integral is based on the identity of part (c) of Problem 3.63. We have

$$\int_0^1 \left(\frac{\ln(1-x)}{x}\right)^2 dx = 2\int_0^1 \sum_{n=1}^{\infty} \frac{H_n}{n+1} x^{n-1} dx$$

$$= 2\sum_{n=1}^{\infty} \frac{H_n}{n+1} \int_0^1 x^{n-1} dx$$

$$= 2\sum_{n=1}^{\infty} \frac{H_n}{n(n+1)}$$

$$= 2\sum_{n=1}^{\infty} \left(\frac{H_n}{n} - \frac{H_n}{n+1}\right)$$

$$= 2\sum_{n=1}^{\infty} \left(\frac{H_n}{n} - \frac{H_{n+1}}{n+1}\right) + 2\sum_{n=1}^{\infty} \frac{1}{(n+1)^2}$$

$$= 2 + 2\zeta(2) - 2$$

$$= 2\zeta(2).$$

3.57. We have

$$\int_0^1 \frac{\ln(1+x)}{x} dx = \int_0^1 \left(\sum_{n=1}^{\infty} (-1)^{n-1} \frac{x^{n-1}}{n}\right) dx$$

$$= \sum_{n=1}^{\infty} \frac{(-1)^{n-1}}{n} \int_0^1 x^{n-1} dx$$

$$= \sum_{n=1}^{\infty} (-1)^{n-1} \frac{1}{n^2}$$

$$= \frac{\pi^2}{12}.$$

We calculate the second integral by parts and then we use the value of the first integral. So, we have

$$\int_0^1 \left(\frac{\ln(1+x)}{x} \right)^2 dx = -\frac{\ln^2(1+x)}{x} \Big|_0^1 + 2\int_0^1 \frac{\ln(1+x)}{x(1+x)} dx$$

$$= -\ln^2 2 + 2\int_0^1 \left(\frac{\ln(1+x)}{x} - \frac{\ln(1+x)}{1+x} \right) dx$$

$$= -2\ln^2 2 + 2\int_0^1 \frac{\ln(1+x)}{x} dx$$

$$= \frac{\pi^2}{6} - 2\ln^2 2.$$

3.58. (a) We have

$$\int_0^1 \frac{\ln(1+x+x^2+\cdots+x^{n-1})}{x} dx = \int_0^1 \frac{\ln(1-x^n)}{x} dx - \int_0^1 \frac{\ln(1-x)}{x} dx$$

$$\overset{x^n=y}{=} \frac{1}{n}\int_0^1 \frac{\ln(1-y)}{y} dy - \int_0^1 \frac{\ln(1-x)}{x} dx$$

$$= \left(\frac{1}{n} - 1 \right) \int_0^1 \frac{\ln(1-x)}{x} dx$$

$$\overset{3.56}{=} \frac{\pi^2(n-1)}{6n}.$$

Part (b) is solved similarly.
3.59. The integral equals $-\frac{\pi^2}{18}$. We have

$$\int_0^1 \frac{\ln(1-x+x^2)}{x} dx = \int_0^1 \frac{\ln(1+x^3)}{x} dx - \int_0^1 \frac{\ln(1+x)}{x} dx$$

$$\overset{x^3=y}{=} \frac{1}{3}\int_0^1 \frac{\ln(1+y)}{y} dy - \int_0^1 \frac{\ln(1+x)}{x} dx$$

$$= -\frac{2}{3}\int_0^1 \frac{\ln(1+x)}{x} dx$$

$$\overset{3.57}{=} -\frac{\pi^2}{18}.$$

3.60. (b) We have

$$\int_0^1 \frac{\ln^k x \, \ln(1-x+x^2-x^3+\cdots+x^{n-1})}{x}\,dx$$

$$=\int_0^1 \frac{\ln^k x \, (\ln(1+x^n) - \ln(1+x))}{x}\,dx$$

$$=\int_0^1 \frac{\ln^k x \, \ln(1+x^n)}{x}\,dx - \int_0^1 \frac{\ln^k x \, \ln(1+x)}{x}\,dx$$

$$\overset{x^n=t}{=} \left(\frac{1}{n^{k+1}} - 1\right)\int_0^1 \frac{\ln^k x \, \ln(1+x)}{x}\,dx$$

$$=\left(\frac{1}{n^{k+1}} - 1\right)\int_0^1 \ln^k x \left(\sum_{n=1}^\infty (-1)^{n-1}\frac{x^{n-1}}{n}\right)dx$$

$$=\left(\frac{1}{n^{k+1}} - 1\right)\sum_{n=1}^\infty (-1)^{n-1}\frac{1}{n}\int_0^1 x^{n-1}\ln^k x \, dx$$

$$=(-1)^k k!\left(\frac{1}{n^{k+1}} - 1\right)\sum_{n=1}^\infty \frac{(-1)^{n-1}}{n^{k+2}}$$

$$=(-1)^k k!\left(1 - \frac{1}{2^{k+1}}\right)\left(\frac{1}{n^{k+1}} - 1\right)\zeta(k+2).$$

Part (a) is solved similarly.

3.61. (a) We have

$$\int_0^1 \frac{\ln(1-x+x^2)}{x-x^2}\,dx = \int_0^1 \ln(1-x+x^2)\left(\frac{1}{1-x} + \frac{1}{x}\right)dx$$

$$=\int_0^1 \frac{\ln(1-x+x^2)}{1-x}\,dx + \int_0^1 \frac{\ln(1-x+x^2)}{x}\,dx$$

$$=2\int_0^1 \frac{\ln(1-x+x^2)}{x}\,dx,$$

and the result follows based on Problem 3.59.

(b) As A. Stadler has observed [141], part (a) of the problem can be used for proving the surprising series with binomial coefficient giving $\zeta(2)$

$$3\sum_{n=1}^\infty \frac{1}{n^2 C_{2n}^n} = \frac{\pi^2}{6}. \tag{11.3}$$

We have

$$\frac{\pi^2}{9} = -\int_0^1 \frac{\ln(1 - x + x^2)}{x - x^2} \mathrm{d}x = \sum_{k=1}^{\infty} \frac{1}{k} \int_0^1 (x - x^2)^{k-1} \mathrm{d}x$$

$$= \sum_{k=1}^{\infty} \frac{1}{k} \int_0^1 x^{k-1}(1 - x)^{k-1} \mathrm{d}x$$

$$= \sum_{k=1}^{\infty} \frac{\Gamma(k)\Gamma(k)}{k\,\Gamma(2k)}$$

$$= 2 \sum_{k=1}^{\infty} \frac{1}{k^2 C_{2k}^k}.$$

Formula (11.3) has an interesting history: it was mentioned by R. Apéry in his notorious lecture at Marseille, where he announced the irrationality of $\zeta(3)$ (see [118]). It seems that the formula is not due to R. Apéry, since it was recorded in [13], together with other series of this form, at the end of problem 36 on page 89.

11.7 Generating Functions

3.62. (a) We have

$$e^{-x} \sin x = \left(\sum_{n=0}^{\infty} (-1)^n \frac{x^n}{n!} \right) \left(\sum_{n=0}^{\infty} (-1)^n \frac{x^{2n+1}}{(2n+1)!} \right) = \sum_{n=0}^{\infty} c_n x^n, \quad x \in \mathbb{R},$$

where $c_n = \sum_{k=0}^{n} a_k b_{n-k}$, $a_n = (-1)^n \frac{1}{n!}$ and $b_{2n} = 0$, $b_{2n+1} = (-1)^n \frac{1}{(2n+1)!}$.
A calculation shows that

$c_0 = a_0 b_0 = 0;$
$c_1 = a_0 b_1 + a_1 b_0 = 1 \cdot 1 + \frac{-1}{1!} \cdot 0 = 1;$
$c_2 = a_0 b_2 + a_1 b_1 + a_2 b_0 = 1 \cdot 0 + \frac{-1}{1!} \cdot \frac{1}{1!} + \frac{1}{2!} \cdot 0 = -1;$
$c_3 = a_0 b_3 + a_1 b_2 + a_2 b_1 + a_3 b_0 = 1 \cdot \frac{-1}{3!} + \frac{-1}{1!} \cdot 0 + \frac{1}{2!} \cdot 1 + \frac{-1}{3!} \cdot 0 = \frac{1}{3}.$

It follows that

$$e^{-x} \sin x = x - x^2 + \frac{x^3}{3} + \cdots, \quad x \in \mathbb{R}.$$

(b) We have

$$e^{3x} \cos x = \left(\sum_{n=0}^{\infty} \frac{(3x)^n}{n!} \right) \left(\sum_{n=0}^{\infty} (-1)^n \frac{x^{2n}}{(2n)!} \right) = \sum_{n=0}^{\infty} c_n x^n, \quad x \in \mathbb{R},$$

where $c_n = \sum_{k=0}^{n} a_k b_{n-k}$, $a_n = \frac{3^n}{n!}$ and $b_{2n} = (-1)^n \frac{1}{(2n)!}$, $b_{2n+1} = 0$.

A calculation shows that

$c_0 = a_0 b_0 = 1 \cdot 1 = 1;$
$c_1 = a_0 b_1 + a_1 b_0 = 1 \cdot 0 + \frac{3}{1!} \cdot 1 = 3;$
$c_2 = a_0 b_2 + a_1 b_1 + a_2 b_0 = 1 \cdot \frac{-1}{2!} + \frac{3}{1!} \cdot 0 + \frac{3^2}{2!} \cdot 1 = 4.$

It follows that

$$e^{3x} \cos x = 1 + 3x + 4x^2 + \cdots, \quad x \in \mathbb{R}.$$

(c) We have

$$\tan x = \frac{\sin x}{\cos x} = \frac{\sum_{n=0}^{\infty} (-1)^n \frac{x^{2n+1}}{(2n+1)!}}{\sum_{n=0}^{\infty} (-1)^n \frac{x^{2n}}{(2n)!}} = \sum_{n=0}^{\infty} c_n x^n, \quad x \in \left(-\frac{\pi}{2}, \frac{\pi}{2} \right).$$

It follows that

$$\left(\sum_{n=0}^{\infty} (-1)^n \frac{x^{2n}}{(2n)!} \right) \cdot \left(\sum_{n=0}^{\infty} c_n x^n \right) = \sum_{n=0}^{\infty} (-1)^n \frac{x^{2n+1}}{(2n+1)!}.$$

We calculate the coefficients c_n by solving the system of equations

$$0 = 1 \cdot c_0 \quad \Rightarrow \quad c_0 = 0;$$

$$1 = 1 \cdot c_1 + 0 \cdot c_0 \quad \Rightarrow \quad c_1 = 1;$$

$$0 = 1 \cdot c_2 + 0 \cdot c_1 - \frac{1}{2!} \cdot c_0 \quad \Rightarrow \quad c_2 = 0;$$

$$-\frac{1}{3!} = 1 \cdot c_3 + 0 \cdot c_2 - \frac{1}{2!} \cdot c_1 + 0 \cdot c_0 \quad \Rightarrow \quad c_3 = \frac{1}{3};$$

$$0 = 1 \cdot c_4 + 0 \cdot c_3 - \frac{1}{2!} \cdot c_2 + 0 \cdot c_1 + \frac{1}{4!} \cdot c_0 \quad \Rightarrow \quad c_4 = 0;$$

$$\frac{1}{5!} = 1 \cdot c_5 + 0 \cdot c_4 - \frac{1}{2!} \cdot c_3 + 0 \cdot c_2 + \frac{1}{4!} \cdot c_1 + 0 \cdot c_0 \quad \Rightarrow \quad c_5 = \frac{2}{15},$$

and we have that

$$\tan x = x + \frac{1}{3}x^3 + \frac{2}{15}x^5 + \cdots, \quad x \in \left(-\frac{\pi}{2}, \frac{\pi}{2}\right).$$

(d) We have

$$\tanh(2x) = \frac{\sinh(2x)}{\cosh(2x)} = \frac{\sum\limits_{n=0}^{\infty} \frac{2^{2n+1}}{(2n+1)!} x^{2n+1}}{\sum\limits_{n=0}^{\infty} \frac{2^{2n}}{(2n)!} x^{2n}} = \sum_{n=0}^{\infty} c_n x^n, \quad x \in \mathbb{R}.$$

It follows that

$$\left(\sum_{n=0}^{\infty} \frac{2^{2n}}{(2n)!} x^{2n}\right) \cdot \left(\sum_{n=0}^{\infty} c_n x^n\right) = \sum_{n=0}^{\infty} \frac{2^{2n+1}}{(2n+1)!} x^{2n+1}.$$

We calculate the coefficients c_n by solving the system of equations

$$0 = 1 \cdot c_0 \quad \Rightarrow \quad c_0 = 0;$$

$$\frac{2}{1!} = 1 \cdot c_1 + 0 \cdot c_0 \quad \Rightarrow \quad c_1 = 2;$$

$$0 = 1 \cdot c_2 + 0 \cdot c_1 + \frac{2^2}{2!} \cdot c_0 \quad \Rightarrow \quad c_2 = 0;$$

$$\frac{2^3}{3!} = 1 \cdot c_3 + 0 \cdot c_2 + \frac{2^2}{2!} \cdot c_1 + 0 \cdot c_0 \quad \Rightarrow \quad c_3 = -\frac{8}{3};$$

$$0 = 1 \cdot c_4 + 0 \cdot c_3 + \frac{2^2}{2!} \cdot c_2 + 0 \cdot c_1 + \frac{2^4}{4!} \cdot c_0 \quad \Rightarrow \quad c_4 = 0;$$

$$\frac{2^5}{5!} = 1 \cdot c_5 + 0 \cdot c_4 + \frac{2^2}{2!} \cdot c_3 + 0 \cdot c_2 + \frac{2^4}{4!} \cdot c_1 + 0 \cdot c_0 \quad \Rightarrow \quad c_5 = \frac{64}{15},$$

and we have that

$$\tanh(2x) = 2x - \frac{8}{3}x^3 + \frac{64}{15}x^5 + \cdots, \quad x \in \mathbb{R}.$$

3.63. (a) Use the Cauchy product to multiply the power series of $\ln(1-x)$ and $\frac{1}{1-x}$.

(d) One way to prove part (d) is by direct computation. Another method is based on part (c) and the Cauchy product for calculating the power series of $\ln^2(1-x)$.

3.64. (a) We have

$$\int_0^{\frac{1}{2}} \frac{\ln(1-x)}{x} dx = \ln(1-x)\ln x \Big|_0^{\frac{1}{2}} + \int_0^{\frac{1}{2}} \frac{\ln x}{1-x} dx$$

$$= \ln^2 2 + \int_0^1 \frac{\ln x}{1-x} dx - \int_{\frac{1}{2}}^1 \frac{\ln x}{1-x} dx$$

$$= \ln^2 2 - \frac{\pi^2}{6} - \int_{\frac{1}{2}}^1 \frac{\ln x}{1-x} dx$$

$$\overset{1-x=t}{=} \ln^2 2 - \frac{\pi^2}{6} - \int_0^{\frac{1}{2}} \frac{\ln(1-t)}{t} dt,$$

and the result follows.

(b) We have

$$\sum_{n=1}^\infty \frac{H_n}{n2^{n-1}} = 2\sum_{n=1}^\infty H_n \int_0^{\frac{1}{2}} x^{n-1} dx = 2\int_0^{\frac{1}{2}} \frac{1}{x} \sum_{n=1}^\infty H_n x^n\, dx \overset{3.63\,(a)}{=} -2\int_0^{\frac{1}{2}} \frac{\ln(1-x)}{x(1-x)} dx$$

$$= -2\int_0^{\frac{1}{2}} \frac{\ln(1-x)}{x} dx - 2\int_0^{\frac{1}{2}} \frac{\ln(1-x)}{1-x} dx \overset{(a)}{=} -2\left(\frac{\ln^2 2}{2} - \frac{\pi^2}{12}\right) + \ln^2 2$$

$$= \frac{\pi^2}{6}.$$

On the other hand,

$$\ln^2 2 = -2\int_0^{\frac{1}{2}} \frac{\ln(1-x)}{1-x} dx = 2\int_0^{\frac{1}{2}} \sum_{n=1}^\infty H_n x^n\, dx = 2\sum_{n=1}^\infty H_n \int_0^{\frac{1}{2}} x^n dx = \sum_{n=1}^\infty \frac{H_n}{(n+1)2^n}.$$

3.66. Part (a) follows by direct computation. Part (b) can be solved either by using the Cauchy product to multiply the power series of the functions $f(x) = \ln(1+x)$ and $g(x) = \frac{1}{1-x}$ or by using part (a) of the problem.

3.67. We have

$$\frac{e^x}{1-x} = \left(\sum_{n=0}^\infty \frac{x^n}{n!}\right) \cdot \left(\sum_{n=0}^\infty x^n\right) = \sum_{n=0}^\infty c_n x^n, \quad x \in (-1, 1).$$

We calculate $c_n = \sum_{k=0}^{n} a_k b_{n-k}$, with $a_n = \frac{1}{n!}$ and $b_n = 1$, and we have that $c_n = \sum_{k=0}^{n} \frac{1}{k!}$. Therefore,

$$\frac{e^x}{1-x} = \sum_{n=0}^{\infty} \left(1 + \frac{1}{1!} + \frac{1}{2!} + \cdots + \frac{1}{n!} \right) x^n, \quad x \in (-1, 1).$$

Remark. The function $f(x) = \frac{e^x}{1-x}$ is known as the generating function of the sequence $(E_n)_{n \geq 0}$ defined by $E_n = 1 + \frac{1}{1!} + \frac{1}{2!} + \cdots + \frac{1}{n!}$.

3.68. We have

$$x = (e^x - 1) \left(\sum_{n=0}^{\infty} \frac{B_n}{n!} x^n \right),$$

which means

$$x = \left(\sum_{n=1}^{\infty} \frac{x^n}{n!} \right) \left(\sum_{n=0}^{\infty} \frac{B_n}{n!} x^n \right)$$

and, therefore,

$$1 = B_0,$$

$$0 = \frac{1}{1!} \cdot \frac{B_1}{1!} + \frac{1}{2!} B_0 \quad \Rightarrow \quad B_1 = -\frac{1}{2},$$

$$0 = \frac{1}{1!} \cdot \frac{B_2}{2!} + \frac{1}{2!} \cdot \frac{B_1}{1!} + \frac{1}{3!} B_0 \quad \Rightarrow \quad B_2 = \frac{1}{6},$$

$$0 = \frac{1}{1!} \cdot \frac{B_3}{3!} + \frac{1}{2!} \cdot \frac{B_2}{2!} + \frac{1}{3!} \cdot \frac{B_1}{1!} + \frac{1}{4!} B_0 \quad \Rightarrow \quad B_3 = 0,$$

$$\cdots \quad \cdots \quad B_4 = -\frac{1}{30},$$

$$\cdots \quad \cdots \quad B_5 = 0,$$

$$\cdots \quad \cdots \quad B_6 = \frac{1}{42}.$$

3.69. *Solution I.* We have

$$1 = (1 - x - x^2) \sum_{n=1}^{\infty} a_n x^{n-1}$$

and it follows that

$$1 = a_1,$$

$$0 = a_2 - a_1 \quad \Rightarrow \quad a_2 = a_1,$$

$$0 = a_3 - a_2 - a_1 \quad \Rightarrow \quad a_3 = a_2 + a_1,$$

$$0 = a_4 - a_3 - a_2 \quad \Rightarrow \quad a_4 = a_3 + a_2,$$

$$\vdots$$

$$0 = a_{n+1} - a_n - a_{n-1} \quad \Rightarrow \quad a_{n+1} = a_n + a_{n-1}.$$

Solution II. For $x \in \left(\frac{1-\sqrt{5}}{2}, \frac{\sqrt{5}-1}{2} \right)$ we have

$$\frac{1}{1-x-x^2} = \frac{1}{\sqrt{5}} \left(\frac{1}{x + \frac{1+\sqrt{5}}{2}} - \frac{1}{x + \frac{1-\sqrt{5}}{2}} \right)$$

$$= \frac{1}{\sqrt{5}} \left(\frac{2}{1+\sqrt{5}} \cdot \frac{1}{1 + \frac{2x}{1+\sqrt{5}}} - \frac{2}{1-\sqrt{5}} \cdot \frac{1}{1 + \frac{2x}{1-\sqrt{5}}} \right)$$

$$= \frac{1}{\sqrt{5}} \left[\frac{2}{1+\sqrt{5}} \cdot \sum_{n=0}^{\infty} (-1)^n \left(\frac{2x}{1+\sqrt{5}} \right)^n \right.$$

$$\left. - \frac{2}{1-\sqrt{5}} \cdot \sum_{n=0}^{\infty} (-1)^n \left(\frac{2x}{1-\sqrt{5}} \right)^n \right]$$

$$= \sum_{n=1}^{\infty} (-1)^{n-1} \frac{2^n}{\sqrt{5}} \left[\frac{1}{(1+\sqrt{5})^n} - \frac{1}{(1-\sqrt{5})^n} \right] x^{n-1}$$

$$= \sum_{n=1}^{\infty} \frac{1}{\sqrt{5}} \left[\left(\frac{1+\sqrt{5}}{2} \right)^n - \left(\frac{1-\sqrt{5}}{2} \right)^n \right] x^{n-1}.$$

3.70. The problem is solved similarly to Problem 3.69.

3.71. (a) Use that $\sin n = \frac{e^{in} - e^{-in}}{2i}$.

   (b) Divide by x the series in part (a) and integrate term by term.

3.72. (a) Use that $\cos n = \frac{e^{in} + e^{-in}}{2}$.

   (b) Divide by x the series in part (a) and integrate term by term.

3.73. Observe that $a_n = n + 1$, $\forall n \geq 0$. It follows that $m = (n^2 + 5n + 5)^2$.

3.74. Eliminate $\frac{1}{b_{2n}} = (2n)!$.

11.8   Series with Harmonic and Skew-Harmonic Numbers

3.75. Part (a) of Problem 3.66 implies that

$$H_n^- - \ln 2 = - \int_0^1 \frac{(-x)^n}{1+x} dx.$$

(a) We have

$$\sum_{n=1}^\infty H_n^- \left(H_n^- - \ln 2\right) = -\sum_{n=1}^\infty H_n^- \int_0^1 \frac{(-x)^n}{1+x} dx = -\int_0^1 \frac{1}{1+x} \left(\sum_{n=1}^\infty H_n^-(-x)^n\right) dx$$

$$\stackrel{3.66\,(b)}{=} -\int_0^1 \frac{\ln(1-x)}{(1+x)^2} dx = \frac{\ln 2}{2}.$$

(b) We have

$$\sum_{n=1}^\infty \left(H_n^- - \ln 2\right) \left(\zeta(2) - 1 - \frac{1}{2^2} - \cdots - \frac{1}{n^2}\right)$$

$$= -\sum_{n=1}^\infty \int_0^1 \frac{(-x)^n}{1+x} dx \left(\zeta(2) - 1 - \frac{1}{2^2} - \cdots - \frac{1}{n^2}\right)$$

$$= -\int_0^1 \frac{1}{1+x} \sum_{n=1}^\infty \left(\zeta(2) - 1 - \frac{1}{2^2} - \cdots - \frac{1}{n^2}\right)(-x)^n dx$$

$$\stackrel{3.4\,C}{=} \int_0^1 \frac{x\zeta(2) + \text{Li}_2(-x)}{(1+x)^2} dx.$$

Integrating by parts, we have that

$$\int_0^1 \frac{x\zeta(2) + \text{Li}_2(-x)}{(1+x)^2} dx = -\frac{x\zeta(2) + \text{Li}_2(-x)}{1+x} \Big|_0^1 + \int_0^1 \frac{\zeta(2) - \frac{\ln(1+x)}{x}}{1+x} dx$$

$$= -\frac{\zeta(2) + \text{Li}_2(-1)}{2} + \zeta(2)\ln 2 - \int_0^1 \frac{\ln(1+x)}{x} dx + \frac{\ln^2 2}{2}$$

$$= -\frac{\pi^2}{8} + \frac{\pi^2 \ln 2}{6} + \frac{\ln^2 2}{2}.$$

(c) We have, based on part (a) of Problem 3.66, that

$$\sum_{n=1}^{\infty} (-1)^n H_n^- \left(\zeta(2) - 1 - \frac{1}{2^2} - \cdots - \frac{1}{n^2} \right)$$

$$= \sum_{n=1}^{\infty} (-1)^n \int_0^1 \frac{1 - (-x)^n}{1 + x} dx \left(\zeta(2) - 1 - \frac{1}{2^2} - \cdots - \frac{1}{n^2} \right)$$

$$= \int_0^1 \frac{1}{1 + x} \sum_{n=1}^{\infty} (-1)^n \left(\zeta(2) - 1 - \frac{1}{2^2} - \cdots - \frac{1}{n^2} \right) dx$$

$$- \int_0^1 \frac{1}{1 + x} \sum_{n=1}^{\infty} x^n \left(\zeta(2) - 1 - \frac{1}{2^2} - \cdots - \frac{1}{n^2} \right) dx$$

$$\stackrel{3.4\,C}{=} \frac{-\zeta(2) - \text{Li}_2(-1)}{2} \ln 2 - \int_0^1 \frac{x\zeta(2) - \text{Li}_2(x)}{(1+x)(1-x)} dx$$

$$= \frac{-\zeta(2) - \text{Li}_2(-1)}{2} \ln 2 - \frac{1}{2} \int_0^1 (x\zeta(2) - \text{Li}_2(x)) \left(\frac{1}{1+x} + \frac{1}{1-x} \right) dx.$$

We calculate the preceding integral using integration by parts, with $f(x) = x\zeta(2) - \text{Li}_2(x)$, $f'(x) = \zeta(2) + \frac{\ln(1-x)}{x}$, $g'(x) = \frac{1}{1+x} + \frac{1}{1-x}$ and $g(x) = \ln(1+x) - \ln(1-x)$, and we obtain that

$$\int_0^1 (x\zeta(2) - \text{Li}_2(x)) \left(\frac{1}{1+x} + \frac{1}{1-x} \right) dx$$

$$= (x\zeta(2) - \text{Li}_2(x)) (\ln(1+x) - \ln(1-x)) \Big|_0^1 - \zeta(2) \int_0^1 (\ln(1+x) - \ln(1-x)) dx$$

$$- \int_0^1 \frac{\ln(1-x)\ln(1+x)}{x} dx + \int_0^1 \frac{\ln^2(1-x)}{x} dx$$

$$= -2\zeta(2) \ln 2 + \frac{21}{8} \zeta(3).$$

It follows that

$$\sum_{n=1}^{\infty} (-1)^n H_n^- \left(\zeta(2) - 1 - \frac{1}{2^2} - \cdots - \frac{1}{n^2} \right)$$

$$= \frac{-\zeta(2) - \text{Li}_2(-1)}{2} \ln 2 + \zeta(2) \ln 2 - \frac{21}{16} \zeta(3) = \frac{\pi^2 \ln 2}{8} - \frac{21}{16} \zeta(3).$$

(d) We have, based on part (a) of Problem 3.66, that

$$\sum_{n=1}^{\infty} H_n^- \left(\zeta(3) - 1 - \frac{1}{2^3} - \cdots - \frac{1}{n^3} \right)$$

$$= \sum_{n=1}^{\infty} \int_0^1 \frac{1 - (-x)^n}{1 + x} dx \left(\zeta(3) - 1 - \frac{1}{2^3} - \cdots - \frac{1}{n^3} \right)$$

$$= \int_0^1 \frac{1}{1 + x} \sum_{n=1}^{\infty} \left(\zeta(3) - 1 - \frac{1}{2^3} - \cdots - \frac{1}{n^3} \right) dx$$

$$- \int_0^1 \frac{1}{1 + x} \sum_{n=1}^{\infty} (-x)^n \left(\zeta(3) - 1 - \frac{1}{2^3} - \cdots - \frac{1}{n^3} \right) dx$$

$$\overset{3.4\,C}{=} \int_0^1 \frac{1}{1 + x} \left(\zeta(2) - \zeta(3) - \frac{-x\zeta(3) - \mathrm{Li}_3(-x)}{1 + x} \right) dx$$

$$= \int_0^1 \frac{\zeta(2) - \zeta(3) + x\zeta(2) + \mathrm{Li}_3(-x)}{(1 + x)^2} dx.$$

We calculate the preceding integral by parts with $f(x) = \zeta(2) - \zeta(3) + x\zeta(2) + \mathrm{Li}_3(-x)$, $f'(x) = \zeta(2) + \frac{\mathrm{Li}_2(-x)}{x}$, $g'(x) = \frac{1}{(1+x)^2}$ and $g(x) = -\frac{1}{1+x}$, and we have that

$$\int_0^1 \frac{\zeta(2) - \zeta(3) + x\zeta(2) + \mathrm{Li}_3(-x)}{(1 + x)^2} dx$$

$$= -\frac{\zeta(2) - \zeta(3) + x\zeta(2) + \mathrm{Li}_3(-x)}{1 + x} \bigg|_0^1 + \int_0^1 \left(\frac{\zeta(2)}{1 + x} + \frac{\mathrm{Li}_2(-x)}{x(1 + x)} \right) dx$$

$$= -\frac{\zeta(3)}{8} + \zeta(2) \ln 2 + \int_0^1 \frac{\mathrm{Li}_2(-x)}{x} dx - \int_0^1 \frac{\mathrm{Li}_2(-x)}{1 + x} dx.$$

A calculation shows that

$$\int_0^1 \frac{\mathrm{Li}_2(-x)}{x} dx = \int_0^1 \sum_{n=1}^{\infty} (-1)^n \frac{x^{n-1}}{n^2} dx = \sum_{n=1}^{\infty} \frac{(-1)^n}{n^3} = \mathrm{Li}_3(-1) = -\frac{3}{4} \zeta(3)$$

and

$$\int_0^1 \frac{\mathrm{Li}_2(-x)}{1 + x} dx = \mathrm{Li}_2(-x) \ln(1 + x) \big|_0^1 + \int_0^1 \frac{\ln^2(1 + x)}{x} dx$$

$$= \mathrm{Li}_2(-1) \ln 2 + \frac{\zeta(3)}{4}$$

$$= -\frac{\zeta(2) \ln 2}{2} + \frac{\zeta(3)}{4}.$$

Putting all these together, we get that

$$\sum_{n=1}^{\infty} H_n^- \left(\zeta(3) - 1 - \frac{1}{2^3} - \cdots - \frac{1}{n^3} \right) = \frac{\pi^2 \ln 2}{4} - \frac{9}{8} \zeta(3).$$

3.76. (a) We have, based on part (a) of Problem 3.66, that

$$\sum_{n=1}^{\infty} (-1)^n \frac{H_n^-}{n} = \sum_{n=1}^{\infty} \frac{(-1)^n}{n} \int_0^1 \frac{1 - (-x)^n}{1 + x} dx$$

$$= \int_0^1 \frac{1}{1 + x} \sum_{n=1}^{\infty} \left(\frac{(-1)^n}{n} - \frac{x^n}{n} \right) dx$$

$$= \int_0^1 \frac{-\ln 2 + \ln(1 - x)}{1 + x} dx$$

$$= -\ln^2 2 + \int_0^1 \frac{\ln(1 - x)}{1 + x} dx$$

$$= -\frac{\pi^2}{12} - \frac{\ln^2 2}{2},$$

where the last equality follows based on part (b) of Problem 7.66.
   (b) We have, based on part (a) of Problem 3.66, that

$$\sum_{n=1}^{\infty} \frac{H_n^-}{n^2} = \sum_{n=1}^{\infty} \frac{1}{n^2} \int_0^1 \frac{1 - (-x)^n}{1 + x} dx$$

$$= \int_0^1 \frac{1}{1 + x} \sum_{n=1}^{\infty} \left(\frac{1}{n^2} - \frac{(-x)^n}{n^2} \right) dx$$

$$= \int_0^1 \frac{\zeta(2) - \text{Li}_2(-x)}{1 + x} dx$$

$$= (\zeta(2) - \text{Li}_2(-x)) \ln(1 + x) \Big|_0^1 - \int_0^1 \frac{\ln^2(1 + x)}{x} dx$$

$$= \frac{\pi^2}{4} \ln 2 - \frac{\zeta(3)}{4},$$

since $\text{Li}_2(-1) = -\frac{\zeta(2)}{2}$ and $\int_0^1 \frac{\ln^2(1+x)}{x} dx = \frac{\zeta(3)}{4}$ (see [6, pp. 291–292]).
   (c) We have, based on part (a) of Problem 3.66, that

$$\sum_{n=1}^{\infty} \frac{H_n^-}{n^3} = \sum_{n=1}^{\infty} \frac{1}{n^3} \int_0^1 \frac{1-(-x)^n}{1+x} dx = \int_0^1 \frac{1}{1+x} \sum_{n=1}^{\infty} \left(\frac{1}{n^3} - \frac{(-x)^n}{n^3} \right) dx$$

$$= \int_0^1 \frac{\zeta(3) - \text{Li}_3(-x)}{1+x} dx$$

$$= (\zeta(3) - \text{Li}_3(-x)) \ln(1+x) \Big|_0^1 + \int_0^1 \frac{\ln(1+x)}{x} \text{Li}_2(-x) dx$$

$$= (\zeta(3) - \text{Li}_3(-1)) \ln 2 - \frac{\text{Li}_2^2(-x)}{2} \Big|_0^1$$

$$= \frac{7}{4}\zeta(3) \ln 2 - \frac{\pi^4}{288}.$$

(d) We have, based on the formula $\int_0^1 x^{n-1} \ln(1-x)dx = -\frac{H_n}{n}, n \geq 1$, (see [26, entry (3.17), p. 206]), that

$$\sum_{n=1}^{\infty} \frac{H_n^-}{n} \cdot \frac{H_{n+1}}{n+1} = -\sum_{n=1}^{\infty} \frac{H_n^-}{n} \int_0^1 x^n \ln(1-x)dx = -\int_0^1 \ln(1-x) \left(\sum_{n=1}^{\infty} \frac{H_n^-}{n} x^n \right) dx$$

$$\overset{3.66\,(b)}{=} -\int_0^1 \ln(1-x) \left(\int_0^x \frac{\ln(1+t)}{t(1-t)} dt \right) dx$$

$$= (1-x)\left[\ln(1-x) - 1\right] \int_0^x \frac{\ln(1+t)}{t(1-t)} dt \Big|_0^1$$

$$- \int_0^1 \frac{\ln(1+x)}{x} (\ln(1-x) - 1) dx$$

$$= -\int_0^1 \frac{\ln(1-x)\ln(1+x)}{x} dx + \int_0^1 \frac{\ln(1+x)}{x} dx$$

$$= \frac{5}{8}\zeta(3) + \frac{\pi^2}{12}.$$

The last equality follows based on formula (10.2) and Problem 3.57.

(e) Using the same idea as in the solution of part (d) we have, after some calculations, that

$$\sum_{n=1}^{\infty}(-1)^n \frac{H_n^-}{n} \cdot \frac{H_{n+1}}{n+1} = -\int_0^1 \frac{1-x}{x(1+x)} \left(\ln^2(1-x) - \ln(1-x)\right) dx$$

$$= -\zeta(2) - 2\zeta(3) - 2\int_0^1 \frac{\ln(1-x)}{1+x} dx + 4\int_0^1 \frac{\ln^2(1-x)}{1+x} dx$$

$$= -\zeta(2) - 2\zeta(3) + 2\mathrm{Li}_2\left(\frac{1}{2}\right) + 4\mathrm{Li}_3\left(\frac{1}{2}\right)$$

$$= \frac{3}{2}\zeta(3) - \frac{\pi^2}{3}\ln 2 - \ln^2 2 + \frac{2}{3}\ln^3 2.$$

3.77. (a) We have, based on the formula $\int_0^1 x^{n-1}\ln(1-x)dx = -\frac{H_n}{n}, n \geq 1$, (see [26, entry (3.17), p. 206]), that

$$\sum_{n=1}^{\infty}(-1)^{n-1}\frac{H_n^- H_{n+1}}{n+1} = \sum_{n=1}^{\infty}(-1)^n H_n^- \int_0^1 x^n \ln(1-x)dx$$

$$= \int_0^1 \ln(1-x)\left(\sum_{n=1}^{\infty} H_n^-(-x)^n\right) dx$$

$$\overset{3.66\,(b)}{=} \int_0^1 \frac{\ln^2(1-x)}{1+x} dx$$

$$= 2\,\mathrm{Li}_3\left(\frac{1}{2}\right).$$

The preceding integral is calculated in the solution of part (b) of Problem 2.81.

(b) We have, based on part (a) of Problem 3.66, that

$$\sum_{n=1}^{\infty}(-1)^n \frac{H_n^- H_n}{n} = \sum_{n=1}^{\infty}(-1)^n \frac{H_n}{n}\int_0^1 \frac{1-(-x)^n}{1+x} dx$$

$$= \int_0^1 \frac{1}{1+x}\left(\sum_{n=1}^{\infty}(-1)^n \frac{H_n}{n} - \sum_{n=1}^{\infty}\frac{H_n}{n}x^n\right) dx$$

$$\overset{(11.9)}{=} \int_0^1 \frac{1}{1+x}\left(S - \mathrm{Li}_2(x) - \frac{1}{2}\ln^2(1-x)\right) dx,$$

where $S \overset{(11.9)}{=} \sum_{n=1}^{\infty}(-1)^n \frac{H_n}{n} = \mathrm{Li}_2(-1) + \frac{1}{2}\ln^2 2 = \frac{\ln^2 2}{2} - \frac{\pi^2}{12}$.

We calculate the integral by parts, with $f(x) = S - \mathrm{Li}_2(x) - \frac{1}{2}\ln^2(1-x)$, $f'(x) = \frac{\ln(1-x)}{x} + \frac{\ln(1-x)}{1-x}$, $g'(x) = \frac{1}{1+x}$ and $g(x) = \ln(1+x) - \ln 2$, and we have that

$$\int_0^1 \frac{1}{1+x} \left(S - \text{Li}_2(x) - \frac{1}{2} \ln^2(1-x) \right) dx$$

$$= \left(S - \text{Li}_2(x) - \frac{1}{2} \ln^2(1-x) \right) (\ln(1+x) - \ln 2) \Big|_0^1$$

$$- \int_0^1 (\ln(1+x) - \ln 2) \left(\frac{\ln(1-x)}{x} + \frac{\ln(1-x)}{1-x} \right) dx$$

$$= S \ln 2 - \int_0^1 \frac{\ln(1-x)\ln(1+x)}{x} dx + \ln 2 \int_0^1 \frac{\ln(1-x)}{x} dx$$

$$- \int_0^1 (\ln(1+x) - \ln 2) \frac{\ln(1-x)}{1-x} dx$$

$$= S \ln 2 + \frac{5}{8}\zeta(3) - \zeta(2) \ln 2 - \int_0^1 (\ln(1+x) - \ln 2) \frac{\ln(1-x)}{1-x} dx.$$

We calculate

$$\int_0^1 (\ln(1+x) - \ln 2) \frac{\ln(1-x)}{1-x} dx - \int_0^1 \frac{\ln y}{y} \ln\left(1 - \frac{y}{2}\right) dy$$

$$= -\int_0^1 \ln y \left(\sum_{n=1}^{\infty} \frac{y^{n-1}}{n 2^n} \right) dy$$

$$= -\sum_{n=1}^{\infty} \frac{1}{n 2^n} \int_0^1 y^{n-1} \ln y \, dy$$

$$= \sum_{n=1}^{\infty} \frac{1}{n^3 2^n}$$

$$= \text{Li}_3\left(\frac{1}{2}\right).$$

It follows that

$$\sum_{n=1}^{\infty} (-1)^n \frac{H_n^- H_n}{n} = S \ln 2 + \frac{5}{8}\zeta(3) - \zeta(2) \ln 2 - \text{Li}_3\left(\frac{1}{2}\right)$$

$$= \frac{\ln^3 2}{3} - \frac{\pi^2}{6} \ln 2 - \frac{\zeta(3)}{4},$$

since $\text{Li}_3\left(\frac{1}{2}\right) = \frac{1}{24}\left(-2\pi^2 \ln 2 + 4\ln^3 2 + 21\zeta(3)\right)$ (see [21, p. 44]).

3.78. (a) and (b) Use that $H_n^- - \ln 2 = -\int_0^1 \frac{(-x)^n}{1+x}dx$, $n \geq 1$ (part (a) of Problem 3.66).

(c) and (d) Add and subtract the series in parts (a) and (b).

3.79. We have, based on the formula $\int_0^1 x^{n-1}\ln(1-x)dx = -\frac{H_n}{n}$, $n \geq 1$, (see [26, entry (3.17), p. 206]), that

$$\sum_{n=1}^{\infty} \frac{H_{2n-1}}{(2n-1)^2} = -\sum_{n=1}^{\infty} \frac{1}{2n-1} \int_0^1 x^{2n-2}\ln(1-x)dx$$

$$= -\int_0^1 \frac{\ln(1-x)}{x} \left(\sum_{n=1}^{\infty} \frac{x^{2n-1}}{2n-1} \right) dx$$

$$= -\frac{1}{2} \int_0^1 \frac{\ln(1-x)}{x} \ln\left(\frac{1+x}{1-x} \right) dx$$

$$= -\frac{1}{2} \int_0^1 \frac{\ln(1-x)\ln(1+x)}{x}dx + \frac{1}{2} \int_0^1 \frac{\ln^2(1-x)}{x}dx$$

$$\overset{(10.2)}{=} \frac{21}{16}\zeta(3).$$

The second part of the problem follows from the formula

$$2\zeta(3) \overset{7.60}{=} \sum_{n=1}^{\infty} \frac{H_n}{n^2} = \sum_{n=1}^{\infty} \frac{H_{2n-1}}{(2n-1)^2} + \frac{1}{4} \sum_{n=1}^{\infty} \frac{H_{2n}}{n^2}.$$

3.80. (a) We have

$$\int_0^1 x^{n-1}\ln(1+x)dx = \int_0^1 x^{n-1} \left(\sum_{i=1}^{\infty} (-1)^{i-1}\frac{x^i}{i} \right) dx$$

$$= \sum_{i=1}^{\infty} \frac{(-1)^{i-1}}{i(i+n)}$$

$$= \frac{1}{n} \left(\sum_{i=1}^{\infty} \frac{(-1)^{i-1}}{i} - \sum_{i=1}^{\infty} \frac{(-1)^{i-1}}{n+i} \right)$$

$$= \frac{1}{n} \left(\ln 2 - (-1)^n \sum_{j=n+1}^{\infty} \frac{(-1)^{j-1}}{j} \right)$$

$$= \frac{1-(-1)^n}{n}\ln 2 + \frac{(-1)^n}{n}H_n^-.$$

(b) We have, based on part (a), that

$$\frac{H_{2n-1}^-}{2n-1} = \frac{2\ln 2}{2n-1} - \int_0^1 x^{2n-2}\ln(1+x)dx$$

and it follows that

$$\sum_{n=1}^\infty (-1)^{n-1}\frac{H_{2n-1}^-}{2n-1} = \sum_{n=1}^\infty (-1)^{n-1}\left(\frac{2\ln 2}{2n-1} - \int_0^1 x^{2n-2}\ln(1+x)dx\right)$$

$$= 2\ln 2\sum_{n=1}^\infty \frac{(-1)^{n-1}}{2n-1} - \int_0^1 \ln(1+x)\left(\sum_{n=1}^\infty (-x^2)^{n-1}\right)dx$$

$$= \frac{\pi\ln 2}{4} - \int_0^1 \frac{\ln(1+x)}{1+x^2}dx$$

$$= \frac{3\pi\ln 2}{8}.$$

The integral $\int_0^1 \frac{\ln(1+x)}{1+x^2}dx = \frac{\pi\ln 2}{8}$ was calculated in the solution of part (b) of Problem 3.25.

(c) Observe that

$$\frac{H_n^-}{n(n+1)} = \frac{H_n^-}{n} - \frac{H_{n+1}^-}{n+1} + \frac{(-1)^n}{(n+1)^2}, \quad n \geq 1. \tag{11.4}$$

It follows that

$$\sum_{n=1}^\infty \frac{H_n^-}{n(n+1)} = \sum_{n=1}^\infty \left(\frac{H_n^-}{n} - \frac{H_{n+1}^-}{n+1}\right) + \sum_{n=1}^\infty \frac{(-1)^n}{(n+1)^2} = 1 + \sum_{n=1}^\infty \frac{(-1)^n}{(n+1)^2} = \frac{\pi^2}{12}.$$

The second series can be calculated based on formula (11.4) and part (a) of Problem 3.76.

(d) We have, based on part (a), that

$$\frac{H_{n+1}^-}{n+1} = (-1)^{n+1}\int_0^1 x^n \ln(1-x)dx + \frac{1-(-1)^{n+1}}{n+1}\ln 2.$$

It follows that

$$\sum_{n=1}^{\infty} \frac{H_n^-}{n} \cdot \frac{H_{n+1}^-}{n+1} = \sum_{n=1}^{\infty} \frac{H_n^-}{n} \left[(-1)^{n+1} \int_0^1 x^n \ln(1-x)dx + \frac{1-(-1)^{n+1}}{n+1} \ln 2 \right]$$

$$= -\int_0^1 \ln(1+x) \left(\sum_{n=1}^{\infty} \frac{H_n^-}{n}(-x)^n \right) dx$$

$$+ \ln 2 \left(\sum_{n=1}^{\infty} \frac{H_n^-}{n(n+1)} - \sum_{n=1}^{\infty} (-1)^{n-1} \frac{H_n^-}{n(n+1)} \right)$$

$$\overset{3.66 \, (b)}{=\!=\!=} -\int_0^1 \ln(1+x) \left(\int_0^{-x} \frac{\ln(1+t)}{t(1-t)} dt \right) dx + \ln 2 \left(\frac{\pi^2}{12} - \ln^2 2 \right).$$

We calculate the preceding integral by parts, and we have

$$\int_0^1 \ln(1+x) \left(\int_0^{-x} \frac{\ln(1+t)}{t(1-t)} dt \right) dx$$

$$= (1+x)\,(\ln(1+x)-1) \int_0^{-x} \frac{\ln(1+t)}{t(1-t)} dt \,\Big|_0^1 - \int_0^1 \frac{\ln(1-x)}{x}(\ln(1+x)-1)dx$$

$$= 2(\ln 2 - 1) \int_0^{-1} \frac{\ln(1+t)}{t(1-t)} dt - \int_0^1 \frac{\ln(1-x)\ln(1+x)}{x} dx + \int_0^1 \frac{\ln(1-x)}{x} dx$$

$$\overset{t=-y}{=\!=\!=} 2(\ln 2 - 1) \int_0^1 \frac{\ln(1-y)}{y(1+y)} dy + \frac{5}{8} \zeta(3) - \frac{\pi^2}{6}$$

$$= 2(\ln 2 - 1) \left[\int_0^1 \frac{\ln(1-y)}{y} dy - \int_0^1 \frac{\ln(1-y)}{1+y} dy \right] + \frac{5}{8} \zeta(3) - \frac{\pi^2}{6}$$

$$\overset{7.66}{=\!=\!=} 2(\ln 2 - 1) \left[-\frac{\pi^2}{6} + \text{Li}_2 \left(\frac{1}{2} \right) \right] + \frac{5}{8} \zeta(3) - \frac{\pi^2}{6}$$

$$= -\frac{\pi^2 \ln 2}{6} - \ln^3 2 + \ln^2 2 + \frac{5}{8} \zeta(3).$$

Putting all these together, we have that part (d) of the problem is solved.

(e) We have, based on part (a), that $\frac{H_n^-}{n} = (-1)^n \int_0^1 x^{n-1} \ln(1+x)dx + \frac{1-(-1)^n}{n} \ln 2$, and it follows that

$$\sum_{n=1}^{\infty} \frac{H_n H_n^-}{n^2} = \sum_{n=1}^{\infty} \frac{H_n}{n} \left[(-1)^n \int_0^1 x^{n-1} \ln(1+x)dx + \frac{1-(-1)^n}{n} \ln 2 \right]$$

$$= \int_0^1 \frac{\ln(1+x)}{x} \left(\sum_{n=1}^{\infty} \frac{H_n}{n}(-x)^n \right) dx + \ln 2 \sum_{n=1}^{\infty} \frac{H_n}{n^2}(1-(-1)^n)$$

$$= \int_0^1 \frac{\ln(1+x)}{x} \left(\text{Li}_2(-x) + \frac{1}{2} \ln^2(1+x) \right) dx + \ln 2 \sum_{n=1}^{\infty} \frac{H_n}{n^2}(1-(-1)^n).$$

On the other hand,

$$\sum_{n=1}^{\infty} \frac{H_n}{n^2}(1-(-1)^n) = 2\sum_{n=1}^{\infty} \frac{H_{2n-1}}{(2n-1)^2} \stackrel{3.79}{=} \frac{21}{8}\zeta(3).$$

It follows that

$$\sum_{n=1}^{\infty} \frac{H_n H_n^-}{n^2} = \int_0^1 \frac{\ln(1+x)}{x}\mathrm{Li}_2(-x)\mathrm{d}x + \frac{1}{2}\int_0^1 \frac{\ln^3(1+x)}{x}\mathrm{d}x + \frac{21}{8}\zeta(3)\ln 2$$

$$= -\frac{\mathrm{Li}_2^2(-x)}{2}\Big|_0^1 + \frac{1}{2}\int_0^1 \frac{\ln^3(1+x)}{x}\mathrm{d}x + \frac{21}{8}\zeta(3)\ln 2$$

$$= \frac{43\pi^4}{1440} + \frac{\pi^2 \ln^2 2}{8} - \frac{\ln^4 2}{8} - 3\mathrm{Li}_4\left(\frac{1}{2}\right).$$

We used a result of V.I. Levin involving the *tetralogarithm* Li_4 ([21, p. 47], [104])

$$\int_0^1 \frac{\ln^3(1+x)}{x}\mathrm{d}x = \frac{\pi^4}{15} + \frac{\pi^2 \ln^2 2}{4} - \frac{\ln^4 2}{4} - \frac{21\ln 2}{4}\zeta(3) - 6\mathrm{Li}_4\left(\frac{1}{2}\right). \quad (11.5)$$

3.81. (a) We apply Abel's summation formula, with $a_n = \frac{(-1)^n}{n}$ and $b_n = \zeta(2) - 1 - \frac{1}{2^2} - \cdots - \frac{1}{n^2}$, and we have that

$$\sum_{n=1}^{\infty} \frac{(-1)^n}{n}\left(\zeta(2) - 1 - \frac{1}{2^2} - \cdots - \frac{1}{n^2}\right)$$

$$= -\lim_{n\to\infty} H_n^-\left(\zeta(2) - 1 - \frac{1}{2^2} - \cdots - \frac{1}{(n+1)^2}\right)$$

$$+ \sum_{n=1}^{\infty}\left(\frac{-1}{1} + \frac{(-1)^2}{2} + \cdots + \frac{(-1)^n}{n}\right)\frac{1}{(n+1)^2}$$

$$= \sum_{n=1}^{\infty}\left(\frac{-1}{1} + \frac{(-1)^2}{2} + \cdots + \frac{(-1)^n}{n} + \frac{(-1)^{n+1}}{n+1} - \frac{(-1)^{n+1}}{n+1}\right)\frac{1}{(n+1)^2}$$

$$= \sum_{n=1}^{\infty}\left(-\frac{H_{n+1}^-}{(n+1)^2} - \frac{(-1)^{n+1}}{(n+1)^3}\right) = \sum_{j=2}^{\infty}\left(\frac{(-1)^{j-1}}{j^3} - \frac{H_j^-}{j^2}\right)$$

$$= \sum_{j=1}^{\infty}\left(\frac{(-1)^{j-1}}{j^3} - \frac{H_j^-}{j^2}\right)$$

$$= \sum_{j=1}^{\infty}\frac{(-1)^{j-1}}{j^3} - \sum_{j=1}^{\infty}\frac{H_j^-}{j^2}$$

$$\stackrel{3.76\,(b)}{=} \zeta(3) - \frac{\pi^2}{4}\ln 2.$$

(b) Use Abel's summation formula, with $a_n = \frac{(-1)^n}{n}$ and $b_n = \zeta(3) - 1 - \frac{1}{2^3} - \cdots - \frac{1}{n^3}$, combined with part (c) of Problem 3.76.

(c) Since $\frac{H_n^-}{n} = (-1)^n \int_0^1 x^{n-1} \ln(1+x)dx + \frac{1-(-1)^n}{n} \ln 2$ (see part (a) of Problem 3.80), we have that

$$\sum_{n=1}^{\infty} \frac{H_n^-}{n} \left(\zeta(2) - 1 - \frac{1}{2^2} - \cdots - \frac{1}{n^2} \right)$$

$$= \sum_{n=1}^{\infty} \left[(-1)^n \int_0^1 x^{n-1} \ln(1+x)dx + \frac{1-(-1)^n}{n} \ln 2 \right] \left(\zeta(2) - 1 - \frac{1}{2^2} - \cdots - \frac{1}{n^2} \right)$$

$$= \int_0^1 \frac{\ln(1+x)}{x} \left[\sum_{n=1}^{\infty} \left(\zeta(2) - 1 - \frac{1}{2^2} - \cdots - \frac{1}{n^2} \right) (-x)^n \right] dx$$

$$+ \ln 2 \sum_{n=1}^{\infty} \frac{1-(-1)^n}{n} \left(\zeta(2) - 1 - \frac{1}{2^2} - \cdots - \frac{1}{n^2} \right)$$

$$\stackrel{3.4\,C}{=} \int_0^1 \frac{\ln(1+x)}{x} \cdot \frac{-x\zeta(2) - Li_2(-x)}{1+x} dx$$

$$+ \ln 2 \sum_{n=1}^{\infty} \frac{1-(-1)^n}{n} \left(\zeta(2) - 1 - \frac{1}{2^2} - \cdots - \frac{1}{n^2} \right).$$

$$\tag{11.6}$$

Using that

$$\sum_{n=1}^{\infty} \frac{1}{n} \left(\zeta(2) - 1 - \frac{1}{2^2} - \cdots - \frac{1}{n^2} \right) = \zeta(3)$$

(see [26, problem 3.20, p. 142]) and part (a) of the problem, we have

$$\sum_{n=1}^{\infty} \frac{1-(-1)^n}{n} \left(\zeta(2) - 1 - \frac{1}{2^2} - \cdots - \frac{1}{n^2} \right) = \frac{\pi^2}{4} \ln 2. \tag{11.7}$$

We calculate

$$I = \int_0^1 \frac{\ln(1+x)}{x} \cdot \frac{-x\zeta(2) - \mathrm{Li}_2(-x)}{1+x} dx$$

$$= -\zeta(2)\frac{\ln^2 2}{2} - \int_0^1 \frac{\ln(1+x)}{x}\mathrm{Li}_2(-x)dx + \int_0^1 \frac{\ln(1+x)}{1+x}\mathrm{Li}_2(-x)dx$$

$$= -\frac{\pi^2 \ln^2 2}{12} + \frac{\mathrm{Li}_2^2(-x)}{2}\Big|_0^1 + \int_0^1 \frac{\ln(1+x)}{1+x}\mathrm{Li}_2(-x)dx$$

$$= -\frac{\pi^2 \ln^2 2}{12} + \frac{\pi^4}{288} + \int_0^1 \frac{\ln(1+x)}{1+x}\mathrm{Li}_2(-x)dx.$$

Integrating by parts, with $f(x) = \mathrm{Li}_2(-x)$, $f'(x) = -\frac{\ln(1+x)}{x}$, $g'(x) = \frac{\ln(1+x)}{1+x}$ and $g(x) = \frac{\ln^2(1+x)}{2}$, we have that

$$\int_0^1 \frac{\ln(1+x)}{1+x}\mathrm{Li}_2(-x)dx = \frac{\mathrm{Li}_2(-x)\ln^2(1+x)}{2}\Big|_0^1 + \frac{1}{2}\int_0^1 \frac{\ln^3(1+x)}{x}dx$$

$$= \frac{\mathrm{Li}_2(-1)\ln^2 2}{2} + \frac{1}{2}\int_0^1 \frac{\ln^3(1+x)}{x}dx$$

$$= \frac{\pi^2 \ln^2 2}{12} + \frac{\pi^4}{30} - \frac{\ln^4 2}{8} - \frac{21\ln 2}{8}\zeta(3) - 3\mathrm{Li}_4\left(\frac{1}{2}\right),$$

where the last equality follows based on formula (11.5).

It follows that

$$I = \frac{53\pi^4}{1440} - \frac{\ln^4 2}{8} - \frac{21\ln 2}{8}\zeta(3) - 3\mathrm{Li}_4\left(\frac{1}{2}\right). \tag{11.8}$$

Combining (11.6), (11.7), and (11.8), we have that part (c) of the problem is solved.

11.9 Remarkable Numerical and Function Series

3.82. We notice that $u'(x) = w(x)$, $v'(x) = u(x)$, and $w'(x) = v(x)$. Let

$$f(x) = u^2(x) + v^2(x) + w^2(x) - u(x)v(x) - v(x)w(x) - w(x)u(x).$$

Then

$$f'(x) = 2u(x)u'(x) + 2v(x)v'(x) + 2w(x)w'(x)$$

$$-u'(x)v(x) - u(x)v'(x) - v'(x)w(x) - v(x)w'(x) - w'(x)u(x) - w(x)u'(x)$$

$$= -f(x).$$

It follows that $f(x) = \mathscr{C}e^{-x}$ and, since $f(0) = 1$, we obtain that $f(x) = e^{-x}$.

To solve the second part of the problem we observe that

$$u(x) + v(x) + w(x) = e^x$$

and we use the identity

$$x^3 + y^3 + z^3 - 3xyz = (x + y + z)(x^2 + y^2 + z^2 - xy - yz - zx).$$

We have

$$
\begin{aligned}
&(u^3(x) + v^3(x) + w^3(x) - 3u(x)v(x)w(x))' \\
&= 3u^2(x)u'(x) + 3v^2(x)v'(x) + 3w^2(x)w'(x) \\
&\quad - 3(u'(x)v(x)w(x) + u(x)v'(x)w(x) + u(x)v(x)w'(x)) \\
&= 3u^2(x)w(x) + 3v^2(x)u(x) + 3w^2(x)v(x) \\
&\quad - 3(v(x)w^2(x) + u^2(x)w(x) + u(x)v^2(x)) \\
&= 0,
\end{aligned}
$$

and it follows that $u^3(x) + v^3(x) + w^3(x) - 3u(x)v(x)w(x) = \mathscr{C}$. When $x = 0$ we have $u(0) = 1$, $v(0) = 0$, $w(0) = 0$, and we obtain that

$$u^3(x) + v^3(x) + w^3(x) - 3u(x)v(x)w(x) = 1.$$

3.83. Part (a) can be solved by direct computation.

(b) We have, based on part (a), that

$$
\begin{aligned}
\sum_{n=1}^{\infty} \frac{1}{n} \left(\ln \frac{1}{1-x} - x - \frac{x^2}{2} - \cdots - \frac{x^n}{n} \right) &= \sum_{n=1}^{\infty} \frac{1}{n} \int_0^x \frac{t^n}{1-t} \, dt \\
&= \int_0^x \frac{1}{1-t} \left(\sum_{n=1}^{\infty} \frac{t^n}{n} \right) dt \\
&= -\int_0^x \frac{\ln(1-t)}{1-t} \, dt \\
&= \frac{\ln^2(1-x)}{2}.
\end{aligned}
$$

3.84. (a) $\ln(1-x) + \frac{x}{1-x}$; (b) $\frac{x}{2(1+x^2)} - \frac{\arctan x}{2}$.

3.85. (a) Use mathematical induction.

(b) *Solution I.* Apply Abel's summation formula with $a_n = H_n$ and $b_n = \ln \frac{1}{1-x} - x - \frac{x^2}{2} - \cdots - \frac{x^n}{n}$.

Solution II. We have, based on part (a) of Problem 3.83, that

$$
\sum_{n=1}^{\infty} H_n \left(\ln \frac{1}{1-x} - x - \frac{x^2}{2} - \cdots - \frac{x^n}{n} \right) = \sum_{n=1}^{\infty} H_n \int_0^x \frac{t^n}{1-t} dt
$$

$$
= \int_0^x \frac{1}{1-t} \left(\sum_{n=1}^{\infty} H_n t^n \right) dt
$$

$$
\overset{3.63\,(a)}{=} -\int_0^x \frac{\ln(1-t)}{(1-t)^2} dt
$$

$$
= -\frac{\ln(1-x)}{1-x} - \frac{x}{1-x}.
$$

(c) Divide by $x \neq 0$ the series in part (b) and integrate from 0 to x. Another method is based on the application of Abel's summation formula with $a_n = H_n$ and $b_n = \text{Li}_2(x) - x - \frac{x^2}{2^2} - \cdots - \frac{x^n}{n^2}$. Parts (d) and (e) are solved similarly.

3.87. (a) We have, based on part (a) of Problem 3.83, that

$$
\sum_{n=1}^{\infty} \left(\ln \frac{1}{1-x} - x - \frac{x^2}{2} - \cdots - \frac{x^n}{n} \right) \left(\zeta(2) - 1 - \frac{1}{2^2} - \cdots - \frac{1}{n^2} \right)
$$

$$
= \sum_{n=1}^{\infty} \int_0^x \frac{t^n}{1-t} dt \left(\zeta(2) - 1 - \frac{1}{2^2} - \cdots - \frac{1}{n^2} \right)
$$

$$
= \int_0^x \frac{1}{1-t} \sum_{n=1}^{\infty} \left(\zeta(2) - 1 - \frac{1}{2^2} - \cdots - \frac{1}{n^2} \right) t^n dt
$$

$$
\overset{3.4\,C}{=} \int_0^x \frac{t\zeta(2) - \text{Li}_2(t)}{(1-t)^2} dt
$$

$$
= \frac{t\zeta(2) - \text{Li}_2(t)}{1-t} \Big|_0^x - \int_0^x \left(\frac{\zeta(2)}{1-t} + \frac{\ln(1-t)}{t} + \frac{\ln(1-t)}{1-t} \right) dt
$$

$$
= \frac{x}{1-x} (\zeta(2) - \text{Li}_2(x)) + \zeta(2)\ln(1-x) + \frac{\ln^2(1-x)}{2}.
$$

(b) Let

$$
f(x) = \sum_{n=1}^{\infty} \left(\text{Li}_2(x) - x - \frac{x^2}{2^2} - \cdots - \frac{x^n}{n^2} \right) \left(\zeta(2) - 1 - \frac{1}{2^2} - \cdots - \frac{1}{n^2} \right).
$$

We have, based on part (a), that

$$f'(x) = \frac{1}{x}\sum_{n=1}^{\infty}\left(\ln\frac{1}{1-x} - x - \frac{x^2}{2} - \cdots - \frac{x^n}{n}\right)\left(\zeta(2) - 1 - \frac{1}{2^2} - \cdots - \frac{1}{n^2}\right)$$

$$= \frac{\zeta(2) - \text{Li}_2(x)}{1-x} + \zeta(2)\frac{\ln(1-x)}{x} + \frac{\ln^2(1-x)}{2x}.$$

It follows that

$$f(x) = \int_0^x \frac{\zeta(2) - \text{Li}_2(t)}{1-t}dt + \zeta(2)\int_0^x \frac{\ln(1-t)}{t}dt + \frac{1}{2}\int_0^x \frac{\ln^2(1-t)}{t}dt$$

$$= -(\zeta(2) - \text{Li}_2(t))\ln(1-t)\Big|_0^x + \int_0^x \frac{\ln^2(1-t)}{t}dt$$

$$- \zeta(2)\text{Li}_2(x) + \frac{1}{2}\int_0^x \frac{\ln^2(1-t)}{t}dt$$

$$= -\ln(1-x)(\zeta(2) - \text{Li}_2(x)) - \zeta(2)\text{Li}_2(x) + \frac{3}{2}\int_0^x \frac{\ln^2(1-t)}{t}dt.$$

3.88. See [4].

3.89. Let $x \in (-1, 1)$. We apply Abel's summation formula with $a_n = x^n$ and $b_n = f(1) - a_1 - a_2 - \cdots - a_n$ and we have, since $A_n = \frac{x - x^{n+1}}{1-x}$, that

$$\sum_{n=1}^{\infty}(f(1) - a_1 - a_2 - \cdots - a_n)x^n = \lim_{n\to\infty}\frac{x - x^{n+1}}{1-x}(f(1) - a_1 - a_2 - \cdots - a_{n+1})$$

$$+ \frac{x}{1-x}\sum_{n=1}^{\infty}(1 - x^n)a_{n+1}$$

$$= \frac{x}{1-x}\sum_{n=1}^{\infty}a_{n+1} - \frac{1}{1-x}\sum_{n=1}^{\infty}a_{n+1}x^{n+1}$$

$$= \frac{x}{1-x}(f(1) - a_1) - \frac{f(x) - a_1 x}{1-x}$$

$$= \frac{f(1)x - f(x)}{1-x}.$$

Now we consider the case $x = 1$. Since the series $\sum_{n=1}^{\infty}(f(1) - a_1 - a_2 - \cdots - a_n)$ converges, we have, based on Abel's theorem for power series, that

$$\sum_{n=1}^{\infty}(f(1)-a_1-a_2-\cdots-a_n) = \lim_{x\to1^-}\sum_{n=1}^{\infty}(f(1)-a_1-a_2-\cdots-a_n)x^n$$

$$= \lim_{x\to1^-}\frac{f(1)x-f(x)}{1-x}$$

$$= f'(1)-f(1).$$

We mention that, when $x\neq1$, the power series $\sum_{n=1}^{\infty}(f(1)-a_1-a_2-\cdots-a_n)x^n$ can be calculated by shifting the index of summation. We have

$$S = \sum_{n=1}^{\infty}(f(1)-a_1-a_2-\cdots-a_n)x^n$$

$$= (f(1)-a_1)x + \sum_{n=2}^{\infty}(f(1)-a_1-a_2-\cdots-a_n)x^n$$

$$\stackrel{n-1=i}{=} (f(1)-a_1)x + \sum_{i=1}^{\infty}(f(1)-a_1-a_2-\cdots-a_{i+1})x^{i+1}$$

$$= (f(1)-a_1)x + x\sum_{i=1}^{\infty}(f(1)-a_1-a_2-\cdots-a_i)x^i - \sum_{i=1}^{\infty}a_{i+1}x^{i+1}$$

$$= f(1)x + xS - f(x).$$

3.90. (a) Let $k\in\mathbb{N}$. We have

$$\sum_{i=1}^{k}\frac{(-1)^{i-1}}{n+i} = \sum_{i=1}^{k}(-1)^{i-1}\int_0^1 x^{n+i-1}dx = \int_0^1 x^n\sum_{i=1}^{k}(-x)^{i-1}dx = \int_0^1 x^n\frac{1-(-x)^k}{1+x}dx.$$

It follows that $\sum_{i=1}^{\infty}\frac{(-1)^{i-1}}{n+i} = \lim_{k\to\infty}\sum_{i=1}^{k}\frac{(-1)^{i-1}}{n+i} = \int_0^1\frac{x^n}{1+x}dx$, since

$$0 < \left|\int_0^1(-1)^k\frac{x^{n+k}}{1+x}dx\right| < \int_0^1 x^{n+k}dx = \frac{1}{n+k+1}.$$

Therefore,

$$\frac{1}{n} - \frac{2}{n+1} + \frac{2}{n+2} - \frac{2}{n+3} + \cdots = \int_0^1 x^{n-1}dx - \int_0^1\frac{2x^n}{1+x}dx = \int_0^1\frac{x^{n-1}(1-x)}{1+x}dx.$$

(b) We calculate the radius of convergence of the power series by the formula $R = \dfrac{1}{\limsup \sqrt[n]{|a_n|}}$. We have

$$\frac{1}{\sqrt[n]{2n(n+1)}} = \sqrt[n]{\int_0^1 \frac{x^{n-1} - x^n}{2}\,dx} < \sqrt[n]{\int_0^1 \frac{x^{n-1}(1-x)}{1+x}\,dx} < \sqrt[n]{\int_0^1 (x^{n-1} - x^n)\,dx}$$

$$= \frac{1}{\sqrt[n]{n(n+1)}}$$

and this implies that $\limsup \sqrt[n]{\int_0^1 \frac{x^{n-1}(1-x)}{1+x}\,dx} = 1$. Therefore, $R = 1$.

Let $x \in \mathbb{R}$, $|x| < 1$, and let $N \in \mathbb{N}$. We have

$$\sum_{n=1}^{N} x^n \left(\frac{1}{n} - \frac{2}{n+1} + \frac{2}{n+2} - \frac{2}{n+3} + \cdots \right) = \sum_{n=1}^{N} x^n \int_0^1 y^{n-1} \frac{1-y}{1+y}\,dy$$

$$= \int_0^1 \frac{(1-y)x}{1+y} \cdot \frac{1 - (xy)^N}{1 - xy}\,dy = \int_0^1 \frac{(1-y)x}{(1+y)(1-xy)}\,dy - \int_0^1 \frac{(1-y)x(xy)^N}{(1+y)(1-xy)}\,dy.$$

Passing to the limit, when $N \to \infty$, we get that

$$\sum_{n=1}^{\infty} x^n \left(\frac{1}{n} - \frac{2}{n+1} + \frac{2}{n+2} - \frac{2}{n+3} + \cdots \right) = \int_0^1 \frac{(1-y)x}{(1+y)(1-xy)}\,dy,$$

since

$$0 < \left| \int_0^1 \frac{(1-y)x(xy)^N}{(1+y)(1-xy)}\,dy \right| \le \frac{1}{1-|x|} \int_0^1 y^N\,dy = \frac{1}{N+1} \cdot \frac{1}{1-|x|} \to 0 \quad (N \to \infty).$$

On the other hand,

$$\int_0^1 \frac{(1-y)x}{(1+y)(1-xy)}\,dy = \frac{x}{1+x} \int_0^1 \left(\frac{2}{1+y} - \frac{1}{1-xy} + \frac{x}{1-xy} \right) dy$$

$$= \frac{x}{1+x} \left(2\ln 2 + \frac{\ln(1-x)}{x} - \ln(1-x) \right)$$

$$= \frac{2x\ln 2 + (1-x)\ln(1-x)}{1+x}.$$

Case $x = -1$. Let $a_n = \frac{1}{n} - \frac{2}{n+1} + \frac{2}{n+2} - \frac{2}{n+3} + \cdots$. We have, based on part (a), that $a_{n-1} = \int_0^1 \frac{y^{n-2}(1-y)}{1+y}\,dy > \int_0^1 \frac{y^{n-1}(1-y)}{1+y}\,dy = a_n$, which implies that the series $\sum_{n=1}^{\infty} (-1)^n a_n$ is a Leibniz series, hence it converges. It follows, based on

Abel's theorem for power series, that

$$\sum_{n=1}^{\infty}(-1)^n a_n = \lim_{x\to-1}\sum_{n=1}^{\infty}x^n\left(\frac{1}{n}-\frac{2}{n+1}+\frac{2}{n+2}-\frac{2}{n+3}+\cdots\right)$$

$$= \lim_{x\to-1}\frac{2x\ln 2+(1-x)\ln(1-x)}{1+x}$$

$$= \ln 2 - 1.$$

Case $x = 1$. The series formula can be proved either by using part (a) or by observing that $S_{2n} = 1 - \frac{1}{2} + \frac{1}{3} - \cdots - \frac{1}{2n}$, $n \geq 1$.

(c) This part is solved similarly to part (d).

(d) We have, based on part (a), that

$$S = \sum_{n=1}^{\infty}(-1)^n\left(\frac{1}{n}-\frac{2}{n+1}+\frac{2}{n+2}-\frac{2}{n+3}+\cdots\right)^2$$

$$= \sum_{n=1}^{\infty}(-1)^n\int_0^1 x^{n-1}\frac{1-x}{1+x}dx\int_0^1 y^{n-1}\frac{1-y}{1+y}dy$$

$$= -\int_0^1\int_0^1\frac{(1-x)(1-y)}{(1+x)(1+y)}\sum_{n=1}^{\infty}(-xy)^{n-1}dxdy$$

$$= -\int_0^1\int_0^1\frac{(1-x)(1-y)}{(1+x)(1+y)(1+xy)}dxdy$$

$$= -\int_0^1\frac{1-x}{1+x}\left(\int_0^1\frac{1-y}{(1+y)(1+xy)}dy\right)dx$$

$$= -\int_0^1\frac{1-x}{1+x}\left[\int_0^1\frac{1-y}{1-x}\left(\frac{1}{1+y}-\frac{x}{1+xy}\right)dy\right]dx$$

$$= -\int_0^1\frac{1}{1+x}\left(\int_0^1\frac{1-y}{1+y}dy-x\int_0^1\frac{1-y}{1+xy}dy\right)dx$$

$$= -\int_0^1\frac{1}{1+x}\left(2\ln 2-1-x\int_0^1\frac{1-y}{1+xy}dy\right)dx.$$

A calculation shows that

$$x\int_0^1\frac{1-y}{1+xy}dy = \frac{1+x}{x}\ln(1+x)-1.$$

It follows that

$$S = -\int_0^1 \frac{1}{1+x} \left(2\ln 2 - \frac{1+x}{x} \ln(1+x) \right) dx$$

$$= -2\ln^2 2 + \int_0^1 \frac{\ln(1+x)}{x} dx$$

$$= \frac{\pi^2}{12} - 2\ln^2 2.$$

3.92. We have

$$S = \sum_{n=1}^{\infty} \left(\frac{1}{n^2} - \frac{1}{(n+1)^2} + \frac{1}{(n+2)^2} - \cdots \right)^2$$

$$= \left(\frac{1}{1^2} - \frac{1}{2^2} + \frac{1}{3^2} - \cdots \right)^2$$

$$+ \sum_{n=2}^{\infty} \left(\frac{1}{n^2} - \frac{1}{(n+1)^2} + \frac{1}{(n+2)^2} - \cdots \right)^2$$

$$\stackrel{n-1=i}{=} \frac{\pi^4}{144} + \sum_{i=1}^{\infty} \left(\frac{1}{(i+1)^2} - \frac{1}{(i+2)^2} + \frac{1}{(i+3)^2} - \cdots \right)^2$$

$$= \frac{\pi^4}{144} + \sum_{i=1}^{\infty} \left[\frac{1}{i^2} - \left(\frac{1}{i^2} - \frac{1}{(i+1)^2} + \frac{1}{(i+2)^2} - \frac{1}{(i+3)^2} + \cdots \right) \right]^2$$

$$= \frac{\pi^4}{144} + \frac{\pi^4}{90} - 2\sum_{i=1}^{\infty} \frac{1}{i^2} \left(\frac{1}{i^2} - \frac{1}{(i+1)^2} + \frac{1}{(i+2)^2} - \cdots \right) + S,$$

and it follows that

$$\sum_{i=1}^{\infty} \frac{1}{i^2} \left(\frac{1}{i^2} - \frac{1}{(i+1)^2} + \frac{1}{(i+2)^2} - \cdots \right) = \frac{1}{2} \left(\frac{\pi^4}{144} + \frac{\pi^4}{90} \right) = \frac{13}{1440} \pi^4.$$

3.93. (a) Let $x \in (-1, 1)$ and let

$$f(x) = \sum_{n=1}^{\infty} \left(\ln \frac{1}{1-x} - x - \frac{x^2}{2} - \cdots - \frac{x^n}{n} \right) \left(\ln \frac{1}{2} - (-1) - \frac{(-1)^2}{2} - \cdots - \frac{(-1)^n}{n} \right).$$

We have

$$f'(x) = \sum_{n=1}^{\infty} \left(\frac{1}{1-x} - 1 - x - \cdots - x^{n-1} \right) \left(\ln \frac{1}{2} - (-1) - \frac{(-1)^2}{2} - \cdots - \frac{(-1)^n}{n} \right)$$

$$= \sum_{n=1}^{\infty} \frac{x^n}{1-x} \left(\ln \frac{1}{2} - (-1) - \frac{(-1)^2}{2} - \cdots - \frac{(-1)^n}{n} \right)$$

$$\underset{3.88 \ (c)}{=} \frac{\ln(1+x) - x \ln 2}{(1-x)^2}.$$

It follows that

$$f(x) = \int_0^x \frac{\ln(1+t) - t \ln 2}{(1-t)^2} dt$$

$$= \frac{\ln(1+t) - t \ln 2}{1-t} \Big|_0^x - \int_0^x \left(\frac{1}{1-t^2} - \frac{\ln 2}{1-t} \right) dt$$

$$= \frac{\ln(1+x) - x \ln 2}{1-x} + \frac{1}{2} \ln \frac{1-x}{1+x} - \ln 2 \ln(1-x).$$

Letting $x \to -1$, we get that

$$\sum_{n=1}^{\infty} \left(\ln \frac{1}{2} - (-1) - \frac{(-1)^2}{2} - \cdots - \frac{(-1)^n}{n} \right)^2$$

$$= \lim_{x \to -1} \left[\frac{\ln(1+x) - x \ln 2}{1-x} + \frac{1}{2} \ln \frac{1-x}{1+x} - \ln 2 \ln(1-x) \right]$$

$$= \ln 2 - \ln^2 2.$$

The case when $x = -1$ also follows from Problem 3.94.

(b) Let $x \in (-1, 1)$ and let $S(x)$ be the function

$$\sum_{n=1}^{\infty} (-1)^n \left(\ln \frac{1}{1-x} - x - \frac{x^2}{2} - \cdots - \frac{x^n}{n} \right) \left(\ln \frac{1}{2} - (-1) - \frac{(-1)^2}{2} - \cdots - \frac{(-1)^n}{n} \right).$$

We have that $S'(x)$ is equal to

$$\sum_{n=1}^{\infty}(-1)^n\left(\frac{1}{1-x}-1-x-\cdots-x^{n-1}\right)\left(\ln\frac{1}{2}-(-1)-\frac{(-1)^2}{2}-\cdots-\frac{(-1)^n}{n}\right)$$

$$=\sum_{n=1}^{\infty}(-1)^n\frac{x^n}{1-x}\left(\ln\frac{1}{2}-(-1)-\frac{(-1)^2}{2}-\cdots-\frac{(-1)^n}{n}\right)$$

$$\underset{\underset{=}{3.88\,(c)}}{}\frac{\ln(1-x)+x\ln 2}{1-x^2}.$$

It follows that

$$S(x)=\int_0^x\frac{\ln(1-t)+t\ln 2}{1-t^2}dt$$

$$=\frac{1}{2}\int_0^x(\ln(1-t)+t\ln 2)\left(\frac{1}{1-t}+\frac{1}{1+t}\right)dt$$

$$=\frac{1}{2}\left(-\frac{1}{2}\ln^2(1-x)+\int_0^x\frac{\ln(1-t)}{1+t}dt-\ln 2\ln(1-x^2)\right).$$

We calculate the preceding integral. We have

$$\int_0^x\frac{\ln(1-t)}{1+t}dt\overset{1+t=y}{=}\int_1^{1+x}\frac{\ln(2-y)}{y}dy$$

$$=\int_1^{1+x}\frac{\ln 2+\ln\left(1-\frac{y}{2}\right)}{y}dy$$

$$\overset{y=2u}{=}\ln 2\ln(1+x)+\int_{\frac{1}{2}}^{\frac{1+x}{2}}\frac{\ln(1-u)}{u}du$$

$$=\ln 2\ln(1+x)+\text{Li}_2\left(\frac{1}{2}\right)-\text{Li}_2\left(\frac{1+x}{2}\right).$$

Putting all these together, we get that

$$S(x)=-\frac{\ln^2(1-x)}{4}+\frac{1}{2}\text{Li}_2\left(\frac{1}{2}\right)-\frac{1}{2}\text{Li}_2\left(\frac{1+x}{2}\right)-\frac{\ln 2\ln(1-x)}{2}.$$

Letting $x \to -1$, we get that

$$\sum_{n=1}^{\infty} (-1)^n \left(\ln \frac{1}{2} - (-1) - \frac{(-1)^2}{2} - \cdots - \frac{(-1)^n}{n} \right)^2$$

$$= \lim_{x \to -1} \left[-\frac{\ln^2(1-x)}{4} + \frac{1}{2}\text{Li}_2\left(\frac{1}{2}\right) - \frac{1}{2}\text{Li}_2\left(\frac{1+x}{2}\right) - \frac{\ln 2 \ln(1-x)}{2} \right]$$

$$= \frac{\pi^2}{24} - \ln^2 2.$$

3.94. We have, based on part (a) of Problem 3.83, that

$$S = \sum_{n=1}^{\infty} \left(\ln \frac{1}{1-x} - x - \frac{x^2}{2} - \cdots - \frac{x^n}{n} \right) \left(\ln \frac{1}{1-y} - y - \frac{y^2}{2} - \cdots - \frac{y^n}{n} \right)$$

$$= \sum_{n=1}^{\infty} \int_0^x \frac{t^n}{1-t} dt \int_0^y \frac{u^n}{1-u} du$$

$$= \int_0^x \int_0^y \frac{1}{(1-t)(1-u)} \sum_{n=1}^{\infty} (ut)^n dt\, du$$

$$= \int_0^x \int_0^y \frac{ut}{(1-t)(1-u)(1-ut)} dt\, du$$

$$= \int_0^x \frac{t}{1-t} \left(\int_0^y \frac{u}{(1-u)(1-tu)} du \right) dt.$$

A calculation shows that

$$\int_0^y \frac{u}{(1-u)(1-tu)} du = \int_0^y \frac{1}{1-t} \left(\frac{u}{1-u} - \frac{ut}{1-ut} \right) du$$

$$= \frac{1}{1-t} \int_0^y \left(\frac{1}{1-u} - \frac{1}{1-ut} \right) du$$

$$= \frac{1}{1-t} \left(-\ln(1-y) + \frac{1}{t}\ln(1-ty) \right).$$

It follows that

$$S = \int_0^x \frac{t}{(1-t)^2} \left(-\ln(1-y) + \frac{1}{t} \ln(1-ty) \right) dt$$

$$= -\ln(1-y) \int_0^x \frac{t}{(1-t)^2} dt + \int_0^x \frac{\ln(1-ty)}{(1-t)^2} dt$$

$$= -\ln(1-y) \left(\frac{x}{1-x} + \ln(1-x) \right) + \int_0^x \frac{\ln(1-ty)}{(1-t)^2} dt$$

$$= -\ln(1-x)\ln(1-y) - \frac{x}{1-x} \ln(1-y)$$

$$- \frac{y}{1-y} \ln(1-x) + \frac{1-xy}{(1-x)(1-y)} \ln(1-xy).$$

The last integral has been calculated by parts as follows:

$$\int_0^x \frac{\ln(1-ty)}{(1-t)^2} dt = \frac{\ln(1-yt)}{1-t} \Big|_0^x + y \int_0^x \frac{1}{(1-ty)(1-t)} dt$$

$$= \frac{\ln(1-xy)}{1-x} + \frac{y}{1-y} \int_0^x \left(\frac{1}{1-t} - \frac{y}{1-ty} \right) dt$$

$$= \frac{1-xy}{(1-x)(1-y)} \ln(1-xy) - \frac{y}{1-y} \ln(1-x).$$

11.10 Multiple Series with the Riemann Zeta Function

3.95. We have

$$\sum_{n=1}^{\infty} \sum_{m=1}^{\infty} \frac{(-1)^{n+m}}{nm(n+m)} = \sum_{n=1}^{\infty} \sum_{m=1}^{\infty} \frac{(-1)^{n+m}}{nm} \int_0^1 x^{n+m-1} dx$$

$$= \int_0^1 \frac{1}{x} \sum_{n=1}^{\infty} \frac{(-x)^n}{n} \sum_{m=1}^{\infty} \frac{(-x)^m}{m} dx$$

$$= \int_0^1 \frac{\ln^2(1+x)}{x} dx$$

$$= \frac{\zeta(3)}{4}.$$

The last integral is calculated in [6, pp. 291–292].
3.96. (a) We have

$$\sum_{i=1}^{\infty}\sum_{j=1}^{\infty}(\zeta(i+j)-1)=\sum_{i=1}^{\infty}\sum_{j=1}^{\infty}\sum_{k=2}^{\infty}\frac{1}{k^{i+j}}=\sum_{k=2}^{\infty}\sum_{i=1}^{\infty}\frac{1}{k^i}\sum_{j=1}^{\infty}\frac{1}{k^j}=\sum_{k=2}^{\infty}\frac{1}{(k-1)^2}=\zeta(2).$$

Part (b) is solved similarly.
3.97. (a) We have, based on symmetry reasons, that

$$\sum_{i=1}^{\infty}\sum_{j=1}^{\infty}i\frac{\zeta(i+j)-1}{i+j}=\sum_{i=1}^{\infty}\sum_{j=1}^{\infty}j\frac{\zeta(i+j)-1}{i+j}$$

and it follows, based on part (a) of Problem 3.96, that

$$\sum_{i=1}^{\infty}\sum_{j=1}^{\infty}i\frac{\zeta(i+j)-1}{i+j}=\frac{1}{2}\sum_{i=1}^{\infty}\sum_{j=1}^{\infty}(\zeta(i+j)-1)=\frac{\zeta(2)}{2}.$$

Part (b) is solved similarly.
3.98. (b) We have

$$\sum_{i=1}^{\infty}\sum_{j=1}^{\infty}\left[2^{2i+j}\left(\zeta(2i+j)-1\right)-1\right]=\sum_{i=1}^{\infty}\sum_{j=1}^{\infty}\left(2^{2i+j}\sum_{n=2}^{\infty}\frac{1}{n^{2i+j}}-1\right)$$

$$=\sum_{i=1}^{\infty}\sum_{j=1}^{\infty}\sum_{n=3}^{\infty}\left(\frac{2}{n}\right)^{2i+j}$$

$$=\sum_{n=3}^{\infty}\sum_{i=1}^{\infty}\left(\frac{2}{n}\right)^{2i}\sum_{j=1}^{\infty}\left(\frac{2}{n}\right)^{j}$$

$$=8\sum_{n=3}^{\infty}\frac{1}{(n^2-4)(n-2)}$$

$$=\frac{\pi^2}{3}-\frac{25}{24}.$$

Part (a) is solved similarly.

3.99. Part (a) follows from part (b), with $k = 1$ and $n = 2$.

(b) We have

$$\sum_{i_1=1}^{\infty} \cdots \sum_{i_n=1}^{\infty} i_1 \cdots i_k \left(\zeta(i_1 + \cdots + i_n) - 1 \right)$$

$$= \sum_{i_1=1}^{\infty} \cdots \sum_{i_n=1}^{\infty} i_1 \cdots i_k \sum_{m=2}^{\infty} \frac{1}{m^{i_1 + \cdots + i_n}}$$

$$= \sum_{m=2}^{\infty} \sum_{i_1=1}^{\infty} \frac{i_1}{m^{i_1}} \cdots \sum_{i_k=1}^{\infty} \frac{i_k}{m^{i_k}} \sum_{i_{k+1}=1}^{\infty} \frac{1}{m^{i_{k+1}}} \cdots \sum_{i_n=1}^{\infty} \frac{1}{m^{i_n}}$$

$$= \sum_{m=2}^{\infty} \left[\frac{\frac{1}{m}}{\left(1 - \frac{1}{m}\right)^2} \right]^k \left(\frac{\frac{1}{m}}{1 - \frac{1}{m}} \right)^{n-k}$$

$$= \sum_{m=2}^{\infty} \frac{m^k}{(m-1)^{n+k}}$$

$$= \sum_{m=2}^{\infty} \frac{1}{(m-1)^{n+k}} \sum_{j=0}^{k} C_k^j (m-1)^{k-j}$$

$$= \sum_{j=0}^{k} C_k^j \sum_{m=2}^{\infty} \frac{1}{(m-1)^{j+n}}$$

$$= \sum_{j=0}^{k} C_k^j \zeta(j + n).$$

3.100. See the solution of Problem 3.99.

3.101. (b) We have

$$\sum_{i_1=1}^{\infty} \cdots \sum_{i_k=1}^{\infty} \left[2^{i_1+\cdots+i_k} (\zeta(i_1 + \cdots + i_k) - 1) - 1 \right]$$

$$= \sum_{i_1=1}^{\infty} \cdots \sum_{i_k=1}^{\infty} \left(2^{i_1+\cdots+i_k} \sum_{n=2}^{\infty} \frac{1}{n^{i_1+\cdots+i_k}} - 1 \right)$$

$$= \sum_{i_1=1}^{\infty} \cdots \sum_{i_k=1}^{\infty} \sum_{n=3}^{\infty} \left(\frac{2}{n} \right)^{i_1+\cdots+i_k}$$

$$= \sum_{n=3}^{\infty} \sum_{i_1=1}^{\infty} \left(\frac{2}{n} \right)^{i_1} \cdots \sum_{i_k=1}^{\infty} \left(\frac{2}{n} \right)^{i_k}$$

$$= \sum_{n=3}^{\infty} \left(\frac{\frac{2}{n}}{1 - \frac{2}{n}} \right)^{k}$$

$$- \sum_{n=3}^{\infty} \frac{2^k}{(n-2)^k}$$

$$= 2^k \zeta(k).$$

3.102. (a) We have

$$\sum_{i=1}^{\infty} \sum_{j=1}^{\infty} \left[3^{i+j} \left(\zeta(i+j) - 1 - \frac{1}{2^{i+j}} \right) - 1 \right] = \sum_{i=1}^{\infty} \sum_{j=1}^{\infty} \left(3^{i+j} \sum_{n=3}^{\infty} \frac{1}{n^{i+j}} - 1 \right)$$

$$= \sum_{i=1}^{\infty} \sum_{j=1}^{\infty} \sum_{n=4}^{\infty} \left(\frac{3}{n} \right)^{i+j}$$

$$= \sum_{n=4}^{\infty} \sum_{i=1}^{\infty} \left(\frac{3}{n} \right)^{i} \sum_{j=1}^{\infty} \left(\frac{3}{n} \right)^{j}$$

$$= \sum_{n=4}^{\infty} \left(\frac{\frac{3}{n}}{1 - \frac{3}{n}} \right)^{2}$$

$$= \sum_{n=4}^{\infty} \frac{9}{(n-3)^2}$$

$$= 9\zeta(2).$$

Part (b) is solved similarly.

The challenge. The sum of the series is $(n+1)^n \zeta(n)$.

3.103. (b) We have

$$\sum_{i=1}^{\infty}\sum_{j=1}^{\infty}(-1)^{i-1}\left[2^{i+j}(\zeta(i+j)-1)-1\right] = \sum_{i=1}^{\infty}\sum_{j=1}^{\infty}(-1)^{i-1}\left(2^{i+j}\sum_{n=2}^{\infty}\frac{1}{n^{i+j}}-1\right)$$

$$= \sum_{i=1}^{\infty}\sum_{j=1}^{\infty}(-1)^{i-1}\sum_{n=3}^{\infty}\left(\frac{2}{n}\right)^{i+j}$$

$$= -\sum_{n=3}^{\infty}\sum_{i=1}^{\infty}\left(-\frac{2}{n}\right)^{i}\sum_{j=1}^{\infty}\left(\frac{2}{n}\right)^{j}$$

$$= \sum_{n=3}^{\infty}\frac{4}{n^2-4}$$

$$= \frac{25}{12}.$$

Part (a) is solved similarly.

3.104. (b) We have

$$\sum_{i_1=1}^{\infty}\cdots\sum_{i_k=1}^{\infty}(-1)^{i_1+\cdots+i_k}\left(\zeta(i_1+\cdots+i_k)-1\right)$$

$$= \sum_{i_1=1}^{\infty}\cdots\sum_{i_k=1}^{\infty}(-1)^{i_1+\cdots+i_k}\sum_{n=2}^{\infty}\frac{1}{n^{i_1+\cdots+i_k}}$$

$$= \sum_{n=2}^{\infty}\sum_{i_1=1}^{\infty}\left(-\frac{1}{n}\right)^{i_1}\cdots\sum_{i_k=1}^{\infty}\left(-\frac{1}{n}\right)^{i_k}$$

$$= \sum_{n=2}^{\infty}\left(\frac{-\frac{1}{n}}{1+\frac{1}{n}}\right)^{k}$$

$$= \sum_{n=2}^{\infty}\frac{(-1)^k}{(n+1)^k}$$

$$= (-1)^k\left(\zeta(k)-1-\frac{1}{2^k}\right).$$

11.11   Series Involving Products of Harmonic Numbers

3.105. *Solution I.* (a) See Problem 7.61.

Solution II. We use the formula (see [26, entry (3.17), p. 206])

$$\int_0^1 x^{n-1} \ln(1-x)dx = -\frac{H_n}{n}, \quad n \geq 1,$$

and the power series

$$\sum_{n=1}^{\infty} \frac{H_n}{n} x^n = \text{Li}_2(x) + \frac{1}{2}\ln^2(1-x), \quad x \in [-1, 1). \tag{11.9}$$

The power series formula can be proved by dividing by x the power series of part (a) of Problem 3.63 and then integrating the series from 0 to x.

(a) We have

$$\sum_{n=1}^{\infty} \frac{H_n}{n} \cdot \frac{H_{n+1}}{n+1} = -\sum_{n=1}^{\infty} \frac{H_n}{n} \int_0^1 x^n \ln(1-x)dx$$

$$= -\int_0^1 \ln(1-x)\left(\sum_{n=1}^{\infty} \frac{H_n}{n} x^n\right) dx$$

$$= -\int_0^1 \ln(1-x)\left(\text{Li}_2(x) + \frac{1}{2}\ln^2(1-x)\right) dx$$

$$= -\int_0^1 \ln(1-x)\text{Li}_2(x)dx + 3.$$

We calculate the preceding integral by parts, and we have that

$$\int_0^1 \ln(1-x)\text{Li}_2(x)dx = -(1-x)\left[\ln(1-x) - 1\right]\text{Li}_2(x)\Big|_0^1$$

$$-\int_0^1 \frac{1-x}{x}\left[\ln^2(1-x) - \ln(1-x)\right]dx$$

$$= -\int_0^1 \frac{1-x}{x}\left[\ln^2(1-x) - \ln(1-x)\right]dx$$

$$= -\int_0^1 \frac{\ln^2(1-x)}{x}dx + \int_0^1 \ln^2(1-x)dx$$

$$+ \int_0^1 \frac{\ln(1-x)}{x}dx - \int_0^1 \ln(1-x)dx$$

$$= -2\zeta(3) - \zeta(2) + 3.$$

Putting all these together, we get that part (a) of the problem is solved.

We mention that the series in part (a) of Problem 3.105, as well as a generalization, are calculated by a different method in [25].

(b) We have

$$
\sum_{n=1}^{\infty} (-1)^n \frac{H_n}{n} \cdot \frac{H_{n+1}}{n+1} = -\sum_{n=1}^{\infty} (-1)^n \frac{H_n}{n} \int_0^1 x^n \ln(1-x) dx
$$

$$
= -\int_0^1 \ln(1-x) \left(\sum_{n=1}^{\infty} \frac{H_n}{n} (-x)^n \right) dx
$$

$$
= -\int_0^1 \ln(1-x) \left(\mathrm{Li}_2(-x) + \frac{1}{2} \ln^2(1+x) \right) dx
$$

$$
= -\int_0^1 \ln(1-x) \mathrm{Li}_2(-x) dx - \frac{1}{2} \int_0^1 \ln(1-x) \ln^2(1+x) dx.
$$
(11.10)

We calculate the first integral by parts and we have

$$
\int_0^1 \ln(1-x) \mathrm{Li}_2(-x) dx = -(1-x) \left[\ln(1-x) - 1 \right] \mathrm{Li}_2(-x) \Big|_0^1
$$

$$
- \int_0^1 \frac{1-x}{x} \ln(1+x) \left[\ln(1-x) - 1 \right] dx
$$

$$
= -\int_0^1 \frac{1-x}{x} \ln(1+x) \left[\ln(1-x) - 1 \right] dx
$$

$$
= -\int_0^1 \frac{\ln(1-x) \ln(1+x)}{x} dx + \int_0^1 \frac{\ln(1+x)}{x} dx
$$

$$
+ \int_0^1 \ln(1-x) \ln(1+x) dx - \int_0^1 \ln(1+x) dx
$$

$$
= \frac{5}{8} \zeta(3) - \frac{\pi^2}{12} + \ln^2 2 - 4 \ln 2 + 3.
$$
(11.11)

We calculate the second integral as follows:

$$
\int_0^1 \ln(1-x) \ln^2(1+x) dx = \frac{1}{3} \int_0^1 \ln^3 \left(\frac{1-x}{1+x} \right) dx - \frac{1}{3} \int_0^1 \ln^3(1-x) dx
$$

$$
+ \int_0^1 \ln^2(1-x) \ln(1+x) dx + \frac{1}{3} \int_0^1 \ln^3(1+x) dx
$$

$$
= \frac{\zeta(3)}{2} + 8 \ln 2 - 4 \ln^2 2 + \frac{4}{3} \ln^3 2 - \frac{\pi^2 \ln 2}{3} - 6 + \frac{\pi^2}{3}.
$$
(11.12)

We used that

$$\int_0^1 \ln^3\left(\frac{1-x}{1+x}\right)dx \overset{x=\frac{1-t}{1+t}}{=} 2\int_0^1 \frac{\ln^3 t}{(1+t)^2}dt = -9\zeta(3)$$

and

$$\int_0^1 \ln^2(1-x)\ln(1+x)dx = -6+\frac{\pi^2}{3}+4\ln 2-2\ln^2 2+\frac{7}{2}\zeta(3)-\frac{\pi^2\ln 2}{3}+\frac{2\ln^3 2}{3}.$$

We sketch below the calculation of the preceding integral. We have

$$\int_0^1 \ln^2(1-x)\ln(1+x)dx = \int_0^1 \ln^2 t \ln(2-t)dt$$

$$= \int_0^1 \ln^2 t\left[\ln 2 + \ln\left(1-\frac{t}{2}\right)\right]dt$$

$$= 2\ln 2 - \sum_{n=1}^{\infty}\frac{1}{2^n n}\int_0^1 t^n \ln^2 t\, dt$$

$$= 2\ln 2 - 2\sum_{n=1}^{\infty}\frac{1}{2^n n(n+1)^3}.$$

The calculation of the last series, which can be expressed in terms of $\text{Li}_2\left(\frac{1}{2}\right)$ and $\text{Li}_3\left(\frac{1}{2}\right)$, is left as a challenge to the interested reader.

Combining (11.10), (11.11), and (11.12), part (b) of the problem is solved.

3.106. Use that

$$\frac{H_n}{n^2}\cdot\frac{H_{n+1}}{n+1} = \frac{H_n^2}{n^2}+\frac{H_n}{n^2}-\frac{H_n}{n}+\frac{H_{n+1}}{n+1}-\frac{1}{(n+1)^2}-\frac{H_n H_{n+1}}{n(n+1)}.$$

3.108. Subtract the series in parts (d) and (b) of Problem 3.107.

Derivatives and Applications 12

12.1 Apéritif

4.1. $J = (-2, \infty)$. Since f is strictly increasing we get that f is injective. We also have that f is surjective and it follows that f is bijective. Since $f(2) = 2$ we have that $g'(2) = \frac{1}{f'(2)} = \frac{1}{9}$. From $f(x) = x^3 - 3x$, $x \in I$, we can write $g^3(y) - 3g(y) = y$, $y \in J$. Taking successive derivatives, one has that $3g^2(y)g'(y) - 3g'(y) = 1$, $6g(y)g'^2(y) + 3g^2(y) \cdot g''(y) - 3g''(y) = 0$, $y \in J$. When $y = 2$ we obtain that $6 \cdot 2 \cdot \frac{1}{81} + 3 \cdot 4 \cdot g''(2) - 3g''(2) = 0$. Therefore, $g''(2) = -\frac{4}{243}$.

4.2. We have

$$\frac{f(x+y) - f(x)}{y} = a \Rightarrow \lim_{y \to 0} \frac{f(x+y) - f(x)}{y} = a,$$

and it follows that $f'(x) = a$, which implies $f(x) = ax + b$, $b \in \mathbb{R}$.

4.3. $f(x) = \frac{1}{2}e^{2x} + 1$.

4.4. $g'(0) = (2021!)^3$.

4.5. The equation of the tangent to the parabola at the point (x_0, y_0) is given by $y - ax_0^2 = 2ax_0(x - x_0)$, and it follows that $x_1 = \frac{x_0}{2}$. Let A_1 be the area of the domain bounded by the tangent to the parabola at (x_0, y_0), parabola and the vertical line $x = \frac{x_0}{2}$, and let A_2 be the area of the domain bounded by the parabola, the x axis and the vertical line $x = \frac{x_0}{2}$. We have

$$A_1 = \int_{\frac{x_0}{2}}^{x_0} \left(ax^2 - 2ax_0(x - x_0) - ax_0^2 \right) dx = \frac{ax_0^3}{24}$$

and

$$A_2 = \int_0^{\frac{x_0}{2}} ax^2 dx = \frac{ax_0^3}{24}.$$

A. Sîntămărian, O. Furdui, *Sharpening Mathematical Analysis Skills*, Problem Books in Mathematics, https://doi.org/10.1007/978-3-030-77139-3_12

Remark. More generally, if f is a twice differentiable function such that f'' is continuous and positive for all x, then the graph of f is a parabola if and only if the area of the domains bounded by the graph of f, the tangents to the graph of f at two distinct points on the graph, and the vertical line that passes through the intersection point of the two tangents are equal [152].

4.6. It suffices to prove that $\ln(1+x-x^2) \geq (1-x)\ln(1+x)$, for all $x \in [0, 1]$. Let $f : [0, 1] \to \mathbb{R}$, $f(y) = \ln(1 + y)$. We have $f''(y) = -\frac{1}{(1+y)^2} < 0$, and it follows that f is concave down. We have, based on the definition of the concavity of f, that $f(\alpha \cdot x + (1 - \alpha) \cdot y) \geq \alpha f(x) + (1 - \alpha) f(y)$, for all $x, y, \alpha \in [0, 1]$. Therefore, $\ln(1 + x(1 - x)) = \ln(1 + (1 - x)x + x \cdot 0) \geq (1 - x)\ln(1 + x) + x\ln(1 + 0) = (1 - x)\ln(1 + x)$.

Remark. The same idea can be used for solving Problem 3.44. It suffices to prove that $\ln(x^2 - x + 1) \geq x \ln x$, for all $x \in (0, 1]$. Let $f : (0, 1] \to \mathbb{R}$, $f(x) = \ln x$. We have $f''(x) = -\frac{1}{x^2} < 0$, and it follows that f is concave down. This implies, based on the definition of the concavity of f, that $f(\alpha \cdot x + (1 - \alpha) \cdot y) \geq \alpha f(x) + (1 - \alpha) f(y)$, for all $x, y \in (0, 1]$ and $\alpha \in [0, 1]$. Therefore, $\ln(x^2 - x + 1) = \ln(x \cdot x + (1 - x) \cdot 1) \geq x \ln x + (1 - x) \ln 1 = x \ln x$.

4.7.

(a) Let $S_1(x) = 1 + x + 2x^2 + 3x^3 + \cdots + nx^n$, $x \in \mathbb{R}$. Using the formula

$$1 + x + x^2 + \cdots + x^n = \frac{1 - x^{n+1}}{1 - x}, \quad x \neq 1,$$

we obtain by differentiation that

$$1 + 2x + 3x^2 + \cdots + nx^{n-1} = \left(\frac{1 - x^{n+1}}{1 - x}\right)'.$$

It follows that

$$S_1(x) = 1 + x(1 + 2x + 3x^2 + \cdots + nx^{n-1}) = 1 + x\left(\frac{1 - x^{n+1}}{1 - x}\right)'$$

$$= 1 + x \cdot \frac{nx^{n+1} - (n + 1)x^n + 1}{(1 - x)^2}, \quad x \neq 1.$$

When $x = 1$ we have $S_1(1) = 1 + 1 + 2 + \cdots + n = 1 + \frac{n(n+1)}{2} = \frac{n^2+n+2}{2}$.

(b) Let $S_2(x) = 1 + 2^2x + 3^2x^2 + \cdots + n^2x^{n-1}$, $x \in \mathbb{R}$. We have

$$S_2(x) = S_1'(x) = \left(1 + x \cdot \frac{nx^{n+1} - (n+1)x^n + 1}{(1-x)^2}\right)'$$

$$= \frac{-n^2 x^{n+2} + (2n^2 + 2n - 1)x^{n+1} - (n+1)^2 x^n + x + 1}{(1-x)^3}, \qquad x \neq 1.$$

When $x = 1$ we obtain that $S_2(1) = 1 + 2^2 + 3^2 + \cdots + n^2 = \frac{n(n+1)(2n+1)}{6}$.

(c) Let $S_3(x) = 2 + 3 \cdot 2x + 4 \cdot 3x^2 + \cdots + n(n-1)x^{n-2}$, $x \in \mathbb{R}$. We have that

$$S_3(x) = (1 + 2x + 3x^2 + \cdots + nx^{n-1})' = \left(\frac{nx^{n+1} - (n+1)x^n + 1}{(1-x)^2}\right)'$$

$$= \frac{-n(n-1)x^{n+1} + 2(n^2 - 1)x^n - n(n+1)x^{n-1} + 2}{(1-x)^3}, \qquad x \neq 1.$$

When $x = 1$ we have

$$S_3(1) = 2 + 2 \cdot 3 + 3 \cdot 4 + \cdots + (n-1) \cdot n = \sum_{k=2}^{n}(k-1)k = \sum_{k=2}^{n}k^2 - \sum_{k=2}^{n}k$$

$$= \frac{n(n+1)(2n+1)}{6} - 1 - \left(\frac{n(n+1)}{2} - 1\right)$$

$$= \frac{(n-1)n(n+1)}{3}.$$

4.8.

(a) The first inequality is equivalent to

$$\frac{e}{2n+2} < e - \left(1 + \frac{1}{n}\right)^n \Leftrightarrow \left(1 + \frac{1}{2n+1}\right)\left(1 + \frac{1}{n}\right)^n < e.$$

Let $f : [1, \infty) \to \mathbb{R}$ be the function defined by

$$f(x) = \left(1 + \frac{1}{2x+1}\right)\left(1 + \frac{1}{x}\right)^x.$$

We have

$$f'(x) = \left(1 + \frac{1}{x}\right)^x \cdot \frac{2x+2}{2x+1} \cdot \left(-\frac{2}{2x+1} + \ln\frac{x+1}{x}\right).$$

Let $g : [1, \infty) \to \mathbb{R}$, $g(x) = -\frac{2}{2x+1} + \ln\frac{x+1}{x}$. A calculation shows that $g'(x) = -\frac{1}{x(x+1)(2x+1)^2} < 0$, $\forall x \geq 1$. It follows that g is strictly decreasing on $[1, \infty)$ and, since $\lim\limits_{x \to \infty} g(x) = 0$, we get that $g(x) > 0$, for all $x \geq 1$. This implies that $f'(x) > 0$, for $x \geq 1$, hence f is strictly increasing on $[1, \infty)$ and, since $\lim\limits_{x \to \infty} f(x) = e$, we obtain that $f(x) < e$.

The second inequality is equivalent to

$$e - \left(1 + \frac{1}{n}\right)^n < \frac{e}{2n+1} \Leftrightarrow e < \left(1 + \frac{1}{2n}\right)\left(1 + \frac{1}{n}\right)^n.$$

Let $u : [1, \infty) \to \mathbb{R}$ be the function defined by

$$u(x) = \left(1 + \frac{1}{2x}\right)\left(1 + \frac{1}{x}\right)^x.$$

We have

$$u'(x) = \left(1 + \frac{1}{x}\right)^x \cdot \frac{2x+1}{2x} \cdot \left(-\frac{1}{x} + \frac{2}{2x+1} + \ln\frac{x+1}{x} - \frac{1}{x+1}\right).$$

Let $v : [1, \infty) \to \mathbb{R}$, $v(x) = -\frac{1}{x} + \frac{2}{2x+1} + \ln\frac{x+1}{x} - \frac{1}{x+1}$. A calculation shows that $v'(x) = \frac{5x^2+5x+1}{x^2(x+1)^2(2x+1)^2} > 0$, $x \in [1, \infty)$. It follows that v is strictly increasing on $[1, \infty)$ and, since $\lim\limits_{x \to \infty} v(x) = 0$, we obtain that $v(x) < 0$, for $x \in [1, \infty)$. This implies $u'(x) < 0$, $x \in [1, \infty)$, hence u is strictly decreasing on $[1, \infty)$ and, since $\lim\limits_{x \to \infty} u(x) = e$, we get that $u(x) > e$, for $x \in [1, \infty)$.

(b) We have, based on part (a), that

$$\lim\limits_{n \to \infty} n\left(e - \left(1 + \frac{1}{n}\right)^n\right) = \frac{e}{2}.$$

The limit can also be calculated directly using l'Hôpital's rule.

4.9. The first inequality is equivalent to $\gamma < x_n - \frac{1}{2n+1}$. Let

$$u_n = 1 + \frac{1}{2} + \cdots + \frac{1}{n} - \ln n - \frac{1}{2n+1}, \quad n \geq 1.$$

We have $u_{n+1} - u_n = \frac{1}{n+1} - \ln\frac{n+1}{n} - \frac{1}{2n+3} + \frac{1}{2n+1}$. Let $u : [1, \infty) \to \mathbb{R}$ be the function defined by $u(x) = \frac{1}{x+1} - \ln\frac{x+1}{x} - \frac{1}{2x+3} + \frac{1}{2x+1}$. We have

$$u'(x) = \frac{16x^3 + 40x^2 + 32x + 9}{x(x+1)^2(2x+1)^2(2x+3)^2} > 0, \quad x \geq 1,$$

from which it follows that u is strictly increasing on $[1, \infty)$ and, since $\lim_{x \to \infty} u(x) = 0$, we obtain that $u(x) < 0$, for all $x \geq 1$. This implies $u_{n+1} < u_n$, $n \geq 1$ and, since $\lim_{n \to \infty} u_n = \gamma$, we have that $u_n > \gamma$.

The second inequality is equivalent to $x_n - \frac{1}{2n} < \gamma$. Let

$$v_n = 1 + \frac{1}{2} + \cdots + \frac{1}{n} - \ln n - \frac{1}{2n}, \quad n \geq 1.$$

We have $v_{n+1} - v_n = \frac{1}{2(n+1)} + \frac{1}{2n} - \ln \frac{n+1}{n}$. Let $v : [1, \infty) \to \mathbb{R}$ be the function defined by $v(x) = \frac{1}{2(x+1)} + \frac{1}{2x} - \ln \frac{x+1}{x}$. A calculation shows that

$$v'(x) = -\frac{1}{2x^2(x+1)^2} < 0, \quad x \in [1, \infty).$$

It follows that v is strictly decreasing and, since $\lim_{x \to \infty} v(x) = 0$, we have that $v(x) > 0$, for all $x \in [1, \infty)$. This implies that $v_{n+1} > v_n$, $n \geq 1$, which shows that the sequence $(v_n)_{n \geq 1}$ is strictly increasing and, since $\lim_{n \to \infty} v_n = \gamma$, we get that $v_n < \gamma$, for $n \geq 1$.

Remark. We recover the inequalities in Problem 4.9 by taking $a = 1$ in [129, Theorem 1.7.2, p. 29] and [129, Theorem 1.9.1, p. 35], where different proofs are given. See [129] for other inequalities and interesting results regarding Euler's constant γ.

12.2 Integral Equations

4.11. We change variables according to $x - t = u$ and we have that

$$f(x) = x + e^{-x} \int_0^x e^u f(u) \, du.$$

Since f is continuous, we have that the function $x \mapsto \int_0^x e^u f(u) \, du$ is differentiable, therefore f is differentiable. We multiply both sides of the preceding equation by e^x and we have

$$e^x f(x) = x e^x + \int_0^x e^u f(u) \, du.$$

Let $g(x) = e^x f(x)$. A calculation shows that

$$g'(x) = e^x + x e^x + g(x).$$

This implies $(g(x)e^{-x})' = 1 + x$. It follows that

$$g(x) = e^x \left(x + \frac{x^2}{2} + \mathscr{C} \right),$$

and

$$f(x) = x + \frac{x^2}{2} + \mathscr{C}, \quad \mathscr{C} \in \mathbb{R}.$$

Since $f(0) = 0$, we obtain that $f(x) = x + \frac{x^2}{2}$.

4.12. (a) $f(x) = \frac{a}{2}x^2 + a, \ x \in \mathbb{R}$; (b) $f(x) = \frac{a}{2}\cos(\sqrt{2}x) + \frac{a}{2}, \ x \in \mathbb{R}$.

4.13.

(a) $f(x) = -x^2 - x + 1, \ x \in \mathbb{R}$;
(b) $f(x) = -x - \frac{x^2}{2}, \ x \in \mathbb{R}$;
(c) $f(x) = \cos x - \sin x - 1, \ x \in \mathbb{R}$.

4.14.

(a) $f(x) = \frac{x^2}{2} + 1, \ x \in \mathbb{R}$;
(b) $f(x) = -\frac{\sin(\sqrt{2}x)}{2\sqrt{2}} - \frac{x}{2}, \ x \in \mathbb{R}$.

4.15. (a) $f(x) = \frac{a}{2} + \frac{a}{2}\cos(\sqrt{2}x), \ x \in \mathbb{R}$; (b) $f(x) = -x - \frac{x^3}{6}, \ x \in \mathbb{R}$.

4.16. $f(x) = a - \frac{2a}{\sqrt{3}}e^{-\frac{x}{2}} \sin\left(\frac{\sqrt{3}}{2}x\right), \ x \in \mathbb{R}$.

4.17.

(a) $f(x) = a \cosh x, \ x \in \mathbb{R}$;
(b) $f(x) = a \cos x, \ x \in \mathbb{R}$;
(c) $f(x) = a \cos x, \ x \in \mathbb{R}$.

12.3   Differential Equations

4.18.

(a) We replace x by $-x$ and we have

$$f'(x) - 2xf(-x) = x, \quad \forall x \in \mathbb{R};$$
$$f'(-x) + 2xf(x) = -x, \quad \forall x \in \mathbb{R}.$$

We add the preceding equations and we obtain that

$$f'(x) + f'(-x) + 2x(f(x) - f(-x)) = 0, \quad \forall x \in \mathbb{R}. \tag{12.1}$$

Let $g : \mathbb{R} \to \mathbb{R}$, $g(x) = f(x) - f(-x)$. We have, based on equation (12.1), that

$$g'(x) + 2xg(x) = 0, \quad \forall x \in \mathbb{R}.$$

We multiply the preceding equation by e^{x^2} and we obtain that

$$\left(g(x)e^{x^2}\right)' = 0, \quad \forall x \in \mathbb{R} \quad \Longrightarrow \quad g(x) = \mathscr{C}e^{-x^2}, \quad \forall x \in \mathbb{R},$$

where $\mathscr{C}$ is a real arbitrary constant. Therefore,

$$f(x) - f(-x) = \mathscr{C}e^{-x^2}, \quad \forall x \in \mathbb{R}. \tag{12.2}$$

When $x = 0$ we obtain, based on (12.2), that $\mathscr{C} = 0$. Therefore, $f(-x) = f(x)$, $\forall x \in \mathbb{R}$. The initial differential equation implies that

$$f'(x) - 2xf(x) = x, \quad \forall x \in \mathbb{R}.$$

It follows

$$\left(f(x)e^{-x^2}\right)' = xe^{-x^2} = \left(-\frac{1}{2}e^{-x^2}\right)', \quad \forall x \in \mathbb{R},$$

and we have $f(x) = -\frac{1}{2} + \mathscr{C}e^{x^2}, x \in \mathbb{R}$.

(b) The solution of this part of the problem is similar to the solution of part (a). We obtain that

$$f(x) = \frac{1}{2} + \mathscr{C}e^{-x^2}, \quad x \in \mathbb{R}.$$

4.19. $f(x) = \frac{x^2}{3} + \mathscr{C}x, \ x \in \mathbb{R}$.

4.20. We have

$$f(x) = \begin{cases} \dfrac{x^2}{k+2} & \text{if } k \text{ is even} \\[2mm] \mathscr{C}x^k + \dfrac{x^2}{k+2}, & \mathscr{C} \in \mathbb{R} \quad \text{if } k \text{ is odd.} \end{cases}$$

4.21.

(a) We prove that $f(x) = \sqrt{x^2 + 1}$. We have, by multiplying by $-\frac{1}{x^2}$, that

$$-\frac{1}{x^2}f'\left(\frac{1}{x}\right)f(x) = -\frac{1}{x^2}, \quad x > 0. \tag{12.3}$$

We replace x by $\frac{1}{x}$ in $f'\left(\frac{1}{x}\right) = \frac{1}{f(x)}$ and we obtain that

$$f'(x)f\left(\frac{1}{x}\right) = 1, \quad x > 0. \tag{12.4}$$

We add equations (12.3) and (12.4) and we have that

$$-\frac{1}{x^2}f'\left(\frac{1}{x}\right)f(x) + f'(x)f\left(\frac{1}{x}\right) = 1 - \frac{1}{x^2}, \quad x > 0.$$

It follows that

$$\left(f\left(\frac{1}{x}\right)f(x)\right)' = 1 - \frac{1}{x^2}, \quad x > 0.$$

Thus,

$$f\left(\frac{1}{x}\right)f(x) = x + \frac{1}{x} + \mathscr{C}.$$

Since $f(1) = \sqrt{2}$ we have that $\mathscr{C} = 0 \Rightarrow f\left(\frac{1}{x}\right)f(x) = x + \frac{1}{x}$, $\forall x > 0$. Using equation (12.4) we obtain that

$$\frac{f'(x)}{f(x)} = \frac{x}{1 + x^2}, \quad x > 0.$$

Therefore, $(\ln f(x))' - (\ln\sqrt{1+x^2})' = 0$, $\forall x > 0$. It follows $\ln f(x) - \ln\sqrt{1+x^2} = \ln\mathscr{C}_1$, where $\mathscr{C}_1 > 0$. We obtain $f(x) = \mathscr{C}_1\sqrt{1+x^2}$. Since $f(1) = \sqrt{2}$ we have that $\mathscr{C}_1 = 1$ and $f(x) = \sqrt{1+x^2}$.

(b) $f(x) = (1+x)e^{\frac{1-x}{2(1+x)}}$.

4.22. The functions are of the following form:

$$f(x) = \sqrt{\mathscr{C}}\left(\frac{x}{\sqrt{a}}\right)^{\frac{ab}{\mathscr{C}}}, \quad \mathscr{C} > 0.$$

4.23. The functions are of the following form:

$$f(x) = \sqrt{\mathscr{C}}e^{\frac{1}{2\mathscr{C}}\left(x^2 - \frac{1}{2}\right)}, \quad \mathscr{C} > 0.$$

4.24. The functions are of the following form:

$$f(x) = \sqrt{\mathscr{C}} e^{\frac{1}{\mathscr{C}}(x - \frac{\pi}{4})}, \quad \mathscr{C} > 0.$$

4.25. $f(x) = 0$, $f(x) = 2x$, $f(x) = \frac{2}{b} \sin(bx)$, $b \in \mathbb{R}^*$, $f(x) = \frac{2}{a} \sinh(ax)$, $a \in \mathbb{R}^*$.

4.26.

(a) We prove that these functions are:

$f(x) = x;$
$f(x) = \frac{\sin(\alpha x)}{\alpha}, \alpha \in \mathbb{R}^*;$
$f(x) = \frac{\sinh(\beta x)}{\beta}, \beta \in \mathbb{R}^*.$

We let $y = 0$ in

$$f(x + y) - f(x - y) = 2f'(x)f(y), \quad x, y \in \mathbb{R} \tag{12.5}$$

and we get that $2f'(x)f(0) = 0$, for all $x \in \mathbb{R}$, and since f is not constant we get that $f(0) = 0$. We differentiate (12.5) with respect to x and then with respect to y and we get that

$$\begin{cases} f'(x + y) - f'(x - y) = 2f''(x)f(y), & x, y \in \mathbb{R} \\ f'(x + y) + f'(x - y) = 2f'(x)f'(y), & x, y \in \mathbb{R}. \end{cases} \tag{12.6}$$

We let $y = 0$ in the second equation of (12.6) and we get that $2f'(x) = 2f'(x)f'(0)$. It follows that $f'(x)(1 - f'(0)) = 0$, for all $x \in \mathbb{R}$. This implies, since f is not constant, that $f'(0) = 1$.

Now we differentiate the equations in (12.6), the first one with respect to x and the second one with respect to y and we get that

$$\begin{cases} f''(x + y) - f''(x - y) = 2f'''(x)f(y), & x, y \in \mathbb{R} \\ f''(x + y) - f''(x - y) = 2f'(x)f''(y), & x, y \in \mathbb{R}. \end{cases}$$

We subtract the preceding equations and we get that $f'''(x)f(y) - f'(x)f''(y) = 0$, for all $x, y \in \mathbb{R}$. Since f is not constant, there is $x_0 \in \mathbb{R}$ such that $f'(x_0) \neq 0$. The preceding equation implies

$$f''(y) - \frac{f'''(x_0)}{f'(x_0)} f(y) = 0, \quad \forall y \in \mathbb{R}.$$

Let $k = \frac{f'''(x_0)}{f'(x_0)}$. Now we consider the following three cases:

Case $k = 0$. This implies that $f''(y) = 0$, for all $y \in \mathbb{R}$, which implies that $f(y) = ay + b$. Since $f(0) = 0$ and $f'(0) = 1$, we get that $f(y) = y$.

Case $k < 0$. Let $k = -\alpha^2$, with $\alpha \neq 0$. We have that $f''(y) + \alpha^2 f(y) = 0$ and this implies $f(y) = \mathscr{C}_1 \cos(\alpha y) + \mathscr{C}_2 \sin(\alpha y)$. Since $f(0) = 0$ and $f'(0) = 1$, we get that $f(y) = \frac{\sin(\alpha y)}{\alpha}$.

Case $k > 0$. Let $k = \beta^2$, with $\beta \neq 0$. We have that $f''(y) - \beta^2 f(y) = 0$ and this implies that $f(y) = \mathscr{C}_3 e^{\beta y} + \mathscr{C}_4 e^{-\beta y}$. Since $f(0) = 0$ and $f'(0) = 1$, we get that $f(y) = \frac{\sinh(\beta y)}{\beta}$.

(b) These functions are:

$$f(x) = \frac{x^2}{2} + \mathscr{C}, \ \mathscr{C} \in \mathbb{R};$$
$$f(x) = \mathscr{C} - \frac{\cos(\alpha x)}{\alpha^2}, \ \mathscr{C} \in \mathbb{R}, \ \alpha \in \mathbb{R}^*;$$
$$f(x) = \mathscr{C} + \frac{\cosh(\beta x)}{\beta^2}, \ \mathscr{C} \in \mathbb{R}, \ \beta \in \mathbb{R}^*.$$

Use a similar technique to that given in part (a) of the problem.

4.27.

(a) $f(x) = \mathscr{C}e^x + \frac{x^2}{2}e^x, \ \mathscr{C} \in \mathbb{R}, \ x \in \mathbb{R}$.

(b) Let $y(x) = \displaystyle\sum_{n=1}^{\infty} n \left(e^x - 1 - \frac{x}{1!} - \frac{x^2}{2!} - \cdots - \frac{x^n}{n!} \right), \ x \in \mathbb{R}$. A calculation shows that $y'(x) - y(x) = xe^x$, with $y(0) = 0$. It follows, based on part (a), that $y(x) = \frac{x^2}{2}e^x$.

4.29.

(a) Calculate the $2n$th partial sum of the series.
(b) We have, based on part (a), that

$$S = \sum_{n=1}^{\infty} \left(\frac{x^n}{n!} - 2\frac{x^{n+1}}{(n+1)!} + 3\frac{x^{n+2}}{(n+2)!} - \cdots \right)$$

$$= \frac{x}{1!} - 2\frac{x^2}{2!} + 3\frac{x^3}{3!} - 4\frac{x^4}{4!} + \cdots + \sum_{n=2}^{\infty} \left(\frac{x^n}{n!} - 2\frac{x^{n+1}}{(n+1)!} + 3\frac{x^{n+2}}{(n+2)!} - \cdots \right)$$

$$\overset{n-1=m}{=} xe^{-x} + \sum_{m=1}^{\infty} \left(\frac{x^{m+1}}{(m+1)!} - 2\frac{x^{m+2}}{(m+2)!} + 3\frac{x^{m+3}}{(m+3)!} + \cdots \right)$$

$$= xe^{-x}$$

$$- \sum_{m=1}^{\infty} \left[\left(\frac{x^m}{m!} - 2\frac{x^{m+1}}{(m+1)!} + 3\frac{x^{m+2}}{(m+2)!} + \cdots \right) - \left(\frac{x^m}{m!} - \frac{x^{m+1}}{(m+1)!} + \frac{x^{m+2}}{(m+2)!} + \cdots \right) \right]$$

$$\overset{(a)}{=} xe^{-x} - S + \sinh x.$$

4.30. Let

$$y(x) = \sum_{n=1}^{\infty} n \left(\frac{x^n}{n!} - \frac{x^{n+1}}{(n+1)!} + \frac{x^{n+2}}{(n+2)!} - \cdots \right), \quad x \in \mathbb{R},$$

and observe that y satisfies the differential equation $y'(x) + y(x) = (x+1)e^x$, $x \in \mathbb{R}$.

4.31. Let

$$f(x) = \sum_{n=k}^{\infty} (-1)^n C_n^k \left(e^x - 1 - x - \frac{x^2}{2!} - \cdots - \frac{x^n}{n!} \right).$$

Then

$$f'(x) = \sum_{n=k}^{\infty} (-1)^n C_n^k \left(e^x - 1 - x - \frac{x^2}{2!} - \cdots - \frac{x^{n-1}}{(n-1)!} \right)$$

$$= \sum_{n=k}^{\infty} (-1)^n C_n^k \left(e^x - 1 - x - \frac{x^2}{2!} - \cdots - \frac{x^{n-1}}{(n-1)!} - \frac{x^n}{n!} \right)$$

$$+ \sum_{n=k}^{\infty} (-1)^n C_n^k \frac{x^n}{n!}$$

$$= f(x) + \frac{(-x)^k}{k!} \sum_{n=k}^{\infty} \frac{(-x)^{n-k}}{(n-k)!}$$

$$= f(x) + \frac{(-x)^k}{k!} e^{-x}.$$

This implies that

$$(f(x)e^{-x})' = (-1)^k \frac{x^k}{k!} e^{-2x} \quad \Rightarrow \quad f(x)e^{-x} = \frac{(-1)^k}{k!} \int_0^x t^k e^{-2t}\,dt + \mathscr{C}, \quad \mathscr{C} \in \mathbb{R}.$$

It follows, since $f(0) = 0$, that $\mathscr{C} = 0$ and this implies that

$$f(x)e^{-x} = \frac{(-1)^k}{k!} \int_0^x t^k e^{-2t}\,dt.$$

Therefore,

$$\lim_{x \to \infty} e^{-x} f(x) = \frac{(-1)^k}{k!} \int_0^{\infty} t^k e^{-2t}\,dt.$$

The substitution $2t = y$ shows that

$$\int_0^\infty t^k e^{-2t}\, dt = \frac{1}{2^{k+1}} \int_0^\infty y^k e^{-y}\, dy = \frac{\Gamma(k+1)}{2^{k+1}} = \frac{k!}{2^{k+1}}.$$

It follows that $\lim_{x\to\infty} e^{-x} f(x) = \frac{(-1)^k}{2^{k+1}}$ and the problem is solved.

4.32. (a) Let $y(x) = \sum_{n=1}^\infty \left\lfloor \frac{n}{2} \right\rfloor \left(e^x - 1 - \frac{x}{1!} - \frac{x^2}{2!} - \cdots - \frac{x^n}{n!} \right)$, $x \in \mathbb{R}$.

We have

$$y'(x) = \sum_{n=1}^\infty \left\lfloor \frac{n}{2} \right\rfloor \left(e^x - 1 - \frac{x}{1!} - \frac{x^2}{2!} - \cdots - \frac{x^{n-1}}{(n-1)!} \right)$$

$$= \sum_{n=1}^\infty \left\lfloor \frac{n}{2} \right\rfloor \left(e^x - 1 - \frac{x}{1!} - \frac{x^2}{2!} - \cdots - \frac{x^n}{n!} \right) + \sum_{n=1}^\infty \left\lfloor \frac{n}{2} \right\rfloor \frac{x^n}{n!}$$

$$= y(x) + \sum_{k=1}^\infty k \frac{x^{2k}}{(2k)!} + \sum_{k=1}^\infty (k-1) \frac{x^{2k-1}}{(2k-1)!}$$

$$= y(x) + \frac{x-1}{2} \sum_{k=1}^\infty \frac{x^{2k-1}}{(2k-1)!} + \frac{x}{2} \sum_{k=1}^\infty \frac{x^{2k-2}}{(2k-2)!}$$

$$= y(x) + \frac{x-1}{2} \sinh x + \frac{x}{2} \cosh x.$$

We obtain the linear differential equation $y'(x) = y(x) + \frac{x-1}{2} \sinh x + \frac{x}{2} \cosh x$, $x \in \mathbb{R}$, with initial condition $y(0) = 0$. A calculation shows that $y(x) = \frac{1}{4} \sinh x + \frac{1}{4}(x^2 - x)e^x$.

The other parts of the problem can be solved similarly.

12.4   Higher Order Derivatives

4.33. Prove the formulae by mathematical induction.

4.34.

(a) We assume that $\sin x$ is a polynomial function, i.e. $\sin x = P(x)$, $\forall x \in \mathbb{R}$, where P is a polynomial of degree n. We differentiate the preceding equality $n + 1$ times and we get, based on part (c) of Problem 4.33, that $\sin\left(x + \frac{n+1}{2}\pi\right) = 0$, $\forall x \in \mathbb{R}$, which is a contradiction.

(b) We assume that e^x is a polynomial function, i.e. $e^x = P(x)$, $\forall x \in \mathbb{R}$. Let $n = \deg P$. We differentiate the preceding equality $n + 1$ times and we get that $e^x = 0$, $\forall x \in \mathbb{R}$, which contradicts $e^x > 0$, $\forall x \in \mathbb{R}$.

(c) We assume that e^x is a rational function. We have $e^x = \frac{P(x)}{Q(x)}$, $\forall x \in \mathbb{R}$, where P and Q are polynomials. It follows that $Q(x)e^x = P(x)$, $\forall x \in \mathbb{R}$. Let $n = $ degree P. We differentiate n times the preceding equality and we have that $\sum_{k=0}^{n} C_n^k Q^{(n-k)}(x)e^x = n!a_n$, where $a_n \neq 0$ is the leading coefficient of P.

It follows that $e^x = \frac{1}{n!a_n} \sum_{k=0}^{n} C_n^k Q^{(n-k)}(-x)$, $\forall x \in \mathbb{R}$, which contradicts the fact that e^x is not a polynomial function.

4.35. We have $g'(x) - g(x) = -f(x) \leq 0$, $\forall x \in \mathbb{R}$. It follows that $(e^{-x}g(x))' \leq 0$, $\forall x \in \mathbb{R}$. This implies that the function $x \to e^{-x}g(x)$ decreases on $\mathbb{R}$ and it follows that $e^{-x}g(x) \geq \lim_{x\to\infty} e^{-x}g(x) = 0 \Rightarrow g(x) \geq 0$, $\forall x \in \mathbb{R}$.

4.36. (a) $f(x) = \mathscr{C}e^{\frac{x}{2}}$, $\mathscr{C} \in \mathbb{R}$; (b) $f(x) = \mathscr{C}e^{-\frac{x}{2}}$, $\mathscr{C} \in \mathbb{R}$; (c) $f(x) = \mathscr{C}e^{\frac{x}{2a}}$, $\mathscr{C} \in \mathbb{R}$; (d) $f(x) = \mathscr{C}e^{\frac{3-\sqrt{5}}{2}x}$, $\mathscr{C} \in \mathbb{R}$; (e) $f(x) = \mathscr{C}e^{\frac{\sqrt{5}-1}{2}x}$, $\mathscr{C} \in \mathbb{R}$; (f) $f(x) = \mathscr{C}_1e^{-\frac{x}{\sqrt{2}}} + \mathscr{C}_2e^{\frac{x}{\sqrt{2}}}$, $\mathscr{C}_1, \mathscr{C}_2 \in \mathbb{R}$; (g) $f(x) = \mathscr{C}\cosh\left(\frac{x}{\sqrt{2}}\right)$, $\mathscr{C} \in \mathbb{R}$.

4.37.

(a) $\left(\dfrac{3x}{1+x}\right)^{(n)} = (-1)^{n+1}\dfrac{3 \cdot n!}{(x+1)^{n+1}}$, $n \geq 1$;

(b) $\left(\dfrac{1}{2x^2+5x-3}\right)^{(n)} = (-1)^n\dfrac{n!}{7}\left(\dfrac{2^{n+1}}{(2x-1)^{n+1}} - \dfrac{1}{(x+3)^{n+1}}\right)$, $n \geq 0$;

(c) $\left(\dfrac{1+x}{\sqrt{x}}\right)^{(n)} = (-1)^{n-1}\dfrac{1 \cdot 3 \cdot 5 \cdots (2n-3)}{2^n}x^{-\frac{2n+1}{2}}(x-2n+1)$, $n \geq 2$;

(d) $\left(\dfrac{1+x}{\sqrt[3]{x}}\right)^{(n)} = (-1)^{n-1}\dfrac{1 \cdot 4 \cdot 7 \cdots (3n-5)}{3^n}x^{-\frac{3n+1}{3}}(2x-3n+2)$, $n \geq 2$;

(e) $(\sin^2 x)^{(n)} = -2^{n-1}\cos\left(2x+\dfrac{n\pi}{2}\right)$, $n \in \mathbb{N}$;

(f) $(\cos^2 x)^{(n)} = 2^{n-1}\cos\left(2x+\dfrac{n\pi}{2}\right)$, $n \in \mathbb{N}$;

(g) use the formula $\sin(3x) = 3\sin x - 4\sin^3 x$ and it follows that
$$(\sin^3 x)^{(n)} = \frac{3}{4}\sin\left(x+\frac{n\pi}{2}\right) - \frac{3^n}{4}\sin\left(3x+\frac{n\pi}{2}\right), \quad n \geq 0;$$

(h) use the formula $\cos(3x) = 4\cos^3 x - 3\cos x$ and it follows that
$$(\cos^3 x)^{(n)} = \frac{3}{4}\cos\left(x+\frac{n\pi}{2}\right) + \frac{3^n}{4}\cos\left(3x+\frac{n\pi}{2}\right), \quad n \geq 0;$$

(i) $\left(\dfrac{e^x - e^{-x}}{2}\right)^{(n)} = \dfrac{1}{2}\left(e^x - (-1)^n e^{-x}\right)$, $n \geq 0$;

(j) $\left(\dfrac{e^x + e^{-x}}{2}\right)^{(n)} = \dfrac{1}{2}\left(e^x + (-1)^n e^{-x}\right)$, $n \geq 0$.

4.38. We have

$$\sin^4 x+\cos^4 x = (\sin^2 x+\cos^2 x)^2-2\sin^2 x \cos^2 x = 1-\frac{1}{2}\sin^2(2x) = \frac{3}{4}+\frac{\cos(4x)}{4},$$

and it follows that $f^{(n)}(x) = 4^{n-1}\cos\left(4x + \frac{n\pi}{2}\right)$.

4.39. *Solution due to Andreea Ilieş, computer science student 2017/2018.*
We have

$$f'(x) = e^x (\cos x + \sin x) = f(x) + g(x),$$

$$g'(x) = e^x (\cos x - \sin x) = g(x) - f(x).$$

It follows that

$$\left(f^{(n)}(x)\right)^2 + \left(g^{(n)}(x)\right)^2 = \left[(f'(x))^{(n-1)}\right]^2 + \left[(g'(x))^{(n-1)}\right]^2$$

$$= \left[(f(x)+g(x))^{(n-1)}\right]^2 + \left[(g(x)-f(x))^{(n-1)}\right]^2$$

$$= \left[f^{(n-1)}(x)+g^{(n-1)}(x)\right]^2 + \left[g^{(n-1)}(x)-f^{(n-1)}(x)\right]^2$$

$$= 2\left[\left(f^{(n-1)}(x)\right)^2 + \left(g^{(n-1)}(x)\right)^2\right]$$

$$= 2^2\left[\left(f^{(n-2)}(x)\right)^2 + \left(g^{(n-2)}(x)\right)^2\right]$$

$$= \cdots$$

$$= 2^{n-1}\left[(f'(x))^2 + (g'(x))^2\right]$$

$$= 2^n\left[f^2(x)+g^2(x)\right]$$

$$= 2^n e^{2x}.$$

4.40.

(a) Since $\frac{1}{1+x^2} = \frac{1}{2i}\left(\frac{1}{x-i} - \frac{1}{x+i}\right)$, we have

$$f^{(n)}(x)=(-1)^n\frac{n!}{2i}\left(\frac{1}{(x-i)^{n+1}} - \frac{1}{(x+i)^{n+1}}\right)=(-1)^n\frac{n!}{2i}\cdot\frac{(x+i)^{n+1}-(x-i)^{n+1}}{(x^2+1)^{n+1}},$$

and it follows that

$$\frac{f^{(2nk)}(1)}{(2nk)!} = \frac{1}{2i}\left[\left(\frac{1+i}{2}\right)^{2nk+1} - \left(\frac{1-i}{2}\right)^{2nk+1}\right].$$

Because $1+i = \sqrt{2}e^{\frac{i\pi}{4}}$ and $1-i = \sqrt{2}e^{-\frac{i\pi}{4}}$, we have that

$$\sum_{n=0}^{\infty}\frac{f^{(2nk)}(1)}{(2nk)!} = \frac{1}{2i}\left[\frac{1+i}{2}\sum_{n=0}^{\infty}\left(\frac{1+i}{2}\right)^{2nk} - \frac{1-i}{2}\sum_{n=0}^{\infty}\left(\frac{1-i}{2}\right)^{2nk}\right]$$

$$= \frac{1}{2i}\left[\frac{1+i}{2}\cdot\frac{2^{2k}}{2^{2k}-(1+i)^{2k}} - \frac{1-i}{2}\cdot\frac{2^{2k}}{2^{2k}-(1-i)^{2k}}\right]$$

$$= 2^{k-1}\frac{2^k - \cos\frac{k\pi}{2} + \sin\frac{k\pi}{2}}{2^{2k} - 2^{k+1}\cos\frac{k\pi}{2} + 1}.$$

(b) When $k=1$ and $k=2$, we obtain the formulae

$$\sum_{n=0}^{\infty}\frac{f^{(2n)}(1)}{(2n)!} = \frac{3}{5} \quad\text{and}\quad \sum_{n=0}^{\infty}\frac{f^{(4n)}(1)}{(4n)!} = \frac{2}{5}.$$

4.41.

(a) $f: \mathbb{R} \to \mathbb{R}$, $f(x) = (x^2 - 3x - 1)e^{2x}$. Using Leibniz's formula, we have

$$f^{(n)}(x) = C_n^0(x^2 - 3x - 1)\cdot 2^n \cdot e^{2x} + C_n^1(2x - 3)\cdot 2^{n-1}e^{2x} + C_n^2 \cdot 2 \cdot 2^{n-2}e^{2x}$$

$$= 2^{n-2}e^{2x}\left(4x^2 + 4(n-3)x + n^2 - 7n - 4\right).$$

(b) $f: \mathbb{R} \to \mathbb{R}$, $f(x) = \frac{x-1}{e^x}$. We use Leibniz's formula and we have

$$f^{(n)}(x) = C_n^0(x-1)(-1)^n e^{-x} + C_n^1(-1)^{n-1}e^{-x} = (-1)^n\frac{x - n - 1}{e^x}.$$

(c) $f: \mathbb{R} \to \mathbb{R}$, $f(x) = e^{ax}\cos(bx + c)$, $a, b, c \in \mathbb{R}$. We have, based on Leibniz's formula, that

$$f^{(n)}(x) = \sum_{k=0}^{n} C_n^k a^{n-k} b^k e^{ax} \cos\left(bx + c + \frac{k\pi}{2}\right)$$

$$= e^{ax}\sum_{k=0}^{n} C_n^k a^{n-k} b^k \cos\left(bx + c + \frac{k\pi}{2}\right).$$

4.42. $x = \frac{\pi}{4} + k\pi$, $k \in \mathbb{Z}$.

4.43. We use that

$$f'(x) = \frac{2}{x+1}\ln(x+1) \quad \text{and} \quad (\ln(1+x))^{(k)} = (-1)^{k-1}\frac{(k-1)!}{(x+1)^k}, \quad k \ge 1.$$

We write $f(x) = \ln^2(x+1) = \ln(x+1) \cdot \ln(x+1)$, we apply Leibniz's formula and we have that

$$f^{(n)}(x) = C_n^0(-1)^{n-1}\frac{(n-1)!}{(x+1)^n} \cdot \ln(x+1)$$

$$+ \sum_{k=1}^{n-1} C_n^k(-1)^{n-k-1}\frac{(n-k-1)!}{(x+1)^{n-k}} \cdot (-1)^{k-1}\frac{(k-1)!}{(x+1)^k}$$

$$+ C_n^n \ln(x+1) \cdot (-1)^{n-1}\frac{(n-1)!}{(x+1)^n}$$

$$= 2(-1)^{n-1}(n-1)!\frac{1}{(x+1)^n}\ln(x+1)$$

$$+ \sum_{k=1}^{n-1}(-1)^{n-2}\frac{n!}{k!(n-k)!} \cdot (k-1)!(n-k-1)! \cdot \frac{1}{(x+1)^n}$$

$$= 2(-1)^{n-1}(n-1)!\frac{1}{(x+1)^n}\ln(x+1)$$

$$+ \sum_{k=1}^{n-1}(-1)^{n-2}(n-1)!\frac{n}{k(n-k)} \cdot \frac{1}{(x+1)^n}$$

$$= 2(-1)^{n-1}(n-1)!\frac{1}{(x+1)^n}\ln(x+1)$$

$$+ \sum_{k=1}^{n-1}(-1)^{n-2}(n-1)!\left(\frac{1}{k} + \frac{1}{n-k}\right)\frac{1}{(x+1)^n}$$

$$= 2(-1)^{n-1}(n-1)!\frac{1}{(x+1)^n}\ln(x+1)$$

$$- 2(-1)^{n-1}(n-1)!\frac{1}{(x+1)^n}\sum_{k=1}^{n-1}\frac{1}{k}$$

$$= 2(-1)^{n-1}(n-1)!\left[\ln(x+1) - \sum_{k=1}^{n-1}\frac{1}{k}\right]\frac{1}{(x+1)^n}, \quad n \ge 2.$$

4.44. We use that

$$\left(\frac{1}{x}\right)^{(k)} = (-1)^k \frac{k!}{x^{k+1}}, \quad k \geq 0 \quad \text{and} \quad (\ln x)^{(k)} = (-1)^{k-1}\frac{(k-1)!}{x^k}, \quad k \in \mathbb{N}.$$

We have

$$f^{(n)}(x) = C_n^0(-1)^n \frac{n!}{x^{n+1}}\ln x + \sum_{k=1}^n C_n^k(-1)^{n-k}\frac{(n-k)!}{x^{n-k+1}}\cdot(-1)^{k-1}\frac{(k-1)!}{x^k}$$

$$= (-1)^{n-1}\frac{n!}{x^{n+1}}\left(1 + \frac{1}{2} + \frac{1}{3} + \cdots + \frac{1}{n} - \ln x\right), \quad n \in \mathbb{N}.$$

4.45. *Solution I. Mathematical induction.* Let $n \geq 1$ and let $P(n)$ be the statement $(x^n \ln x)^{(n)} = n!(H_n + \ln x)$. One can check that $(x \ln x)' = 1 + \ln x = 1!(H_1 + \ln x)$, so $P(1)$ is true. We assume that $P(n-1)$ is true and we prove that $P(n)$ holds true. Since $f(x) = x^n \ln x$, we have that $f'(x) = nx^{n-1}\ln x + x^{n-1}$ and by differentiation $n-1$ times we obtain that

$$f^{(n)}(x) = n(x^{n-1}\ln x)^{(n-1)} + (x^{n-1})^{(n-1)} \quad (P(n-1) \text{ holds true})$$

$$= n(n-1)!(H_{n-1} + \ln x) + (n-1)!$$

$$= n!\left(H_{n-1} + \ln x + \frac{1}{n}\right)$$

$$= n!(H_n + \ln x).$$

Solution II. Leibniz's formula. We have

$$(x^n \ln x)^{(n)} = (x^n)^{(n)}\ln x + \sum_{k=1}^n C_n^k(x^n)^{(n-k)}(\ln x)^{(k)}$$

$$= n!\ln x + \sum_{k=1}^n C_n^k \cdot \frac{n!}{k!}\cdot x^k \cdot (-1)^{k-1}\frac{(k-1)!}{x^k}$$

$$= n!\ln x + n!\sum_{k=1}^n (-1)^{k-1}C_n^k \cdot \frac{1}{k}.$$

Now we prove that the following binomial identity holds true.

A Binomial Identity Let $n \in \mathbb{N}$. Then

$$\sum_{k=1}^n (-1)^{k-1}\frac{1}{k}C_n^k = 1 + \frac{1}{2} + \frac{1}{3} + \cdots + \frac{1}{n}.$$

Proof. We have

$$\sum_{k=1}^{n}(-1)^{k-1}\frac{1}{k}C_n^k = \sum_{k=1}^{n}C_n^k\int_0^1(-x)^{k-1}\mathrm{d}x$$

$$= -\int_0^1\frac{1}{x}\sum_{k=1}^{n}C_n^k(-x)^k\mathrm{d}x$$

$$= \int_0^1\frac{1-(1-x)^n}{x}\mathrm{d}x \quad (1-x=t)$$

$$= \int_0^1\frac{1-t^n}{1-t}\mathrm{d}t$$

$$= \int_0^1(1+t+t^2+\cdots+t^{n-1})\mathrm{d}t$$

$$= 1+\frac{1}{2}+\frac{1}{3}+\cdots+\frac{1}{n},$$

and the identity is proved.   ∎

4.46. Use Problem 4.44.

4.47. Let $f(x) = x^{n-1}e^{\frac{1}{x}}$. We prove that

$$f^{(n)}(x) = (-1)^n\frac{f(x)}{x^{2n}}, \quad \text{i.e.} \quad \left(x^{n-1}e^{\frac{1}{x}}\right)^{(n)} = (-1)^n\frac{1}{x^{n+1}}e^{\frac{1}{x}}.$$

We solve the problem by mathematical induction. Let $n \geq 1$ and let $P(n)$ be the statement

$$\left(x^{n-1}e^{\frac{1}{x}}\right)^{(n)} = (-1)^n\frac{1}{x^{n+1}}e^{\frac{1}{x}}.$$

Since $\left(e^{\frac{1}{x}}\right)' = -\frac{1}{x^2}e^{\frac{1}{x}}$, we have that $P(1)$ holds true. We assume that $P(k)$ is true for all $k < n$ and we prove that $P(n)$ holds true.

We have

$$f'(x) = (n-1)x^{n-2}e^{\frac{1}{x}} - x^{n-3}e^{\frac{1}{x}}.$$

We differentiate the preceding equality $n-1$ times and we have that

$$f^{(n)}(x) = (f'(x))^{(n-1)}(x)$$

$$= (n-1)\left(x^{n-2}e^{\frac{1}{x}}\right)^{(n-1)} - \left(x^{n-3}e^{\frac{1}{x}}\right)^{(n-1)} \qquad (P(n-1) \text{ holds true})$$

$$= (n-1)\left((-1)^{n-1}\frac{1}{x^n}e^{\frac{1}{x}}\right) - \left[\left(x^{n-3}e^{\frac{1}{x}}\right)^{(n-2)}\right]' \qquad (P(n-2) \text{ holds true})$$

$$= (n-1)(-1)^{n-1}\frac{1}{x^n}e^{\frac{1}{x}} - \left((-1)^{n-2}\frac{1}{x^{n-1}}e^{\frac{1}{x}}\right)'$$

$$= (n-1)(-1)^{n-1}\frac{1}{x^n}e^{\frac{1}{x}} + (-1)^{n-1}\left(-\frac{n-1}{x^n}e^{\frac{1}{x}} - \frac{1}{x^{n+1}}e^{\frac{1}{x}}\right)$$

$$= (-1)^n\frac{1}{x^{n+1}}e^{\frac{1}{x}}.$$

4.48. We have

$$f'(x) = \frac{2\arcsin x}{\sqrt{1-x^2}}$$

$$f''(x) = \frac{2}{1-x^2} + \frac{xf'(x)}{1-x^2}.$$

Therefore,

$$(1-x^2)f''(x) - xf'(x) - 2 = 0.$$

We differentiate $n-1$ times the preceding formula, using Leibniz's formula, and we have that

$$(1-x^2)f^{(n+1)}(x) - (2n-1)xf^{(n)}(x) - (n-1)^2 f^{(n-1)}(x) = 0, \quad n \geq 2.$$

When $x = 0$ we obtain the recurrence formula

$$f^{(n+1)}(0) = (n-1)^2 f^{(n-1)}(0), \quad n \geq 2.$$

When $n = 2p$, $p \in \mathbb{N}$, we have that

$$f^{(2p+1)}(0) = (2p-1)^2(2p-3)^2 \cdots 1^2 \cdot f'(0).$$

Since $f'(0) = 0$, we obtain that $f^{(2p+1)}(0) = 0$, for all $p \in \mathbb{N}$.
When $n = 2p+1$, $p \in \mathbb{N}$, we obtain from the recurrence formula that

$$f^{(2p+2)}(0) = (2p)^2(2p-2)^2 \cdots 2^2 \cdot f''(0).$$

Since $f''(0) = 2$, we have that $f^{(2p+2)}(0) = 2^{2p+1}(p!)^2$, for all $p \in \mathbb{N}$.

4.49. We have $(1 + x^2)f(x) = 1$. We differentiate n times the preceding equality, using Leibniz's formula, and we have

$$(x^2 + 1)f^{(n)}(x) + 2nxf^{(n-1)}(x) + n(n-1)f^{(n-2)}(x) = 0, \quad n \geq 2.$$

When $x = 0$ we obtain the recurrence formula

$$f^{(n)}(0) = -n(n-1)f^{(n-2)}(0), \quad n \geq 2.$$

When $n = 2p$, $p \in \mathbb{N}$, it follows that

$$f^{(2p)}(0) = (-1)^p (2p)(2p-1) \cdots 2 \cdot 1 \cdot f^{(0)}(0).$$

Since $f^{(0)}(0) = f(0) = 1$, we obtain that $f^{(2p)}(0) = (-1)^p (2p)!$, for all $p \in \mathbb{N}$. When $n = 2p + 1$, $p \in \mathbb{N}$, we obtain from the recurrence formula that

$$f^{(2p+1)}(0) = (-1)^p (2p+1)(2p) \cdots 3 \cdot 2 \cdot f'(0).$$

Since $f'(0) = 0$, we have that $f^{(2p+1)}(0) = 0$, for all $p \in \mathbb{N}$.

For the function g we have

$$g'(x) = \frac{1}{1 + x^2} = f(x).$$

Therefore, $g^{(2n)}(0) = f^{(2n-1)}(0) = 0$, $n \in \mathbb{N}$. Since $g^{(0)}(0) = g(0) = 0$, it follows that $g^{(2n)}(0) = 0$, $\forall n \geq 0$. Also, $g^{(2n+1)}(0) = f^{(2n)}(0) = (-1)^n (2n)!$, $\forall n \geq 0$.

For the function h, using Leibniz's formula and the derivatives (already obtained) of the function g, we have that

$$h^{(n)}(0) = \sum_{k=0}^{n} C_n^k g^{(n-k)}(0) g^{(k)}(0), \quad n \geq 0.$$

Therefore,

$$h^{(2n)}(0) = \sum_{k=0}^{2n} C_{2n}^k g^{(2n-k)}(0) g^{(k)}(0)$$

$$= \sum_{k=1}^{n} C_{2n}^{2k-1} g^{(2n-2k+1)}(0) g^{(2k-1)}(0)$$

$$= \sum_{k=1}^{n} C_{2n}^{2k-1} (-1)^{n-k} (2n-2k)! (-1)^{k-1} (2k-2)!$$

$$= (-1)^{n-1}(2n)! \sum_{k=1}^{k} \frac{1}{(2n - 2k + 1)(2k - 1)}$$

$$= (-1)^{n-1}\frac{(2n)!}{2n} \sum_{k=1}^{n} \left(\frac{1}{2n - 2k + 1} + \frac{1}{2k - 1} \right)$$

$$= (-1)^{n-1}2(2n - 1)! \sum_{k=1}^{n} \frac{1}{2k - 1}, \quad \forall n \in \mathbb{N}.$$

When $n = 0$ we have $h^{(0)}(0) = h(0) = 0$.
On the other hand,

$$h^{(2n+1)}(0) = \sum_{k=0}^{2n+1} C_{2n+1}^k g^{(2n-k+1)}(0) g^{(k)}(0) = 0, \quad \forall n \geq 0.$$

4.50. We have

$$f^{(3n)}(0) = (-1)^n(3n)!, \quad n \geq 0$$
$$f^{(3n+1)}(0) = 0, \quad n \geq 0$$
$$f^{(3n+2)}(0) = 0, \quad n \geq 0.$$

4.51. We have

$$f^{(2n)}\left(-\frac{1}{2}\right) = (-1)^n \left(\frac{4}{3}\right)^{n+1} (2n)!, \quad n \geq 0$$

$$f^{(2n+1)}\left(-\frac{1}{2}\right) = 0, \quad n \geq 0.$$

4.52. We have

$$f^{(3n+1)}(0) = 0, \quad n \geq 0$$
$$f^{(3n+2)}(0) = 0, \quad n \geq 0$$
$$f^{(3n+3)}(0) = -3(3n + 2)!, \quad n \geq 0.$$

4.53.

- We have

$$e^a \int_0^1 e^{(x-a)t} dt = \frac{e^x - e^a}{x - a}, \quad x \neq a.$$

We obtain, by differentiating n times the preceding formula, that

$$e^a \int_0^1 e^{(x-a)t} t^n dt = \left(\frac{e^x - e^a}{x - a}\right)^{(n)}.$$

It follows that

$$n \left(\frac{e^x - e^a}{x - a}\right)^{(n)} = ne^a \int_0^1 e^{(x-a)t} t^n dt = e^a \int_0^1 e^{(x-a)\sqrt[n]{y}} \sqrt[n]{y}\, dy.$$

Thus,

$$\lim_{n\to\infty} n \left(\frac{e^x - e^a}{x - a}\right)^{(n)} = \lim_{n\to\infty} e^a \int_0^1 e^{(x-a)\sqrt[n]{y}} \sqrt[n]{y}\, dy = e^a \int_0^1 e^{x-a} dy = e^x.$$

- We have

$$\frac{\sin x}{x} = \int_0^1 \cos(xt)dt, \quad x \in \mathbb{R}^*.$$

We differentiate $2n + 1$ times the preceding equality and we have that

$$\left(\frac{\sin x}{x}\right)^{(2n+1)} = \int_0^1 t^{2n+1} \cos\left(xt + \frac{2n+1}{2}\pi\right) dt = (-1)^{n-1} \int_0^1 t^{2n+1} \sin(xt)\, dt.$$

It follows that

$$(-1)^{n-1} n \left(\frac{\sin x}{x}\right)^{(2n+1)} = n \int_0^1 t^{2n+1} \sin(xt)\, dt$$

$$= \frac{n}{2n+1} \int_0^1 {}^{2n+1}\!\sqrt{y} \sin\left(x\ {}^{2n+1}\!\sqrt{y}\right) dy$$

and

$$\lim_{n\to\infty} (-1)^{n-1} n \left(\frac{\sin x}{x}\right)^{(2n+1)} = \lim_{n\to\infty} \frac{n}{2n+1} \int_0^1 {}^{2n+1}\!\sqrt{y} \sin\left(x\ {}^{2n+1}\!\sqrt{y}\right) dy$$

$$= \frac{1}{2} \int_0^1 \sin x\, dy$$

$$= \frac{\sin x}{2}.$$

- We have

$$\frac{\ln(1 + y)}{y} = \int_1^\infty \frac{dx}{x(1 + (1 + y)(x - 1))}, \quad y \in (-1, 1), \ y \neq 0$$

and it follows that

$$\left(\frac{\ln(1+y)}{y}\right)^{(n)} = \int_1^\infty \frac{(-1)^n n! (x-1)^n}{[1+(1+y)(x-1)]^{n+1}} \cdot \frac{dx}{x}$$

$$\overset{x-1=t}{=} (-1)^n n! \int_0^\infty \frac{t^n}{[1+(1+y)t]^{n+1}} \cdot \frac{dt}{1+t}$$

$$\overset{t=\frac{u}{1-u(1+y)}}{=} (-1)^n n! \int_0^{\frac{1}{1+y}} \frac{u^n}{1-uy} du$$

$$\overset{u=\frac{v}{1+y}}{=} \frac{(-1)^n n!}{(1+y)^{n+1}} \int_0^1 \frac{v^n}{1-\frac{vy}{1+y}} dv$$

$$\overset{v^n=t}{=} \frac{(-1)^n (n-1)!}{(1+y)^{n+1}} \int_0^1 \frac{\sqrt[n]{t}}{1-\frac{y\sqrt[n]{t}}{1+y}} dt.$$

This implies that

$$\lim_{n\to\infty} (-1)^n \frac{(1+y)^n}{(n-1)!} \left(\frac{\ln(1+y)}{y}\right)^{(n)} = \frac{1}{1+y} \lim_{n\to\infty} \int_0^1 \frac{\sqrt[n]{t}}{1-\frac{y\sqrt[n]{t}}{1+y}} dt$$

$$= \frac{1}{1+y} \int_0^1 \frac{1}{1-\frac{y}{1+y}} dt$$

$$= 1.$$

4.54. *Solution I due to H. Ricardo* [123]. The crucial observation is that $f_n^{(n)}(0)$ is $n!$ times the coefficient of x^n in the Taylor expansion of $f_n(x)$ about the origin. We have $\sin(kx) = kx + O(x^3)$, which gives

$$f_n(x) = \prod_{k=1}^n (kx + O(x^3)) = n! x^n + O\left(x^{n+2}\right).$$

Thus, $f_n^{(n)}(0) = n!^2$.

Solution II. We apply the generalized Leibniz formula. We have, since $\sin^{(k)}(ax+b) = a^k \sin\left(ax + b + \frac{k\pi}{2}\right)$, $k \geq 0$, $a, b \in \mathbb{R}$, that

$$f_n^{(n)}(x) = \sum_{k_1+k_2+\cdots+k_n=n} \frac{n!}{k_1! k_2! \cdots k_n!} 1^{k_1} 2^{k_2} \cdots n^{k_n} \sin\left(x + \frac{k_1\pi}{2}\right) \cdots \sin\left(nx + \frac{k_n\pi}{2}\right)$$

and when $x = 0$ one has that

$$f_n^{(n)}(0) = \sum_{k_1+k_2+\cdots+k_n=n} \frac{n!}{k_1!k_2!\cdots k_n!} 1^{k_1} 2^{k_2} \cdots n^{k_n} \sin\left(\frac{k_1\pi}{2}\right) \cdots \sin\left(\frac{k_n\pi}{2}\right),$$

$$(12.7)$$

where $k_i \geq 0$ and $k_1 + k_2 + \cdots + k_n = n$.

If k_i is an even integer, i.e. $k_i = 2m_i$, $m_i \geq 0$, one has that $\sin\frac{k_i\pi}{2} = \sin m_i\pi = 0$. If k_i are odd integers, i.e. $k_i = 2m_i - 1$, $m_i \geq 1$, then $\sin\frac{k_i\pi}{2} = \sin\frac{(2m_i-1)\pi}{2} = (-1)^{m_i-1}$. The equation $k_1 + k_2 + \cdots + k_n = n$ implies that $2m_1 - 1 + 2m_2 - 1 + \cdots + 2m_n - 1 = n \Rightarrow$

$$m_1 + m_2 + \cdots + m_n = n. \tag{12.8}$$

However, $m_i \geq 1, \forall i = 1, 2, \ldots, n$ and this implies that $m_1 + m_2 + \cdots + m_n \geq n$. Thus, the only case when one has equality in (12.8) is when $m_1 = m_2 = \cdots = m_n = 1 \Rightarrow k_1 = k_2 = \cdots = k_n = 1$. It follows, from formula (12.7), that

$$f_n^{(n)}(0) = \frac{n!}{1! \cdot 1! \cdots 1!} \cdot 1 \cdot 2 \cdot 3 \cdots n = n!^2.$$

4.55. If n is odd, then $g_n^{(n)}(0) = 0$. If n is even, then we have

$$g_{2n}^{(2n)}(0) = (-1)^n \frac{(2n)!^3}{2^n} \sum_{1 \leq i_1 < i_2 < \cdots < i_n \leq 2n} \frac{1}{i_1^2 i_2^2 \cdots i_n^2}.$$

The preceding formula is obtained by calculating the coefficient of x^n in the Taylor expansion of $g_n(x)$ about the origin. See the first solution of Problem 4.54.

4.56. $f_n^{(n)}(0) = n!^2$. See the first solution of Problem 4.54.

4.57. If n is odd, then $g_n^{(n)}(0) = 0$. If n is even, then we have

$$g_{2n}^{(2n)}(0) = \frac{(2n)!^3}{2^n} \sum_{1 \leq i_1 < i_2 < \cdots < i_n \leq 2n} \frac{1}{i_1^2 i_2^2 \cdots i_n^2}.$$

The preceding formula is obtained by calculating the coefficient of x^n in the Taylor expansion of $g_n(x)$ about the origin. See the first solution of Problem 4.54.

12.5   Taylor's Formula

4.58.

(a) $\forall x \in (-3, \infty)$, $\exists \theta_{-1,x} \in (0, 1)$ such that

$$\frac{1}{x+3} = \sum_{k=0}^{n}(-1)^k \frac{(x+1)^k}{2^{k+1}} + (-1)^{n+1}\frac{(x+1)^{n+1}}{[2+\theta_{-1,x}(x+1)]^{n+2}};$$

(b) $\forall x \in \mathbb{R}, \exists \theta_{2,x} \in (0,1)$ such that

$$e^{2x-1} = e^3\left[\sum_{k=0}^{n}\frac{2^k}{k!}(x-2)^k + \frac{2^{n+1}}{(n+1)!}e^{2\theta_{2,x}(x-2)}(x-2)^{n+1}\right];$$

(c) when $n = 2m$, $m \geq 0$, $\forall x \in \mathbb{R}, \exists \theta_{\frac{\pi}{2},x} \in (0,1)$ such that

$$\sin(3x) = \sum_{k=0}^{m}(-1)^{k+1}\frac{3^{2k}}{(2k)!}\left(x-\frac{\pi}{2}\right)^{2k}$$

$$+ (-1)^{m+2}\frac{3^{2m+1}}{(2m+1)!}\sin\left(3\theta_{\frac{\pi}{2},x}\left(x-\frac{\pi}{2}\right)\right)\left(x-\frac{\pi}{2}\right)^{2m+1},$$

and, when $n = 2m+1$, $m \geq 0$, $\forall x \in \mathbb{R}, \exists \theta_{\frac{\pi}{2},x} \in (0,1)$ such that

$$\sin(3x) = \sum_{k=0}^{m}(-1)^{k+1}\frac{3^{2k}}{(2k)!}\left(x-\frac{\pi}{2}\right)^{2k}$$

$$+ (-1)^{m+2}\frac{3^{2m+2}}{(2m+2)!}\cos\left(3\theta_{\frac{\pi}{2},x}\left(x-\frac{\pi}{2}\right)\right)\left(x-\frac{\pi}{2}\right)^{2m+2};$$

(d) when $n = 2m$, $m \geq 0$, $\forall x \in \mathbb{R}, \exists \theta_{1,x} \in (0,1)$ such that

$$\cosh(x-1) = 1 + \frac{(x-1)^2}{2!} + \frac{(x-1)^4}{4!} + \cdots + \frac{(x-1)^{2m}}{(2m)!}$$

$$+ \frac{(x-1)^{2m+1}}{(2m+1)!}\sinh(\theta_{1,x}(x-1)),$$

and, when $n = 2m+1$, $m \geq 0$, $\forall x \in \mathbb{R}, \exists \theta_{1,x} \in (0,1)$ such that

$$\cosh(x-1) = 1 + \frac{(x-1)^2}{2!} + \frac{(x-1)^4}{4!} + \cdots + \frac{(x-1)^{2m}}{(2m)!}$$

$$+ \frac{(x-1)^{2m+2}}{(2m+2)!}\cosh(\theta_{1,x}(x-1)).$$

4.59.

(a) $f(x) = x^5 + 7x^4 - 2x^3 - 46x^2 + 65x - 25$;
(b) $f(x) = x^4 - 5x^3 - 31x^2 - 43x - 18$.

4.60. $f(x) = (x-1)^5 + 3(x-1)^2 + 1.$
4.61. $f(x) = (x+1)^5 - 7(x+1)^3 - 1.$
4.62.

(a) $(M_n f)(x) = \frac{1}{2} \sum_{k=0}^{n} (-1)^k \left(1 - \frac{1}{3^{k+1}}\right) x^k;$

(b) $(M_n f)(x) = 1 + \frac{1}{2}x + \sum_{k=2}^{n} (-1)^{k-1} \frac{(2k-3)!!}{(2k)!!} x^k;$

(c) $(M_n f)(x) = -\ln 2 - \sum_{k=1}^{n} \left(1 - \frac{1}{2^k}\right) \frac{x^k}{k}.$

4.63.

(a) $\forall x \in \mathbb{R}, \exists \theta_x \in (0,1)$ such that

$$e^x = 1 + \frac{x}{1!} + \frac{x^2}{2!} + \frac{x^3}{3!} + \cdots + \frac{x^n}{n!} + \frac{x^{n+1}}{(n+1)!} e^{\theta_x x};$$

(b) when $n = 2m, m \geq 0, \forall x \in \mathbb{R}, \exists \theta_x \in (0,1)$ such that

$$\sin x = x - \frac{x^3}{3!} + \frac{x^5}{5!} - \cdots + (-1)^{m-1} \frac{x^{2m-1}}{(2m-1)!} + (-1)^m \frac{x^{2m+1}}{(2m+1)!} \cos(\theta_x x),$$

and, when $n = 2m - 1, m \in \mathbb{N}, \forall x \in \mathbb{R}, \exists \theta_x \in (0,1)$ such that

$$\sin x = x - \frac{x^3}{3!} + \frac{x^5}{5!} - \cdots + (-1)^{m-1} \frac{x^{2m-1}}{(2m-1)!} + (-1)^m \frac{x^{2m}}{(2m)!} \sin(\theta_x x);$$

(c) when $n = 2m, m \geq 0, \forall x \in \mathbb{R}, \exists \theta_x \in (0,1)$ such that

$$\cos x = 1 - \frac{x^2}{2!} + \frac{x^4}{4!} - \cdots + (-1)^m \frac{x^{2m}}{(2m)!} + (-1)^{m+1} \frac{x^{2m+1}}{(2m+1)!} \sin(\theta_x x),$$

and, when $n = 2m + 1, m \geq 0, \forall x \in \mathbb{R}, \exists \theta_x \in (0,1)$ such that

$$\cos x = 1 - \frac{x^2}{2!} + \frac{x^4}{4!} - \cdots + (-1)^m \frac{x^{2m}}{(2m)!} + (-1)^{m+1} \frac{x^{2m+2}}{(2m+2)!} \cos(\theta_x x);$$

(d) $\forall x \in (-1, \infty), \exists \theta_x \in (0,1)$ such that

$$(1+x)^\alpha = 1 + \frac{\alpha}{1!}x + \frac{\alpha(\alpha-1)}{2!}x^2 + \frac{\alpha(\alpha-1)(\alpha-2)}{3!}x^3 + \cdots$$

$$+ \frac{\alpha(\alpha-1)(\alpha-2)\cdots(\alpha-n+1)}{n!}x^n$$

(continued)

$$+\frac{\alpha(\alpha-1)(\alpha-2)\cdots(\alpha-n)}{(n+1)!}(1+\theta_x x)^{\alpha-n-1}x^{n+1};$$

(e) $\forall x \in (-1, \infty)$, $\exists \theta_x \in (0, 1)$ such that

$$\ln(1+x) = x - \frac{x^2}{2} + \frac{x^3}{3} - \cdots + (-1)^{n-1}\frac{x^n}{n} + (-1)^n \frac{x^{n+1}}{n+1}\frac{1}{(1+\theta_x x)^{n+1}};$$

(f) when $n = 2m$, $m \geq 0$, $\forall x \in \mathbb{R}$, $\exists \theta_x \in (0, 1)$ such that

$$\sinh x = x + \frac{x^3}{3!} + \frac{x^5}{5!} + \cdots + \frac{x^{2m-1}}{(2m-1)!} + \frac{x^{2m+1}}{(2m+1)!}\cosh(\theta_x x),$$

and, when $n = 2m - 1$, $m \in \mathbb{N}$, $\forall x \in \mathbb{R}$, $\exists \theta_x \in (0, 1)$ such that

$$\sinh x = x + \frac{x^3}{3!} + \frac{x^5}{5!} + \cdots + \frac{x^{2m-1}}{(2m-1)!} + \frac{x^{2m}}{(2m)!}\sinh(\theta_x x);$$

(g) when $n = 2m$, $m \geq 0$, $\forall x \in \mathbb{R}$, $\exists \theta_x \in (0, 1)$ such that

$$\cosh x = 1 + \frac{x^2}{2!} + \frac{x^4}{4!} + \cdots + \frac{x^{2m}}{(2m)!} + \frac{x^{2m+1}}{(2m+1)!}\sinh(\theta_x x),$$

and, when $n = 2m + 1$, $m \geq 0$, $\forall x \in \mathbb{R}$, $\exists \theta_x \in (0, 1)$ such that

$$\cosh x = 1 + \frac{x^2}{2!} + \frac{x^4}{4!} + \cdots + \frac{x^{2m}}{(2m)!} + \frac{x^{2m+2}}{(2m+2)!}\cosh(\theta_x x).$$

4.64. It suffices to consider only the case when $x < 0$. We have, based on Maclaurin formula of order $2n$ for the exponential function that, $\forall x \in \mathbb{R}$, $\exists \theta \in (0, 1)$ such that

$$e^x = 1 + \frac{x}{1!} + \frac{x^2}{2!} + \cdots + \frac{x^{2n}}{(2n)!} + \frac{x^{2n+1}}{(2n+1)!}e^{\theta x}.$$

When $x < 0 \Rightarrow x^{2n+1} < 0 \Rightarrow 1 + \frac{x}{1!} + \frac{x^2}{2!} + \cdots + \frac{x^{2n}}{(2n)!} = e^x - \frac{x^{2n+1}}{(2n+1)!}e^{\theta x} > 0.$

4.65. e is not a rational number

We assume that e is a rational number, i.e. $\exists\, a, b \in \mathbb{N}$ such that $e = \frac{a}{b}$. We have, based on Maclaurin's formula of order n for the exponential function, that $\forall x \in \mathbb{R}, \exists\, \theta \in (0, 1)$ such that

$$e^x = 1 + \frac{x}{1!} + \cdots + \frac{x^n}{n!} + \frac{x^{n+1}}{(n+1)!} e^{\theta x}.$$

When $x = 1$ we have that

$$\frac{a}{b} = e = 1 + \frac{1}{1!} + \cdots + \frac{1}{n!} + \frac{e^{\theta}}{(n+1)!}.$$

The preceding equality implies that

$$n!\left(\frac{a}{b} - 1 - \frac{1}{1!} - \cdots - \frac{1}{n!}\right) = \frac{e^{\theta}}{n+1}. \tag{12.9}$$

We note that $0 < \frac{e^{\theta}}{n+1} < 1$, for $n \geq 2$. Let $n > \max\{2, b\}$. Equality (12.9) is a contradiction since the left hand side of (12.9) is an integer, while the right hand side of (12.9) is a real number between 0 and 1.

$e^{\sqrt{2}}$ is not a rational number

We assume that $e^{\sqrt{2}}$ is a rational number. This implies that $e^{-\sqrt{2}}$ is also a rational number and it follows that $\cosh \sqrt{2}$ is a rational number, i.e. $\exists\, a, b \in \mathbb{N}$ such that $\cosh \sqrt{2} = \frac{a}{b}$. We have, based on Maclaurin's formula for cosh, that $\forall x \in \mathbb{R}, \exists\, \theta \in (0, 1)$ such that

$$\cosh x = 1 + \frac{x^2}{2!} + \cdots + \frac{x^{2n}}{(2n)!} + \frac{x^{2n+2}}{(2n+2)!} \cosh(\theta x).$$

When $x = \sqrt{2}$ we have that

$$\frac{a}{b} = \cosh \sqrt{2} = 1 + \frac{2}{2!} + \cdots + \frac{2^n}{(2n)!} + \frac{2^{n+1}}{(2n+2)!} \cosh\left(\theta \sqrt{2}\right).$$

Since $\frac{2^k}{(2k)!} = \frac{1}{(2k-1)!!k!}$, $k \geq 1$, the preceding equality becomes

$$\frac{a}{b} = \cosh \sqrt{2} = 1 + \frac{1}{1!!1!} + \cdots + \frac{1}{(2n-1)!!n!} + \frac{\cosh\left(\theta \sqrt{2}\right)}{(2n+1)!!(n+1)!}.$$

(continued)

It follows that

$$(2n-1)!!n!\left(\frac{a}{b}-1-\frac{1}{1!!1!}-\cdots-\frac{1}{(2n-1)!!(n-1)!}\right)=\frac{\cosh\left(\theta\sqrt{2}\right)}{(2n+1)(n+1)}.$$

For n large enough, the preceding equality contradicts the fact that the left hand side of the equality is an integer, while the right hand side is a real number between 0 and 1.

To prove that $\sin 1$ and $\cos 1$ are irrational numbers, one can apply the same technique as above.

4.66.

(a) We have

$$\sum_{n=1}^{\infty}\frac{n^3}{n!}=\sum_{n=1}^{\infty}\frac{n^2}{(n-1)!}=\sum_{k=0}^{\infty}\frac{k^2+2k+1}{k!}=\sum_{k=1}^{\infty}\frac{k^2}{k!}+2\sum_{k=1}^{\infty}\frac{k}{k!}+e$$

$$=\sum_{k=1}^{\infty}\frac{k}{(k-1)!}+2\sum_{k=1}^{\infty}\frac{1}{(k-1)!}+e$$

$$=\sum_{k=2}^{\infty}\frac{1}{(k-2)!}+\sum_{k=1}^{\infty}\frac{1}{(k-1)!}+3e$$

$$=5e.$$

(b) We have

$$f(x)=\sum_{n=0}^{\infty}\frac{f^{(n)}(0)}{n!}x^n,\quad |x|<R.$$

It follows that

$$\left(x\left(xf'(x)\right)'\right)'=\sum_{n=1}^{\infty}\frac{n^3}{n!}f^{(n)}(0)x^{n-1}.$$

This implies that

$$\left(x\left(xf'(x)\right)'\right)'(1)=\sum_{n=1}^{\infty}\frac{n^3}{n!}f^{(n)}(0).$$

A calculation shows that $\left(x\left(xf'(x)\right)'\right)'(1) = f'(1) + 3f''(1) + f'''(1)$.

4.67. We have, based on Maclaurin's formula of order 1, that $\forall x \in \mathbb{R}$ there exists $\theta \in (0, 1)$ such that

$$f(x) = 1 + \frac{1}{2}f''(\theta x)x^2$$

and it follows that

$$f\left(\frac{x}{\sqrt{n}}\right) = 1 + \frac{1}{2}f''\left(\theta\frac{x}{\sqrt{n}}\right)\frac{x^2}{n}.$$

This implies that

$$\lim_{n\to\infty}\left(f\left(\frac{x}{\sqrt{n}}\right)\right)^n = \lim_{n\to\infty}\left(1 + \frac{1}{2}f''\left(\theta\frac{x}{\sqrt{n}}\right)\frac{x^2}{n}\right)^n$$

$$= e^{\lim_{n\to\infty}\frac{x^2}{2}f''\left(\theta\frac{x}{\sqrt{n}}\right)}$$

$$= e^{p\frac{x^2}{2}}.$$

12.6   Series with the Maclaurin Remainder of a Function f

4.68.

(a) Use that

$$\frac{x^n}{n} + \frac{x^{n+1}}{n+1} + \frac{x^{n+2}}{n+2} + \cdots = \int_0^x \frac{t^{n-1}}{1-t}\,dt, \quad x \in [-1, 1).$$

(b) Use Taylor's formula with the remainder in integral form or Abel's summation formula.

(c) We have

$$S = \sum_{n=0}^{\infty} a^n \left(\frac{f^{(n)}(0)}{n!} x^n + \frac{f^{(n+1)}(0)}{(n+1)!} x^{n+1} + \cdots \right)$$

$$= f(0) + \frac{f'(0)}{1!} x + \frac{f''(0)}{2!} x^2 + \cdots + \sum_{n=1}^{\infty} a^n \left(\frac{f^{(n)}(0)}{n!} x^n + \frac{f^{(n+1)}(0)}{(n+1)!} x^{n+1} + \cdots \right)$$

$$= f(x) + \sum_{n=1}^{\infty} a^n \left(\frac{f^{(n)}(0)}{n!} x^n + \frac{f^{(n+1)}(0)}{(n+1)!} x^{n+1} + \cdots \right)$$

$$\overset{n-1=m}{=} f(x) + \sum_{m=0}^{\infty} a^{m+1} \left(\frac{f^{(m+1)}(0)}{(m+1)!} x^{m+1} + \frac{f^{(m+2)}(0)}{(m+2)!} x^{m+2} + \cdots \right)$$

$$= f(x) + a \sum_{m=0}^{\infty} a^m \left[\left(\frac{f^{(m)}(0)}{m!} x^m + \frac{f^{(m+1)}(0)}{(m+1)!} x^{m+1} + \cdots \right) - \frac{f^{(m)}(0)}{m!} x^m \right]$$

$$= f(x) + aS - a \sum_{m=0}^{\infty} \frac{f^{(m)}(0)}{m!} (ax)^m$$

$$= f(x) + aS - f(ax).$$

4.69.

(a) Integrate by parts.

(b) We have

$$\sum_{n=0}^{\infty} \left(e^x - 1 - \frac{x}{1!} - \frac{x^2}{2!} - \cdots - \frac{x^n}{n!} \right) = \sum_{n=0}^{\infty} \int_0^x \frac{(x-t)^n}{n!} e^t \, dt$$

$$= \int_0^x e^t \left(\sum_{n=0}^{\infty} \frac{(x-t)^n}{n!} \right) dt$$

$$= \int_0^x e^t e^{x-t} \, dt$$

$$= \int_0^x e^x \, dt$$

$$= x e^x.$$

(c) This part of the problem is solved similarly to part (b).

4.70. Use Taylor's formula with the remainder in integral form or apply Abel's summation formula.

4.71.

(a) Use mathematical induction.
(b) Apply Abel's summation formula with $a_n = (-1)^{n-1} n^2$ and $b_n = e^x - 1 - \frac{x}{1!} - \frac{x^2}{2!} - \cdots - \frac{x^n}{n!}$.

4.72. We have

$$S = \sum_{n=0}^{\infty} \cos \frac{n\pi}{2} \left(f(x) - f(0) - \frac{f'(0)}{1!}x - \frac{f''(0)}{2!}x^2 - \cdots - \frac{f^{(n)}(0)}{n!}x^n \right)$$

$$= \sum_{k=0}^{\infty} (-1)^k \left(f(x) - f(0) - \frac{f'(0)}{1!}x - \frac{f''(0)}{2!}x^2 - \cdots - \frac{f^{(2k)}(0)}{(2k)!}x^{2k} \right)$$

$$= f(x) - f(0) + \sum_{k=1}^{\infty} (-1)^k \left(f(x) - f(0) - \frac{f'(0)}{1!}x - \cdots - \frac{f^{(2k)}(0)}{(2k)!}x^{2k} \right)$$

$$\overset{k-1=i}{=} f(x) - f(0) + \sum_{i=0}^{\infty} (-1)^{i+1} \left(f(x) - f(0) - \frac{f'(0)}{1!}x - \cdots - \frac{f^{(2i+2)}(0)}{(2i+2)!}x^{2i+2} \right)$$

$$= f(x) - f(0) - \sum_{i=0}^{\infty} (-1)^i \left(f(x) - f(0) - \frac{f'(0)}{1!}x - \cdots - \frac{f^{(2i)}(0)}{(2i)!}x^{2i} \right)$$

$$+ \sum_{i=0}^{\infty} (-1)^i \frac{f^{(2i+1)}(0)}{(2i+1)!}x^{2i+1} + \sum_{i=0}^{\infty} (-1)^i \frac{f^{(2i+2)}(0)}{(2i+2)!}x^{2i+2}$$

$$= f(x) - f(0) - S + \frac{f(xi) - f(-xi)}{2i} + f(0) - \frac{f(xi) + f(-xi)}{2}.$$

We used in the previous calculations the following formulae:

- $\displaystyle \sum_{i=0}^{\infty} (-1)^i \frac{f^{(2i+2)}(0)}{(2i+2)!}x^{2i+2} = f(0) - \frac{f(xi) + f(-xi)}{2}, \quad |x| < R;$

- $\displaystyle \sum_{i=0}^{\infty} (-1)^i \frac{f^{(2i+1)}(0)}{(2i+1)!}x^{2i+1} = \frac{f(xi) - f(-xi)}{2i}, \quad |x| < R.$

The second series formula can be proved similarly.

4.73.

(a) First, we prove that the series converges absolutely. We apply the Maclaurin formula of order n to the function $f(x) = \sin x$ and we have, since $f(x) = M_n(x) + R_n(x)$, that

$$\sin x - x + \frac{1}{3!}x^3 - \cdots - \frac{\sin \frac{n\pi}{2}}{n!}x^n = R_n(x).$$

Since $\sin^{(k)}(ax) = a^k \sin \left(ax + \frac{k\pi}{2}\right)$, $k \geq 0$, $a, x \in \mathbb{R}$, we have, by using the formula for the remainder R_n, that

$$R_n(x) = \frac{\sin^{(n+1)}(\theta_x x)}{(n+1)!}x^{n+1} = \frac{\sin \left(\theta_x x + \frac{(n+1)\pi}{2}\right)}{(n+1)!}x^{n+1},$$

for some $\theta_x \in (0, 1)$. It follows that

$$\sum_{n=0}^{\infty} \left| 3^n \left(\sin x - x + \frac{1}{3!}x^3 - \cdots - \frac{\sin \frac{n\pi}{2}}{n!}x^n \right) \right| = \sum_{n=0}^{\infty} \left| 3^n \frac{\sin \left(\theta_x x + \frac{(n+1)\pi}{2}\right)}{(n+1)!}x^{n+1} \right|$$

$$\leq \frac{1}{3} \sum_{n=0}^{\infty} \frac{(3|x|)^{n+1}}{(n+1)!}$$

$$= \frac{1}{3} \left(e^{3|x|} - 1 \right),$$

which shows that the series converges absolutely.

Now, we calculate its sum. We have

$$f(x) = \sum_{n=0}^{\infty} 3^n \left(\sin x - x + \frac{1}{3!}x^3 - \cdots - \frac{\sin \frac{n\pi}{2}}{n!}x^n \right)$$

$$= \sin x + \sum_{n=1}^{\infty} 3^n \left(\sin x - x + \frac{1}{3!}x^3 - \cdots - \frac{\sin \frac{n\pi}{2}}{n!}x^n \right)$$

$$\overset{n-1=m}{=} \sin x + \sum_{m=0}^{\infty} 3^{m+1} \left(\sin x - x + \frac{1}{3!}x^3 - \cdots - \frac{\sin \frac{(m+1)\pi}{2}}{(m+1)!}x^{m+1} \right)$$

$$= \sin x + 3\sum_{m=0}^{\infty} 3^m \left(\sin x - x + \frac{1}{3!}x^3 - \cdots - \frac{\sin \frac{m\pi}{2}}{m!}x^m \right)$$

$$-\sum_{m=0}^{\infty} 3^{m+1} \frac{\sin \frac{(m+1)\pi}{2}}{(m+1)!} x^{m+1}$$

$$= \sin x + 3f(x) - \sum_{m=0}^{\infty} \frac{\sin^{(m+1)}(0)}{(m+1)!} (3x)^{m+1}$$

$$= \sin x + 3f(x) - \sin(3x),$$

and it follows that $2f(x) = \sin(3x) - \sin x \Rightarrow f(x) = \sin x \cos(2x)$.

(b) See [144].

4.74. We have, based on part (a) of Problem 4.69, that

$$\sum_{n=k}^{\infty} C_n^k f^{(n)}(0) \left(e^x - 1 - \frac{x}{1!} - \frac{x^2}{2!} - \cdots - \frac{x^n}{n!} \right) = \sum_{n=k}^{\infty} C_n^k f^{(n)}(0) \int_0^x \frac{(x-t)^n}{n!} e^t \, dt$$

$$= \frac{1}{k!} \int_0^x e^t \sum_{n=k}^{\infty} \frac{f^{(n)}(0)}{(n-k)!} (x-t)^n \, dt \overset{n-k=i}{=} \frac{1}{k!} \int_0^x e^t \sum_{i=0}^{\infty} \frac{f^{(k+i)}(0)}{i!} (x-t)^{k+i} \, dt$$

$$= \frac{1}{k!} \int_0^x e^t (x-t)^k \sum_{i=0}^{\infty} \frac{\left(f^{(k)} \right)^{(i)}(0)}{i!} (x-t)^i \, dt = \frac{1}{k!} \int_0^x e^t (x-t)^k f^{(k)}(x-t) \, dt$$

$$\overset{x-t=y}{=} \frac{1}{k!} \int_0^x e^{x-y} y^k f^{(k)}(y) \, dy.$$

4.75. Apply Abel's summation formula with $a_n = (-1)^n$ and $b_n = f(x) - (T_n f)(x)$. The series equals $f(0) - \frac{f(x)+f(-x)}{2}$.

4.76.

(a) *Solution I.* Calculate the $4n$th partial sum of the series.

 Solution II. We have

$$y(x) = \sum_{n=1}^{\infty} (-1)^{\lfloor \frac{n}{2} \rfloor} \left(e^x - 1 - \frac{x}{1!} - \frac{x^2}{2!} - \cdots - \frac{x^n}{n!} \right)$$

$$= e^x - 1 - \frac{x}{1!} + \sum_{n=2}^{\infty} (-1)^{\lfloor \frac{n}{2} \rfloor} \left(e^x - 1 - \frac{x}{1!} - \frac{x^2}{2!} - \cdots - \frac{x^n}{n!} \right)$$

and it follows that

$$y''(x) = e^x + \sum_{n=2}^{\infty}(-1)^{\lfloor\frac{n}{2}\rfloor}\left(e^x - 1 - \frac{x}{1!} - \frac{x^2}{2!} - \cdots - \frac{x^{n-2}}{(n-2)!}\right)$$

$$\overset{n-2=m}{=\!=\!=} e^x + \sum_{m=0}^{\infty}(-1)^{\lfloor\frac{m+2}{2}\rfloor}\left(e^x - 1 - \frac{x}{1!} - \frac{x^2}{2!} - \cdots - \frac{x^m}{m!}\right)$$

$$= e^x - \sum_{m=0}^{\infty}(-1)^{\lfloor\frac{m}{2}\rfloor}\left(e^x - 1 - \frac{x}{1!} - \frac{x^2}{2!} - \cdots - \frac{x^m}{m!}\right)$$

$$= e^x - (e^x - 1) - y(x)$$

$$= 1 - y(x).$$

We obtained the differential equation with constant coefficients $y'' + y = 1$. The solution of this equation is $y(x) = \mathscr{C}_1 \cos x + \mathscr{C}_2 \sin x + 1$, where $\mathscr{C}_1, \mathscr{C}_2 \in \mathbb{R}$. Since $y(0) = 0$ and $y'(0) = 0$, we get that $\mathscr{C}_1 = -1$ and $\mathscr{C}_2 = 0$ and these imply that $y(x) = 1 - \cos x$.

(b) Calculate the $4n$th partial sum of the series or differentiate the series in part (a).

The general case

Let $S_{4n} = \sum_{k=1}^{4n}(-1)^{\lfloor\frac{k}{2}\rfloor}\left(f(x) - f(0) - \frac{f'(0)}{1!}x - \frac{f''(0)}{2!}x^2 - \cdots - \frac{f^{(k)}(0)}{k!}x^k\right)$

be the $4n$th partial sum of the first series.

A calculation shows that

$$S_{4n} = \frac{f''(0)}{2!}x^2 - \frac{f^{(4)}(0)}{4!}x^4 + \cdots + \frac{f^{(4n-2)}(0)}{(4n-2)!}x^{4n-2} - \frac{f^{(4n)}(0)}{(4n)!}x^{4n}$$

and it follows that

$$\lim_{n\to\infty} S_{4n} = f(0) - \frac{f(ix) + f(-ix)}{2}.$$

The second series can be calculated similarly.

4.77. *Solution I.*

(a) Prove the formula by mathematical induction.
(b) We apply Taylor's formula with the remainder in integral form and we have

$$\sum_{n=k}^{\infty} C_n^k \left(f(x) - f(0) - \frac{f'(0)}{1!}x - \frac{f''(0)}{2!}x^2 - \cdots - \frac{f^{(n)}(0)}{n!}x^n \right)$$

$$= \sum_{n=k}^{\infty} C_n^k \int_0^x \frac{(x-t)^n}{n!} f^{(n+1)}(t)dt$$

$$= \frac{1}{k!} \int_0^x \sum_{n=k}^{\infty} \frac{(x-t)^n}{(n-k)!} f^{(n+1)}(t)dt$$

$$\stackrel{n-k=i}{=} \frac{1}{k!} \int_0^x (x-t)^k \sum_{i=0}^{\infty} \frac{(x-t)^i}{i!} \left(f^{(k+1)} \right)^{(i)}(t)dt$$

$$\stackrel{x-t=y}{=} \frac{1}{k!} \int_0^x y^k \sum_{i=0}^{\infty} \frac{y^i}{i!} \left(f^{(k+1)} \right)^{(i)}(x-y)dy$$

$$= \frac{1}{k!} \int_0^x y^k f^{(k+1)}(x)dy$$

$$= \frac{f^{(k+1)}(x)}{(k+1)!} x^{k+1}.$$

Solution II. Use Abel's summation formula combined to part (a) of the problem. The challenging part of this solution would be to prove that

$$\lim_{n\to\infty} C_{n+1}^{k+1} \left(f(x) - f(0) - \frac{f'(0)}{1!}x - \frac{f''(0)}{2!}x^2 - \cdots - \frac{f^{(n+1)}(0)}{(n+1)!}x^{n+1} \right) = 0.$$

This limit can be proved using an idea of R. Mabry, which appears in the solution given by him [107] for a problem proposed by O. Furdui [33].

4.78. We integrate by parts twice and we have

$$I_n = \int_0^\infty \frac{T_n(x) - \sin x}{x^{2n+1}} dx = -\frac{T_n(x) - \sin x}{2nx^{2n}} \Big|_0^\infty + \frac{1}{2n} \int_0^\infty \frac{T_n'(x) - \cos x}{x^{2n}} dx$$

$$= \frac{1}{2n} \int_0^\infty \frac{T_n'(x) - \cos x}{x^{2n}} dx = -\frac{T_n'(x) - \cos x}{2n(2n-1)x^{2n-1}} \Big|_0^\infty$$

$$+ \frac{1}{2n(2n-1)} \int_0^\infty \frac{T_n''(x) + \sin x}{x^{2n-1}} dx$$

$$= -\frac{1}{2n(2n-1)} \int_0^\infty \frac{T_{n-1}(x) - \sin x}{x^{2n-1}} dx$$

$$= -\frac{1}{2n(2n-1)} I_{n-1},$$

since $T_n''(x) = -T_{n-1}(x)$. It follows that

$$I_n = -\frac{1}{2n(2n-1)}I_{n-1} \quad \text{and} \quad I_n = \frac{(-1)^{n-1}}{2n(2n-1)\cdots 4\cdot 3}I_1.$$

We calculate I_1 integrating by parts twice and we have

$$I_1 = \int_0^\infty \frac{x-\sin x}{x^3}dx = -\frac{x-\sin x}{2x^2}\Big|_0^\infty + \frac{1}{2}\int_0^\infty \frac{1-\cos x}{x^2}dx$$

$$= \frac{1}{2}\int_0^\infty \frac{1-\cos x}{x^2}dx = -\frac{1-\cos x}{2x}\Big|_0^\infty + \frac{1}{2}\int_0^\infty \frac{\sin x}{x}dx$$

$$= \frac{1}{2}\int_0^\infty \frac{\sin x}{x}dx$$

$$= \frac{\pi}{4}.$$

12.7   Series with Fractional Part Function

4.80, **4.81**, and **4.82**. See [59].

12.8   Extrema of One Variable Functions

4.83. (a) min $f(\mathbb{R}) = -17$, max $f(\mathbb{R}) = 17$; (b) max $f(\mathbb{R}) = f\left(\ln\frac{\sqrt{5}+1}{2}\right) = -\frac{5}{4}$;

(c) min $f((-\infty, 0)) = f(-1) = \frac{3\pi^2}{16}$; (d) min $f(\mathbb{R}) = -625$.
4.84. min $f(C) = f(-2) = -30$, max $f(C) = f(2) = 30$.
4.85.

(a) $x = -1$ is a global minimum point and $f(-1) = -\frac{1}{2}$ is the global minimum value, $x = 1$ is a global maximum point, and $f(1) = \frac{1}{2}$ is the global maximum value;

(b) $x = -1$ is a global maximum point and $f(-1) = \frac{1}{2}$ is the global maximum value, $x = 0$ is a global minimum point and $f(0) = 0$ is the global minimum value, $x = 1$ is a global maximum point, and $f(1) = \frac{1}{2}$ is the global maximum value;

(c) $x = -1$ is a local maximum point and $f(-1) = 8$ is the local maximum value, $x = 0$ is not a local extremum point of f (it is an inflection point), $x = 5$ is a local minimum point, and $f(5) = -3124$ is the local minimum value;

(d) $x = 0$ is not a local extremum point of f (it is an inflection point), $x = 2$ is a global minimum point, and $f(2) = -65$ is the global minimum value of f;

(e) $x = -\frac{\sqrt{2}}{2}$ is a global minimum point of f and $f\left(-\frac{\sqrt{2}}{2}\right) = -\frac{\sqrt{2}}{2}e^{-\frac{1}{2}}$ is the global minimum value of f, $x = \frac{\sqrt{2}}{2}$ is a global maximum point of f, and $f(\frac{\sqrt{2}}{2}) = \frac{\sqrt{2}}{2}e^{-\frac{1}{2}}$ is the global maximum value of f;

(f) $x = -1$ is a global maximum point of f and $f(-1) = \frac{1}{e}$ is the global maximum value of f, $x = 0$ is a global minimum point of f and $f(0) = 0$ is the global minimum value of f, $x = 1$ is a global maximum point of f, and $f(1) = \frac{1}{e}$ is the global maximum value of f;

(g) $x = 0$ is a global minimum point of f and $f(0) = 1$ is the global minimum value of f;

(h) $x = \frac{1}{e^2}$ is a local maximum point of f and $f\left(\frac{1}{e^2}\right) = \frac{4}{e^2}$ is the local maximum value of f, $x = 1$ is a global minimum point of f, and $f(1) = 0$ is the global minimum value of f;

(i) $x = \frac{1}{e}$ is a global minimum point of f and $f\left(\frac{1}{e}\right) = \left(\frac{1}{e}\right)^{\frac{1}{e}}$ is the global minimum value of f;

(j) $x = \frac{\pi}{4} + 2k\pi$, $k \in \mathbb{Z}$, are local maximum points of f and $f\left(\frac{\pi}{4} + 2k\pi\right) = \frac{\sqrt{2}}{2}e^{\frac{\pi}{4}+2k\pi}$, $k \in \mathbb{Z}$, are local maximum values of f, $x = \frac{\pi}{4} + (2k+1)\pi$, $k \in \mathbb{Z}$, are local minimum points of f, and $f\left(\frac{\pi}{4} + (2k+1)\pi\right) = -\frac{\sqrt{2}}{2}e^{\frac{\pi}{4}+(2k+1)\pi}$, $k \in \mathbb{Z}$, are local minimum values of f;

(k) $x = -1$ is a global minimum point of f and $f(-1) = -\arctan\frac{1}{e}$ is the global minimum value of f;

(l) $x = -\frac{\pi}{2}$ is a global minimum point of f and $f\left(-\frac{\pi}{2}\right) = -\frac{\pi^2}{4} - 2$ is the global minimum value of f;

(m) $x = -\frac{\pi}{2}$ is a global maximum point of f and $f\left(-\frac{\pi}{2}\right) = \frac{\pi^2}{4} - 2$ is the global maximum value of f.

We have $f'(x) = 2\cos x - 2x - \pi$. Therefore, $f'(x) = 0$ if and only if $\cos x = x + \frac{\pi}{2}$. Since $y = x + \frac{\pi}{2}$ is the equation of the tangent line to the graph of the cosine function at $-\frac{\pi}{2}$, it follows that the only critical point of f is $-\frac{\pi}{2}$. Since $f''\left(-\frac{\pi}{2}\right) = f'''\left(-\frac{\pi}{2}\right) = 0$, $f^{IV}\left(-\frac{\pi}{2}\right) = -2 < 0$, we have that $x = -\frac{\pi}{2}$ is a local maximum point of f and $f\left(-\frac{\pi}{2}\right) = \frac{\pi^2}{4} - 2$ is the local maximum value of f.

Now we prove that $x = -\frac{\pi}{2}$ is in fact a global maximum point of f, since

$$f(x) \le \frac{\pi^2}{4} - 2 \iff 2\sin x - x^2 - \pi x \le \frac{\pi^2}{4} - 2$$

$$\iff 2(1 + \sin x) \le \left(x + \frac{\pi}{2}\right)^2$$

$$\overset{y=x+\frac{\pi}{2}}{\iff} 2(1 - \cos y) \le y^2$$

$$\iff \sin^2 \frac{y}{2} \le \left(\frac{y}{2}\right)^2.$$

4.86. Let $M(0, 33)$ and let $P(x, y)$ be a point on the graph of the parabola $x = 4y^2$. The distance from M to P is $d(M, P) = \sqrt{x^2 + (y - 33)^2}$. Let

$$f(x, y) = x^2 + (y-33)^2 \quad \text{and} \quad g(y) = f(4y^2, y) = 16y^4 + y^2 - 66y + 1089, \quad y \in \mathbb{R}.$$

Solution I. We observe that the function g can be written as

$$g(y) = 16(y^2 - 1)^2 + 33(y - 1)^2 + 1040, \quad y \in \mathbb{R}.$$

It follows that the minimum value of g is attained at $y = 1$, which shows that the distance from M to the graph of the parabola $x = 4y^2$ is obtained at the point $P_0(4, 1)$ and the distance equals $d(M, P_0) = 4\sqrt{65}$.

Solution II. The only critical point of g is $y = 1$ and, since $g''(1) = 194 > 0$, it follows that this is a local minimum point of g. In fact, this point is a global minimum point of g, since $g(y) - g(1) = 16(y^2 - 1)^2 + 33(y - 1)^2 \geq 0$, for all $y \in \mathbb{R}$. Therefore, the distance from M to the graph of the parabola $x = 4y^2$ is attained at $P_0(4, 1)$ and the distance equals $d(M, P_0) = 4\sqrt{65}$.

4.87. Let $M(0, b)$ and $P(x, y)$ be a point on the graph of the parabola $x^2 = 4y$. The distance from M to P is $d(M, P) = \sqrt{x^2 + (y - b)^2}$. Let

$$f(x, y) = x^2 + (y-b)^2 \quad \text{and} \quad g(y) = f(\pm 2\sqrt{y}, y) = y^2 + 2(2-b)y + b^2, \quad y \geq 0.$$

Solution I. We write the function g as follows:

$$g(y) = (y - b + 2)^2 + 4b - 4, \quad y \geq 0.$$

When $b \geq 2$, the minimum value of g is attained at $y = b - 2$, which implies that the distance from M to the graph of the parabola $x^2 = 4y$ is obtained for the points $P_{1,2}(\pm 2\sqrt{b - 2}, b - 2)$ and the distance is $d(M, P_1) = d(M, P_2) = 2\sqrt{b - 1}$. We mention that, when $b = 2$ the points P_1 and P_2 are identical to $O(0, 0)$.

When $b < 2$, the minimum value of g is attained at $y = 0$, which implies that the distance from M to the graph of the parabola $x^2 = 4y$ is obtained for the point $P_0(0, 0) = O(0, 0)$ and the distance is $d(M, O) = |b|$.

Solution II. We distinguish several cases.

Case $b \geq 2$. The only critical point of g is $y = b - 2$ and since $g''(b-2) = 2 > 0$, it follows that this is a local minimum point of g. In fact, this point is a global minimum point of g, since $g(y) - g(b - 2) = (y - b + 2)^2 \geq 0$, for all $y \geq 0$. Therefore, the distance from M to the graph of the parabola $x^2 = 4y$ is attained at $P_{1,2}(\pm 2\sqrt{b - 2}, b-2)$ and the distance is $d(M, P_1) = d(M, P_2) = 2\sqrt{b - 1}$. We mention that, when $b = 2$ points P_1 and P_2 are identical to $O(0, 0)$.

Case $0 < b < 2$. In triangle $OM'P'$, where $O(0, 0)$, $M'(0, 2)$, and $P'(x, y)$ is a point on the parabola, with $y > 0$, we have $OM' = 2$ and $M'P' > 2$. From here, it follows that $\widehat{M'OP'} > \widehat{M'P'O}$. Therefore, in triangle OMP' we have $\widehat{MOP'} > \widehat{MP'O}$, which implies $OM < MP'$. Thus, the distance from M to

the graph of the parabola is attained at $P_0(0, 0) = O(0, 0)$ and the distance is
$d(M, O) = b$.

Case $b = 0$.   The point $M(0, 0) = O(0, 0)$ is on the parabola and the distance is
$d(O, O) = 0$.

Case $b < 0$.   The distance is $d(M, O) = -b$.

Therefore, the distance from M to the graph of the parabola $x^2 = 4y$ is $2\sqrt{b-1}$,
for $b \geq 2$, and $|b|$, for $b < 2$.

4.88. The critical points of f are 0 and $\pm\sqrt{-\frac{3b}{5a}}$. Since $f''(0) = 0$ and $f'''(0) = 6b \neq 0$, it follows that 0 is not a local extremum point of f.

Let $\alpha = \sqrt{-\frac{3b}{5a}}$. Since $f''(\mp\alpha) = \pm 6b\alpha \neq 0$, it follows that points $-\alpha$ and α
are local extremum points of f. We have

$$M + m = f(-\alpha) + f(\alpha) = -a\alpha^5 - ba^3 + c + a\alpha^5 + ba^3 + c = 2c.$$

On the other hand,

$$
\begin{aligned}
Mm &= f(-\alpha)f(\alpha) \\
&= (-a\alpha^5 - ba^3 + c)(a\alpha^5 + ba^3 + c) \\
&= [-\alpha^3(a\alpha^2 + b) + c][\alpha^3(a\alpha^2 + b) + c] \\
&= \left(-\alpha^3\frac{2b}{5} + c\right)\left(\alpha^3\frac{2b}{5} + c\right) \\
&= -\alpha^6\frac{4b^2}{25} + c^2 \\
&= \frac{108b^5}{3125a^3} + c^2.
\end{aligned}
$$

4.89. First, we consider the case $n = 1$. We have $x - xy + y - 1 = (x - 1)(1 - y) = (x - 1)^2 \geq 0$.

Now, we consider the case $n \geq 2$. Let $f : [0, 2] \to \mathbb{R}$, $f(x) = x^n - 2x + x^2 + (2 - x)^n$. A calculation shows that

$$
\begin{aligned}
f'(x) &= nx^{n-1} - 2 + 2x - n(2 - x)^{n-1} \\
f''(x) &= n(n-1)x^{n-2} + 2 + n(n-1)(2 - x)^{n-2}.
\end{aligned}
$$

Since $f''(x) > 0, \forall x \in [0, 2]$, we have that f' is a strictly increasing function and
it follows, since $f'(1) = 0$, that $f'(x) \leq 0, \forall x \in [0, 1]$ and $f'(x) \geq 0, \forall x \in [1, 2]$.
This shows that $f(x) \geq f(1) = 1, \forall x \in [0, 2]$.

Partial Derivatives and Applications

13

13.1 Partial Derivatives, the Jacobian and the Hessian Matrices, Differential Operators

5.1.

(a) $\dfrac{\partial f}{\partial x}(x, y) = -(y^2 + 2) \cos(\cos(xy^2 + 2x - y + 10)) \sin(xy^2 + 2x - y + 10),$

$\dfrac{\partial f}{\partial y}(x, y) = -(2xy - 1) \cos(\cos(xy^2 + 2x - y + 10)) \sin(xy^2 + 2x - y + 10);$

(b) $\dfrac{\partial f}{\partial x}(x, y) = \dfrac{2y \sin(xy) \cos(xy)}{\sin^2(xy) + 1}, \quad \dfrac{\partial f}{\partial y}(x, y) = \dfrac{2x \sin(xy) \cos(xy)}{\sin^2(xy) + 1};$

(c) $\dfrac{\partial f}{\partial x}(x, y) = \dfrac{y}{x} + \ln z, \quad \dfrac{\partial f}{\partial y}(x, y) = \dfrac{z}{y} + \ln x, \quad \dfrac{\partial f}{\partial z}(x, y) = \dfrac{x}{z} + \ln y;$

(d) $\dfrac{\partial f}{\partial x}(x, y, z) = e^{xyz} \cos(x - yz) \, [yz \cos(x - yz) - 2 \sin(x - yz)],$

$\dfrac{\partial f}{\partial y}(x, y, z) = e^{xyz} \cos(x - yz) \, [xz \cos(x - yz) + 2z \sin(x - yz)],$

$\dfrac{\partial f}{\partial z}(x, y, z) = e^{xyz} \cos(x - yz) \, [xy \cos(x - yz) + 2y \sin(x - yz)];$

(e) $\dfrac{\partial f}{\partial x}(x, y, z) = zx^{z-1} y^z, \quad \dfrac{\partial f}{\partial y}(x, y, z) = zx^z y^{z-1}, \quad \dfrac{\partial f}{\partial z}(x, y, z) = (xy)^z \ln(xy);$

(f) $\dfrac{\partial f}{\partial x}(x, y, z) = yz \cdot x^{yz-1}, \quad \dfrac{\partial f}{\partial y}(x, y, z) = x^{yz} \cdot z \ln x, \quad \dfrac{\partial f}{\partial z}(x, y, z) = x^{yz} \cdot y \ln x;$

(g) $\dfrac{\partial f}{\partial x}(x, y, z) = y^z \cdot x^{y^z - 1}, \quad \dfrac{\partial f}{\partial y}(x, y, z) = z \cdot y^{z-1} \cdot x^{y^z} \ln x,$

$\dfrac{\partial f}{\partial z}(x, y, z) = y^z \cdot x^{y^z} \ln x \ln y.$

© The Author(s), under exclusive license to Springer Nature Switzerland AG 2021
A. Sîntămărian, O. Furdui, *Sharpening Mathematical Analysis Skills*, Problem Books
in Mathematics, https://doi.org/10.1007/978-3-030-77139-3_13

5.2.

(a) $\dfrac{\partial f}{\partial x}(x, y) = \dfrac{y}{xy + 1}$, $\dfrac{\partial f}{\partial y}(x, y) = \dfrac{x}{xy + 1}$, $\dfrac{\partial^2 f}{\partial x^2}(x, y) = -\dfrac{y^2}{(xy + 1)^2}$,

$\dfrac{\partial^2 f}{\partial y \partial x}(x, y) = \dfrac{1}{(xy + 1)^2}$, $\dfrac{\partial^2 f}{\partial y^2}(x, y) = -\dfrac{x^2}{(xy + 1)^2}$;

(b) $\dfrac{\partial f}{\partial x}(x, y) = \cos(xy) - y(x - y)\sin(xy)$,

$\dfrac{\partial f}{\partial y}(x, y) = -\cos(xy) - x(x - y)\sin(xy)$,

$\dfrac{\partial^2 f}{\partial x^2}(x, y) = -2y\sin(xy) - y^2(x - y)\cos(xy)$,

$\dfrac{\partial^2 f}{\partial y \partial x}(x, y) = -2(x - y)\sin(xy) - xy(x - y)\cos(xy)$,

$\dfrac{\partial^2 f}{\partial y^2}(x, y) = 2x\sin(xy) - x^2(x - y)\cos(xy)$;

(c) $\dfrac{\partial f}{\partial x}(x, y) = 2x\arctan\dfrac{x}{y} + y$, $\dfrac{\partial f}{\partial y}(x, y) = 2y\arctan\dfrac{x}{y} - x$,

$\dfrac{\partial^2 f}{\partial x^2}(x, y) = 2\arctan\dfrac{x}{y} + \dfrac{2xy}{x^2 + y^2}$, $\dfrac{\partial^2 f}{\partial y \partial x}(x, y) = \dfrac{-x^2 + y^2}{x^2 + y^2}$,

$\dfrac{\partial^2 f}{\partial y^2}(x, y) = 2\arctan\dfrac{x}{y} - \dfrac{2xy}{x^2 + y^2}$;

(d) $\dfrac{\partial f}{\partial x}(x, y, z) = e^{x^2 y}[2xy\sin(x - z) + \cos(x - z)]$,

$\dfrac{\partial f}{\partial y}(x, y, z) = x^2 e^{x^2 y}\sin(x - z)$, $\dfrac{\partial f}{\partial z}(x, y, z) = -e^{x^2 y}\cos(x - z)$,

$\dfrac{\partial^2 f}{\partial x^2}(x, y, z) = (4x^2 y^2 + 2y - 1)e^{x^2 y}\sin(x - z) + 4xye^{x^2 y}\cos(x - z)$,

$\dfrac{\partial^2 f}{\partial y \partial x}(x, y, z) = 2x(x^2 y + 1)e^{x^2 y}\sin(x - z) + x^2 e^{x^2 y}\cos(x - z)$,

$\dfrac{\partial^2 f}{\partial z \partial x}(x, y, z) = e^{x^2 y}\sin(x - z) - 2xye^{x^2 y}\cos(x - z)$,

$\dfrac{\partial^2 f}{\partial y^2}(x, y, z) = x^4 e^{x^2 y}\sin(x - z)$,

$\dfrac{\partial^2 f}{\partial z \partial y}(x, y, z) = -x^2 e^{x^2 y}\cos(x - z)$,

$\dfrac{\partial^2 f}{\partial z^2}(x, y, z) = -e^{x^2 y}\sin(x - z)$;

(e) $\dfrac{\partial f}{\partial x}(x, y, z) = -\dfrac{y^2 e^{-z}}{(xy - 1)^2}$, $\quad \dfrac{\partial f}{\partial y}(x, y, z) = -\dfrac{e^{-z}}{(xy - 1)^2}$,

$\dfrac{\partial f}{\partial z}(x, y, z) = -\dfrac{y e^{-z}}{xy - 1}$,

$\dfrac{\partial^2 f}{\partial x^2}(x, y, z) = \dfrac{2 y^3 e^{-z}}{(xy - 1)^3}$, $\quad \dfrac{\partial^2 f}{\partial y \partial x}(x, y, z) = \dfrac{2 y e^{-z}}{(xy - 1)^3}$,

$\dfrac{\partial^2 f}{\partial z \partial x}(x, y, z) = \dfrac{y^2 e^{-z}}{(xy - 1)^2}$, $\quad \dfrac{\partial^2 f}{\partial y^2}(x, y, z) = \dfrac{2 x e^{-z}}{(xy - 1)^3}$,

$\dfrac{\partial^2 f}{\partial z \partial y}(x, y, z) = \dfrac{e^{-z}}{(xy - 1)^2}$, $\quad \dfrac{\partial^2 f}{\partial z^2}(x, y, z) = \dfrac{y e^{-z}}{xy - 1}$;

(f) $\dfrac{\partial f}{\partial x}(x, y, z) = \dfrac{yz - 1}{(xz - 1)^2}$, $\quad \dfrac{\partial f}{\partial y}(x, y, z) = -\dfrac{1}{xz - 1}$,

$\dfrac{\partial f}{\partial z}(x, y, z) = -\dfrac{x(x - y)}{(xz - 1)^2}$,

$\dfrac{\partial^2 f}{\partial x^2}(x, y, z) = -\dfrac{2z(yz - 1)}{(xz - 1)^3}$, $\quad \dfrac{\partial^2 f}{\partial y \partial x}(x, y, z) = \dfrac{z}{(xz - 1)^2}$,

$\dfrac{\partial^2 f}{\partial z \partial x}(x, y, z) = \dfrac{2x - y - xyz}{(xz - 1)^3}$, $\quad \dfrac{\partial^2 f}{\partial y^2}(x, y, z) = 0$,

$\dfrac{\partial^2 f}{\partial z \partial y}(x, y, z) = \dfrac{x}{(xz - 1)^2}$, $\quad \dfrac{\partial^2 f}{\partial z^2}(x, y, z) = \dfrac{2x^2(x - y)}{(xz - 1)^3}$;

(g) $\dfrac{\partial f}{\partial x}(x, y, z) = \dfrac{z}{yz - 1}$, $\quad \dfrac{\partial f}{\partial y}(x, y, z) = -\dfrac{xz^2}{(yz - 1)^2}$,

$\dfrac{\partial f}{\partial z}(x, y, z) = -\dfrac{x}{(yz - 1)^2}$,

$\dfrac{\partial^2 f}{\partial x^2}(x, y, z) = 0$, $\quad \dfrac{\partial^2 f}{\partial y \partial x}(x, y, z) = -\dfrac{z^2}{(yz - 1)^2}$,

$\dfrac{\partial^2 f}{\partial z \partial x}(x, y, z) = -\dfrac{1}{(yz - 1)^2}$, $\quad \dfrac{\partial^2 f}{\partial y^2}(x, y, z) = \dfrac{2xz^3}{(yz - 1)^3}$,

$\dfrac{\partial^2 f}{\partial z \partial y}(x, y, z) = \dfrac{2xz}{(yz - 1)^3}$, $\quad \dfrac{\partial^2 f}{\partial z^2}(x, y, z) = \dfrac{2xy}{(yz - 1)^3}$;

(h) $\dfrac{\partial f}{\partial x}(x, y, z) = \dfrac{\cos(x - y)}{\cos(y - z)}$, $\quad \dfrac{\partial f}{\partial y}(x, y, z) = -\dfrac{\cos(x - z)}{\cos^2(y - z)}$,

$\dfrac{\partial f}{\partial z}(x, y, z) = -\dfrac{\sin(x - y)\sin(y - z)}{\cos^2(y - z)}$,

$\dfrac{\partial^2 f}{\partial x^2}(x, y, z) = -\dfrac{\sin(x - y)}{\cos(y - z)}$, $\quad \dfrac{\partial^2 f}{\partial y \partial x}(x, y, z) = \dfrac{\sin(x - z)}{\cos^2(y - z)}$,

$$\frac{\partial^2 f}{\partial z \partial x}(x, y, z) = -\frac{\cos(x - y)\sin(y - z)}{\cos^2(y - z)},$$

$$\frac{\partial^2 f}{\partial y^2}(x, y, z) = -\frac{2\cos(x - z)\sin(y - z)}{\cos^3(y - z)},$$

$$\frac{\partial^2 f}{\partial z \partial y}(x, y, z) = \frac{\cos(x - z)\sin(y - z) - \sin(x - y)}{\cos^3(y - z)},$$

$$\frac{\partial^2 f}{\partial z^2}(x, y, z) = \sin(x - y) \cdot \frac{1 + \sin^2(y - z)}{\cos^3(y - z)}.$$

5.3.

(a) We have

$$\frac{\partial f}{\partial x}(x, y) = \begin{cases} \frac{x^4 y + 4x^2 y^3 - y^5}{(x^2 + y^2)^2} & \text{if } (x, y) \neq (0, 0) \\ 0 & \text{if } (x, y) = (0, 0) \end{cases}$$

and

$$\frac{\partial f}{\partial y}(x, y) = \begin{cases} \frac{x^5 - 4x^3 y^2 - xy^4}{(x^2 + y^2)^2} & \text{if } (x, y) \neq (0, 0) \\ 0 & \text{if } (x, y) = (0, 0). \end{cases}$$

Evidently, the function $\frac{\partial f}{\partial x}$ is continuous on $\mathbb{R}^2 \setminus \{(0, 0)\}$. Since

$$\left| \frac{\partial f}{\partial x}(x, y) \right| \leq |y| \frac{x^4 + 4x^2 y^2 + y^4}{(x^2 + y^2)^2} \leq 2|y|,$$

for all $(x, y) \neq (0, 0)$, it follows that $\lim\limits_{(x,y) \to (0,0)} \frac{\partial f}{\partial x}(x, y) = 0 = \frac{\partial f}{\partial x}(0, 0)$, so $\frac{\partial f}{\partial x}$ is continuous at $(0, 0)$. We proved that the partial derivative $\frac{\partial f}{\partial x}$ is continuous on $\mathbb{R}^2$. Similarly, one can prove that the partial derivative $\frac{\partial f}{\partial y}$ is continuous on $\mathbb{R}^2$, so f is of class C^1.

(b) We have

$$\frac{\partial^2 f}{\partial y \partial x}(0, 0) = \frac{\partial}{\partial y}\left(\frac{\partial f}{\partial x}\right)(0, 0) = \lim_{y \to 0} \frac{\frac{\partial f}{\partial x}(0, y) - \frac{\partial f}{\partial x}(0, 0)}{y - 0} = \lim_{y \to 0} \frac{-y}{y} = -1$$

and

$$\frac{\partial^2 f}{\partial x \partial y}(0, 0) = \frac{\partial}{\partial x}\left(\frac{\partial f}{\partial y}\right)(0, 0) = \lim_{x \to 0} \frac{\frac{\partial f}{\partial y}(x, 0) - \frac{\partial f}{\partial y}(0, 0)}{x - 0} = \lim_{x \to 0} \frac{x}{x} = 1.$$

Therefore, $\frac{\partial^2 f}{\partial y \partial x}(0, 0) \neq \frac{\partial^2 f}{\partial x \partial y}(0, 0)$.

(c) We prove that f does not satisfy the conditions of Schwarz's theorem. We have

$$\frac{\partial^2 f}{\partial y \partial x}(x, y) = \begin{cases} \frac{x^6 + 9x^4 y^2 - 9x^2 y^4 - y^6}{(x^2 + y^2)^3} & \text{if} \quad (x, y) \neq (0, 0) \\ -1 & \text{if} \quad (x, y) = (0, 0). \end{cases}$$

The sequence $\left(\left(\frac{1}{k}, 0 \right) \right)_{k \in \mathbb{N}}$ converges to $(0, 0)$, but

$$\lim_{k \to \infty} \frac{\partial^2 f}{\partial y \partial x}\left(\frac{1}{k}, 0 \right) = 1 \neq \frac{\partial^2 f}{\partial y \partial x}(0, 0).$$

Therefore, the function $\frac{\partial^2 f}{\partial y \partial x}$ is not continuous at $(0, 0)$. Similarly, one can prove that the other partial derivative $\frac{\partial^2 f}{\partial x \partial y}$ is not continuous at $(0, 0)$.

5.4.

(a) For all $(x, y) \in \mathbb{R}^2$, with $xy \neq 0$, we have

$$\frac{\partial f}{\partial x}(x, y) = 2x \arctan \frac{y}{x} - y \quad \text{and} \quad \frac{\partial f}{\partial y}(x, y) = -2y \arctan \frac{x}{y} + x.$$

Let $a \in \mathbb{R}$ be arbitrary. We have

$$\lim_{x \to a} \frac{f(x, 0) - f(a, 0)}{x - a} = \lim_{x \to a} \frac{0}{x - a} = 0,$$

so, f is partially derivable with respect to the variable x at $(a, 0)$ and $\frac{\partial f}{\partial x}(a, 0) = 0$. Also, if $a = 0$, then

$$\lim_{y \to 0} \frac{f(a, y) - f(a, 0)}{y - 0} = \lim_{y \to 0} \frac{0}{y} = 0,$$

and if $a \neq 0$, then

$$\lim_{y \to 0} \frac{f(a, y) - f(a, 0)}{y - 0} = \lim_{y \to 0} \left(a^2 \frac{\arctan \frac{y}{a}}{y} - y \arctan \frac{a}{y} \right) = a.$$

Thus, f is partially derivable with respect to the variable y at $(a, 0)$ and $\frac{\partial f}{\partial y}(a, 0) = a$.

Similarly, one can prove that f is partially derivable at $(0, a)$ and $\frac{\partial f}{\partial x}(0, a) = -a$, $\frac{\partial f}{\partial y}(0, a) = 0$.

In conclusion, f is partially derivable on $\mathbb{R}^2$ and we have

$$\frac{\partial f}{\partial x}(x, y) = \begin{cases} 2x \arctan \frac{y}{x} - y & \text{if } x \neq 0 \\ -y & \text{if } x = 0 \end{cases}$$

$$\frac{\partial f}{\partial y}(x, y) = \begin{cases} -2y \arctan \frac{x}{y} + x & \text{if } y \neq 0 \\ x & \text{if } y = 0. \end{cases}$$

One can prove, based on the expressions of the partial derivatives of f written above, that these are continuous functions on $\mathbb{R}^2$, so f is of class C^1.

(b) We have

$$\frac{\partial^2 f}{\partial y \partial x}(0, 0) = \lim_{y \to 0} \frac{\frac{\partial f}{\partial x}(0, y) - \frac{\partial f}{\partial x}(0, 0)}{y - 0} = \lim_{y \to 0} \frac{-y}{y} = -1$$

and

$$\frac{\partial^2 f}{\partial x \partial y}(0, 0) = \lim_{x \to 0} \frac{\frac{\partial f}{\partial y}(x, 0) - \frac{\partial f}{\partial y}(0, 0)}{x - 0} = \lim_{x \to 0} \frac{x}{x} = 1.$$

(c) We prove that f does not satisfy the conditions of Schwarz's theorem. We have

$$\frac{\partial^2 f}{\partial y \partial x}(x, y) = \begin{cases} \frac{x^2 - y^2}{x^2 + y^2} & \text{if } x \neq 0 \\ -1 & \text{if } x = 0. \end{cases}$$

Let $(x_n)_{n \in \mathbb{N}}$ be a sequence of real numbers which converges to 0. The sequence $((x_n, 0))_{n \in \mathbb{N}}$ converges to $(0, 0)$ and

$$\lim_{n \to \infty} \frac{\partial^2 f}{\partial y \partial x}(x_n, 0) = 1 \neq \frac{\partial^2 f}{\partial y \partial x}(0, 0).$$

Therefore, the function $\frac{\partial^2 f}{\partial y \partial x}$ is not continuous at $(0, 0)$. Similarly, one can prove that the function $\frac{\partial^2 f}{\partial x \partial y}$ is not continuous at $(0, 0)$.

5.6. $L(f) = -2f$ and $L(L(f)) = 4f$.

5.9. $\frac{\partial^{2n} f}{\partial y^n \partial x^n}(x, y) = e^{x+y} \sum_{k=0}^{n} \left(C_n^k\right)^2 (n - k)!(x^k + y^k)$.

5.10. When $n = 2p$, $p \geq 0$ is an integer, we have $\frac{\partial^{4p} f}{\partial y^{2p} \partial x^{2p}}(0, 0) = 0$, and when $n = 2p + 1$, $p \geq 0$ is an integer, we have $\frac{\partial^{2(2p+1)} f}{\partial y^{2p+1} \partial x^{2p+1}}(0, 0) = \left[\frac{(2p+2)!}{2}\right]^2$.

5.13. We have

$$J(f)(x, y) = \begin{pmatrix} \dfrac{1}{y} & \dfrac{\sin y}{x} \\ \dfrac{1+x^2y^2} & \dfrac{1+x^2y^2} \\ \pi\cos(\pi x - y) & -\cos(\pi x - y) \end{pmatrix}, \quad (x, y) \in \mathbb{R}^2,$$

and

$$J(f)(1, 0) = \begin{pmatrix} 1 & 0 \\ 0 & 1 \\ -\pi & 1 \end{pmatrix}.$$

5.14. We have

$$J(f)(x, y, z) = \begin{pmatrix} \dfrac{ye^{xy-z}}{e^{xy-z}+1} & \dfrac{xe^{xy-z}}{e^{xy-z}+1} & -\dfrac{e^{xy-z}}{e^{xy-z}+1} \\ 2xyz(x^2+1)^{yz-1} & z(x^2+1)^{yz}\ln(x^2+1) & y(x^2+1)^{yz}\ln(x^2+1) \end{pmatrix},$$

for all $(x, y, z) \in \mathbb{R}^3$, and

$$J(f)(-1, 1, -1) = \frac{1}{2}\begin{pmatrix} 1 & -1 & -1 \\ 1 & -\ln 2 & \ln 2 \end{pmatrix}.$$

5.15. $\dfrac{D(x,y)}{D(\rho,\theta)} = \rho.$

5.16. $\dfrac{D(x,y,z)}{D(\rho,\varphi,\theta)} = \rho^2\sin\varphi.$

5.17. $\dfrac{D(x,y)}{D(u,v)} = \dfrac{1}{(1-u^2-v^2)^2}.$

5.18. $\dfrac{D(x,y,z)}{D(u,v,w)} = 24uvw(w-u)(v-u)(w-v).$

5.19. We have

$$H(f)(x, y) = \begin{pmatrix} y(y-1)x^{y-2} & x^{y-1}(1+y\ln x) \\ x^{y-1}(1+y\ln x) & x^y\ln^2 x \end{pmatrix}, \quad (x, y) \in (0, \infty) \times \mathbb{R},$$

and it follows that

$$H(f)(1, -1) = \begin{pmatrix} 2 & 1 \\ 1 & 0 \end{pmatrix}.$$

5.20. We have

$$H(f)(x, y, z) = \begin{pmatrix} \dfrac{2(y-z)}{(z-x)^3} & \dfrac{1}{(z-x)^2} & \dfrac{x-2y+z}{(z-x)^3} \\[2mm] \dfrac{1}{(z-x)^2} & 0 & -\dfrac{1}{(z-x)^2} \\[2mm] \dfrac{x-2y+z}{(z-x)^3} & -\dfrac{1}{(z-x)^2} & -\dfrac{2(x-y)}{(z-x)^3} \end{pmatrix},$$

for all $(x, y, z) \in A$, and

$$H(f)(1, -1, 0) = \begin{pmatrix} 2 & 1 & -3 \\ 1 & 0 & -1 \\ -3 & -1 & 4 \end{pmatrix}.$$

5.21. $\nabla f(x, y, z) = \left(\dfrac{yz}{1+x^2y^2z^2}, \dfrac{xz}{1+x^2y^2z^2}, \dfrac{xy}{1+x^2y^2z^2} \right);$

$\Delta f(x, y, z) = -\dfrac{2xyz(x^2y^2 + y^2z^2 + x^2z^2)}{(1+x^2y^2z^2)^2}.$

5.22. $\nabla f(x, y, z) = \left(-\dfrac{x}{x^2+y^2+z^2}, -\dfrac{y}{x^2+y^2+z^2}, -\dfrac{z}{x^2+y^2+z^2} \right);$

$\Delta f(x, y, z) = -\dfrac{1}{x^2+y^2+z^2}.$

5.24. The only point which satisfies the property is $(3, 3, 3)$.

5.25. The only point which satisfies the property is $(5, \frac{5}{3}, \frac{1}{2})$.

5.26. div $F(x, y, z) = 1 + x - \dfrac{y}{1+(x-yz)^2};$

rot $F(x, y, z) = \left(\dfrac{-z}{1+(x-yz)^2} + e^z, 1 - \dfrac{1}{1+(x-yz)^2}, y + 1 \right).$

5.27. div $F(x, y, z) = \cos(yz) + e^{xz} + xy\cos(xyz);$

rot $F(x, y, z) = (xz\cos(xyz) - xy\, e^{xz}, -xy\sin(yz) - yz\cos(xyz),$
$\qquad\qquad\qquad yz\, e^{xz} + xz\sin(yz)).$

13.2   The Chain Rule

5.28.

(a) Let $u(x) = \sin x$ and $v(x) = \arctan x$. Then $F(x) = f(u(x), v(x))$ and we have

$$F'(x) = \cos x \cdot f'_u(u(x), v(x)) + \dfrac{1}{1+x^2} \cdot f'_v(u(x), v(x)),$$

$$F''(x) = \cos^2 x \cdot f''_{u^2}(u(x), v(x)) + \dfrac{2\cos x}{1+x^2} \cdot f''_{uv}(u(x), v(x))$$

$$+ \frac{1}{(1+x^2)^2} \cdot f''_{v^2}(u(x), v(x)) - \sin x \cdot f'_u(u(x), v(x))$$

$$- \frac{2x}{(1+x^2)^2} \cdot f'_v(u(x), v(x)).$$

(b) Let $u(x) = \ln(x^2 + 1)$ and $v(x) = \cos x$. Then $F(x) = f(u(x), v(x))$ and we have

$$F'(x) = \frac{2x}{x^2+1} \cdot f'_u(u(x), v(x)) - \sin x \cdot f'_v(u(x), v(x)),$$

$$F''(x) = \frac{4x^2}{(x^2+1)^2} \cdot f''_{u^2}(u(x), v(x)) - \frac{4x \sin x}{x^2+1} \cdot f''_{uv}(u(x), v(x))$$

$$+ \sin^2 x \cdot f''_{v^2}(u(x), v(x)) - \frac{2(x^2-1)}{(x^2+1)^2} \cdot f'_u(u(x), v(x))$$

$$- \cos x \cdot f'_v(u(x), v(x)).$$

5.29.

(a) Let $u(x) = -x$ and $v(x) = 3x$. Then $F(x) = f(u(x), v(x))$ and we have that

$$F'''(x) = -f'''_{u^3}(u(x), v(x)) + 9 f'''_{u^2 v}(u(x), v(x))$$

$$- 27 f'''_{uv^2}(u(x), v(x)) + 27 f'''_{v^3}(u(x), v(x)).$$

(b) Let $u(x) = x$ and $v(x) = x^2$. Then $F(x) = f(u(x), v(x))$ and we have that

$$F'''(x) = f'''_{u^3}(u(x), v(x)) + 6x f'''_{u^2 v}(u(x), v(x)) + 12x^2 f'''_{uv^2}(u(x), v(x))$$

$$+ 8x^3 f'''_{v^3}(u(x), v(x)) + 6 f''_{uv}(u(x), v(x)) + 12x f''_{v^2}(u(x), v(x)).$$

5.30.

(a) Let $u(x) = e^x$, $v(x) = \ln(x + \sqrt{x^2 + 1})$ and $w(x) = x^2$. Then $F(x) = f(u(x), v(x), w(x))$ and we have that

$$F'(x) = e^x \cdot f'_u(u(x), v(x), w(x)) + \frac{1}{\sqrt{x^2+1}} \cdot f'_v(u(x), v(x), w(x))$$

$$+ 2x \cdot f'_w(u(x), v(x), w(x)),$$

$$F''(x) = e^{2x} \cdot f''_{u^2}(u(x), v(x), w(x)) + \frac{1}{x^2+1} \cdot f''_{v^2}(u(x), v(x), w(x))$$

$$+ 4x^2 \cdot f''_{w^2}(u(x), v(x), w(x)) + \frac{2 e^x}{\sqrt{x^2+1}} \cdot f''_{uv}(u(x), v(x), w(x))$$

$$+\frac{4x}{\sqrt{x^2+1}} \cdot f''_{vw}(u(x), v(x), w(x)) + 4x\, e^x \cdot f''_{uw}(u(x), v(x), w(x))$$

$$+e^x \cdot f'_u(u(x), v(x), w(x)) - \frac{x}{(x^2+1)\sqrt{x^2+1}} \cdot f'_v(u(x), v(x), w(x))$$

$$+2 \cdot f'_w(u(x), v(x), w(x)).$$

(b) Let $u(x) = 2^x$, $v(x) = \operatorname{arccot} x$ and $w(x) = e^{x^2}$. Then $F(x) = f(u(x), v(x), w(x))$ and we have that

$$F'(x) = 2^x \ln 2 \cdot f'_u(u(x), v(x), w(x)) - \frac{1}{1+x^2} \cdot f'_v(u(x), v(x), w(x))$$

$$+2x\, e^{x^2} \cdot f'_w(u(x), v(x), w(x)),$$

$$F''(x) = 2^{2x} \ln^2 2 \cdot f''_{u^2}(u(x), v(x), w(x)) + \frac{1}{(1+x^2)^2} \cdot f''_{v^2}(u(x), v(x), w(x))$$

$$+4x^2\, e^{2x^2} \cdot f''_{w^2}(u(x), v(x), w(x)) - \frac{2^{x+1} \ln 2}{1+x^2} \cdot f''_{uv}(u(x), v(x), w(x))$$

$$-\frac{4x\, e^{x^2}}{1+x^2} \cdot f''_{vw}(u(x), v(x), w(x)) + x\, e^{x^2} 2^{x+2} \ln 2 \cdot f''_{uw}(u(x), v(x), w(x))$$

$$+2^x \ln^2 2 \cdot f'_u(u(x), v(x), w(x)) + \frac{2x}{(1+x^2)^2} \cdot f'_v(u(x), v(x), w(x))$$

$$+2(2x^2+1)\, e^{x^2} \cdot f'_w(u(x), v(x), w(x)).$$

5.31.

(a) Let $u(x, y) = x^2 - 2xy + 3y$. Then $F(x, y) = f(u(x, y))$ and we have that

$$F'_x(x, y) = 2(x - y) \cdot f'(u(x, y)),$$

$$F'_y(x, y) = -(2x - 3) \cdot f'(u(x, y)),$$

$$F''_{x^2}(x, y) = 4(x - y)^2 \cdot f''(u(x, y)) + 2 \cdot f'(u(x, y)),$$

$$F''_{xy}(x, y) = -2(x - y)(2x - 3) \cdot f''(u(x, y)) - 2 \cdot f'(u(x, y)),$$

$$F''_{y^2}(x, y) = (2x - 3)^2 \cdot f''(u(x, y)).$$

(b) Let $u(x, y) = x \sin y - y \cos x$. Then $F(x, y) = f(u(x, y))$ and we have that

$$F'_x(x, y) = (\sin y + y \sin x) \cdot f'(u(x, y)),$$

$$F'_y(x, y) = (x \cos y - \cos x) \cdot f'(u(x, y)),$$

$$F''_{x^2}(x, y) = (\sin y + y \sin x)^2 \cdot f''(u(x, y)) + y \cos x \cdot f'(u(x, y)),$$

$$F''_{xy}(x, y) = (\sin y + y \sin x)(x \cos y - \cos x) \cdot f''(u(x, y))$$

$$+ (\cos y + \sin x) \cdot f'(u(x, y)),$$

$$F''_{y^2}(x, y) = (x \cos y - \cos x)^2 \cdot f''(u(x, y)) - x \sin y \cdot f'(u(x, y)).$$

5.32.

(a) Let $u(x, y, z) = x - y - z$. Then $F(x, y, z) = f(u(x, y, z))$ and we have that

$$F'''_{xyz}(x, y, z) = f'''(u(x, y, z)),$$

$$F'''_{yz^2}(x, y, z) = -f'''(u(x, y, z)).$$

(b) Let $u(x, y, z) = xyz$. Then $F(x, y, z) = f(u(x, y, z))$ and we have that

$$F'''_{x^3}(x, y, z) = y^3 z^3 f'''(u(x, y, z)),$$

$$F'''_{x^2 z}(x, y, z) = xy^3 z^2 f'''(u(x, y, z)) + 2y^2 z f''(u(x, y, z)).$$

(c) Let $u(x, y, z) = x^2 - xy + z$. Then $F(x, y, z) = f(u(x, y, z))$ and we have that

$$F'''_{xy^2}(x, y, z) = x^2(2x - y) f'''(u(x, y, z)) + 2x f''(u(x, y, z)),$$

$$F'''_{xyz}(x, y, z) = -x(2x - y) f'''(u(x, y, z)) - f''(u(x, y, z)).$$

(d) Let $u(x, y, z, t) = xyt - z + t$. Then $F(x, y, z, t) = f(u(x, y, z, t))$ and we have

$$F'''_{xyt}(x, y, z, t) = x(xy + 1)yt^2 f'''(u(x, y, z, t)) + (3xy + 1)t f''(u(x, y, z, t))$$

$$+ f'(u(x, y, z, t)),$$

$$F'''_{xzt}(x, y, z, t) = -(xy + 1)yt f'''(u(x, y, z, t)) - y f''(u(x, y, z, t)),$$

$$F'''_{xt^2}(x, y, z, t) = (xy + 1)^2 yt f'''(u(x, y, z, t)) + 2(xy + 1)y f''(u(x, y, z, t)),$$

$$F'''_{yzt}(x, y, z, t) = -x(xy + 1)t f'''(u(x, y, z, t)) - x f''(u(x, y, z, t)),$$

$$F^{IV}_{xyzt}(x, y, z, t) = -x(xy + 1)yt^2 f^{IV}(u(x, y, z, t)) - (3xy + 1)t f'''(u(x, y, z, t))$$

$$- f''(u(x, y, z, t)).$$

5.33.

(a) Let $u(x, y) = xy - x + y$ and $v(x, y) = x \arctan y$. We can write $F(x, y) = f(u(x, y), v(x, y))$ and we have

$$F'_x(x, y) = (y - 1) \cdot f'_u(u(x, y), v(x, y)) + \arctan y \cdot f'_v(u(x, y), v(x, y)),$$

$$F'_y(x, y) = (x+1) \cdot f'_u(u(x, y), v(x, y)) + \frac{x}{1+y^2} \cdot f'_v(u(x, y), v(x, y)),$$

$$F''_{x^2}(x, y) = (y-1)^2 \cdot f''_{u^2}(u(x, y), v(x, y)) + 2(y-1) \arctan y \cdot f''_{uv}(u(x, y), v(x, y))$$
$$+ \arctan^2 y \cdot f''_{v^2}(u(x, y), v(x, y)),$$

$$F''_{xy}(x, y) = (x+1)(y-1) \cdot f''_{u^2}(u(x, y), v(x, y))$$
$$+ \left[\frac{x(y-1)}{1+y^2} + (x+1) \arctan y \right] \cdot f''_{uv}(u(x, y), v(x, y))$$
$$+ \frac{x \arctan y}{1+y^2} \cdot f''_{v^2}(u(x, y), v(x, y))$$
$$+ f'_u(u(x, y), v(x, y)) + \frac{1}{1+y^2} \cdot f'_v(u(x, y), v(x, y)),$$

$$F''_{y^2}(x, y) = (x+1)^2 \cdot f''_{u^2}(u(x, y), v(x, y)) + \frac{2x(x+1)}{1+y^2} \cdot f''_{uv}(u(x, y), v(x, y))$$
$$+ \frac{x^2}{(1+y^2)^2} \cdot f''_{v^2}(u(x, y), v(x, y)) - \frac{2xy}{(1+y^2)^2} \cdot f'_v(u(x, y), v(x, y)).$$

(b) Let $u(x, y) = \sin(x-y)$ and $v(x, y) = xy - \ln(y^2+1)$. We can write $F(x, y) = f(u(x, y), v(x, y))$ and it follows that

$$F'_x(x, y) = \cos(x-y) \cdot f'_u(u(x, y), v(x, y)) + y \cdot f'_v(u(x, y), v(x, y)),$$

$$F'_y(x, y) = -\cos(x-y) \cdot f'_u(u(x, y), v(x, y)) + \left(x - \frac{2y}{y^2+1}\right) \cdot f'_v(u(x, y), v(x, y)),$$

$$F''_{x^2}(x, y) = \cos^2(x-y) \cdot f''_{u^2}(u(x, y), v(x, y)) + 2y \cos(x-y) \cdot f''_{uv}(u(x, y), v(x, y))$$
$$+ y^2 \cdot f''_{v^2}(u(x, y), v(x, y)) - \sin(x-y) \cdot f'_u(u(x, y), v(x, y)),$$

$$F''_{xy}(x, y) = -\cos^2(x-y) \cdot f''_{u^2}(u(x, y), v(x, y))$$
$$+ \left(x - y - \frac{2y}{y^2+1}\right) \cos(x-y) \cdot f''_{uv}(u(x, y), v(x, y))$$
$$+ y \left(x - \frac{2y}{y^2+1}\right) \cdot f''_{v^2}(u(x, y), v(x, y))$$
$$+ \sin(x-y) \cdot f'_u(u(x, y), v(x, y)) + f'_v(u(x, y), v(x, y)),$$

$$F''_{y^2}(x, y) = \cos^2(x-y) \cdot f''_{u^2}(u(x, y), v(x, y))$$
$$- 2 \left(x - \frac{2y}{y^2+1}\right) \cos(x-y) \cdot f''_{uv}(u(x, y), v(x, y))$$
$$+ \left(x - \frac{2y}{y^2+1}\right)^2 \cdot f''_{v^2}(u(x, y), v(x, y))$$

$$-\sin(x-y)\cdot f_u'(u(x,y),v(x,y))+\frac{2(y^2-1)}{(y^2+1)^2}\cdot f_v'(u(x,y),v(x,y)).$$

5.34. To simplify the writing, we drop out the argument of the functions that appears in the calculations.

(a) Let $u(x,y,z)=x-yz$ and $v(x,y,z)=x-y$. Then

$$F(x,y,z)=f(u(x,y,z),v(x,y,z))$$

and we have that

$$F_{y^2}''=z^2f_{u^2}''+2zf_{uv}''+f_{v^2}'',$$
$$F_{yz}''=yzf_{u^2}''+yf_{uv}''-f_u',$$
$$F_{xyz}'''=yzf_{u^3}'''+y(z+1)f_{u^2v}'''+yf_{uv^2}'''-f_{u^2}''-f_{uv}''.$$

(b) Let $u(x,y,z)=x-z$ and $v(x,y,z)=y-z$. Then

$$F(x,y,z)=f(u(x,y,z),v(x,y,z))$$

and we have that

$$F_{xz}''=-f_{u^2}''-f_{uv}'',$$
$$F_{yz}''=-f_{uv}''-f_{v^2}'',$$
$$F_{xyz}'''=-f_{u^2v}'''-f_{uv^2}''',$$
$$F_{xz^2}'''=f_{u^3}'''+2f_{u^2v}'''+f_{uv^2}'''.$$

(c) Let $u(x,y,z)=xy-z$ and $v(x,y,z)=xz-y$. Then

$$F(x,y,z)=f(u(x,y,z),v(x,y,z))$$

and we have that

$$F_{xy}''=xyf_{u^2}''+(xz-y)f_{uv}''-zf_{v^2}''+f_u',$$
$$F_{yz}''=-xf_{u^2}''+(x^2+1)f_{uv}''-xf_{v^2}'',$$
$$F_{xyz}'''=-xyf_{u^3}'''+(x^2y-xz+y)f_{u^2v}'''+(x^2z-xy+z)f_{uv^2}'''$$
$$-xzf_{v^3}'''-f_{u^2}''+2xf_{uv}''-f_{v^2}'',$$

$$F'''_{y^2z} = -x^2 f'''_{u^3} + x(x^2+2) f'''_{u^2v} - (2x^2+1) f'''_{uv^2} + x f'''_{v^3}.$$

(d) Let $u(x, y, z, t) = x - t$ and $v(x, y, z, t) = yz - t$. Then

$$F(x, y, z, t) = f(u(x, y, z, t), v(x, y, z, t))$$

and we have that

$$F''_{yt} = -z f''_{uv} - z f''_{v^2},$$

$$F''_{zt} = -y f''_{uv} - y f''_{v^2},$$

$$F'''_{xyz} = yz f'''_{uv^2} + f''_{uv},$$

$$F'''_{xyt} = -z f'''_{u^2v} - z f'''_{uv^2},$$

$$F'''_{xzt} = -y f'''_{u^2v} - y f'''_{uv^2},$$

$$F'''_{xt^2} = f'''_{u^3} + 2 f'''_{u^2v} + f'''_{uv^2},$$

$$F^{IV}_{xyzt} = -yz f^{IV}_{u^2v^2} - yz f^{IV}_{uv^3} - f'''_{u^2v} - f'''_{uv^2}.$$

5.35. To simplify the writing, we drop out the argument of the functions that appears in the calculations.

(a) Let $u(x, y, z) = x - y$, $v(x, y, z) = y - z$ and $w(x, y, z) = xy - z$. Then

$$F(x, y, z) = f(u(x, y, z), v(x, y, z), w(x, y, z))$$

and we have that

$$F''_{xy} = -f''_{u^2} + xy f''_{w^2} + f''_{uv} + (x - y) f''_{uw} + y f''_{vw} + f'_{w},$$

$$F''_{xz} = -y f''_{w^2} - f''_{uv} - f''_{uw} - y f''_{vw},$$

$$F''_{yz} = -f''_{v^2} - x f''_{w^2} + f''_{uv} + f''_{uw} - (x + 1) f''_{vw},$$

$$F'''_{xyz} = -xy f'''_{w^3} + f'''_{u^2v} + f'''_{u^2w} - f'''_{uv^2} - (x - y + 1) f'''_{uvw} - (x - y) f'''_{uw^2}$$
$$\quad - y f'''_{v^2w} - (x + 1) y f'''_{vw^2} - f''_{vw} - f''_{w^2},$$

$$F'''_{yz^2} = f'''_{v^3} + x f'''_{w^3} - f'''_{uv^2} - 2 f'''_{uvw} - f'''_{uw^2} + (x + 2) f'''_{v^2w} + (2x + 1) f'''_{vw^2}.$$

(b) Let $u(x, y, z) = y$, $v(x, y, z) = x - z$ and $w(x, y, z) = x - yz$. Then

$$F(x, y, z) = f(u(x, y, z), v(x, y, z), w(x, y, z))$$

and we have that

$$F''_{xy} = -z f''_{w^2} + f''_{uv} + f''_{uw} - z f''_{vw},$$

$$F''_{xz} = -f''_{v^2} - y f''_{w^2} - (y+1) f''_{vw},$$

$$F''_{yz} = yz f''_{w^2} - f''_{uv} - y f''_{uw} + z f''_{vw} - f'_w,$$

$$F''_{z^2} = f''_{v^2} + y^2 f''_{w^2} + 2y f''_{vw},$$

$$F'''_{xy^2} = z^2 f'''_{w^3} + f'''_{u^2 v} + f'''_{u^2 w} - 2z f'''_{uvw} - 2z f'''_{uw^2} + z^2 f'''_{vw^2},$$

$$F'''_{xyz} = yz f'''_{w^3} - f'''_{uv^2} - (y+1) f'''_{uvw} - y f'''_{uw^2} + z f'''_{v^2 w} + (y+1)z f'''_{vw^2}$$
$$- f''_{vw} - f''_{w^2},$$

$$F'''_{y^2 z} = -yz^2 f'''_{w^3} - f'''_{u^2 v} - y f'''_{u^2 w} + 2z f'''_{uvw} + 2yz f'''_{uw^2} - z^2 f'''_{vw^2}$$
$$+ 2z f''_{w^2} - 2 f''_{uw}.$$

(c) Let $u(x, y, z) = z$, $v(x, y, z) = x - y - z$ and $w(x, y, z) = x$. Then

$$F(x, y, z) = f(u(x, y, z), v(x, y, z), w(x, y, z))$$

and we have that

$$F''_{xy} = -f''_{v^2} - f''_{vw},$$

$$F''_{xz} = -f''_{v^2} + f''_{uv} + f''_{uw} - f''_{vw},$$

$$F''_{yz} = f''_{v^2} - f''_{uv},$$

$$F''_{z^2} = f''_{u^2} + f''_{v^2} - 2 f''_{uv},$$

$$F'''_{xyz} = f'''_{v^3} - f'''_{uv^2} - f'''_{uvw} + f'''_{v^2 w},$$

$$F'''_{xz^2} = f'''_{v^3} + f'''_{u^2 v} + f'''_{u^2 w} - 2 f'''_{uv^2} - 2 f'''_{uvw} + f'''_{v^2 w},$$

$$F^{IV}_{xy^3} = -f^{IV}_{v^4} - f^{IV}_{v^3 w}.$$

(d) Let $u(x, y, z, t) = xt$, $v(x, y, z, t) = y - z$ and $w(x, y, z, t) = x - t$. Then

$$F(x, y, z, t) = f(u(x, y, z, t), v(x, y, z, t), w(x, y, z, t))$$

and we have that

$$F''_{xy} = t f''_{uv} + f''_{vw},$$

$$F''_{yt} = x f''_{uv} - f''_{vw},$$

$$F''_{zt} = -x f''_{uv} + f''_{vw},$$

$$F''_{t^2} = x^2 f''_{u^2} + f''_{w^2} - 2x f''_{uw},$$

$$F'''_{xyz} = -t f'''_{uv^2} - f'''_{v^2 w},$$

$$F'''_{xyt} = xt f'''_{u^2 v} + (x-t) f'''_{uvw} - f'''_{vw^2} + f''_{uv},$$

$$F'''_{yzt} = -x f'''_{uv^2} + f'''_{v^2 w},$$

$$F'''_{t^3} = x^3 f'''_{u^3} - f'''_{w^3} - 3x^2 f'''_{u^2 w} + 3x f'''_{uw^2},$$

$$F^{IV}_{xyzt} = -xt f^{IV}_{u^2 v^2} - (x-t) f^{IV}_{uv^2 w} + f^{IV}_{v^2 w^2} - f'''_{uv^2}.$$

5.36. $a - b = 0$.

5.37. $a + b = 0$.

13.3   Homogeneous Functions. Euler's Identity

5.38. Homogeneous functions of degree. (a) $-\frac{1}{2}$; (b) -1; (c) 1; (d) -2; (e) 1.

13.4   Taylor's Formula for Real Functions of Two Real Variables

5.40.

(a) $f(x, y) = xy^2 - 2xy + 2y^2 + x - 4y + 2$;

(b) $f(x, y) = x^2 y + x^2 - 2xy - 2x + y + 1$.

5.41. $f(x, y) = (x-1)^3 + (x-1)^2(y+1) + 3(x-1)^2 - (x-1)(y+1) + 3(x-1) + 7$.

5.42. $f(x, y) = (x + 1)^2 (y - 2) + 4(y - 2)^3 + 2(x + 1)(y - 2) + 4(y - 2)^2 + 8$.

5.43. (a) $(M_2 f)(x, y) = xy$;

(b) $(M_3 f)(x, y) = 1 + x + \frac{1}{2!}(x^2 - y^2) + \frac{1}{3!}(x^3 - 3xy^2)$;

(c) $(M_4 f)(x, y) = 1 + xy + \frac{x^2 y^2}{2}$.

5.44. We have

(a) $(T_2 f)(x, y) = 1 + [(x - 1) + (y + 1)]$
$$+ \frac{1}{2!} \left[3(x - 1)^2 + 4(x - 1)(y + 1) + (y + 1)^2 \right];$$

(b) $(T_2 f)(x, y) = 1 + \frac{1}{2!} \left[-(x - 1)^2 - 2(x - 1)(y - 1) - (y - 1)^2 \right]$.

5.45. Use the function $f(x, y) = \sqrt{x} \sqrt[3]{y}$ and the point $(x_0, y_0) = (1, 1)$.

5.46. Use the function $f(x, y) = x^y$ and the point $(x_0, y_0) = (1, 3)$.

5.47. The coefficient of $x^n y^n$ is 5^n.

5.48. The coefficient of $(x - 1)^n (y - 1)^n$ is C_{2n}^n.

5.49. The coefficient of $x^n y^n$ is C_{2n+1}^n. We have

$$f(x, y) = \frac{1}{1-x} \cdot \frac{1}{1-x-y} = \left(\sum_{n=0}^{\infty} x^n\right)\left(\sum_{m=0}^{\infty} (x+y)^m\right)$$

$$= \left(\sum_{n=0}^{\infty} x^n\right)\left(\sum_{m=0}^{\infty} \sum_{k=0}^{m} C_m^k x^{m-k} y^k\right).$$

The coefficient of $x^n y^n$ is $C_n^n + C_{n+1}^n + C_{n+2}^n + \cdots + C_{2n}^n$. We observe that this is the coefficient of x^n from the expression $(1+x)^n + (1+x)^{n+1} + \cdots + (1+x)^{2n}$. This sum can be written in the following way:

$$(1+x)^n + (1+x)^{n+1} + \cdots + (1+x)^{2n}$$

$$= (1+x)^n [1 + (1+x) + (1+x)^2 + \cdots + (1+x)^n]$$

$$= (1+x)^n \cdot \frac{(1+x)^{n+1} - 1}{x}$$

$$= \frac{1}{x}[(1+x)^{2n+1} - (1+x)^n].$$

Therefore, the coefficient we are looking for is the coefficient of x^{n+1} from the expression

$$(1+x)^{2n+1} - (1+x)^n,$$

which is $C_{2n+1}^{n+1} = C_{2n+1}^n$.

13.5 The Differential of Several Real Variable Functions

5.50. $df(0, 1)(1, 2) = 2e + 1$ and $d^2 f(0, 1)(1, 2) = 6e + 8$.

5.51. $df(1, 0)(2, 1) = 4e + 1$ and $d^2 f(1, 0)(2, 1) = 22e + 4$.

5.52. $df(\pi, -1, 1)(-1, 2, 1) = -1 - \pi$ and $d^2 f(\pi, -1, 1)(-1, 2, 1) = 2 - 4\pi$.

5.53. $df\left(\frac{\pi}{2}, 1, 0\right)(1, -2, 2) = -3$ and $d^2 f\left(\frac{\pi}{2}, 1, 0\right)(1, -2, 2) = 8$.

5.54. $d^4 f(2, 1, 2)(-1, 1, 1) = \frac{5}{2}$.

5.55. -28.

5.56. $v = (1, -2, 2)$, $\frac{df}{dv}(1, 1, -1) = 1$ and $\frac{df}{dv}(-1, 1, -1) = 0$.

13.6 Extrema of Several Real Variable Functions

5.57. We have $x + 1 \geq 2\sqrt{x}$, $y + 1 \geq 2\sqrt{y}$ and $x + y \geq 2\sqrt{xy}$. It follows that

$$f(x, y) \leq \frac{xy}{2\sqrt{x} \cdot 2\sqrt{y} \cdot 2\sqrt{xy}} = \frac{1}{8} = f(1, 1), \quad \forall (x, y) \in (0, \infty)^2.$$

Therefore, $(1, 1)$ is a global maximum point of f.

5.58. (a) $(a, a), a > 0$, points of global minimum;   (b) $(0, 0)$ point of local minimum, $\left(\frac{\sqrt{3}}{3}, -\frac{\sqrt{3}}{3}\right)$ and $\left(-\frac{\sqrt{3}}{3}, \frac{\sqrt{3}}{3}\right)$ saddle points;   (c) $(0, 0)$ saddle point, $\left(\frac{1}{2}, 1\right)$ point of global maximum;   (d) $(-2, 1)$ point of local maximum;   (e) $\left(\frac{2\pi}{3}, \frac{2\pi}{3}\right)$ point of local minimum;   (f) $(a, a, a), a > 0$, points of global minimum;   (g) $(-1, 0, 0)$ point of local minimum;   (h) $(1, 1, 1)$ point of global minimum;   (i) $(a, 0, a + 1)$, $a \in \mathbb{R}$, points of global minimum;   (j) $(0, 0, 0)$ saddle point, $\left(-\frac{\pi}{4}, -\frac{\pi}{4}, -\frac{\pi}{4}\right)$ and $\left(\frac{\pi}{4}, \frac{\pi}{4}, \frac{\pi}{4}\right)$ points of local maximum.

5.59. If $k \leq 0$ or $l \leq 0$, then f does not have critical points. If $k > 0$ and $l > 0$, then $\left(\sqrt[3]{\frac{k^2}{l}}, \sqrt[3]{\frac{l^2}{k}}\right)$ is a point of global minimum.

5.60.

(a) *Solution I.* The minimum value is $\frac{1}{180}$. We have that

$$\int_0^1 (x^2 - a - bx)^2 \mathrm{d}x = a^2 + \frac{b^2}{3} - \frac{2a}{3} - \frac{b}{2} + ab + \frac{1}{5}.$$

Let $f(a, b) = a^2 + \frac{b^2}{3} - \frac{2a}{3} - \frac{b}{2} + ab + \frac{1}{5}$. A calculation shows that the only critical point of f is $\left(-\frac{1}{6}, 1\right)$ and Sylvester's criterion implies that this is a local minimum point of f. We prove that this point is a global minimum point of f, i.e., $f(a, b) \geq f\left(-\frac{1}{6}, 1\right)$, $\forall (a, b) \in \mathbb{R}^2$. To prove that the previous inequality holds we observe, since f is a polynomial of degree 2 in two variables, that f coincides with its Taylor's polynomial of degree 2 corresponding to f at the point $\left(-\frac{1}{6}, 1\right)$. A calculation shows that

$$f(a, b) = (T_2 f)(a, b) = \left(a + \frac{1}{6}\right)^2 + \left(a + \frac{1}{6}\right)(b - 1) + \frac{1}{3}(b - 1)^2 + \frac{1}{180}$$

$$\geq f\left(-\frac{1}{6}, 1\right) = \frac{1}{180}, \quad \forall (a, b) \in \mathbb{R}^2.$$

The last inequality follows since $x^2 + xy + \frac{1}{3}y^2 \geq 0$, for all $x, y \in \mathbb{R}$.

Solution II. We consider the vector space $C([0, 1], \mathbb{R})$, of continuous real valued functions on the interval $[0, 1]$, together with the inner product defined by $\langle f, g \rangle = \int_0^1 f(x)g(x)\mathrm{d}x$.

(continued)

Let $V = \text{span}\{1, x\}$ and observe that $\min\limits_{a, b \in \mathbb{R}} \int_0^1 (x^2 - a - bx)^2 dx$ is the square of the distance, in $C([0, 1], \mathbb{R})$, from the vector x^2 to the subspace V. It follows that

$$\min_{a, b \in \mathbb{R}} \int_0^1 (x^2 - a - bx)^2 dx = d^2(x^2, V) = \frac{G(1, x, x^2)}{G(1, x)} = \frac{1}{180},$$

where $G(v_1, v_2, \ldots, v_n)$ denotes the *Gram determinant* of vectors $v_1, v_2, \ldots, v_n$.

(b) $\frac{8}{175}$.

5.61. $\max f(C) = f\left(2, \frac{3}{2}\right) = 3$.

5.62. $\min f(C) = f(4, 1) = 5$.

5.63. Let $L : \mathbb{R}^2 \times \mathbb{R} \to \mathbb{R}$, $L(x, y, \lambda) = x^2 + 2y^2 + \lambda(2x + 4y - 3)$ be the Lagrange function. The only critical point of L is $\left(\frac{1}{2}, \frac{1}{2}, -\frac{1}{2}\right)$. We obtain that $\left(\frac{1}{2}, \frac{1}{2}\right)$ is a conditional local minimum point of f.

5.64. Since C is compact and f is continuous, it follows, based on the Weierstrass theorem, that $f|_C$ is bounded and it attains its minimum and maximum values on C. The points where $f|_C$ attains its minimum and maximum values are located between the conditional critical points of f relative to the set C.

Let $L : \mathbb{R}^2 \times \mathbb{R} \to \mathbb{R}$, $L(x, y, \lambda) = x - y + \lambda(x^2 + y^2 - 2)$ be the Lagrange function. The critical points of L are $\left(-1, 1, \frac{1}{2}\right)$ and $\left(1, -1, -\frac{1}{2}\right)$. Since $f(-1, 1) = -2$ and $f(1, -1) = 2$, it follows that $\min f(C) = -2$ and $\max f(C) = 2$.

5.65. $\min f(C) = -\frac{1}{2}$ and $\max f(C) = \frac{1}{2}$.

5.66. Let $L : \mathbb{R}^2 \times \mathbb{R} \to \mathbb{R}$, $L(x, y, \lambda) = xy + \lambda(x + 2y - 4)$ be the Lagrange function. The only critical point of L is $(2, 1, -1)$. We obtain that $(2, 1)$ is a conditional local maximum point of f.

5.69. The set C is the triangular surface with vertices $(\pi, 0, 0)$, $(0, \pi, 0)$, and $(0, 0, \pi)$. Since C is a compact set and f is continuous, it follows, based on the Weierstrass theorem, that the function $f|_C$ is bounded and it attains its bounds on C. The points where $f|_C$ attains its bounds are located between:

- The conditional critical points of f relative to the set

$$C_0 = \left\{ (x, y, z) \in (0, \infty)^3 \mid x + y + z = \pi \right\}$$

(the interior of the triangular surface);
- The conditional critical points of f relative to the set

$$C_1 = \{ (x, y, z) \mid x = 0, y, z > 0, y + z = \pi \}$$

(the side of the triangle, without endpoints, situated in the plane yOz);
- The conditional critical points of f relative to the set

$$C_2 = \{ (x, y, z) \mid y = 0, x, z > 0, x + z = \pi \}$$

(the side of the triangle, without the endpoints, situated in the plane xOz);
- The conditional critical points of f relative to the set

$$C_3 = \{ (x, y, z) \mid z = 0, x, y > 0, x + y = \pi \}$$

(the side of triangle, without the endpoints, situated in the plane xOy);
- The points $(\pi, 0, 0)$, $(0, \pi, 0)$, $(0, 0, \pi)$ (the triangle vertices).

To determine the conditional critical points of f relative to the set C_0, we consider the Lagrange function $L : (0, \infty)^3 \times \mathbb{R} \to \mathbb{R}$, $L(x, y, z, \lambda) = \sin x + \sin y + \sin z + \lambda(x + y + z - \pi)$. The only critical point of L is $\left(\frac{\pi}{3}, \frac{\pi}{3}, \frac{\pi}{3}, -\frac{1}{2} \right)$.

To determine the conditional critical points of f relative to the set C_1, we consider the function $g(y) = f(0, y, \pi - y) = 2 \sin y$, whose only critical point is $y = \frac{\pi}{2}$. Therefore, $\left(0, \frac{\pi}{2}, \frac{\pi}{2} \right)$ is the conditional critical point of f relative to the set C_1. Similarly, one can obtain the conditional critical points of f relative to the other two sides of the triangle: $\left(\frac{\pi}{2}, 0, \frac{\pi}{2} \right)$, $\left(\frac{\pi}{2}, \frac{\pi}{2}, 0 \right)$. Since $f\left(\frac{\pi}{3}, \frac{\pi}{3}, \frac{\pi}{3} \right) = \frac{3\sqrt{3}}{2}$, $f\left(0, \frac{\pi}{2}, \frac{\pi}{2} \right) = f\left(\frac{\pi}{2}, 0, \frac{\pi}{2} \right) = f\left(\frac{\pi}{2}, \frac{\pi}{2}, 0 \right) = 2$, and $f(\pi, 0, 0) = f(0, \pi, 0) = f(0, 0, \pi) = 0$, it follows that $\min f(C) = 0$ and $\max f(C) = \frac{3\sqrt{3}}{2}$.

Remark. In any triangle ABC the following inequality holds true:

$$\sin A + \sin B + \sin C \leq \frac{3\sqrt{3}}{2}.$$

The equality is valid if and only if the triangle ABC is equilateral.

5.70. Let $L : \mathbb{R}^3 \times \mathbb{R} \to \mathbb{R}$, $L(x, y, z, \lambda) = x + y^2 + z^3 + \lambda(x + 2y + 3z - a)$ be the Lagrange function. The critical points of L are $(a - 5, 1, 1, -1)$ and $(a + 1, 1, -1, -1)$. We obtain that $(a - 5, 1, 1)$ is a conditional local minimum point of f, and $(a + 1, 1, -1)$ is not a conditional local extremum point of f.

5.71. Let $L : (0, \infty)^3 \times \mathbb{R} \to \mathbb{R}$, $L(x, y, z, \lambda) = xyz^2 + \lambda(x + y + 2z - a)$ be the Lagrange function. The only critical point of L is $\left(\frac{a}{4}, \frac{a}{4}, \frac{a}{4}, -\frac{a^3}{4^3}\right)$. We obtain that $\left(\frac{a}{4}, \frac{a}{4}, \frac{a}{4}\right)$ is a conditional local maximum point of f.

5.72. Let $L : \mathbb{R}^3 \times \mathbb{R} \to \mathbb{R}$, $L(x, y, z, \lambda) = x(y + z) + 3yz + \lambda(xyz - 3)$ be the Lagrange function. The only critical point of L is $(3, 1, 1, -2)$. We obtain that $(3, 1, 1)$ is a conditional local minimum point of f.

5.73. Since C is compact and f is continuous, it follows, based on the Weierstrass theorem, that $f|_C$ is bounded and attains its bounds on C.

The points where $f|_C$ attains its bounds are located between the conditional critical points of f relative to the set C.

Let $L : \mathbb{R}^3 \times \mathbb{R}^2 \to \mathbb{R}$, $L(x, y, z, \lambda, \mu) = x + y + z + \lambda(x - 2y + z - 2) + \mu(x^2 + y^2 + z^2 - 1)$ be the Lagrange function. The critical points of L are $\left(0, -1, 0, -1, \frac{3}{2}\right)$ and $\left(\frac{2}{3}, -\frac{1}{3}, \frac{2}{3}, 1, -\frac{3}{2}\right)$. Since $f(0, -1, 0) = -1$ and $f\left(\frac{2}{3}, -\frac{1}{3}, \frac{2}{3}\right) = 1$, it follows that $\min f(C) = -1$ and $\max f(C) = 1$.

5.74. Let $L : \mathbb{R}^3 \times \mathbb{R}^2 \to \mathbb{R}$, $L(x, y, z, \lambda, \mu) = xyz + \lambda(x + y + z - 4) + \mu(xy + yz + xz - 5)$ be the Lagrange function. The critical points of L are $(1, 1, 2, 1, -1)$, $(1, 2, 1, 1, -1)$, $(2, 1, 1, 1, -1)$, $\left(\frac{5}{3}, \frac{5}{3}, \frac{2}{3}, \frac{25}{9}, -\frac{5}{3}\right)$, $\left(\frac{5}{3}, \frac{2}{3}, \frac{5}{3}, \frac{25}{9}, -\frac{5}{3}\right)$, $\left(\frac{2}{3}, \frac{5}{3}, \frac{5}{3}, \frac{25}{9}, -\frac{5}{3}\right)$. We obtain that $(1, 1, 2)$, $(1, 2, 1)$, and $(2, 1, 1)$ are the conditional local maximum points of f, and $\left(\frac{5}{3}, \frac{5}{3}, \frac{2}{3}\right)$, $\left(\frac{5}{3}, \frac{2}{3}, \frac{5}{3}\right)$, and $\left(\frac{2}{3}, \frac{5}{3}, \frac{5}{3}\right)$ are the conditional local minimum points of f.

Remark. The set C is bounded, since

$$x^2 + y^2 + z^2 = (x + y + z)^2 - 2(xy + yz + xz) = 6,$$

for all $(x, y, z) \in C$, and it is closed, being the intersection of closed sets. Thus, C, which is *a circle* in space, is compact and f is continuous and it follows, based on the Weierstrass theorem, that f is bounded and attains its bounds on C. The points where $f|_C$ attains its bounds are located between the conditional critical points of f relative to the set C. Since

$$f(1, 1, 2) = f(1, 2, 1) = f(2, 1, 1) = 2,$$

$$f\left(\frac{5}{3}, \frac{5}{3}, \frac{2}{3}\right) = f\left(\frac{5}{3}, \frac{2}{3}, \frac{5}{3}\right) = f\left(\frac{2}{3}, \frac{5}{3}, \frac{5}{3}\right) = \frac{50}{27},$$

it follows that $(1, 1, 2)$, $(1, 2, 1)$, and $(2, 1, 1)$ are conditional global maximum points of f, and $\left(\frac{5}{3}, \frac{5}{3}, \frac{2}{3}\right)$, $\left(\frac{5}{3}, \frac{2}{3}, \frac{5}{3}\right)$, and $\left(\frac{2}{3}, \frac{5}{3}, \frac{5}{3}\right)$ are the conditional global minimum points of f.

5.75. C is compact and f is continuous. We have, based on the Weierstrass theorem, that $f|_C$ is bounded and attains its bounds on C. The points where $f|_C$ attains its bounds are located between:

– The critical points of f situated in the interior of C;
– The conditional critical points of f relative to the set

$$C_{\text{sphere}} = \left\{(x, y, z) \in \mathbb{R}^3 \mid x^2 + y^2 + z^2 = 1\right\}.$$

Since f does not have critical points, we determine the conditional critical points of f relative to the set C_{sphere}.

Let $L : \mathbb{R}^3 \times \mathbb{R} \to \mathbb{R}$, $L(x, y, z, \lambda) = 1 - x - y - z + \lambda(x^2 + y^2 + z^2 - 1)$ be the Lagrange function. The critical points of L are

$$\left(\frac{\sqrt{3}}{3}, \frac{\sqrt{3}}{3}, \frac{\sqrt{3}}{3}, \frac{\sqrt{3}}{2}\right) \quad \text{and} \quad \left(-\frac{\sqrt{3}}{3}, -\frac{\sqrt{3}}{3}, -\frac{\sqrt{3}}{3}, -\frac{\sqrt{3}}{2}\right).$$

Since $f\left(\frac{\sqrt{3}}{3}, \frac{\sqrt{3}}{3}, \frac{\sqrt{3}}{3}\right) = 1 - \sqrt{3}$ and $f\left(-\frac{\sqrt{3}}{3}, -\frac{\sqrt{3}}{3}, -\frac{\sqrt{3}}{3}\right) = 1 + \sqrt{3}$, we have that min $f(C) = 1 - \sqrt{3}$ and max $f(C) = 1 + \sqrt{3}$.

5.76. The set C is compact and the function f is continuous. We have, based on the Weierstrass theorem, that $f|_C$ is bounded and attains its bounds on C. The points where $f|_C$ attains its bounds are located between:

– The critical points of f situated in the interior of C;
– The conditional critical points of f relative to the set
   $C_1 = \left\{(x, y, z) \in \mathbb{R}^3 \mid z = x^2 + y^2, z < 1\right\}$;
– The conditional critical points of f relative to the set
   $C_2 = \left\{(x, y, z) \in \mathbb{R}^3 \mid x^2 + y^2 < 1, z = 1\right\}$;
– The conditional critical points of f relative to the set
   $C_3 = \left\{(x, y, z) \in \mathbb{R}^3 \mid x^2 + y^2 = 1, z = 1\right\}$.

The only critical point of f is $(0, 0, 0)$, which is not located in the interior of C.

To determine the conditional critical point of f relative to the set C_1, we consider the function $g(x, y) = f(x, y, x^2 + y^2) = x^3 + x^2 y + xy^2 + y^3 + 2xy$. The critical points of g are $(0, 0)$ and $\left(-\frac{1}{3}, -\frac{1}{3}\right)$. It follows that the conditional critical points of f relative to the set C_1 are $(0, 0, 0)$ and $\left(-\frac{1}{3}, -\frac{1}{3}, \frac{2}{9}\right)$.

To determine the conditional critical points of f relative to the set C_2, we consider the function $h(x, y) = f(x, y, 1) = 2xy + x + y$. The only critical point of h is $\left(-\frac{1}{2}, -\frac{1}{2}\right)$. We have $\left(-\frac{1}{2}\right)^2 + \left(-\frac{1}{2}\right)^2 < 1$ and it follows that $\left(-\frac{1}{2}, -\frac{1}{2}, 1\right)$ is the unique conditional critical point of f relative to the set C_2.

To determine the conditional critical points of f relative to the set C_3, we consider the Lagrange function $L(x, y, \lambda) = 2xy + x + y + \lambda(x^2 + y^2 - 1)$. The critical points of L are $\left(\frac{\sqrt{2}}{2}, \frac{\sqrt{2}}{2}, -1 - \frac{\sqrt{2}}{2}\right)$, $\left(-\frac{\sqrt{2}}{2}, -\frac{\sqrt{2}}{2}, -1 + \frac{\sqrt{2}}{2}\right)$, $\left(\frac{\sqrt{7}-1}{4}, -\frac{\sqrt{7}+1}{4}, 1, 1\right)$, and $\left(-\frac{\sqrt{7}+1}{4}, \frac{\sqrt{7}-1}{4}, 1, 1\right)$, which implies that the conditional critical points of f relative to the set C_3 are $\left(\frac{\sqrt{2}}{2}, \frac{\sqrt{2}}{2}, 1\right)$, $\left(-\frac{\sqrt{2}}{2}, -\frac{\sqrt{2}}{2}, 1\right)$, $\left(\frac{\sqrt{7}-1}{4}, -\frac{\sqrt{7}+1}{4}, 1\right)$, and $\left(-\frac{\sqrt{7}+1}{4}, \frac{\sqrt{7}-1}{4}, 1\right)$.

Since

$$f(0, 0, 0) = 0, \quad f\left(-\frac{1}{3}, -\frac{1}{3}, \frac{2}{9}\right) = \frac{2}{27}, \quad f\left(-\frac{1}{2}, -\frac{1}{2}, 1\right) = -\frac{1}{2},$$

$$f\left(\frac{\sqrt{2}}{2}, \frac{\sqrt{2}}{2}, 1\right) = 1 + \sqrt{2}, \quad f\left(-\frac{\sqrt{2}}{2}, -\frac{\sqrt{2}}{2}, 1\right) = 1 - \sqrt{2},$$

$$f\left(\frac{\sqrt{7}-1}{4}, -\frac{\sqrt{7}+1}{4}, 1\right) = -\frac{5}{4} \quad \text{and} \quad f\left(-\frac{\sqrt{7}+1}{4}, \frac{\sqrt{7}-1}{4}, 1\right) = -\frac{5}{4},$$

it follows that $\min f(C) = -\frac{5}{4}$ and $\max f(C) = 1 + \sqrt{2}$.

5.77. [132] The minimum value of the expression $x + yz$ is 50.

Let $a = x + yz$. We analyze the following cases:

| $y = 1$ | Then $z = 160 - x$ and $a = x + 160 - x = 160$. |

| $y = 2$ | Because $2x + z = 160$, it follows that $x \leq 79$. Then |

$$a = x + 2(160 - 2x) \implies x = \frac{2 \cdot 160 - a}{3} \leq 79 \implies a \geq 83.$$

For $x = 79$ and $z = 2$, we have $a = 83$.

| $y = 3$ | Because $3x + z = 160$, it follows that $x \leq 53$. From $a = x + 3(160 - 3x)$ we obtain that $a \geq 56$. For $x = 53$ and $z = 1$ we have $a = 56$. |

| $y = 4$ | Because $4x + z = 160$, it follows that $x \leq 39$. From $a = x + 4(160 - 4x)$ we obtain that $a \geq 55$. For $x = 39$ and $z = 4$ we have $a = 55$. |

| $y = 5$ | Because $5x + z = 160$, it follows that $x \leq 31$. From $a = x + 5(160 - 5x)$ we obtain that $a \geq 56$. For $x = 31$ and $z = 5$ we have $a = 56$. |

| $y = 6$ | Because $6x + z = 160$, it follows that $x \leq 26$. From $a = x + 6(160 - 6x)$ we obtain that $a \geq 50$. For $x = 26$ and $z = 4$ we have $a = 50$. |

$\boxed{y = 7}$ Because $7x + z = 160$, it follows that $x \leq 22$. From $a = x + 7(160 - 7x)$
we obtain that $a \geq 64$. For $x = 22$ and $z = 6$ we have $a = 64$.

$\boxed{y = 8}$ Because $8x + z = 160$, it follows that $x \leq 19$. From $a = x + 8(160 - 8x)$
we obtain that $a \geq 83$. For $x = 19$ and $z = 8$ we have $a = 83$.

$\boxed{y = 9}$ Because $9x + z = 160$, it follows that $x \leq 17$. From $a = x + 9(160 - 9x)$
we obtain that $a \geq 80$. For $x = 17$ and $z = 7$ we have $a = 80$.

We also analyze the following cases:

$\boxed{z = 1}$ Then $xy = 159 = 3 \cdot 53$. Therefore, we can have

$$
\begin{aligned}
x &= 1,   & y &= 159 \implies a = 160, \\
x &= 3,   & y &= 53 \implies a = 56, \\
x &= 53, & y &= 3   \qquad \text{was analyzed,} \\
x &= 159, & y &= 1   \qquad \text{was analyzed.}
\end{aligned}
$$

$\boxed{z = 2}$ Then $xy = 158 = 2 \cdot 79$. Therefore, we can have

$$
\begin{aligned}
x &= 1,   & y &= 158 \implies a = 517, \\
x &= 2,   & y &= 79 \implies a = 160, \\
x &= 79, & y &= 2   \qquad \text{was analyzed,} \\
x &= 158, & y &= 1   \qquad \text{was analyzed.}
\end{aligned}
$$

$\boxed{z = 3}$ Then $xy = 157$. Therefore, we can have

$$
\begin{aligned}
x &= 1,   & y &= 157 \implies a = 472, \\
x &= 157, & y &= 1   \qquad \text{was analyzed.}
\end{aligned}
$$

$\boxed{z = 4}$ Then $xy = 156 = 2^2 \cdot 3 \cdot 13$. Therefore, we can have

$$
\begin{aligned}
x &= 1, & y &= 156 \implies a = 625, \\
x &= 2, & y &= 78 \implies a = 314, \\
x &= 3, & y &= 52 \implies a = 211, \\
x &= 4, & y &= 39 \implies a = 160, \\
x &= 6, & y &= 26 \implies a = 110, \\
x &= 12, & y &= 13 \implies a = 64, \\
x &= 13, & y &= 12 \implies a = 61,
\end{aligned}
$$

$$x = 26, \quad y = 6 \quad \text{was analyzed,}$$
$$x = 39, \quad y = 4 \quad \text{was analyzed,}$$
$$x = 52, \quad y = 3 \quad \text{was analyzed,}$$
$$x = 78, \quad y = 2 \quad \text{was analyzed,}$$
$$x = 156, \, y = 1 \quad \text{was analyzed.}$$

If $y \geq 10$ and $z \geq 5$, then $a = x + yz > 50$. Therefore, the minimum value of the expression $x + yz$ is 50.

5.78. *Solution I.* Since $2(x^2 + y^2) \geq (x + y)^2$, for all $x, y \in \mathbb{R}$, we have that

$$\left| (x + y)e^{-x^2 - y^2} \right| \leq \frac{|x + y|}{e^{\frac{(x+y)^2}{2}}} = \frac{|x + y|}{e^{\frac{|x+y|^2}{2}}}.$$

Using the substitution $|x + y| = t\sqrt{2}$, we have to prove that $\sqrt{2}te^{-t^2} \leq \frac{1}{\sqrt{e}}$, $\forall t \geq 0$.

Solution II. We determine the global extremum values of the function $f : \mathbb{R}^2 \to \mathbb{R}$, $f(x, y) = (x + y)e^{-x^2 - y^2}$. We consider the compact set

$$D(0, r) = \left\{ (x, y) \in \mathbb{R}^2 \mid x^2 + y^2 \leq r^2 \right\}, \quad r > 0,$$

and we determine the extremum values of f on $D(0, r)$. We have

$$f'_x(x, y) = e^{-x^2 - y^2}(1 - 2x(x + y)) \quad \text{and} \quad f'_y(x, y) = e^{-x^2 - y^2}(1 - 2y(x + y)).$$

The critical points of f are $\left(\frac{1}{2}, \frac{1}{2} \right)$ and $\left(-\frac{1}{2}, -\frac{1}{2} \right)$, which are local extremum points of f and

$$f\left(\frac{1}{2}, \frac{1}{2} \right) = \frac{1}{\sqrt{e}}, \quad f\left(-\frac{1}{2}, -\frac{1}{2} \right) = -\frac{1}{\sqrt{e}}.$$

We determine the extremum values of f on the boundary of $D(0, r)$. Let

$$L(x, y, \lambda) = (x + y)^{-(x^2 + y^2)} - \lambda(x^2 + y^2 - r^2), \quad \lambda \in \mathbb{R}.$$

The conditional critical points can be obtained by solving the system $L'_x = 0$, $L'_y = 0$, $x^2 + y^2 = r^2$. It follows that $x = y = \pm\frac{r\sqrt{2}}{2}$. Since

$$f\left(\frac{r\sqrt{2}}{2}, \frac{r\sqrt{2}}{2} \right) = r\sqrt{2}e^{-r^2} \leq \frac{1}{\sqrt{e}},$$

it follows that $\max_{(x,y) \in \mathbb{R}^2} |f(x, y)| = \frac{1}{\sqrt{e}}$.

Implicit Functions

14

14.1 Implicit Functions of One Real Variable Defined by an Equation

6.1. $y'(0) = 1$, $y''(0) = \frac{1}{2}$, $y'''(0) = \frac{11}{4}$.

6.2. $y'(0) = -\pi^2$, $y''(0) = 4\pi^3$.

6.3. $y'(0) = 0$, $y''(0) = -\frac{2}{3}$, $y'''(0) = -\frac{2}{3}$.

6.4. $y'(0) = 0$, $y''(0) = \pi + 1$.

6.5. $y'(0) = 2$, $y''(0) = 4$.

6.6. $y'(1) = 1$, $y''(1) = 0$.

6.7. Let (x_0, y_0) be an arbitrary point on the curve, with $x_0 y_0 \neq 0$. We have $\sqrt{x_0} + \sqrt{y_0} = \sqrt{c}$. We differentiate the equation $\sqrt{x} + \sqrt{y(x)} = \sqrt{c}$ with respect to x, and we obtain that $y'(x) = -\sqrt{\frac{y(x)}{x}}$. The equation of the tangent to the curve at the point (x_0, y_0) is given by

$$y - y_0 = -\sqrt{\frac{y_0}{x_0}}(x - x_0).$$

A calculation shows that the lengths of the segments are

$$x = x_0 + \sqrt{x_0 y_0} \quad \text{and} \quad y = y_0 + \sqrt{x_0 y_0}.$$

It follows that $x + y = x_0 + 2\sqrt{x_0 y_0} + y_0 = (\sqrt{x_0} + \sqrt{y_0})^2 = c$.

When $x_0 = c$ and $y_0 = 0$, we have $y'_-(c) = 0$. In this case $x = c$ and $y = 0$, which implies $x + y = c$.

When $x_0 = 0$ and $y_0 = c$, we have $y'_+(0) = -\infty$. In this case $x = 0$ and $y = c$, which implies $x + y = c$.

6.8. $y'\left(\frac{3a}{2}\right) = -1$, $y''\left(\frac{3a}{2}\right) = -\frac{32}{3a}$, $y'''\left(\frac{3a}{2}\right) = -\frac{512}{3a^2}$.

© The Author(s), under exclusive license to Springer Nature Switzerland AG 2021
A. Sîntămărian, O. Furdui, *Sharpening Mathematical Analysis Skills*, Problem Books in Mathematics, https://doi.org/10.1007/978-3-030-77139-3_14

6.9. $y'\left(\frac{a\sqrt{2}}{2}\right) = -1,\ \ y''\left(\frac{a\sqrt{2}}{2}\right) = -\frac{6\sqrt{2}}{a}.$

6.10. $\left(\frac{a\sqrt{3}}{2}, \frac{a}{2}\right).$

6.13.

(a) $y'(1) = -\frac{4}{3}, y''(1) = -\frac{68}{9};$

(b) $y'(-1) = \frac{4}{3}, y''(-1) = -\frac{68}{9}.$

6.14. $y'(1) = -1, y''(1) = -4.$

14.2 Implicit Functions of Two Real Variables Defined by an Equation

6.15. $z'_x(-1, 1) = 0,\ \ z'_y(-1, 1) = 1,\ \ z''_{x^2}(-1, 1) = 0,\ \ z''_{xy}(-1, 1) = 1,$
$z''_{y^2}(-1, 1) = -1.$

6.16. $z'_x(1, -1) = \frac{1}{2},\ \ z'_y(1, -1) = -1,\ \ z''_{x^2}(1, -1) = -\frac{11}{8},\ \ z''_{xy}(1, -1) = \frac{1}{2},$
$z''_{y^2}(1, -1) = -\frac{5}{2}.$

6.17. $z'_x(-1, 2) = -4,\ \ z'_y(-1, 2) = -1,\ \ z''_{x^2}(-1, 2) = -12,\ \ z''_{xy}(-1, 2) = 2,$
$z''_{y^2}(-1, 2) = 2.$

14.3 Implicit Functions of One Real Variable Defined by a System of Equations

6.18. $y'(1) = -2,\ \ z'(1) = -3,\ \ y''(1) = -9,\ \ z''(1) = 2.$
6.19. $y'(1) = 4,\ \ z'(1) = -14,\ \ y''(1) = -32,\ \ z''(1) = 293.$
6.20. $y'(-1) = -3,\ \ z'(-1) = -3,\ \ y''(-1) = 16,\ \ z''(-1) = 1.$

14.4 Implicit Functions of Two Real Variables Defined by a System of Equations

6.21. $u'_x(1, -1) = -1,\ \ u'_y(1, -1) = 1,\ \ v'_x(1, -1) = -2\pi + 2,\ \ v'_y(1, -1) = \pi - 2.$
6.22. $u'_x(-1, 2) = 1,\ \ u'_y(-1, 2) = 0,\ \ v'_x(-1, 2) = 3,\ \ v'_y(-1, 2) = 1.$

Challenges, Gems, and Mathematical Beauties

15.1 Limits of Sequences

7.1. (a) x^2; (b) xy; (c) 0.
7.2. The sequence diverges.
7.3. k^k. Use Stirling's formula.
7.4. $\frac{1}{2}$. Use that $\lim_{n\to\infty} n(H_n - \ln n - \gamma) - \frac{1}{2}$ (see Problem 1.31).
7.5. (b) Use the inequality in part (a) with $x = \sqrt[n]{k}$, $k = 1, 2, \ldots, n$.
7.6. e^{a-2b}.
7.7. 1.
7.8.

(a) The limit equals 2. Let $x_n = \sqrt[n]{2^n \sin 1 + 2^n \sin 2 + \cdots + 2^n \sin n}$. Since $2^n \sin i \le 2^n$, we get that $x_n \le 2\sqrt[n]{n}$. Fix $i = 1, \ldots, n$. We have $2^{\sin i} \le x_n$ and it follows that $2^{\sin i} \le \liminf_{n\to\infty} x_n \le \limsup_{n\to\infty} x_n \le 2$. Therefore, $2 = \limsup_{i\to\infty} 2^{\sin i} \le \liminf_{n\to\infty} x_n \le \limsup_{n\to\infty} x_n \le 2$.
(b) The limit equals $2^{\sup x_n}$.

7.9. (a) $\frac{1}{2}$; (b) $\frac{1}{16}$; (c) $\frac{a+\sqrt{a^2-1}}{2}$; (d) $\frac{\sqrt{a^2+b^2}}{4}$.
7.10. We have

$$\lim_{n\to\infty} \left(\frac{1 + \frac{1}{3} + \cdots + \frac{1}{2n-1}}{\frac{1}{2} + \frac{1}{4} + \cdots + \frac{1}{2n}} \right)^{\ln n} = \lim_{n\to\infty} \left(\frac{H_{2n} - \frac{1}{2}H_n}{\frac{1}{2}H_n} \right)^{\ln n}$$

$$= \lim_{n\to\infty} \left(2\frac{H_{2n}}{H_n} - 1 \right)^{\ln n}$$

$$= \lim_{n\to\infty} \left[1 + 2\left(\frac{H_{2n}}{H_n} - 1 \right) \right]^{\ln n}$$

$$= e^{2 \lim_{n \to \infty} \ln n \left(\frac{H_{2n}}{H_n} - 1 \right)}$$

$$= e^{2 \lim_{n \to \infty} \frac{\ln n}{H_n} (H_{2n} - H_n)}$$

$$= 4,$$

since $\lim_{n \to \infty} \frac{\ln n}{H_n} = 1$ and $\lim_{n \to \infty} (H_{2n} - H_n) = \ln 2$.

7.12. A. (a), (c), and (d) Use Taylor's formula. **B.** (a) Use Cauchy–d'Alembert's criterion. **A.** (b) and **B.** (b) use the technique for calculating limits of sequences of indeterminate form 1^∞. In our case, we have the following.

Let $(x_n)_{n \in \mathbb{N}}$, $x_n \in \mathbb{R}$, $\forall n \in \mathbb{N}$, be such that $\lim_{n \to \infty} x_n = x$ and $\lim_{n \to \infty} \left(\frac{x_n}{x} \right)^n = L$. If $\lim_{n \to \infty} n(x_n - x)$ exists, then $\lim_{n \to \infty} n(x_n - x) = x \ln L$.

A. (e) *The challenge*: Use a similar technique to that given in the solution of problem 1.14 in [26].

7.13. We use Abel's summation formula, with $a_k = 1$ and $b_k = \frac{1}{k^2} + \frac{1}{(k+1)^2} + \cdots$, and we have that

$$\sum_{k=1}^{n} \left(\frac{1}{k^2} + \frac{1}{(k+1)^2} + \cdots \right) = n \left(\frac{1}{(n+1)^2} + \frac{1}{(n+2)^2} + \cdots \right) + \sum_{k=1}^{n} \frac{1}{k}.$$

It follows that

$$\sum_{k=1}^{n} \left(\frac{1}{k^2} + \frac{1}{(k+1)^2} + \cdots \right) - \ln n = n \left(\frac{1}{(n+1)^2} + \frac{1}{(n+2)^2} + \cdots \right) + \sum_{k=1}^{n} \frac{1}{k} - \ln n$$

and the result follows since $\lim_{n \to \infty} n \left(\frac{1}{(n+1)^2} + \frac{1}{(n+2)^2} + \cdots \right) = 1$.

7.14.

(a) Let $f_n(x) = x^n + x^{n-1} + \cdots + x - 1 = \frac{x^{n+1} - 2x + 1}{x - 1}$. Since $f_n'(x) = nx^{n-1} + \cdots + 1 > 0$, $x \geq 0$, we have that f_n strictly increases on $(0, \infty)$. Since $f_n \left(\frac{1}{2} \right) = -\frac{1}{2^n} < 0$ and $f_n(1) = n - 1 > 0$, $n \geq 2$, we have that $x_n \in \left(\frac{1}{2}, 1 \right)$.

(b) We have

$$f_{n+1}(x_n) = \frac{x_n^{n+2} - 2x_n + 1}{x_n - 1} = \frac{x_n(2x_n - 1) - 2x_n + 1}{x_n - 1} = 2x_n - 1 > 0,$$

since $x_n > \frac{1}{2}$. Also, $f_{n+1} \left(\frac{1}{2} \right) = -\frac{1}{2^{n+1}} < 0$ and this implies $x_{n+1} \in \left(\frac{1}{2}, x_n \right)$.

(c) We prove that $x_n \in \left(\frac{1}{2}, \frac{1}{2} + \frac{1}{2^{n+1}} \right)$. It suffices to show that $f_n \left(\frac{1}{2} + \frac{1}{2^{n+1}} \right) > 0$. We have

$$f_n\left(\frac{1}{2}+\frac{1}{2^{n+1}}\right)=\frac{1}{2^{n-1}}\cdot\frac{\frac{1}{2}\left(1+\frac{1}{2^n}\right)^{n+1}-1}{\frac{1}{2^n}-1}>0,$$

since $\left(1+\frac{1}{2^n}\right)^{n+1}<2, n\geq 2$. The preceding inequality is equivalent to $\sqrt[n+1]{2}-1>\frac{1}{2^n}, \forall n\geq 2$. The inequality is verified for $n=2$. Let $n\geq 3$. We have, based on Lagrange's Mean Value Theorem, that $2^{\frac{1}{n+1}}-2^0=\frac{\ln 2}{n+1}2^{\theta_n}$, for some $\theta_n\in\left(0,\frac{1}{n+1}\right)$. It follows that

$$2^{\frac{1}{n+1}}-1=\frac{\ln 2}{n+1}2^{\theta_n}>\frac{\ln 2}{n+1}>\frac{1}{2^n},\quad n\geq 3,$$

since $2^n\ln 2=2^{n-1}\ln 4>2^{n-1}\geq n+1, \forall n\geq 3$.

(d) We have

$$\lim_{n\to\infty}2^n\left(x_n-\frac{1}{2}\right)=\lim_{n\to\infty}2^{n-1}(2x_n-1)=\lim_{n\to\infty}2^{n-1}x_n^{n+1}$$

$$=\frac{1}{4}\lim_{n\to\infty}(2x_n)^{n+1}=\frac{1}{4}e^{\lim_{n\to\infty}(n+1)(2x_n-1)}=\frac{1}{4},$$

since (see the solution of part (c)) $0<(n+1)(2x_n-1)<\frac{n+1}{2^n}$.

7.15. Use that:

(a) $\dfrac{n^2}{k^2(n-k)^2}=\dfrac{1}{k^2}+\dfrac{1}{(n-k)^2}+\dfrac{2}{n}\left(\dfrac{1}{k}+\dfrac{1}{n-k}\right)$;

(b) $\dfrac{n^3}{k^3(n-k)^3}=\dfrac{1}{k^3}+\dfrac{1}{(n-k)^3}+\dfrac{3}{n}\left(\dfrac{1}{k^2}+\dfrac{1}{(n-k)^2}\right)+\dfrac{6}{n^2}\left(\dfrac{1}{k}+\dfrac{1}{n-k}\right)$.

7.16. (a) $e^{\frac{1}{n+2}}$; (b) $e^{\frac{n+1}{n+2}}$.

7.17. (a) $-\frac{x^2}{2}$; (b) $\frac{x^2}{2}\left(f''(0)-(f'(0))^2\right)e^{f'(0)x}$.

7.18. (a) $\frac{f(1)}{2}$; (b) $\frac{2}{3}(1-\ln 2)f(1)$.

7.19.

(i) Let $x_n=\dfrac{a_{qn+k+1}a_{qn+k+1+p}\cdots a_{qn+k+1+s(n-1)p}}{a_{qn+k}a_{qn+k+p}\cdots a_{qn+k+s(n-1)p}}, n\in\mathbb{N}$.

We have $a_na_{n+2}<a_{n+1}^2$, for all $n\in\mathbb{N}$. Hence

$$\prod_{i=1}^{p-1}a_{n+i-1}^ia_{n+i+1}^i<\prod_{i=1}^{p-1}a_{n+i}^{2i}\quad\text{and}\quad\prod_{i=1}^{p-1}a_{n+i-1}^{p-i}a_{n+i+1}^{p-i}<\prod_{i=1}^{p-1}a_{n+i}^{2(p-i)},$$

for all $n \in \mathbb{N}$. Therefore,

$$a_n a_{n+p}^{p-1} < a_{n+p-1}^p \quad \text{and} \quad a_n^{p-1} a_{n+p} < a_{n+1}^p, \quad n \in \mathbb{N}.$$

Taking into account these inequalities, we get

$$\frac{a_{qn+k+1+s(n-1)p}^p}{a_{qn+k} a_{qn+k+s(n-1)p}^{p-1}} < x_n^p < \frac{a_{qn+k+1}^{p-1} a_{qn+k+1+s(n-1)p}}{a_{qn+k}^p},$$

for all $n \in \mathbb{N}\setminus\{1\}$. It follows, based on Squeeze Theorem, that $\lim\limits_{n\to\infty} x_n^p = \frac{ps+q}{q}$.

Therefore, $\lim\limits_{n\to\infty} x_n = \sqrt[p]{\frac{ps+q}{q}}$.

(ii) Let $y_n = \dfrac{a_{qn+k} a_{qn+k+p} \cdots a_{qn+k+s(n-1)p}}{(n!)^s}$, $n \in \mathbb{N}$. Using the limit from part (i), it follows that

$$\lim\limits_{n\to\infty} \frac{y_{n+1}}{y_n} = \left(\sqrt[p]{\frac{ps+q}{q}}\right)^q [(ps+q)r]^s.$$

Now, according to the Cauchy–d'Alembert criterion, we are able to write

$$\lim\limits_{n\to\infty} \sqrt[n]{y_n} = \lim\limits_{n\to\infty} \frac{y_{n+1}}{y_n} = \left(\sqrt[p]{\frac{ps+q}{q}}\right)^q [(ps+q)r]^s.$$

(iii) Let $z_n = \dfrac{a_{qn+k} a_{qn+k+p} \cdots a_{qn+k+s(n-1)p}}{n^{ns}}$, $n \in \mathbb{N}$. We have

$$\frac{z_{n+1}}{z_n} = \frac{y_{n+1}}{y_n} \cdot \frac{1}{\left[\left(1+\frac{1}{n}\right)^n\right]^s}, \quad n \in \mathbb{N}.$$

It follows that

$$\lim\limits_{n\to\infty} \frac{z_{n+1}}{z_n} = \left(\sqrt[p]{\frac{ps+q}{q}}\right)^q \left[\frac{(ps+q)r}{e}\right]^s.$$

Now, according to the Cauchy–d'Alembert criterion, we are able to write

$$\lim\limits_{n\to\infty} \sqrt[n]{z_n} = \lim\limits_{n\to\infty} \frac{z_{n+1}}{z_n} = \left(\sqrt[p]{\frac{ps+q}{q}}\right)^q \left[\frac{(ps+q)r}{e}\right]^s.$$

7.20. We have

$$x_n = \frac{5^{5n} \left(C_{2n}^n\right)^3}{C_{10n}^{5n} C_{5n}^n C_{4n}^{2n}} = \frac{[(5n+5)(5n+10)\cdots(10n)]^5}{(5n+1)(5n+2)\cdots(10n)}$$

$$= \left[\frac{(5n+5)(5n+10)\cdots(10n)}{(5n+4)(5n+9)\cdots(10n-1)}\right]^4 \cdot \left[\frac{(5n+4)(5n+9)\cdots(10n-1)}{(5n+3)(5n+8)\cdots(10n-2)}\right]^3$$

$$\times \left[\frac{(5n+3)(5n+8)\cdots(10n-2)}{(5n+2)(5n+7)\cdots(10n-3)}\right]^2 \cdot \frac{(5n+2)(5n+7)\cdots(10n-3)}{(5n+1)(5n+6)\cdots(10n-4)},$$

for all $n \in \mathbb{N}$.

Choosing $a = r = 1$, $p = 5$, $q = 5$, and $s = 1$ in part (i) of Problem 7.19, we obtain

$$\lim_{n\to\infty} \frac{(5n+k+1)(5n+k+6)\cdots(10n+k-4)}{(5n+k)(5n+k+5)\cdots(10n+k-5)} = \sqrt[5]{2}, \quad k \in \{1,2,3,4\}.$$

So,

$$\lim_{n\to\infty} x_n = (\sqrt[5]{2})^4 (\sqrt[5]{2})^3 (\sqrt[5]{2})^2 \sqrt[5]{2} = 4.$$

7.21. Having in view the recurrence relation $nx_{n+1} = \left(n + \dfrac{1}{p}\right) x_n$, for $n \in \mathbb{N}$, we can write that

$$\prod_{j=1}^n pjx_{j+1} = \prod_{j=1}^n (pj+1)x_j$$

and

$$\sum_{j=1}^n pjx_{j+1} = \sum_{j=1}^n [p(j-1)x_j + (p+1)x_j],$$

for all $n \in \mathbb{N}$. We get

$$x_{n+1} = \frac{(p+1)(2p+1)\cdots(np+1)}{p(2p)\cdots(np)} x_1$$

and

$$x_1 + x_2 + \cdots + x_n = \frac{pn}{p+1} x_{n+1},$$

for all $n \in \mathbb{N}$. Therefore,

$$x_{\alpha n} + x_{\alpha n+1} + \cdots + x_{\beta n}$$

$$= \frac{p\beta n}{p+1} x_{\beta n+1} - \frac{p(\alpha n - 1)}{p+1} x_{\alpha n}$$

$$= \frac{p\beta n}{p+1} \cdot \frac{(p+1)(2p+1)\cdots(\beta np+1)}{p(2p)\cdots(\beta np)} x_1 - \frac{p(\alpha n - 1)}{p+1} x_{\alpha n}$$

$$= \frac{p\beta n}{p+1} \cdot \frac{(\alpha np + 1)((\alpha n + 1)p + 1)\cdots(\beta np + 1)}{(\alpha np)((\alpha n + 1)p)\cdots(\beta np)} x_{\alpha n} - \frac{p(\alpha n - 1)}{p+1} x_{\alpha n},$$

for all $n \in \mathbb{N}$.

The above relations, obtained for the sequence $(x_n)_{n\in\mathbb{N}}$, hold for $p = 1$ too.
Hence

$$y_{n+1} = (n+1)y_1$$

and

$$y_{\alpha n} + y_{\alpha n+1} + \cdots + y_{\beta n} = \frac{(\beta^2 - \alpha^2)n + \alpha + \beta}{2\alpha} y_{\alpha n},$$

for all $n \in \mathbb{N}$.

It follows that

$$\frac{y_{\alpha n}(x_{\alpha n} + x_{\alpha n+1} + \cdots + x_{\beta n})}{x_{\alpha n}(y_{\alpha n} + y_{\alpha n+1} + \cdots + y_{\beta n})}$$

$$= \frac{2p\alpha\beta n}{(p+1)[(\beta^2 - \alpha^2)n + \alpha + \beta]}$$

$$\times \frac{(\alpha np + 1)((\alpha n + 1)p + 1)\cdots(\beta np - (\beta - \alpha)p + 1)}{(\alpha np)((\alpha n + 1)p)\cdots(\beta np - (\beta - \alpha)p)}$$

$$\times \frac{(\beta np - (\beta - \alpha)p + p + 1)\cdots(\beta np + 1)}{(\beta np - (\beta - \alpha)p + p)\cdots(\beta np)}$$

$$- \frac{2p\alpha(\alpha n - 1)}{(p+1)[(\beta^2 - \alpha^2)n + \alpha + \beta]},$$

for all $n \in \mathbb{N}$.

Choosing $a = r = 1, k = 0, q = \alpha p$, and $s = \beta - \alpha$ in part (i) of Problem 7.19, we obtain

$$\lim_{n\to\infty} \frac{(\alpha np + 1)((\alpha n + 1)p + 1)\cdots(\beta np - (\beta - \alpha)p + 1)}{(\alpha np)((\alpha n + 1)p)\cdots(\beta np - (\beta - \alpha)p)} = \sqrt[p]{\frac{\beta}{\alpha}}.$$

So,

$$\lim_{n\to\infty} \frac{y_{\alpha n}(x_{\alpha n} + x_{\alpha n+1} + \cdots + x_{\beta n})}{x_{\alpha n}(y_{\alpha n} + y_{\alpha n+1} + \cdots + y_{\beta n})} = \frac{2p}{p+1} \cdot \frac{\frac{\beta}{\alpha}\sqrt[p]{\frac{\beta}{\alpha}} - 1}{\left(\frac{\beta}{\alpha}\right)^2 - 1}.$$

7.22. Let $(z_n)_{n\in\mathbb{N}}$ be a sequence, with $z_1 \neq 0$, defined by the recurrence relation $nz_{n+1} = (n+1)z_n$, for $n \in \mathbb{N}$. Using Problem 7.21, we can write that

$$\lim_{n\to\infty} \frac{y_{\alpha n}(x_{\alpha n} + x_{\alpha n+1} + \cdots + x_{\beta n})}{x_{\alpha n}(y_{\alpha n} + y_{\alpha n+1} + \cdots + y_{\beta n})}$$

$$= \lim_{n\to\infty} \left[\frac{z_{\alpha n}(x_{\alpha n} + x_{\alpha n+1} + \cdots + x_{\beta n})}{x_{\alpha n}(z_{\alpha n} + z_{\alpha n+1} + \cdots + z_{\beta n})} \cdot \frac{y_{\alpha n}(z_{\alpha n} + z_{\alpha n+1} + \cdots + z_{\beta n})}{z_{\alpha n}(y_{\alpha n} + y_{\alpha n+1} + \cdots + y_{\beta n})} \right]$$

$$= \frac{2\lambda}{\lambda + 1} \cdot \frac{\left(\frac{\beta}{\alpha}\right)^{\frac{\lambda+1}{\lambda}} - 1}{\left(\frac{\beta}{\alpha}\right)^2 - 1} \cdot \frac{\mu + 1}{2\mu} \cdot \frac{\left(\frac{\beta}{\alpha}\right)^2 - 1}{\left(\frac{\beta}{\alpha}\right)^{\frac{\mu+1}{\mu}} - 1} = \frac{\lambda(\mu + 1)}{\mu(\lambda + 1)} \cdot \frac{\left(\frac{\beta}{\alpha}\right)^{\frac{\lambda+1}{\lambda}} - 1}{\left(\frac{\beta}{\alpha}\right)^{\frac{\mu+1}{\mu}} - 1}.$$

7.23. We use Problem 7.21, and we obtain that the limit is $\frac{p}{p+1}\left(b\sqrt[p]{\frac{b}{a}} - a\right)$.

7.24. For $m \in \{1, 2, \ldots, p\}$, let $(x_n^{(m)})_{n\in\mathbb{N}}$ be the sequence defined by

$$x_n^{(m)} = \prod_{j=0}^{n-1} \frac{n + jp + m}{n + jp + m - 1}.$$

Clearly, $x_n^{(1)} = x_n$. As one can observe

$$x_{n+m}^{(1)} = x_n^{(m+1)} \prod_{k=1}^{m} \frac{n + (n+k-1)p + m + 1}{n + (n+k-1)p + m}, \quad m \in \{1, 2, \ldots, p-1\}.$$

Thus,

$$x_n^{(1)} x_{n+1}^{(1)} \cdots x_{n+p-1}^{(1)} = x_n^{(1)} x_n^{(2)} \cdots x_n^{(p)} \prod_{m=1}^{p-1} \prod_{k=1}^{m} \frac{n + (n+k-1)p + m + 1}{n + (n+k-1)p + m}.$$

However, the product $x_n^{(1)} x_n^{(2)} \cdots x_n^{(p)}$ telescopes as

$$x_n^{(1)} x_n^{(2)} \cdots x_n^{(p)} = \prod_{j=0}^{n-1} \prod_{m=1}^{p} \frac{n + jp + m}{n + jp + m - 1} = \prod_{j=0}^{n-1} \frac{n + (j+1)p}{n + jp} = p + 1.$$

Therefore,

$$x_n^{(1)}x_{n+1}^{(1)}\cdots x_{n+p-1}^{(1)} = (p+1)\prod_{m=1}^{p-1}\prod_{k=1}^{m}\frac{n+(n+k-1)p+m+1}{n+(n+k-1)p+m}.$$

It follows that $(\lim_{n\to\infty} x_n)^p = p+1$, and so $\lim_{n\to\infty} x_n = \sqrt[p]{p+1}$.

7.25. We have

$$\lim_{n\to\infty}\sum_{k=1}^{\infty}\frac{\sqrt[p]{n}}{\sum_{j=1}^{p}\sqrt[p]{k^j(n+k)^{p-j+1}}} = \lim_{n\to\infty}\frac{1}{\sqrt[p]{n^{p-1}}}\sum_{k=1}^{\infty}\left(\frac{1}{\sqrt[p]{k}}-\frac{1}{\sqrt[p]{n+k}}\right)$$

$$= \lim_{n\to\infty}\frac{1}{\sqrt[p]{n^{p-1}}}\sum_{k=1}^{n}\frac{1}{\sqrt[p]{k}}$$

$$= \lim_{n\to\infty}\frac{1}{n}\sum_{k=1}^{n}\frac{1}{\sqrt[p]{\frac{k}{n}}}$$

$$= \int_0^1\frac{1}{\sqrt[p]{x}}\,dx$$

$$= \frac{p}{p-1}.$$

7.26.

(a) The limit equals $1/4$. It is known that $\lim_{n\to\infty} S_n = \ln(4/\pi)$ (see [140]). We have, based on Stolz–Cesàro lemma, the $0/0$ case, that

$$\lim_{n\to\infty}\frac{S_{2n}-\ln\frac{4}{\pi}}{\frac{1}{(2n)^2}} = \lim_{n\to\infty}\frac{S_{2n+2}-S_{2n}}{\frac{1}{(2n+2)^2}-\frac{1}{(2n)^2}}$$

$$= \lim_{n\to\infty}\frac{\frac{1}{2n+1}-\frac{1}{2n+2}+\ln\frac{(2n+1)(2n+3)}{(2n+2)^2}}{\frac{1}{(2n+2)^2}-\frac{1}{(2n)^2}}$$

$$= -\frac{1}{4},$$

where the last equality follows by l'Hôpital's rule.

Also, an application of Stolz–Cesàro lemma, the $0/0$ case, gives

$$\lim_{n\to\infty}\frac{S_{2n+1}-\ln\frac{4}{\pi}}{\frac{1}{(2n+1)^2}} = \lim_{n\to\infty}\frac{S_{2n+3}-S_{2n+1}}{\frac{1}{(2n+3)^2}-\frac{1}{(2n+1)^2}}$$

$$= \lim_{n \to \infty} \frac{-\frac{1}{2n+2} + \frac{1}{2n+3} + \ln \frac{(2n+3)^2}{(2n+2)(2n+4)}}{\frac{1}{(2n+3)^2} - \frac{1}{(2n+1)^2}}$$

$$= \frac{1}{4}.$$

Consequently,

$$\lim_{n \to \infty} n^2 \left| S_n - \ln \frac{4}{\pi} \right| = \lim_{n \to \infty} \frac{\left| S_n - \ln \frac{4}{\pi} \right|}{\frac{1}{n^2}} = \frac{1}{4}.$$

(b) In a similar way as in part (a) we get that the limit is $\frac{5}{12}$.

15.2   Limits of Integrals

7.27.

(a) The limit equals 1. Observe that

$$\int_0^2 \frac{x^n}{1+x^n}dx = \int_0^1 \frac{x^n}{1+x^n}dx + 1 - \int_1^2 \frac{1}{1+x^n}dx.$$

We have

$$0 < \int_0^1 \frac{x^n}{1+x^n}dx \le \int_0^1 x^n dx = \frac{1}{n+1}$$

and it follows that $\lim_{n \to \infty} \int_0^1 \frac{x^n}{1+x^n}dx = 0$.

On the other hand, $1 + x^n \ge 2x^{\frac{n}{2}}$, and we have that

$$\int_1^2 \frac{1}{1+x^n}dx \le \int_1^2 \frac{1}{2x^{\frac{n}{2}}}dx = \frac{1}{n-2}\left(1 - 2^{1-\frac{n}{2}}\right), \quad n > 2,$$

which shows that $\lim_{n \to \infty} \int_1^2 \frac{1}{1+x^n}dx = 0$.

(b) First we observe, since $\lim_{x \to \infty} f(x)$ exists, that f is bounded (prove it!). We have

$$\int_0^2 f(x^n)dx = \int_0^1 f(x^n)dx + \int_1^2 f(x^n)dx.$$

An application of the Bounded Convergence Theorem shows that, the details are left to the reader, $\lim_{n\to\infty} \int_0^1 f(x^n)dx = f(0)$ and $\lim_{n\to\infty} \int_1^2 f(x^n)dx = f(\infty)$.

   A *challenge.* Prove that the preceding two limits hold by using an $\epsilon - \delta$ argument.

7.28. *Solution I.* Let $\epsilon > 0$. Since f is continuous at 1, we have that there exists $\delta > 0$ such that $|f(x) - f(1)| < \epsilon$, for $x \in (\delta, 1]$. We have

$$x \int_a^1 y^x f(y)dy = x \int_a^\delta y^x(f(y) - f(1))dy + x \int_\delta^1 y^x(f(y) - f(1))dy$$
$$+ \frac{x}{x+1}f(1)(1 - a^{x+1}).$$

On the one hand,

$$\left| x \int_\delta^1 y^x(f(y) - f(1))dy \right| \le \epsilon \frac{x}{x+1}\left(1 - \delta^{x+1}\right) \le \epsilon$$

which implies that

$$\lim_{x\to\infty} \left| x \int_\delta^1 y^x(f(y) - f(1))dy \right| \le \epsilon.$$

Since $\epsilon > 0$ was arbitrarily taken, we get that $\lim_{x\to\infty} x \int_\delta^1 y^x(f(y) - f(1))dy = 0$.

   On the other hand,

$$\left| x \int_a^\delta y^x(f(y) - f(1))dy \right| \le \frac{2x}{x+1}\|f\|\left(\delta^{x+1} - a^{x+1}\right) \to 0, \quad x \to \infty.$$

It follows, based on the previous calculations, that $\lim_{x\to\infty} x \int_a^1 y^x f(y)dy = f(1)$ and the problem is solved.

   Solution II. Approximate f by a polynomial function.

7.29. The limit equals 2. We have

$$x \int_0^1 (t^2-t+1)^x dt = x \int_0^{\frac{1}{4}} (t^2-t+1)^x dt + x \int_{\frac{1}{4}}^{\frac{3}{4}} (t^2-t+1)^x dt + x \int_{\frac{3}{4}}^1 (t^2-t+1)^x dt.$$

Using the substitution $t^2 - t + 1 = y$, we get that

$$x \int_0^{\frac{1}{4}} (t^2 - t + 1)^x dt = x \int_{\frac{13}{16}}^1 \frac{y^x}{\sqrt{4y - 3}}dy$$

and it follows, based on Problem 7.28, that

$$\lim_{x\to\infty} x \int_0^{\frac{1}{4}} (t^2 - t + 1)^x \, dt = \lim_{x\to\infty} x \int_{\frac{13}{16}}^1 \frac{y^x}{\sqrt{4y - 3}} \, dy = 1.$$

On the other hand,

$$x \int_{\frac{1}{4}}^{\frac{3}{4}} (t^2 - t + 1)^x \, dt = x \int_{\frac{1}{4}}^{\frac{1}{2}} (t^2 - t + 1)^x \, dt + x \int_{\frac{1}{2}}^{\frac{3}{4}} (t^2 - t + 1)^x \, dt \le 2x \left(\frac{13}{16}\right)^x,$$

which implies that $\lim_{x\to\infty} x \int_{\frac{1}{4}}^{\frac{3}{4}} (t^2 - t + 1)^x \, dt = 0$.
The substitution $t^2 - t + 1 = y$ shows that

$$x \int_{\frac{3}{4}}^1 (t^2 - t + 1)^x \, dt = x \int_{\frac{13}{16}}^1 \frac{y^x}{\sqrt{4y - 3}} \, dy$$

and it follows, based on Problem 7.28, that

$$\lim_{x\to\infty} x \int_{\frac{3}{4}}^1 (t^2 - t + 1)^x \, dt = \lim_{x\to\infty} x \int_{\frac{13}{16}}^1 \frac{y^x}{\sqrt{4y - 3}} \, dy = 1.$$

7.30. Solution due to I. Gavrea. The limit equals 1. We have

$$x \int_0^1 t^{t^x} \, dt = x \int_0^{\frac{1}{2e}} t^{t^x} \, dt + x \int_{\frac{1}{2e}}^{\frac{3}{2e}} t^{t^x} \, dt + x \int_{\frac{3}{2e}}^1 t^{t^x} \, dt.$$

Let $\alpha = \left(\frac{1}{2e}\right)^{\frac{1}{2e}}$ and let $f(t) = t^t$, $t \in \left(0, \frac{1}{2e}\right]$, with $f(0) = 1$. Using the substitution $t^t = y \Rightarrow dt = \frac{1}{y(\ln f^{-1}(y) + 1)} \, dy$, we get that

$$x \int_0^{\frac{1}{2e}} t^{t^x} \, dt = -x \int_\alpha^1 \frac{y^x}{y(\ln f^{-1}(y) + 1)} \, dy$$

and it follows, based on Problem 7.28, that

$$\lim_{x\to\infty} x \int_0^{\frac{1}{2e}} t^{t^x} \, dt = -\lim_{x\to\infty} x \int_\alpha^1 \frac{y^x}{y(\ln f^{-1}(y) + 1)} \, dy = -\frac{1}{1 \cdot (\ln f^{-1}(1) + 1)} = 0.$$

Let $\beta = \left(\frac{3}{2e}\right)^{\frac{3}{2e}}$. We have

$$x \int_{\frac{1}{2e}}^{\frac{3}{2e}} t^{tx} \, dt = x \int_{\frac{1}{2e}}^{\frac{1}{e}} t^{tx} \, dt + x \int_{\frac{1}{e}}^{\frac{3}{2e}} t^{tx} \, dt \le x \left(\alpha^x + \beta^x \right)$$

and this implies that $\displaystyle\lim_{x\to\infty} x \int_{\frac{1}{2e}}^{\frac{3}{2e}} t^{tx} \, dt = 0$.

Let $g(t) = t^t$, $t \in \left[\frac{3}{2e}, 1 \right]$. Using the substitution $t^t = y \Rightarrow dt = \frac{1}{y(\ln g^{-1}(y)+1)} \, dy$, we get that

$$x \int_{\frac{3}{2e}}^{1} t^{tx} \, dt = x \int_{\beta}^{1} \frac{y^x}{y(\ln g^{-1}(y) + 1)} \, dy$$

and it follows, based on Problem 7.28, that

$$\lim_{x\to\infty} x \int_{\frac{3}{2e}}^{1} t^{tx} \, dt = \lim_{x\to\infty} x \int_{\beta}^{1} \frac{y^x}{y(\ln g^{-1}(y) + 1)} \, dy = \frac{1}{1 \cdot (\ln g^{-1}(1) + 1)} = 1.$$

7.31. The limit equals $\frac{2}{\ln a}$. We have

$$x \int_{0}^{1} a^{t(t-1)x} \, dt = 2x \int_{0}^{\frac{1}{2}} \left(\frac{1}{a}\right)^{t(1-t)x} \, dt$$

$$\overset{t(1-t)=y}{=} 2x \int_{0}^{\frac{1}{4}} \left(\frac{1}{a}\right)^{yx} \frac{dy}{\sqrt{1-4y}}$$

$$\overset{1-4y=u}{=} \frac{x}{2} \left(\frac{1}{a}\right)^{\frac{x}{4}} \int_{0}^{1} \frac{a^{\frac{xu}{4}}}{\sqrt{u}} \, du$$

$$\overset{u=v^2}{=} x \left(\frac{1}{a}\right)^{\frac{x}{4}} \int_{0}^{1} a^{\frac{xv^2}{4}} \, dv$$

$$\overset{\frac{xv^2}{4}=t}{=} \frac{\sqrt{x}}{a^{\frac{x}{4}}} \int_{0}^{\frac{x}{4}} \frac{a^t}{\sqrt{t}} \, dt.$$

It follows that

$$\lim_{x\to\infty} x \int_{0}^{1} a^{t(t-1)x} \, dt = \lim_{x\to\infty} \frac{\sqrt{x}}{a^{\frac{x}{4}}} \int_{0}^{\frac{x}{4}} \frac{a^t}{\sqrt{t}} \, dt$$

$$\overset{\frac{x}{4}=b}{=} 2 \lim_{b\to\infty} \frac{\int_{0}^{b} \frac{a^t}{\sqrt{t}} \, dt}{\frac{a^b}{\sqrt{b}}}$$

$$\overset{\text{l'Hôpital's rule } \frac{\infty}{\infty}}{=} 2 \lim_{b \to \infty} \frac{\frac{a^b}{\sqrt{b}}}{a^b \ln a \sqrt{b} - a^b \cdot \frac{1}{2\sqrt{b}}}$$

$$= 2 \lim_{b \to \infty} \frac{\sqrt{b}}{\ln a \sqrt{b} - \frac{1}{2\sqrt{b}}}$$

$$= \frac{2}{\ln a}.$$

When we applied l'Hôpital's rule, we have used that $\lim_{b \to \infty} \int_0^b \frac{a^t}{\sqrt{t}} dt = \int_0^\infty \frac{a^t}{\sqrt{t}} dt = \infty$.

7.32. The limit equals 1. We have

$$\int_0^\infty \{x\}^n \, e^{-x} dx = \sum_{k=0}^\infty \int_k^{k+1} \{x\}^n \, e^{-x} dx$$

$$= \sum_{k=0}^\infty \int_k^{k+1} (x-k)^n e^{-x} dx$$

$$\overset{x-k=y}{=} \sum_{k=0}^\infty \int_0^1 y^n e^{-(k+y)} dy$$

$$= \sum_{k=0}^\infty e^{-k} \int_0^1 y^n e^{-y} dy$$

$$= \frac{e}{e-1} \int_0^1 y^n e^{-y} dy.$$

It follows that

$$\lim_{n \to \infty} \sqrt[n]{\int_0^\infty \{x\}^n \, e^{-x} dx} = \lim_{n \to \infty} \frac{\sqrt[n]{e}}{\sqrt[n]{e-1}} \sqrt[n]{\int_0^1 y^n e^{-y} dy} = 1,$$

since $\lim_{n \to \infty} \sqrt[n]{\int_0^1 y^n e^{-y} dy} = 1$. The preceding limit follows based on the inequalities

$$\frac{1}{e(n+1)} \leq \int_0^1 y^n e^{-y} dy \leq \frac{1}{n+1}.$$

7.33. The limit equals $\frac{1}{e}$. On the one hand,

$$\int_0^\infty \lfloor x \rfloor^n e^{-x} dx < \int_0^\infty x^n e^{-x} dx = n!.$$

On the other hand,

$$\frac{n^n}{e^n}\left(1 - \frac{1}{e}\right) = n^n \int_n^{n+1} e^{-x} dx = \int_n^{n+1} \lfloor x \rfloor^n e^{-x} dx < \int_0^\infty \lfloor x \rfloor^n e^{-x} dx.$$

These imply that

$$\frac{1}{e}\sqrt[n]{1 - \frac{1}{e}} < \frac{1}{n}\sqrt[n]{\int_0^\infty \lfloor x \rfloor^n e^{-x} dx} < \frac{\sqrt[n]{n!}}{n},$$

and the result follows since $\lim\limits_{n\to\infty} \frac{\sqrt[n]{n!}}{n} = \frac{1}{e}$.

7.34. The limit equals $\ln(a + b)$. Integrate by parts and observe that

$$n^2 \int_0^1 x^n \left(\sqrt[n]{ax+b} - 1\right) dx = \frac{n^2}{n+1}\left(\sqrt[n]{a+b} - 1\right) - \frac{an}{n+1}\int_0^1 \frac{x^{n+1}}{\sqrt[n]{(ax+b)^{n-1}}} dx.$$

7.35. Using the substitution $x^n = y$, we have that

$$n^2 \int_0^1 (\sqrt[n]{1 + x^n} - 1) dx = n \int_0^1 \left(e^{\frac{1}{n}\ln(1+y)} - 1\right) y^{\frac{1}{n}-1} dy$$

$$= \int_0^1 \frac{\ln(1 + y)}{y} \sqrt[n]{y}\, e^{\theta_n(y)} dy,$$

where $\theta_n(y) \in \left(0, \frac{1}{n}\ln(1 + y)\right)$.

Let $f_n : [0, 1] \to \mathbb{R}$, $f_n(y) = \frac{\ln(1+y)}{y}\sqrt[n]{y}\,e^{\theta_n(y)}$. We have that $\lim\limits_{n\to\infty} f_n(y) = \frac{\ln(1+y)}{y}$ and $|f_n(y)| \le 2\frac{\ln(1+y)}{y}$, which is integrable over $[0, 1]$. It follows, based on Lebesgue Dominated Convergence Theorem, that

$$\lim_{n\to\infty} \int_0^1 \frac{\ln(1 + y)}{y} \sqrt[n]{y}\, e^{\theta_n(y)} dy = \int_0^1 \frac{\ln(1 + y)}{y} dy = \frac{\pi^2}{12}.$$

Similarly, one can prove that

$$\lim_{n\to\infty} n^2\left(1 - \int_0^1 \sqrt[n]{1 - x^n}\, dx\right) = \zeta(2).$$

For a more general problem, the reader is referred to [26, problem 1.60, p. 11].

7.36. (a) $\left(\sqrt{a} + \sqrt{b}\right)\sqrt{\frac{\pi}{2}}$; (b) $\left(\sqrt{a}f\left(\frac{\pi}{2}\right) + \sqrt{b}f(0)\right)\sqrt{\frac{\pi}{2}}$.

7.37. The limit equals $\int_a^b f(x)dx$. Since f is Riemann integrable, we have that f is bounded, i.e. there exists $M > 0$ such that $|f(x)| \leq M, \forall x \in [a, b]$. We have

$$\left| \int_a^b f(x)dx - \int_a^b \frac{f(x)}{1 + \cos x \cos(x + 1) \cdots \cos(x + n)}dx \right|$$

$$\leq M \int_a^b \frac{|\cos x \cos(x + 1) \cdots \cos(x + n)|}{|1 + \cos x \cos(x + 1) \cdots \cos(x + n)|}dx$$

$$\leq M \int_a^b \frac{|\cos x \cos(x + 1) \cdots \cos(x + n)|}{1 - |\cos x \cos(x + 1) \cdots \cos(x + n)|}dx,$$

since $|1 + \cos x \cos(x + 1) \cdots \cos(x + n)| \geq 1 - |\cos x \cos(x + 1) \cdots \cos(x + n)|$. An application of the $AM - GM$ inequality shows that

$$\cos^2 x \cos^2(x + 1) \cdots \cos^2(x + n) \leq \left(\frac{\cos^2 x + \cos^2(x + 1) + \cdots + \cos^2(x + n)}{n + 1} \right)^{n+1}.$$

A calculation shows

$$\cos^2 x + \cos^2(x + 1) + \cdots + \cos^2(x + n) = \frac{n + 1}{2} + \frac{\sin(n + 1)\cos(n + 2x)}{2\sin 1},$$

and it follows that

$$|\cos x \cos(x + 1) \cdots \cos(x + n)| \leq \left(\frac{1}{2} + \frac{1}{2(n + 1)\sin 1} \right)^{\frac{n+1}{2}}.$$

Since the function $h(x) = \frac{x}{1-x}$, $x \in [0, 1)$ is strictly increasing, we have that

$$\left| \int_a^b f(x)dx - \int_a^b \frac{f(x)}{1 + \cos x \cos(x + 1) \cdots \cos(x + n)}dx \right|$$

$$\leq M \cdot (b - a) \cdot \frac{\left(\frac{1}{2} + \frac{1}{2(n+1)\sin 1} \right)^{\frac{n+1}{2}}}{1 - \left(\frac{1}{2} + \frac{1}{2(n+1)\sin 1} \right)^{\frac{n+1}{2}}}.$$

Passing to the limit, as $n \to \infty$, in the previous inequality, the result follows.

7.38.

(a) The limit equals $\Gamma(k)$, where Γ denotes the Gamma function. See [108].

(b) The limit equals $\Gamma(k + 1)$, where Γ denotes the Gamma function.

We have

$$n^{k+1} \int_0^\infty \frac{x^k \sin x}{e^{(n+1)x} - e^{nx}} dx = n^{k+1} \int_0^\infty \frac{x^k \sin x \, e^{-(n+1)x}}{1 - e^{-x}} dx$$

$$\overset{e^{-x}=t}{=} -n^{k+1} \int_0^1 \frac{(-\ln t)^k \sin(\ln t) t^n}{1 - t} dt$$

$$\overset{t^n=y}{=} \int_0^1 (-\ln y)^{k+1} \cdot \frac{\sin\left(\ln \sqrt[n]{y}\right)}{\ln \sqrt[n]{y}} \cdot \sqrt[n]{y} \cdot \frac{\frac{1}{n}}{1 - \sqrt[n]{y}} dy.$$

Let $f_n(y) = (-\ln y)^{k+1} \cdot \frac{\sin(\ln \sqrt[n]{y})}{\ln \sqrt[n]{y}} \cdot \sqrt[n]{y} \cdot \frac{\frac{1}{n}}{1-\sqrt[n]{y}}$, $y \in (0, 1)$.

We have that $\lim\limits_{n\to\infty} f_n(y) = (-\ln y)^k$, $y \in (0, 1)$, and

$$|f_n(y)| = \left| (-\ln y)^{k+1} \cdot \frac{\sin\left(\ln \sqrt[n]{y}\right)}{\ln \sqrt[n]{y}} \cdot \sqrt[n]{y} \cdot \frac{\frac{1}{n}}{1 - \sqrt[n]{y}} \right| \leq \frac{(-\ln y)^{k+1}}{1 - y},$$

since $\left| \frac{\sin x}{x} \right| \leq 1$, $\forall x \in \mathbb{R}$, and $\frac{\frac{1}{n}}{1-\sqrt[n]{y}} \leq \frac{1}{1-y}$, for all $n \in \mathbb{N}$. The last inequality follows from the fact that the function $g : [0, 1] \to \mathbb{R}$, $g(x) = \frac{x}{1-y^x}$ is an increasing function when $y \in (0, 1)$. Since

$$\int_0^1 \frac{(-\ln y)^{k+1}}{1 - y} dy \overset{y=e^{-t}}{=} \int_0^\infty \frac{t^{k+1} e^{-t}}{1 - e^{-t}} dt = \int_0^\infty t^{k+1} e^{-t} \sum_{i=0}^\infty e^{-it} dt$$

$$= \sum_{i=0}^\infty \int_0^\infty t^{k+1} e^{-(i+1)t} dt \overset{(i+1)t=x}{=} \sum_{i=0}^\infty \frac{1}{(i+1)^{k+2}} \int_0^\infty x^{k+1} e^{-x} dx$$

$$= \Gamma(k+2)\zeta(k+2),$$

we have that the positive function $y \to \frac{(-\ln y)^{k+1}}{1-y}$ is integrable over $[0, 1]$.

It follows, based on Lebesgue Dominated Convergence Theorem, that

$$\lim_{n\to\infty} n^{k+1} \int_0^\infty \frac{x^k \sin x}{e^{(n+1)x} - e^{nx}} dx$$

$$= \lim_{n\to\infty} \int_0^1 (-\ln y)^{k+1} \cdot \frac{\sin\left(\ln \sqrt[n]{y}\right)}{\ln \sqrt[n]{y}} \cdot \sqrt[n]{y} \cdot \frac{\frac{1}{n}}{1 - \sqrt[n]{y}} dy$$

$$= (-1)^k \int_0^1 \ln^k y \, dy$$

$$\overset{y=e^{-t}}{=} \int_0^\infty t^k e^{-t} dt$$

$$= \Gamma(k+1).$$

7.39. Use the substitution $x = \sinh \frac{y}{2n}$ and apply Lebesgue Dominated Convergence Theorem.

7.40.

(a) The limit equals $\frac{2}{\pi}$. Let $k = \lfloor \frac{n}{\pi} \rfloor$. This implies that $k\pi \le n < (k+1)\pi$. We have

$$\int_0^n |\sin x| dx = \int_0^{k\pi} |\sin x| dx + \int_{k\pi}^n |\sin x| dx$$

$$= \sum_{i=0}^{k-1} \int_{i\pi}^{(i+1)\pi} |\sin x| dx + \int_{k\pi}^n |\sin x| dx$$

$$\overset{x-i\pi=y}{=} k \int_0^\pi \sin y \, dy + \int_{k\pi}^n |\sin x| dx$$

$$= 2k + \int_{k\pi}^n |\sin x| dx.$$

It follows that

$$\frac{1}{n} \int_0^n |\sin x| dx = \frac{2k}{n} + \frac{1}{n} \int_{k\pi}^n |\sin x| dx.$$

We have

$$\frac{2k}{n} = \frac{2}{n} \left(\frac{n}{\pi} - \left\{ \frac{n}{\pi} \right\} \right) = \frac{2}{\pi} - \frac{2}{n} \left\{ \frac{n}{\pi} \right\} \to \frac{2}{\pi}$$

and

$$0 < \frac{1}{n} \int_{k\pi}^n |\sin x| dx \le \frac{n - k\pi}{n} \to 0,$$

since $0 \le \frac{n-k\pi}{n} < \frac{\pi}{n}$.

(b) The limit equals $\frac{2}{3\pi}$. Use a technique similar to that in the solution of part (a).

7.41.

(a) Let $k = \lfloor \frac{n}{\pi} \rfloor$. This implies that $k\pi \le n < (k+1)\pi$. Using the substitution $nx = t$, we have that

$$\int_0^1 f\left(|\sin nx|\right) dx = \frac{1}{n} \int_0^n f\left(|\sin t|\right) dt$$

$$= \frac{1}{n} \int_0^{k\pi} f\left(|\sin t|\right) dt + \frac{1}{n} \int_{k\pi}^n f\left(|\sin t|\right) dt$$

$$= \frac{1}{n} \sum_{i=0}^{k-1} \int_{i\pi}^{(i+1)\pi} f\left(|\sin t|\right) dt + \frac{1}{n} \int_{k\pi}^n f\left(|\sin t|\right) dt$$

$$\stackrel{t-i\pi=y}{=} \frac{k}{n} \int_0^\pi f\left(\sin y\right) dy + \frac{1}{n} \int_{k\pi}^n f\left(|\sin t|\right) dt.$$

We have

$$\frac{k}{n} = \frac{1}{n}\left(\frac{n}{\pi} - \left\{\frac{n}{\pi}\right\}\right) = \frac{1}{\pi} - \frac{1}{n}\left\{\frac{n}{\pi}\right\} \to \frac{1}{\pi}.$$

Since f is Riemann integrable, we have that f is bounded and it follows that

$$0 < \left| \frac{1}{n} \int_{k\pi}^n f\left(|\sin x|\right) dx \right| \le \frac{n - k\pi}{n} \|f\| \to 0,$$

since $0 \le \frac{n-k\pi}{n} < \frac{\pi}{n}$.

(b) This part can be solved similarly as in part (a).

The identity

$$\sum_{i=1}^n \sum_{j=1}^n \min\{i, j\} = \sum_{i=1}^n i^2 = \frac{n(n+1)(2n+1)}{6}, \quad n \in \mathbb{N}$$

can be proved by mathematical induction.

7.43. $\ln \sqrt{k}$. Make the substitution $x^n = t$ and apply Stolz–Cesàro Theorem.

7.44. $\frac{\sqrt[k]{2}-1}{2}$.

15.3   Convergence and Evaluation of Series

7.45. The series converges for $\alpha > 1$ and diverges for $\alpha \in (0, 1]$. We have, based on Lagrange Mean Value Theorem, that

$$\left(\sqrt[n]{n} - 1\right)^{\alpha} \ln^{\beta} n = \left(e^{\frac{\ln n}{n}} - 1\right)^{\alpha} \ln^{\beta} n = \frac{\ln^{\alpha+\beta} n}{n^{\alpha}} e^{\alpha \theta_n}, \quad \theta_n \in \left(0, \frac{\ln n}{n}\right).$$

Therefore, the series behaves like $\sum_{n=2}^{\infty} \frac{\ln^{\alpha+\beta} n}{n^{\alpha}}$.

7.46. The series converges. Observe the sequence $\left(\sin \frac{1}{n}\right)_{n \in \mathbb{N}}$ converges decreasingly to 0 and

$$\sin 1 + \sin 2 + \cdots + \sin n = \frac{\sin \frac{n}{2} \sin \frac{n+1}{2}}{\sin \frac{1}{2}}, \quad n \in \mathbb{N}.$$

Now the result follows based on Dirichlet's criterion for the convergence of series.

Dirichlet's Criterion If $(a_n)_{n \geq 1}$ is a decreasing sequence of real numbers, convergent to 0, and $(b_n)_{n \geq 1}$ is a sequence of real numbers with the property that there exists $M > 0$ such that $|\sum_{k=1}^{n} b_k| \leq M, \forall n \geq 1$, then the series $\sum_{n=1}^{\infty} a_n b_n$ converges.

7.47. (b) First we prove that the following identity holds:

A Fractional Part Identity If $n \in \mathbb{N}$, then

$$\left\{(2 + \sqrt{3})^n\right\} = 1 - (2 - \sqrt{3})^n,$$

where $\{x\}$ denotes the fractional part of x.

We have

$$\begin{cases} (2 + \sqrt{3})^n = a_n + \sqrt{3} b_n \\ (2 - \sqrt{3})^n = a_n - \sqrt{3} b_n \end{cases} \quad a_n, b_n \in \mathbb{N}.$$

Now we observe that $(2 + \sqrt{3})^n = 2a_n - (a_n - \sqrt{3} b_n) = 2a_n - (2 - \sqrt{3})^n \Rightarrow$
$\left\{(2 + \sqrt{3})^n\right\} = \left\{-(2 - \sqrt{3})^n\right\} = 1 - (2 - \sqrt{3})^n.$

We are ready to solve part (b) of the problem. We have

$$\left| \sin \pi \left(2 + \sqrt{3}\right)^n \right| = \left| \sin \pi \left(\left\lfloor \left(2 + \sqrt{3}\right)^n \right\rfloor + \left\{ \left(2 + \sqrt{3}\right)^n \right\} \right) \right|$$

$$= \left| \sin \pi \left\{ \left(2 + \sqrt{3}\right)^n \right\} \right|$$

$$= \left| \sin \pi \left(1 - \left(2 - \sqrt{3}\right)^n \right) \right|$$

$$= \left| \sin \pi \left(2 - \sqrt{3}\right)^n \right|$$

$$\leq \pi \left(2 - \sqrt{3}\right)^n.$$

Since $\sum_{n=1}^{\infty} \left(2 - \sqrt{3}\right)^n$ converges, it follows that $\sum_{n=1}^{\infty} \sin \left(\pi (2 + \sqrt{3})^n \right)$ converges absolutely.

7.48. The series converges. We have, based on Lagrange Mean Value Theorem, that

$$1 - n^{\frac{\sin n}{n}} = 1 - e^{\frac{\sin n}{n} \ln n} = -\frac{\sin n}{n} \ln n \, e^{\theta_n},$$

where θ_n is between 0 and $\frac{\sin n}{n} \ln n$.

This implies, based on Lagrange Mean Value Theorem, that

$$1 - n^{\frac{\sin n}{n}} = -\frac{\sin n}{n} \ln n \, e^{\theta_n}$$

$$= -\frac{\sin n}{n} \ln n \left(e^{\theta_n} - 1 \right) - \frac{\sin n}{n} \ln n$$

$$= -\frac{\sin n}{n} \ln n \cdot \theta_n \cdot e^{\theta_n'} - \frac{\sin n}{n} \ln n,$$

where θ_n' is between 0 and θ_n.

It follows that

$$\sum_{n=2}^{\infty} \left(1 - n^{\frac{\sin n}{n}}\right) = -\sum_{n=2}^{\infty} \frac{\sin n}{n} \ln n \cdot \theta_n \cdot e^{\theta_n'} - \sum_{n=2}^{\infty} \frac{\sin n}{n} \ln n.$$

Observe that $|\theta_n'| \leq |\theta_n| \leq \frac{|\sin n|}{n} \ln n \leq \frac{\ln n}{n} < 2$ and we have

$$\sum_{n=2}^{\infty} \left| \frac{\sin n}{n} \ln n \cdot \theta_n \cdot e^{\theta_n'} \right| \le e^2 \sum_{n=2}^{\infty} \frac{\ln^2 n}{n^2},$$

which implies that the series $\sum_{n=2}^{\infty} \frac{\sin n}{n} \ln n \cdot \theta_n \cdot e^{\theta_n'}$ converges absolutely.

On the other hand, the series $\sum_{n=2}^{\infty} \frac{\sin n}{n} \ln n$ converges based on Dirichlet's criterion.

7.49. *Solution I.* The series converges absolutely. Observe that

$$\left| \int_0^1 \frac{x^n}{1+x+x^2+\cdots+x^n} f(x) dx \right| \le \|f\| \int_0^1 \frac{x^n}{1+x+x^2+\cdots+x^n} dx$$

$$\le \|f\| \int_0^1 \frac{x^n}{(n+1)x^{\frac{n}{2}}} dx$$

$$= \frac{2\|f\|}{(n+1)(n+2)},$$

since the $AM - GM$ inequality implies that $1 + x + x^2 + \cdots + x^n \ge (n+1)x^{\frac{n}{2}}$.

Solution II. Prove that (see [26, problem 1.39 (b)])

$$\lim_{n\to\infty} n^2 \int_0^1 \frac{x^n}{1+x+x^2+\cdots+x^n} f(x) dx = f(1)\zeta(2),$$

and compare the series with $\sum_{n=1}^{\infty} \frac{1}{n^2}$.

7.50.

(a) *Solution I.* The series diverges. Using the substitution $x = nt$, we have that

$$\sum_{n=1}^{\infty} \left(\frac{\pi}{2} - \int_0^n \frac{\sin^2 x}{x^2} dx \right) = \sum_{n=1}^{\infty} \int_n^{\infty} \frac{\sin^2 x}{x^2} dx$$

$$= \sum_{n=1}^{\infty} \frac{1}{n} \int_1^{\infty} \frac{\sin^2(nt)}{t^2} dt$$

$$= \int_1^{\infty} \frac{1}{t^2} \left(\sum_{n=1}^{\infty} \frac{\sin^2(nt)}{n} \right) dt$$

$$= \infty.$$

We used that

$$\sum_{n=1}^{\infty} \frac{\sin^2(nt)}{n} = \sum_{n=1}^{\infty} \frac{1 - \cos(2nt)}{2n} = \infty,$$

since the harmonic series diverges and the series $\sum_{n=1}^{\infty} \frac{\cos(2nt)}{n}$ converges, based on Dirichlet's criterion.

Solution II. The series diverges. Let $f : [1, \infty) \to (0, \infty)$, $f(x) = \int_x^{\infty} \frac{\sin^2 y}{y^2} dy$. Observe that f decreases and we have, based on Cauchy's integral criterion, that the series $\sum_{n=1}^{\infty} f(n)$ behaves the same like the integral $\int_1^{\infty} \left(\int_x^{\infty} \frac{\sin^2 y}{y^2} dy \right) dx$. The domain of integration of the preceding integral is the infinite triangle with vertex $(1, 1)$ and sides $x = 1$ and $y = x$. Changing the order of integration, Tonelli's Theorem applies, we get that

$$\int_1^{\infty} \left(\int_x^{\infty} \frac{\sin^2 y}{y^2} dy \right) dx = \int_1^{\infty} \left(\int_1^y \frac{\sin^2 y}{y^2} dx \right) dy$$

$$= \int_1^{\infty} \frac{\sin^2 y}{y^2} (y - 1) dy$$

$$= \int_1^{\infty} \frac{\sin^2 y}{y} dy - \int_1^{\infty} \frac{\sin^2 y}{y^2} dy$$

$$= \infty,$$

since $\int_1^{\infty} \frac{\sin^2 y}{y} dy = \infty$ and $\int_1^{\infty} \frac{\sin^2 y}{y^2} dy < \infty$.

(b) The series converges. See the solution of part (a).

7.51.

(a) We need the following inequality due to Bernoulli:

Bernoulli's inequality. If $x \in [0, 1]$ and $\alpha \in [0, 1]$, then $(1-x)^{\alpha} \leq 1-\alpha x$.

The inequality to prove reads

$$\frac{1}{n+1} < p - p \sqrt[p]{\frac{n}{n+1}}.$$

We have, based on Bernoulli's inequality, that

$$\sqrt[p]{\frac{n}{n+1}} = \left(1 - \frac{1}{n+1} \right)^{\frac{1}{p}} < 1 - \frac{1}{p} \cdot \frac{1}{n+1},$$

and the preceding inequality follows.

7.52. We have

$$\int_0^\infty \{x\}\, e^{-x}\, dx = \int_0^\infty x e^{-x}\, dx - \int_0^\infty \lfloor x \rfloor e^{-x}\, dx$$

$$= 1 - \sum_{k=0}^\infty \int_k^{k+1} \lfloor x \rfloor e^{-x}\, dx$$

$$= 1 - \sum_{k=0}^\infty k \int_k^{k+1} e^{-x}\, dx$$

$$= 1 - \left(1 - \frac{1}{e}\right) \sum_{k=1}^\infty k e^{-k}$$

$$= 1 - \left(1 - \frac{1}{e}\right) \cdot \frac{\frac{1}{e}}{\left(1 - \frac{1}{e}\right)^2}$$

$$= \frac{e-2}{e-1}.$$

Similarly, one can prove that $\displaystyle\int_0^\infty x e^{-\lfloor x \rfloor}\, dx = \frac{e\,(e+1)}{2\,(e-1)^2}$.

7.53. $\displaystyle\sum_{n=1}^\infty \frac{1}{(n-1)! + n!} = 1$ and $\displaystyle\sum_{n=0}^\infty \frac{(-1)^n}{(n+(-1)^n)!} = -e^{-1}$.

7.54. $\displaystyle\sum_{n=1}^\infty \frac{n}{(n+1)!} = 1$ and $\displaystyle\sum_{n=1}^\infty \frac{n}{(n+(-1)^n)!} = 2\cosh 1$.

7.55. Use that

$$\frac{1}{(n+k-1)(n+k)(n+k)!} = \frac{1}{(n+k-1)(n+k-1)!} - \frac{1}{(n+k)(n+k)!} - \frac{1}{(n+k)!}.$$

7.56. Calculate the $2n$th partial sum of the series.

7.57. For the first series, calculate the $2n$th partial sum of the series or use part (b) of Problem 3.23. For the second series, use part (d) of Problem 3.23.

7.58. Calculate the $4n$th partial sum of the series and use Wallis formula.

7.59.

(a) The integral can be calculated using integration by parts. An alternative solution is based on the formula

$$\int_0^1 (1-x)^n x^{k-1}\, dx = B\,(n+1, k),$$

where B denotes the beta function of Euler.

(b) We have

$$\sum_{k=1}^{\infty} \frac{1}{k(k+1)(k+2)\cdots(k+n)} = \sum_{k=1}^{\infty} \frac{1}{n!} \int_0^1 (1-x)^n x^{k-1} dx$$

$$= \frac{1}{n!} \int_0^1 (1-x)^n \left(\sum_{k=1}^{\infty} x^{k-1}\right) dx$$

$$= \frac{1}{n!} \int_0^1 (1-x)^{n-1} dx$$

$$= \frac{1}{n \cdot n!}.$$

15.4   Harmonic Series

7.60. We have

$$\int_0^1 \frac{\ln x \ln(1-x)}{x(1-x)} dx = 2 \int_0^1 \frac{\ln x \ln(1-x)}{x} dx$$

$$= -2 \int_0^1 \frac{\ln x}{x} \sum_{n=1}^{\infty} \frac{x^n}{n} dx$$

$$= -2 \sum_{n=1}^{\infty} \frac{1}{n} \int_0^1 x^{n-1} \ln x \, dx$$

$$= 2 \sum_{n=1}^{\infty} \frac{1}{n^3}$$

$$= 2\zeta(3).$$

On the other hand, using the generating function of the nth harmonic number (see part (a) of Problem 3.63), we have that

$$\int_0^1 \frac{\ln x \ln(1-x)}{x(1-x)} dx = \int_0^1 \frac{\ln x}{x} \left(-\sum_{n=1}^{\infty} H_n x^n\right) dx$$

$$= -\sum_{n=1}^{\infty} H_n \int_0^1 x^{n-1} \ln x \, dx$$

$$= \sum_{n=1}^{\infty} \frac{H_n}{n^2}.$$

7.61.

(a) Observe that $\frac{1}{n(n+k)} = \frac{1}{k}\left(\frac{1}{n} - \frac{1}{n+k}\right)$.

(b) We have, based on part (a), that

$$\sum_{k=1}^{\infty} \frac{H_k}{k^2} = \sum_{k=1}^{\infty}\sum_{n=1}^{\infty} \frac{1}{nk(n+k)}$$

$$= \sum_{k=1}^{\infty}\sum_{n=1}^{\infty} \frac{1}{nk} \int_0^1 x^{n+k-1} dx$$

$$= \int_0^1 \sum_{k=1}^{\infty} \frac{x^k}{k} \sum_{n=1}^{\infty} \frac{x^{n-1}}{n} dx$$

$$= \int_0^1 \frac{\ln^2(1-x)}{x} dx$$

$$= \int_0^1 \frac{\ln^2 x}{1-x} dx$$

$$= 2\zeta(3).$$

(e) Observe that

$$\frac{H_n^2}{n(n+1)} = \frac{H_n^2}{n} - \frac{H_{n+1}^2}{n+1} + 2\frac{H_{n+1}}{(n+1)^2} - \frac{1}{(n+1)^3}.$$

7.62.

(a) Prove the formula by mathematical induction.
    Use part (a) of the problem and Abel's summation formula with:
(b) $a_n = H_n^2$ and $b_n = \zeta(2) - 1 - \frac{1}{2^2} - \cdots - \frac{1}{n^2} - \frac{1}{n}$;
(c) $a_n = H_n^2$ and $b_n = \zeta(3) - 1 - \frac{1}{2^3} - \cdots - \frac{1}{n^3}$.

7.63.

(a) Use Abel's summation formula with
    $a_n = \frac{1}{n}$ and $b_n = \zeta(2) - \frac{1}{1^2} - \frac{1}{2^2} - \cdots - \frac{1}{n^2}$.
(b) We have, based on Abel's summation formula with $a_n = \frac{1}{n}$ and $b_n = \frac{1}{n^2} + \frac{1}{(n+1)^2} + \frac{1}{(n+2)^2} + \cdots$, that

$$\sum_{n=1}^{\infty} \frac{1}{n} \left(\frac{1}{n^2} + \frac{1}{(n+1)^2} + \frac{1}{(n+2)^2} + \cdots \right) = \lim_{n \to \infty} H_n \left(\frac{1}{(n+1)^2} + \frac{1}{(n+2)^2} + \cdots \right)$$

$$+ \sum_{n=1}^{\infty} \frac{H_n}{n^2}$$

$$\stackrel{7.61\,(b)}{=} \sum_{n=1}^{\infty} \frac{H_n}{n^2}$$

$$= 2\zeta(3).$$

An alternative solution is based on part (a) of the problem.

7.64. We apply Abel's summation formula with

$$a_n = H_n \quad \text{and} \quad b_n = \frac{\pi^4}{72} - \frac{H_1}{1^3} - \frac{H_2}{2^3} - \cdots - \frac{H_n}{n^3}$$

and we have, since $\sum_{k=1}^{n} H_k = (n+1)H_{n+1} - (n+1)$, that

$$\sum_{n=1}^{\infty} H_n \left(\frac{\pi^4}{72} - \frac{H_1}{1^3} - \frac{H_2}{2^3} - \cdots - \frac{H_n}{n^3} \right)$$

$$= \lim_{n \to \infty} [(n+1)H_{n+1} - (n+1)] \left(\frac{\pi^4}{72} - \frac{H_1}{1^3} - \frac{H_2}{2^3} - \cdots - \frac{H_{n+1}}{(n+1)^3} \right)$$

$$+ \sum_{n=1}^{\infty} [(n+1)H_{n+1} - (n+1)] \frac{H_{n+1}}{(n+1)^3}$$

$$= \sum_{n=1}^{\infty} \left[\frac{H_{n+1}^2}{(n+1)^2} - \frac{H_{n+1}}{(n+1)^2} \right]$$

$$= \frac{17}{4}\zeta(4) - 2\zeta(3).$$

The last equality follows based on the Sandham–Yeung series, discussed in Chap. 8, and part (b) of Problem 7.61.

7.67. Add and subtract the series in part (a) of Problem 3.76 and part (b) of Problem 7.66.

7.68.

(a) We use the formula (prove it!)

$$\ln \frac{1}{2} + 1 - \frac{1}{2} + \cdots + \frac{(-1)^{n-1}}{n} = (-1)^{n-1} \int_0^1 \frac{x^n}{1+x} dx, \quad n \geq 1.$$

We have

$$\sum_{n=1}^{\infty} H_n \left(\ln \frac{1}{2} + 1 - \frac{1}{2} + \cdots + \frac{(-1)^{n-1}}{n} \right) = \sum_{n=1}^{\infty} (-1)^{n-1} H_n \int_0^1 \frac{x^n}{1+x} dx$$

$$= - \int_0^1 \frac{1}{1+x} \sum_{n=1}^{\infty} (-x)^n H_n dx$$

$$\overset{3.63\,(a)}{=} \int_0^1 \frac{\ln(1+x)}{(1+x)^2} dx$$

$$= \frac{1 - \ln 2}{2}.$$

7.70.

(a) Solve this part of the problem by direct computations.
(b) Use part (a) of the problem and part C of Problem 3.4.

7.71. Calculate the nth partial sum of the series, use part (a) of Problem 7.94 and the definition of the Glaisher–Kinkelin constant.

15.5   Series with Factorials

7.72. (b) We have, based on part (a), that

$$\sum_{n=1}^{\infty} \sum_{m=1}^{\infty} \frac{(n-1)!(m-1)}{(n+m)!} = \sum_{n=1}^{\infty} \frac{(n-1)!}{n} \sum_{m=1}^{\infty} \left(\frac{(m-1)!}{(n+m-1)!} - \frac{m!}{(n+m)!} \right)$$

$$= \sum_{n=1}^{\infty} \frac{(n-1)!}{n} \cdot \frac{1}{n!}$$

$$= \sum_{n=1}^{\infty} \frac{1}{n^2}$$

$$= \zeta(2).$$

7.73. For the solution of this problem see [116].
7.74.

(a) We have, based on part (a) of Problem 7.72, that

$$\sum_{i=1}^{\infty}\sum_{j=1}^{\infty}(-1)^{i-1}\frac{(i-1)!(j-1)!}{(i+j)!}$$

$$=\sum_{i=1}^{\infty}(-1)^{i-1}\frac{(i-1)!}{i}\sum_{j=1}^{\infty}\left(\frac{(j-1)!}{(i+j-1)!}-\frac{j!}{(i+j)!}\right)$$

$$=\sum_{i=1}^{\infty}(-1)^{i-1}\frac{(i-1)!}{i}\cdot\frac{1}{i!}$$

$$=\sum_{i=1}^{\infty}\frac{(-1)^{i-1}}{i^2}$$

$$=\frac{\zeta(2)}{2}.$$

(b) Use symmetry and part (a) of the problem.

   Parts (c) and (d) can be solved using the formula in part (a) of Problem 7.72.

15.6   Series of Functions

7.75. Let $\alpha \in (0, 1)$ and let $p, q > 1$, with $p > \frac{5}{\alpha}$, such that $\frac{1}{p} + \frac{1}{q} = 1$. Let $\epsilon > 0$ and let $\delta > 0$ be given by

$$\delta = \left|\frac{\epsilon^p}{2^{\frac{p}{q}}\zeta^{\frac{p}{q}}\left((2-\alpha)q\right)\zeta(\alpha p - 4)}\right|.$$

Let $x, y \in \mathbb{R}$ such that $|x - y| \le \delta$. We have

$$|f(x) - f(y)| = \left|\sum_{n=1}^{\infty}\frac{\cos(n^4 x) - \cos(n^4 y)}{n^2}\right|$$

$$\le 2\sum_{n=1}^{\infty}\frac{1}{n^2}\left|\sin\frac{n^4(x-y)}{2}\right|$$

$$\le 2\left(\sum_{n=1}^{\infty}\frac{1}{n^{\alpha p}}\left|\sin\frac{n^4(x-y)}{2}\right|^p\right)^{\frac{1}{p}}\left(\sum_{n=1}^{\infty}\frac{1}{n^{(2-\alpha)q}}\right)^{\frac{1}{q}},$$

where the last inequality follows based on Hölder's inequality for series.

   On the other hand,

$$\left|\sin\frac{n^4(x-y)}{2}\right|^p \le \left|\sin\frac{n^4(x-y)}{2}\right| \le \left|\sin\left|\frac{n^4(x-y)}{2}\right|\right| \le \frac{n^4}{2}|x-y| \le \frac{n^4}{2}\delta,$$

since $|\sin x| \le |x|$, $\forall x \in \mathbb{R}$.

It follows that

$$|f(x)-f(y)| \le 2\left(\sum_{n=1}^{\infty}\frac{\delta}{2n^{\alpha p-4}}\right)^{\frac{1}{p}}\left(\sum_{n=1}^{\infty}\frac{1}{n^{(2-\alpha)q}}\right)^{\frac{1}{q}}$$

$$\le \delta^{\frac{1}{p}}2^{\frac{1}{q}}\zeta^{\frac{1}{p}}(\alpha p-4)\zeta^{\frac{1}{q}}((2-\alpha)q)$$

$$\le \epsilon,$$

and this implies that f is uniformly continuous on $\mathbb{R}$.

7.76. For $\epsilon > 0$ there exists $n_0 \in \mathbb{N}$ such that $|a_n - l| < \epsilon$, $\forall n \ge n_0$.

We have

$$e^{-x}\sum_{n=0}^{\infty}a_n\frac{x^n}{n!} - l = e^{-x}\sum_{n=0}^{n_0-1}(a_n-l)\frac{x^n}{n!} + e^{-x}\sum_{n=n_0}^{\infty}(a_n-l)\frac{x^n}{n!}$$

and it follows that

$$\left|e^{-x}\sum_{n=0}^{\infty}a_n\frac{x^n}{n!} - l\right| \le e^{-x}\sum_{n=0}^{n_0-1}|a_n-l|\frac{|x|^n}{n!} + e^{-x}\sum_{n=n_0}^{\infty}|a_n-l|\frac{|x|^n}{n!}$$

$$\le e^{-x}\sum_{n=0}^{n_0-1}|a_n-l|\frac{|x|^n}{n!} + \epsilon e^{-x}\cdot e^{|x|}.$$

Passing to the limit, as $x \to \infty$, in the previous inequality, we have that

$$\lim_{x\to\infty}\left|e^{-x}\sum_{n=0}^{\infty}a_n\frac{x^n}{n!} - l\right| \le \lim_{x\to\infty}e^{-x}\sum_{n=0}^{n_0-1}|a_n-l|\frac{|x|^n}{n!} + \epsilon = \epsilon,$$

and, since $\epsilon > 0$ was arbitrary, it follows that $\lim_{x\to\infty}e^{-x}\sum_{n=0}^{\infty}a_n\frac{x^n}{n!} = l$.

7.77.

(a) We assume that $\sup a_n = 1$. Otherwise, we replace the sequence a_n by $\frac{a_n}{\sup a_n}$.

We assume that the series $\sum_{n=1}^{\infty}a_n^x$ converges. This implies that $a_n \to 0$ and it follows that $\sup a_n$ is finite.

Let $\epsilon > 0$. Since the tail of a convergent series converges to 0, we have that there exists $n_0 \in \mathbb{N}$ such that $\sum_{k=n}^{\infty}a_k^x < \epsilon$, $\forall n \ge n_0$.

We have

$$\left(a_1^x + a_2^x + \cdots + a_n^x\right)^{\frac{1}{x}} \le \left(\sum_{n=1}^{\infty} a_n^x\right)^{\frac{1}{x}} \le \left(a_1^x + a_2^x + \cdots + a_n^x + \epsilon\right)^{\frac{1}{x}}, \quad n \ge n_0.$$

Passing, to the limit as $x \to \infty$, in the previous inequalities, we get that

$$\max_{k=1,\ldots,n} a_k \le \lim_{x\to\infty} \left(\sum_{n=1}^{\infty} a_n^x\right)^{\frac{1}{x}} \le 1, \quad n \ge n_0, \tag{15.1}$$

since

$$\lim_{x\to\infty} \left(a_1^x + a_2^x + \cdots + a_n^x\right)^{\frac{1}{x}} = \max_{k=1,\ldots,n} a_k$$

and $\lim_{x\to\infty} \left(a_1^x + a_2^x + \cdots + a_n^x + \epsilon\right)^{\frac{1}{x}} = 1$, when $a_i \in (0,1)$, $i = 1, \ldots, n$ and $\epsilon > 0$.

Passing to the limit, as $n \to \infty$, in (15.1), we get that

$$1 = \sup a_n = \lim_{n\to\infty} \max_{k=1,\ldots,n} a_k \le \lim_{x\to\infty} \left(\sum_{n=1}^{\infty} a_n^x\right)^{\frac{1}{x}} \le 1,$$

which implies that $\lim_{x\to\infty} \left(\sum_{n=1}^{\infty} a_n^x\right)^{\frac{1}{x}} = 1$.

(b) $\lim_{x\to\infty} \zeta(x)^{\frac{1}{x}} = 1$.

Remark. If the series $\sum_{n=1}^{\infty} a_n^x$ does not converge, the problem is no longer valid. If $a_n = \frac{1}{\ln(n+1)}$, then $\sum_{n=1}^{\infty} a_n^x = \sum_{n=1}^{\infty} \frac{1}{\ln^x(n+1)} = \infty$, $\forall x > 0$ and $\sup_{n\ge 1} \frac{1}{\ln(n+1)} = \frac{1}{\ln 2}$.

7.78. We have, if $x \ne 0$, that

$$\int_0^1 \frac{1}{t^{xt}} dt = \int_0^1 e^{-tx \ln t} dt$$

$$= \int_0^1 \left(\sum_{n=0}^{\infty} \frac{(-tx \ln t)^n}{n!}\right) dt$$

$$= \sum_{n=0}^{\infty} (-1)^n \frac{x^n}{n!} \int_0^1 t^n \ln^n t \, dt$$

$$= \sum_{n=0}^{\infty} \frac{x^n}{(n+1)^{n+1}}$$

$$= \frac{1}{x} \sum_{n=1}^{\infty} \left(\frac{x}{n}\right)^n.$$

We used in our calculations the formula

$$\int_0^1 x^m \ln^n x \, dx = (-1)^n \frac{n!}{(m+1)^{n+1}}, \quad m, n \geq 0.$$

It follows that, for $x > 0$, one has

$$\left(\sum_{n=1}^{\infty} \left(\frac{x}{n}\right)^n\right)^{\frac{1}{x}} = x^{\frac{1}{x}} \left(\int_0^1 \frac{1}{t^{xt}} dt\right)^{\frac{1}{x}} = x^{\frac{1}{x}} \left(\int_0^1 \left(\frac{1}{t^t}\right)^x dt\right)^{\frac{1}{x}}.$$

This implies that

$$\lim_{x \to \infty} \left(\sum_{n=1}^{\infty} \left(\frac{x}{n}\right)^n\right)^{\frac{1}{x}} = \lim_{x \to \infty} \left(\int_0^1 \left(\frac{1}{t^t}\right)^x dt\right)^{\frac{1}{x}} = \max_{t \in [0,1]} f(t),$$

where f is the continuous function defined by

$$f(t) = \begin{cases} 1 & \text{if} \quad t = 0 \\ \frac{1}{t^t} & \text{if} \quad t \in (0, 1]. \end{cases}$$

A calculation shows that f attains its maximum when $t = \frac{1}{e}$ and $\max_{t \in [0,1]} f(t) = e^{e^{-1}}$.

7.79. Part (a) follows by direct computations, see the solution of Problem 7.78. For part (b) of the problem see the solution of Problem 7.30.

7.80. We have

$$\sum_{n=2}^{\infty} \frac{x^n}{(-1)^n n + 1} = \frac{1}{2x} \ln \frac{1+x}{1-x} + \frac{x}{2} \ln(1 - x^2) - 1, \quad x \in (-1, 1).$$

7.81. We have, for $x \in (-1, 1)$, that

$$\sum_{n=1}^{\infty} H_n \left(\frac{1}{1-x} - 1 - x - x^2 - \cdots - x^n \right) = \sum_{n=1}^{\infty} H_n \sum_{m=1}^{\infty} x^{n+m}$$

$$= \sum_{m=1}^{\infty} x^m \sum_{n=1}^{\infty} H_n x^n$$

$$\overset{\text{3.63 (a)}}{=} -\frac{x}{(1-x)^2} \ln(1-x).$$

7.82. *Solution I.* We apply Abel's summation formula with $a_n = 1$ and $b_n = \sum_{m=1}^{\infty} \left(e^x - 1 - \frac{x}{1!} - \cdots - \frac{x^{n+m-2}}{(n+m-2)!} \right)$ and we have, since $b_n - b_{n+1} = \sum_{m=1}^{\infty} \frac{x^{n+m-1}}{(n+m-1)!}$, that

$$\sum_{n=1}^{\infty} \sum_{m=1}^{\infty} \left(e^x - 1 - \frac{x}{1!} - \frac{x^2}{2!} - \cdots - \frac{x^{n+m-2}}{(n+m-2)!} \right)$$

$$= \sum_{n=1}^{\infty} n \sum_{m=1}^{\infty} \frac{x^{n+m-1}}{(n+m-1)!}$$

$$= \sum_{n=1}^{\infty} n \left(\frac{x^n}{n!} + \frac{x^{n+1}}{(n+1)!} + \cdots \right)$$

$$= \sum_{n=1}^{\infty} \frac{n(n+1)}{2} \cdot \frac{x^n}{n!}$$

$$= \frac{x^2 + 2x}{2} e^x,$$

where the third equality follows based on Abel's summation formula with $a_n = n$ and $b_n = \frac{x^n}{n!} + \frac{x^{n+1}}{(n+1)!} + \cdots$.

Solution II. Use part (a) of Problem 4.69.

7.83. We integrate by parts and we have

$$-\text{Li}_2(z) = \int_0^z \frac{\ln(1-t)}{t} dt = \ln t \ln(1-t) \Big|_0^z + \int_0^z \frac{\ln t}{1-t} dt$$

$$= \ln z \ln(1-z) + \int_0^1 \frac{\ln t}{1-t} dt - \int_z^1 \frac{\ln t}{1-t} dt$$

$$= \ln z \ln(1-z) - \frac{\pi^2}{6} - \int_0^{1-z} \frac{\ln(1-u)}{u} du$$

$$= \ln z \ln(1-z) - \frac{\pi^2}{6} + \text{Li}_2(1-z).$$

(b) Let $z = \sin^2 \theta$ and we have, based on part (a), that

$$\sum_{n=1}^{\infty} \frac{\sin^{2n} \theta + \cos^{2n} \theta}{n^2} = \frac{\pi^2}{6} - 4 \ln |\sin \theta| \ln |\cos \theta|.$$

7.85. We have, based on the formula (see part (a) of Problem 4.69)

$$\int_0^x \frac{(x - t)^n}{n!} e^t \, dt = e^x - 1 - \frac{x}{1!} - \frac{x^2}{2!} - \cdots - \frac{x^n}{n!}, \quad \forall n \in \mathbb{N}, \ x \in \mathbb{R},$$

that

$$\sum_{n=0}^{\infty} n! \left(e - 1 - \frac{1}{1!} - \frac{1}{2!} - \cdots - \frac{1}{n!} \right)^2 = \sum_{n=0}^{\infty} \int_0^1 (1 - x)^n e^x dx \int_0^1 \frac{(1 - y)^n}{n!} e^y dy$$

$$= \int_0^1 \int_0^1 e^{x+y} \sum_{n=0}^{\infty} \frac{[(1 - x)(1 - y)]^n}{n!} dx dy$$

$$= \int_0^1 \int_0^1 e^{x+y} \cdot e^{(1-x)(1-y)} dx dy$$

$$= e \int_0^1 \int_0^1 e^{xy} dx dy$$

$$= e \int_0^1 \frac{e^x - 1}{x} dx$$

$$= e \int_0^1 \sum_{n=1}^{\infty} \frac{x^{n-1}}{n!} dx$$

$$= e \sum_{n=1}^{\infty} \frac{1}{n \cdot n!}.$$

7.86. Formula (7.1) can be proved by differentiation and then by using part (a) of Problem 3.63.

7.87. Divide by $x \neq 0$ and integrate from 0 to x the power series in Problem 3.4 C. The power series can also be proved by differentiation and by comparing to the power series in Problem 3.4 C.

7.88. Use Abel's summation formula with

$a_n = 1$ and $b_n = \ln^2(1 - x) - \left(x + \frac{x^2}{2} + \cdots + \frac{x^n}{n} \right)^2.$

15.7   Pearls of Series with Tails of Zeta Function Values

7.90.
(a) Use Abel's summation formula with
$$a_n = H_n \text{ and } b_n = \zeta(2) - 1 - \frac{1}{2^2} - \cdots - \frac{1}{n^2} - \frac{1}{n}.$$
(b) Let $H_n^{(2)} = 1 + \frac{1}{2^2} + \cdots + \frac{1}{n^2}$. Use that

$$H_n^{(2)}\left(\zeta(2) - H_n^{(2)} - \frac{1}{n}\right) = -\left(\zeta(2) - H_n^{(2)}\right)^2 + \zeta(2)\left(\zeta(2) - H_n^{(2)} - \frac{1}{n}\right)$$
$$+ \frac{1}{n}\left(\zeta(2) - H_n^{(2)}\right).$$

7.92. Use Abel's summation formula with:

(a) $a_n = n$ and $b_n = \zeta(3) - 1 - \frac{1}{2^3} - \cdots - \frac{1}{n^3} - \frac{1}{2n^2}$;
(b) $a_n = n^2$ and $b_n = \zeta(4) - 1 - \frac{1}{2^4} - \cdots - \frac{1}{n^4} - \frac{1}{3n^3}$;
(c) $a_n = n^3$ and $b_n = \zeta(5) - 1 - \frac{1}{2^5} - \cdots - \frac{1}{n^5} - \frac{1}{4n^4}$.

7.94.

(a) Prove the formula by mathematical induction.
(b) Use Abel's summation formula with
$$a_n = nH_n \text{ and } b_n = \zeta(2) - 1 - \frac{1}{2^2} - \cdots - \frac{1}{n} + \frac{1}{2n^2}.$$
(c) Use Abel's summation formula with
$$a_n = nH_n \text{ and } b_n = \zeta(3) - 1 - \frac{1}{2^3} - \cdots - \frac{1}{n^3} - \frac{1}{2n^2}.$$
(d) Use Abel's summation formula with
$$a_n = nH_n \text{ and } b_n = \zeta(4) - 1 - \frac{1}{2^4} - \cdots - \frac{1}{n^4}.$$

7.95.

(a) Prove the formula by mathematical induction.
    Use Abel's summation formula with:
(b) $a_n = (-1)^n n^2$ and $b_n = \zeta(3) - 1 - \frac{1}{2^3} - \cdots - \frac{1}{n^3} - \frac{1}{2n^2}$;
(c) $a_n = (-1)^n n^2$ and $b_n = \zeta(4) - 1 - \frac{1}{2^4} - \cdots - \frac{1}{n^4}$.

7.97. Use Abel's summation formula with
$$a_n = 1 \text{ and } b_n = \zeta^2(3) - \left(1 + \frac{1}{2^3} + \cdots + \frac{1}{n^3}\right)^2.$$
7.98. See [61].
7.99.

(a) A solution is given in [27]. An alternative solution is based on the
    application of Abel's summation formula with $a_n = \frac{1}{n}$ and $b_n =$

$\left(\zeta(2) - 1 - \frac{1}{2^2} - \cdots - \frac{1}{n^2}\right)^2$, combined with the formulae $\sum_{n=1}^{\infty} \frac{H_n^{(2)}}{n^3} = 3\zeta(2)\zeta(3) - \frac{9}{2}\zeta(5)$ [22, Theorem 3.1, p. 22] and $\sum_{n=1}^{\infty} \frac{H_n H_n^{(2)}}{n^2} = \zeta(2)\zeta(3) + \zeta(5)$ [109, Theorem 6, p. 210], where $H_n^{(2)} = 1 + \frac{1}{2^2} + \cdots + \frac{1}{n^2}$.

(b) Use part (a).

7.100. Use Abel's summation formula with:

(a) $a_n = n$ and $b_n = \left(\zeta(3) - 1 - \frac{1}{2^3} - \cdots - \frac{1}{n^3}\right)^2$;

(b) $a_n = n^2$ and $b_n = \left(\zeta(3) - 1 - \frac{1}{2^3} - \cdots - \frac{1}{n^3}\right)^2$;

(c) $a_n = n^3$ and $b_n = \left(\zeta(3) - 1 - \frac{1}{2^3} - \cdots - \frac{1}{n^3}\right)^2 - \frac{1}{4n^4}$.

7.102.

(a) We need the following formulae which can be proved by using Abel's summation formula:

$$\bullet \quad \sum_{n=1}^{\infty} \frac{\zeta(2) - 1 - \frac{1}{2^2} - \cdots - \frac{1}{n^2}}{n} = \zeta(3);$$

$$\bullet \quad \sum_{n=1}^{\infty} \frac{\zeta(3) - 1 - \frac{1}{2^3} - \cdots - \frac{1}{n^3}}{n} = \frac{1}{4}\zeta(4);$$

$$\bullet \quad \sum_{n=1}^{\infty} \left(\zeta(3) - 1 - \frac{1}{2^3} - \cdots - \frac{1}{n^3}\right) = \zeta(2) - \zeta(3).$$

Now we are ready to solve part (a) of the problem. We apply Abel's summation formula with $a_n = n$ and $b_n = \left(\zeta(2) - 1 - \frac{1}{2^2} - \cdots - \frac{1}{n^2}\right)\left(\zeta(3) - 1 - \frac{1}{2^3} - \cdots - \frac{1}{n^3}\right)$ and we have that

$$\sum_{n=1}^{\infty} n \left(\zeta(2) - 1 - \frac{1}{2^2} - \cdots - \frac{1}{n^2}\right)\left(\zeta(3) - 1 - \frac{1}{2^3} - \cdots - \frac{1}{n^3}\right)$$

$$= \sum_{n=1}^{\infty} \frac{n(n+1)}{2} \left(\frac{\zeta(2) - 1 - \frac{1}{2^2} - \cdots - \frac{1}{n^2}}{(n+1)^3} + \frac{\zeta(3) - 1 - \frac{1}{2^3} - \cdots - \frac{1}{n^3}}{(n+1)^2} - \frac{1}{(n+1)^5}\right)$$

$$= \frac{1}{2} \sum_{n=1}^{\infty} n \left(\frac{\zeta(2) - 1 - \frac{1}{2^2} - \cdots - \frac{1}{n^2}}{(n+1)^2} + \frac{\zeta(3) - 1 - \frac{1}{2^3} - \cdots - \frac{1}{n^3}}{n+1} - \frac{1}{(n+1)^4}\right)$$

$$= \frac{1}{2} \sum_{n=1}^{\infty} n \left(\frac{\zeta(2) - 1 - \frac{1}{2^2} - \cdots - \frac{1}{(n+1)^2}}{(n+1)^2} + \frac{\zeta(3) - 1 - \frac{1}{2^3} - \cdots - \frac{1}{(n+1)^3}}{n+1} + \frac{1}{(n+1)^4} \right)$$

$$= \frac{1}{2} \sum_{m=2}^{\infty} (m-1) \left(\frac{\zeta(2) - 1 - \frac{1}{2^2} - \cdots - \frac{1}{m^2}}{m^2} + \frac{\zeta(3) - 1 - \frac{1}{2^3} - \cdots - \frac{1}{m^3}}{m} + \frac{1}{m^4} \right)$$

$$= \frac{1}{2} \sum_{m=1}^{\infty} (m-1) \left(\frac{\zeta(2) - 1 - \frac{1}{2^2} - \cdots - \frac{1}{m^2}}{m^2} + \frac{\zeta(3) - 1 - \frac{1}{2^3} - \cdots - \frac{1}{m^3}}{m} + \frac{1}{m^4} \right)$$

$$= \frac{1}{2} \sum_{m=1}^{\infty} \frac{\zeta(2) - 1 - \frac{1}{2^2} - \cdots - \frac{1}{m^2}}{m} - \frac{1}{2} \sum_{m=1}^{\infty} \frac{\zeta(2) - 1 - \frac{1}{2^2} - \cdots - \frac{1}{m^2}}{m^2}$$

$$+ \frac{1}{2} \sum_{m=1}^{\infty} \left(\zeta(3) - 1 - \frac{1}{2^3} - \cdots - \frac{1}{m^3} \right) - \frac{1}{2} \sum_{m=1}^{\infty} \frac{\zeta(3) - 1 - \frac{1}{2^3} - \cdots - \frac{1}{m^3}}{m}$$

$$+ \frac{1}{2} \sum_{m=1}^{\infty} \frac{1}{m^3} - \frac{1}{2} \sum_{m=1}^{\infty} \frac{1}{m^4}$$

$$= \frac{1}{2} \zeta(3) - \frac{3}{8} \zeta(4) + \frac{1}{2} (\zeta(2) - \zeta(3)) - \frac{1}{8} \zeta(4) + \frac{1}{2} \zeta(3) - \frac{1}{2} \zeta(4)$$

$$= \frac{1}{2} (\zeta(2) + \zeta(3) - 2\zeta(4)) .$$

(b) We need the following formulae:

- $$\sum_{n=1}^{\infty} \left(\zeta(4) - 1 - \frac{1}{2^4} - \cdots - \frac{1}{n^4} \right) = \zeta(3) - \zeta(4);$$

- $$\sum_{n=1}^{\infty} \frac{\zeta(4) - 1 - \frac{1}{2^4} - \cdots - \frac{1}{n^4}}{n} = 2\zeta(5) - \zeta(2)\zeta(3);$$

- $$\sum_{n=1}^{\infty} \frac{\zeta(2) - 1 - \frac{1}{2^2} - \cdots - \frac{1}{n^2}}{n^3} = \frac{9}{2} \zeta(5) - 2\zeta(2)\zeta(3).$$

The first two equalities can be proved by using Abel's summation formula, and the last series equality is a consequence of the series formula

$$\sum_{n=1}^{\infty} \frac{1 + \frac{1}{2^2} + \cdots + \frac{1}{n^2}}{n^3} = -\frac{9}{2} \zeta(5) + 3\zeta(2)\zeta(3),$$

which follows based on Theorem 3.1 in [22].

We apply Abel's summation formula with $a_n = n$ and

$$b_n = \left(\zeta(2) - 1 - \frac{1}{2^2} - \cdots - \frac{1}{n^2} \right) \left(\zeta(4) - 1 - \frac{1}{2^4} - \cdots - \frac{1}{n^4} \right),$$

and we have that

$$\sum_{n=1}^{\infty} n \left(\zeta(2) - 1 - \frac{1}{2^2} - \cdots - \frac{1}{n^2} \right) \left(\zeta(4) - 1 - \frac{1}{2^4} - \cdots - \frac{1}{n^4} \right)$$

$$= \sum_{n=1}^{\infty} \frac{n(n+1)}{2} \left(\frac{\zeta(2) - 1 - \frac{1}{2^2} - \cdots - \frac{1}{n^2}}{(n+1)^4} + \frac{\zeta(4) - 1 - \frac{1}{2^4} - \cdots - \frac{1}{n^4}}{(n+1)^2} - \frac{1}{(n+1)^6} \right)$$

$$= \frac{1}{2} \sum_{n=1}^{\infty} n \left(\frac{\zeta(2) - 1 - \frac{1}{2^2} - \cdots - \frac{1}{n^2}}{(n+1)^3} + \frac{\zeta(4) - 1 - \frac{1}{2^4} - \cdots - \frac{1}{n^4}}{n+1} - \frac{1}{(n+1)^5} \right)$$

$$= \frac{1}{2} \sum_{m=1}^{\infty} (m-1) \left(\frac{\zeta(2) - 1 - \frac{1}{2^2} - \cdots - \frac{1}{m^2}}{m^3} + \frac{\zeta(4) - 1 - \frac{1}{2^4} - \cdots - \frac{1}{m^4}}{m} + \frac{1}{m^5} \right)$$

$$= \frac{1}{2} \sum_{m=1}^{\infty} \frac{\zeta(2) - 1 - \frac{1}{2^2} - \cdots - \frac{1}{m^2}}{m^2} - \frac{1}{2} \sum_{m=1}^{\infty} \frac{\zeta(2) - 1 - \frac{1}{2^2} - \cdots - \frac{1}{m^2}}{m^3}$$

$$+ \frac{1}{2} \sum_{m=1}^{\infty} \left(\zeta(4) - 1 - \frac{1}{2^4} - \cdots - \frac{1}{m^4} \right) - \frac{1}{2} \sum_{m=1}^{\infty} \frac{\zeta(4) - 1 - \frac{1}{2^4} - \cdots - \frac{1}{m^4}}{m}$$

$$+ \frac{1}{2} \sum_{m=1}^{\infty} \frac{1}{m^4} - \frac{1}{2} \sum_{m=1}^{\infty} \frac{1}{m^5}$$

$$= \frac{3}{8} \zeta(4) - \frac{1}{2} \left(\frac{9}{2} \zeta(5) - 2\zeta(2)\zeta(3) \right) + \frac{1}{2} \left(\zeta(3) - \zeta(4) \right) - \frac{1}{2} (2\zeta(5) - \zeta(2)\zeta(3))$$

$$+ \frac{1}{2} \zeta(4) - \frac{1}{2} \zeta(5)$$

$$= \frac{3}{2} \zeta(2)\zeta(3) - \frac{15}{4} \zeta(5) + \frac{3}{8} \zeta(4) + \frac{1}{2} \zeta(3).$$

7.105. (b) We have

$$\sum_{n=1}^{\infty} \left(\frac{1}{n^k} + \frac{1}{(n+1)^k} + \cdots \right)^2$$

$$= \zeta^2(k) + \sum_{n=2}^{\infty} \left(\frac{1}{n^k} + \frac{1}{(n+1)^k} + \cdots \right)^2$$

$$\stackrel{n-1=m}{=} \zeta^2(k) + \sum_{m=1}^{\infty} \left(\frac{1}{(m+1)^k} + \frac{1}{(m+2)^k} + \cdots \right)^2$$

$$= \zeta^2(k) + \sum_{m=1}^{\infty} \left[\left(\frac{1}{m^k} + \frac{1}{(m+1)^k} + \cdots \right) - \frac{1}{m^k} \right]^2$$

$$= \zeta^2(k) + \sum_{m=1}^{\infty} \left(\frac{1}{m^k} + \frac{1}{(m+1)^k} + \cdots \right)^2 - 2 \sum_{m=1}^{\infty} \frac{1}{m^k} \left(\frac{1}{m^k} + \frac{1}{(m+1)^k} + \cdots \right)$$

$$+ \zeta(2k)$$

and it follows that

$$\sum_{n=1}^{\infty} \frac{1}{n^k} \left(\frac{1}{n^k} + \frac{1}{(n+1)^k} + \cdots \right) = \frac{\zeta^2(k) + \zeta(2k)}{2}.$$

To prove the equality in Remark 7.7, we note that

$$\sum_{n=1}^{\infty} \frac{1 + \frac{1}{2^k} + \cdots + \frac{1}{n^k}}{n^k} + \sum_{n=1}^{\infty} \frac{1}{n^k} \left(\frac{1}{n^k} + \frac{1}{(n+1)^k} + \cdots \right)$$

$$= \sum_{n=1}^{\infty} \frac{1}{n^{2k}} + \sum_{n=1}^{\infty} \frac{1}{n^k} \left(1 + \frac{1}{2^k} + \frac{1}{3^k} + \cdots \right)$$

$$= \zeta(2k) + \zeta^2(k),$$

and the equality follows based on part (b) of the problem.

7.106. We have

$$\sum_{n=1}^{\infty} \left(\frac{1}{n^2} + \frac{1}{(n+1)^2} + \cdots \right) \left(\frac{1}{(n+1)^2} + \frac{1}{(n+2)^2} + \cdots \right)$$

$$= \sum_{n=1}^{\infty} \left[\left(\frac{1}{n^2} + \frac{1}{(n+1)^2} + \cdots \right)^2 - \frac{1}{n^2} \left(\frac{1}{n^2} + \frac{1}{(n+1)^2} + \cdots \right) \right]$$

$$= \sum_{n=1}^{\infty} \left(\frac{1}{n^2} + \frac{1}{(n+1)^2} + \cdots \right)^2 - \sum_{n=1}^{\infty} \frac{1}{n^2} \left(\frac{1}{n^2} + \frac{1}{(n+1)^2} + \cdots \right)$$

$$= 3\zeta(3) - \frac{7}{4}\zeta(4),$$

where the last equality follows based on formula (7.2) and part (a) of Problem 7.105.

7.107. (b) *Shifting the index of summation.*
  We have

$$\sum_{n=1}^{\infty} \left(\frac{1}{n^k} + \frac{1}{(n+1)^k} + \cdots \right)^3 = \zeta^3(k) + \sum_{n=2}^{\infty} \left(\frac{1}{n^k} + \frac{1}{(n+1)^k} + \cdots \right)^3$$

$$\overset{n-1=m}{=} \zeta^3(k) + \sum_{m=1}^{\infty} \left(\frac{1}{(m+1)^k} + \frac{1}{(m+2)^k} + \cdots \right)^3$$

$$= \zeta^3(k) + \sum_{m=1}^{\infty} \left[\left(\frac{1}{m^k} + \frac{1}{(m+1)^k} + \cdots \right) - \frac{1}{m^k} \right]^3$$

$$\overset{(a-b)^3=a^3-3ab(a-b)-b^3}{=} \zeta^3(k) + \sum_{m=1}^{\infty} \left(\frac{1}{m^k} + \frac{1}{(m+1)^k} + \cdots \right)^3$$

$$- 3 \sum_{m=1}^{\infty} \frac{1}{m^k} \left(\frac{1}{m^k} + \frac{1}{(m+1)^k} + \cdots \right) \left(\frac{1}{(m+1)^k} + \frac{1}{(m+2)^k} + \cdots \right) - \zeta(3k),$$

and the result follows.

7.108.

(a) We need the series formula

$$\sum_{n=1}^{\infty} \frac{1}{n} \left(\frac{1}{n^2} + \frac{1}{(n+1)^2} + \cdots \right) = 2\zeta(3), \tag{15.2}$$

which can be proved using Abel's summation formula with $a_n = \frac{1}{n}$ and $b_n = \frac{1}{n^2} + \frac{1}{(n+1)^2} + \cdots$.
  We have

$$\sum_{n=1}^{\infty} \left[\left(\frac{1}{n^2} + \frac{1}{(n+1)^2} + \cdots \right) - \frac{1}{n} \right]^2$$

$$= \sum_{n=1}^{\infty} \left(\frac{1}{n^2} + \frac{1}{(n+1)^2} + \cdots \right)^2$$

$$-2\sum_{n=1}^{\infty}\frac{1}{n}\left(\frac{1}{n^2}+\frac{1}{(n+1)^2}+\cdots\right)+\zeta(2)$$

$$=\zeta(2)-\zeta(3),$$

where the last equality follows based on formulae (7.2) and (15.2).

(b) We apply Abel's summation formula with $a_n = n$ and

$$b_n = \left[\left(\frac{1}{n^2}+\frac{1}{(n+1)^2}+\cdots\right)-\frac{1}{n}\right]^2,$$

and we have, since

$$b_n - b_{n+1} = \frac{1}{n^2(n+1)}\left[2\left(\frac{1}{n^2}+\frac{1}{(n+1)^2}+\cdots\right)-\frac{1}{n^2}-\frac{1}{n}-\frac{1}{n+1}\right],$$

that

$$\sum_{n=1}^{\infty}n\left[\left(\frac{1}{n^2}+\frac{1}{(n+1)^2}+\cdots\right)-\frac{1}{n}\right]^2$$

$$=\frac{1}{2}\sum_{n=1}^{\infty}\frac{1}{n}\left[2\left(\frac{1}{n^2}+\frac{1}{(n+1)^2}+\cdots\right)-\frac{1}{n^2}-\frac{1}{n}-\frac{1}{n+1}\right]$$

$$=\sum_{n=1}^{\infty}\frac{1}{n}\left(\frac{1}{n^2}+\frac{1}{(n+1)^2}+\cdots\right)-\frac{1}{2}\sum_{n=1}^{\infty}\frac{1}{n^3}-\frac{1}{2}\sum_{n=1}^{\infty}\frac{1}{n^2}-\frac{1}{2}\sum_{n=1}^{\infty}\frac{1}{n(n+1)}$$

$$=\frac{3}{2}\zeta(3)-\frac{1}{2}\zeta(2)-\frac{1}{2},$$

where the last equality follows based on formula (15.2).

7.109. Use Abel's summation formula with:

(a) $a_n = 2n-1$ and $b_n = \left(\frac{1}{n^3}+\frac{1}{(n+1)^3}+\cdots\right)^2$;

(b) $a_n = 2n-1$ and $b_n = \left[\left(\frac{1}{n^2}+\frac{1}{(n+1)^2}+\cdots\right)^2-\frac{1}{n^2}\right]$.

Parts (a) and (b) of the problem are solved in [20] and [80].

7.110.

(a) We use Abel's summation formula with $a_n = 3n^2 - 3n + 1$ and $b_n = \left(\frac{1}{n^3}+\frac{1}{(n+1)^3}+\cdots\right)^2$, and we have, since

$$\sum_{k=1}^{n} a_k = n^3 \quad \text{and} \quad b_n - b_{n+1} = \frac{1}{n^3}\left[\left(\frac{2}{n^3} + \frac{2}{(n+1)^3} + \cdots\right) - \frac{1}{n^3}\right],$$

that

$$\sum_{n=1}^{\infty}(3n^2 - 3n + 1)\left(\frac{1}{n^3} + \frac{1}{(n+1)^3} + \cdots\right)^2 = \sum_{n=1}^{\infty}\left[\left(\frac{2}{n^3} + \frac{2}{(n+1)^3} + \cdots\right) - \frac{1}{n^3}\right]$$

$$= 2\sum_{n=1}^{\infty}\left(\frac{1}{n^3} + \frac{1}{(n+1)^3} + \cdots\right) - \zeta(3)$$

$$= 2\zeta(2) - \zeta(3).$$

We used the formula

$$\sum_{n=1}^{\infty}\left(\frac{1}{n^3} + \frac{1}{(n+1)^3} + \cdots\right) = \zeta(2).$$

More generally, for $k > 2$, one has that

$$\sum_{n=1}^{\infty}\left(\frac{1}{n^k} + \frac{1}{(n+1)^k} + \cdots\right) = \zeta(k-1).$$

(b) Use Abel's summation formula with $a_n = 4n^3 - 6n^2 + 4n - 1$ and $b_n = \left(\frac{1}{n^4} + \frac{1}{(n+1)^4} + \cdots\right)^2$.

7.111. Use Abel's summation formula with:

(a) $a_n = H_n$ and $b_n = \frac{1}{n^3} + \frac{1}{(n+1)^3} + \cdots$;

(b) $a_n = n$ and $b_n = \left(\frac{1}{n^3} + \frac{1}{(n+1)^3} + \cdots\right)^2$.

7.112. We will be using the following formula (see part (a) of Problem 7.66):

$$\int_0^1 x^{n-1}\ln(1-x)dx = -\frac{H_n}{n}, \quad n \geq 1. \tag{15.3}$$

See [26, p. 206] for information about the history of this formula. We have

$$\sum_{n=1}^{\infty} \frac{H_n}{n} \left(\frac{1}{n^2} + \frac{1}{(n+1)^2} + \cdots \right) = \zeta(2) + \sum_{n=2}^{\infty} \frac{H_n}{n} \left(\frac{1}{n^2} + \frac{1}{(n+1)^2} + \cdots \right)$$

$$= \zeta(2) - \sum_{n=2}^{\infty} \int_0^1 x^{n-1} \ln(1-x) \left(\zeta(2) - 1 - \frac{1}{2^2} - \cdots - \frac{1}{(n-1)^2} \right) dx$$

$$= \zeta(2) - \int_0^1 \ln(1-x) \sum_{n=2}^{\infty} \left(\zeta(2) - 1 - \frac{1}{2^2} - \cdots - \frac{1}{(n-1)^2} \right) x^{n-1} dx$$

$$= \zeta(2) - \int_0^1 \ln(1-x) \sum_{m=1}^{\infty} \left(\zeta(2) - 1 - \frac{1}{2^2} - \cdots - \frac{1}{m^2} \right) x^m dx$$

$$= \zeta(2) - \int_0^1 \frac{\ln(1-x)}{1-x} (x\zeta(2) - \text{Li}_2(x))\, dx,$$

$$\tag{15.4}$$

where the last equality follows based on part **C** of Problem 3.4.

We calculate the preceding integral by parts, with $f(x) = x\zeta(2) - \text{Li}_2(x)$, $f'(x) = \zeta(2) + \frac{\ln(1-x)}{x}$, $g'(x) = -\frac{\ln(1-x)}{1-x}$, and $g(x) = \frac{1}{2}\ln^2(1-x)$, and we have that

$$\int_0^1 -\frac{\ln(1-x)}{1-x} (x\zeta(2) - \text{Li}_2(x))\, dx$$

$$= (x\zeta(2) - \text{Li}_2(x)) \frac{\ln^2(1-x)}{2} \bigg|_0^1 - \int_0^1 \left(\zeta(2)\frac{\ln^2(1-x)}{2} + \frac{\ln^3(1-x)}{2x} \right) dx$$

$$= -\frac{\zeta(2)}{2} \int_0^1 \ln^2(1-x)dx - \frac{1}{2} \int_0^1 \frac{\ln^3(1-x)}{x}dx$$

$$= -\zeta(2) + \frac{\pi^4}{30},$$

$$\tag{15.5}$$

since $\int_0^1 \ln^2(1-x)dx = 2$ and $\int_0^1 \frac{\ln^3(1-x)}{x}dx = -\frac{\pi^4}{15}$.

It follows, based on (15.4) and (15.5), that

$$\sum_{n=1}^{\infty} \frac{H_n}{n} \left(\frac{1}{n^2} + \frac{1}{(n+1)^2} + \cdots \right) = \frac{\pi^4}{30},$$

and the problem is solved.

7.113.

(a) Let $x_n = \frac{H_n}{n(n+1)} \left(\zeta(2) - 1 - \frac{1}{2^2} - \cdots - \frac{1}{n^2} \right)$ and $y_n = \zeta(2) - 1 - \frac{1}{2^2} - \cdots - \frac{1}{n^2}$.
    Observe that

$$x_n = \frac{H_n y_n}{n} - \frac{H_{n+1} y_{n+1}}{n+1} - \frac{H_{n+1}}{(n+1)^3} + \frac{y_{n+1}}{(n+1)^2} + \frac{1}{(n+1)^4}, \quad n \geq 1.$$

7.114. Part (a) follows based on the power series expansion of the function $\frac{1}{1-x}$.

(b) We have, based on part (a), that

$$\sum_{n=1}^{\infty} \frac{(-1)^{n-1}}{n} \left(\zeta(2) - 1 - \frac{1}{2^2} - \cdots - \frac{1}{n^2} \right)$$

$$= \sum_{n=1}^{\infty} \frac{(-1)^n}{n} \int_0^1 \frac{x^n}{1-x} \ln x \, dx$$

$$= \int_0^1 \frac{\ln x}{1-x} \sum_{n=1}^{\infty} \frac{(-x)^n}{n} \, dx$$

$$= -\int_0^1 \frac{\ln x \ln(1+x)}{1-x} \, dx$$

$$\overset{1-x=t}{=} -\int_0^1 \frac{\ln(1-t) \ln(2-t)}{t} \, dt$$

$$= -\int_0^1 \frac{\ln(1-t) \left[\ln 2 + \ln \left(1 - \frac{t}{2} \right) \right]}{t} \, dt$$

$$= -\ln 2 \int_0^1 \frac{\ln(1-t)}{t} \, dt - \int_0^1 \frac{\ln(1-t) \ln \left(1 - \frac{t}{2} \right)}{t} \, dt$$

$$= \ln 2 \zeta(2) + \int_0^1 \frac{\ln(1-t)}{t} \sum_{n=1}^{\infty} \frac{\left(\frac{t}{2} \right)^n}{n} \, dt$$

$$= \ln 2 \zeta(2) + \sum_{n=1}^{\infty} \frac{1}{n 2^n} \int_0^1 t^{n-1} \ln(1-t) \, dt$$

$$\overset{(15.3)}{=} \ln 2 \zeta(2) - \sum_{n=1}^{\infty} \frac{H_n}{n^2 2^n}$$

$$= \frac{\pi^2}{4} \ln 2 - \zeta(3),$$

since

$$\sum_{n=1}^{\infty} \frac{H_n}{n^2 2^n} = \zeta(3) - \frac{\pi^2}{12} \ln 2.$$

The previous formula follows directly from the generating function of the sequence $\frac{H_n}{n^2}$ (see [109, formula 28, p. 217])

$$\sum_{n=1}^{\infty} \frac{H_n}{n^2} z^n = \frac{1}{2} \ln^2(1-z) \ln z + \ln(1-z) \text{Li}_2(1-z) - \text{Li}_3(1-z) + \text{Li}_3(z) + \zeta(3),$$

for $|z| < 1$.

For an alternative solution of part (b) see [61, Lemma 4, (d)].

7.115. We have, based on part (a) of Problem 7.114, that

$$\sum_{n=1}^{\infty} (-1)^n H_n \left(\zeta(2) - 1 - \frac{1}{2^2} - \cdots - \frac{1}{n^2} \right)$$

$$= -\sum_{n=1}^{\infty} (-1)^n H_n \int_0^1 \frac{x^n}{1-x} \ln x dx$$

$$= -\int_0^1 \frac{\ln x}{1-x} \sum_{n=1}^{\infty} H_n (-x)^n dx$$

$$\overset{3.63\,(a)}{=} \int_0^1 \frac{\ln x \ln(1+x)}{1-x^2} dx$$

$$= \frac{1}{2} \int_0^1 \frac{\ln x \ln(1+x)}{1-x} dx + \frac{1}{2} \int_0^1 \frac{\ln x \ln(1+x)}{1+x} dx$$

$$= \frac{7\zeta(3)}{16} - \frac{\pi^2 \ln 2}{8},$$

since (see the solution of Problem 7.114)

$$\int_0^1 \frac{\ln x \ln(1+x)}{1-x} dx = \zeta(3) - \frac{\pi^2}{4} \ln 2$$

and

$$\int_0^1 \frac{\ln x \ln(1+x)}{1+x} dx = -\frac{1}{2} \int_0^1 \frac{\ln^2(1+x)}{x} dx = -\frac{\zeta(3)}{8},$$

where the last equality is a formula due to Ramanujan [6, pp. 291–292].

7.116. See [105].

7.119. Use Abel's summation formula with:

(a) $a_n = 1$ and $b_n = \frac{\pi^2}{8} - 1 - \frac{1}{3^2} - \cdots - \frac{1}{(2n-1)^2} - \frac{1}{4n}$;

(b) $a_n = n$ and $b_n = \frac{\pi^2}{8} - 1 - \frac{1}{3^2} - \cdots - \frac{1}{(2n-1)^2} - \frac{1}{4n}$;

(c) $a_n = (-1)^n n$ and $b_n = \frac{\pi^2}{8} - 1 - \frac{1}{3^2} - \cdots - \frac{1}{(2n-1)^2} - \frac{1}{4n}$;

(d) Use that

$$\frac{\pi^2}{8} - 1 - \frac{1}{3^2} - \cdots - \frac{1}{(2n-1)^2} = -\int_0^1 \frac{x^{2n}}{1-x^2} \ln x \, dx, \quad n \geq 1;$$

(e) $a_n = \frac{1}{(2n-1)^2}$ and $b_n = \frac{\pi^2}{8} - 1 - \frac{1}{3^2} - \cdots - \frac{1}{(2n-1)^2}$;

(f) $a_n = 1$ and $b_n = \frac{\pi^4}{64} - \left(1 + \frac{1}{3^2} + \cdots + \frac{1}{(2n-1)^2}\right)^2 - \frac{\pi^2}{16n}$, and parts (d) and (e) of the problem;

(g) $a_n = 1$ and $b_n = \left(\frac{\pi^2}{8} - 1 - \frac{1}{3^2} - \cdots - \frac{1}{(2n-1)^2}\right)^2$, and parts (d) and (e) of the problem.

15.8 Exotic Zeta Series

7.121. (c) Using the formula

$$\int_0^1 x^m \ln^{k-1} x \, dx = (-1)^{k-1} \frac{(k-1)!}{(m+1)^k}, \quad m \in \mathbb{N}, \ k \in \mathbb{N},$$

we have that

$$\sum_{n=0}^{\infty} \left(\frac{1}{(6n+1)^k} + \frac{1}{(6n+5)^k}\right) = \sum_{n=0}^{\infty} \frac{(-1)^{k-1}}{(k-1)!} \int_0^1 \ln^{k-1} x \left(x^{6n} + x^{6n+4}\right) dx$$

$$= \frac{(-1)^{k-1}}{(k-1)!} \int_0^1 \ln^{k-1} x \sum_{n=0}^{\infty} \left(x^{6n} + x^{6n+4}\right) dx$$

$$= \frac{(-1)^{k-1}}{(k-1)!} \int_0^1 \frac{1+x^4}{1-x^6} \ln^{k-1} x \, dx$$

$$= \frac{(-1)^{k-1}}{(k-1)!} \int_0^1 \ln^{k-1} x \left(\frac{1}{1-x^2} - \frac{x^2}{1-x^6}\right) dx$$

$$= \frac{(-1)^{k-1}}{(k-1)!} \left(1 - \frac{1}{3^k}\right) \int_0^1 \frac{\ln^{k-1} x}{1-x^2} dx$$

$$= \frac{(-1)^{k-1}}{(k-1)!} \left(1 - \frac{1}{3^k}\right) \int_0^1 \ln^{k-1} x \sum_{n=0}^{\infty} x^{2n} dx$$

$$= \left(1 - \frac{1}{3^k}\right) \sum_{n=0}^{\infty} \frac{1}{(2n+1)^k}$$

$$= \left(1 - \frac{1}{2^k} \right) \left(1 - \frac{1}{3^k} \right) \zeta(k).$$

The other parts of the problem can be solved similarly.

7.122. (e) We have

$$\zeta(3) + \zeta(5) + \cdots + \zeta(2n+1) = \sum_{i=1}^{n} \sum_{k=1}^{\infty} \frac{1}{k^{2i+1}}$$

$$= n + \sum_{i=1}^{n} \sum_{k=2}^{\infty} \frac{1}{k^{2i+1}}$$

$$= n + \sum_{k=2}^{\infty} \sum_{i=1}^{n} \frac{1}{k^{2i+1}}$$

$$= n + \sum_{k=2}^{\infty} \frac{1}{k} \cdot \frac{1 - \frac{1}{k^{2n}}}{k^2 - 1}$$

$$= n + \sum_{k=2}^{\infty} \frac{1}{k(k^2 - 1)} - \sum_{k=2}^{\infty} \frac{1}{(k^2 - 1)k^{2n+1}}$$

$$= n + \frac{1}{4} - \sum_{k=2}^{\infty} \frac{1}{(k^2 - 1)k^{2n+1}}.$$

It follows that

$$n + \frac{1}{4} - \zeta(3) - \zeta(5) - \cdots - \zeta(2n+1) = \sum_{k=2}^{\infty} \frac{1}{(k^2 - 1)k^{2n+1}} \qquad (15.6)$$

and

$$\lim_{n \to \infty} \left(n + \frac{1}{4} - \zeta(3) - \zeta(5) - \cdots - \zeta(2n+1) \right) = \lim_{n \to \infty} \sum_{k=2}^{\infty} \frac{1}{(k^2 - 1)k^{2n+1}} = 0.$$

We have, based on formula (15.6), that

$$\sum_{n=1}^{\infty} \left(n + \frac{1}{4} - \zeta(3) - \zeta(5) - \cdots - \zeta(2n+1) \right) = \sum_{n=1}^{\infty} \sum_{k=2}^{\infty} \frac{1}{k(k^2 - 1)k^{2n}}$$

$$= \sum_{k=2}^{\infty} \frac{1}{k(k^2 - 1)} \sum_{n=1}^{\infty} \frac{1}{k^{2n}}$$

$$= \sum_{k=2}^{\infty} \frac{1}{k(k^2-1)^2}$$

$$= \frac{1}{16}.$$

The other parts of the problem can be solved either by using Abel's summation formula, or by using a similar technique to the one given in the solution of part (e).
7.123. Use Abel's summation formula with $a_n = 1$ and $b_n = \frac{1}{2} - \zeta(2) - \zeta(4) - \cdots - \zeta(2n) + \zeta(3) + \zeta(5) + \cdots + \zeta(2n+1)$.

15.9 Special Differential Equations

7.124. (a) $f(x) = f(0)e^{\frac{k}{f^2(0)}x}$, $x \in \mathbb{R}$; (b) $f(x) = f(0)e^{\frac{x^{k+1}}{(k+1)f^2(0)}}$, $x \in \mathbb{R}$.
7.125.

(a) $f(x) = 2f(0)\frac{e^{2f(0)x}}{1+e^{2f(0)x}}$, $x \in \mathbb{R}$.

(b) $f(x) = f(0)e^{f^2(0)x}$, $x \in \mathbb{R}$. We have

$$f''(x) = f^2(x)f(-x), \ x \in \mathbb{R}$$
$$f'(-x) = f^2(-x)f(x), \ x \in \mathbb{R}.$$

We multiply the first equation by $f(-x)$ and the second equation by $f(x)$ and we have that $f'(x)f(-x) - f'(-x)f(x) = 0$, $\forall x \in \mathbb{R}$. It follows that $(f(x)f(-x))' = 0$, $\forall x \in \mathbb{R}$, and this shows that $f(x)f(-x) = \mathscr{C}$, $\forall x \in \mathbb{R}$. When $x = 0$ we have $\mathscr{C} = f^2(0)$, and it follows that $f(x)f(-x) = f^2(0)$. Thus, $f'(x) = f^2(x)f(-x) = f^2(0)f(x)$, $\forall x \in \mathbb{R}$. The solution of this differential equation with separable variables is $f(x) = f(0)e^{f^2(0)x}$, $x \in \mathbb{R}$.

7.126. (a) $f(x) = \sqrt{2}\frac{e^{2x}}{\sqrt{1+e^{4x}}}$, $x \in \mathbb{R}$; (b) $f(x) = \sqrt{2}\frac{e^{\frac{2}{3}x^3}}{\sqrt{1+e^{\frac{4}{3}x^3}}}$, $x \in \mathbb{R}$.

7.127. $f(x) = \sqrt[3]{2}f(0)\frac{e^{2f^3(0)x}}{\sqrt[3]{1+e^{6f^3(0)x}}}$, $x \in \mathbb{R}$.
7.128. $f(x) = ax + b$, $\forall x \in \mathbb{R}$, $a, b \in \mathbb{R}$.
7.129. $f(x) = ax^2 + bx + c$, $\forall x \in \mathbb{R}$, $a, b, c \in \mathbb{R}$.
7.130. $f(x) = ax^2 + bx + c$, $\forall x \in \mathbb{R}$, $a, b, c \in \mathbb{R}$.
7.131. $f(x) = ax + b$, $\forall x \in \mathbb{R}$, $a, b \in \mathbb{R}$.
7.132. $f(x) = \sin x + \cos x$, $\forall x \in \mathbb{R}$.

15.10   Inequalities

7.133. Use that

$$\frac{1}{a_i^2 + a_j^2} = \int_0^\infty e^{-(a_i^2 + a_j^2)x}\mathrm{d}x.$$

7.134. We have

$$S = \sum_{i=1}^n \sum_{j=1}^n \frac{ij}{i+j-1} a_i a_j$$

$$= \sum_{i=1}^n \sum_{j=1}^n ij a_i a_j \int_0^1 x^{i+j-2}\mathrm{d}x$$

$$= \int_0^1 \left(\sum_{i=1}^n i a_i x^{i-1}\right)\left(\sum_{j=1}^n j a_j x^{j-1}\right)\mathrm{d}x$$

$$= \int_0^1 \left(\sum_{i=1}^n i a_i x^{i-1}\right)^2 \mathrm{d}x$$

$$= \int_0^1 f^2(x)\mathrm{d}x,$$

where $f(x) = \sum_{i=1}^n i a_i x^{i-1}$. It follows, based on Hölder's inequality for integrals, that

$$\sqrt{S} = \sqrt{\int_0^1 f^2(x)\mathrm{d}x} \geq \int_0^1 |f(x)|\mathrm{d}x \geq \left|\int_0^1 f(x)\mathrm{d}x\right| = \left|\sum_{i=1}^n a_i\right|.$$

15.11   Fabulous Integrals

7.135. We have

$$\int_0^1 \frac{\ln(x^a + (1-x)^a)}{x}\mathrm{d}x = \int_0^{\frac{1}{2}} \frac{\ln(1-x)^a\left(1 + \frac{x^a}{(1-x)^a}\right)}{x}\mathrm{d}x + \int_{\frac{1}{2}}^1 \frac{\ln x^a\left(1 + \frac{(1-x)^a}{x^a}\right)}{x}\mathrm{d}x$$

(continued)

$$= a \int_0^{\frac{1}{2}} \frac{\ln(1-x)}{x} dx + \int_0^{\frac{1}{2}} \frac{\ln\left(1 + \frac{x^a}{(1-x)^a}\right)}{x} dx$$

$$+ a \int_{\frac{1}{2}}^1 \frac{\ln x}{x} dx + \int_{\frac{1}{2}}^1 \frac{\ln\left(1 + \frac{(1-x)^a}{x^a}\right)}{x} dx$$

(15.7)

$$= -a\mathrm{Li}_2\left(\frac{1}{2}\right) - a\frac{\ln^2 2}{2} + \int_0^{\frac{1}{2}} \frac{\ln\left(1 + \frac{x^a}{(1-x)^a}\right)}{x} dx$$

$$+ \int_{\frac{1}{2}}^1 \frac{\ln\left(1 + \frac{(1-x)^a}{x^a}\right)}{x} dx.$$

Using the substitution $\frac{x}{1-x} = t$, we have that

$$\int_0^{\frac{1}{2}} \frac{\ln\left(1 + \frac{x^a}{(1-x)^a}\right)}{x} dx = \int_0^1 \frac{\ln(1 + t^a)}{t(1+t)} dt$$

(15.8)

$$= \int_0^1 \frac{\ln(1 + t^a)}{t} dt - \int_0^1 \frac{\ln(1 + t^a)}{1+t} dt.$$

On the other hand, the substitution $\frac{1-x}{x} = t$ shows that

$$\int_{\frac{1}{2}}^1 \frac{\ln\left(1 + \frac{(1-x)^a}{x^a}\right)}{x} dx = \int_0^1 \frac{\ln(1 + t^a)}{1+t} dt.$$

(15.9)

It follows, based on (15.7), (15.8), and (15.9), that

$$\int_0^1 \frac{\ln(x^a + (1-x)^a)}{x} dx = -a\mathrm{Li}_2\left(\frac{1}{2}\right) - a\frac{\ln^2 2}{2} + \int_0^1 \frac{\ln(1 + t^a)}{t} dt$$

$$\overset{t^a=y}{=} -a\frac{\pi^2}{12} + \frac{1}{a} \int_0^1 \frac{\ln(1 + y)}{y} dy$$

$$= \frac{\pi^2}{12}\left(\frac{1}{a} - a\right).$$

7.136. A more general version of this integral was proposed by O. Furdui as a problem in [47] and two solutions by A. Kotronis and O. Kouba have been published in [101].

7.138. Using the substitution $\frac{1}{x} = t$, we have that

$$\int_0^1 \frac{x^m}{\left\lfloor \frac{1}{x} \right\rfloor}\,dx = \int_1^\infty \frac{1}{t^{m+2} \lfloor t \rfloor}\,dt = \sum_{k=1}^\infty \int_k^{k+1} \frac{1}{t^{m+2} \lfloor t \rfloor}\,dt$$

$$= \frac{\zeta(m+2)}{m+1} - \frac{1}{m+1} \sum_{k=1}^\infty \frac{1}{k(k+1)^{m+1}}.$$

Let $S_{m+1} = \sum_{k=1}^\infty \frac{1}{k(k+1)^{m+1}}$. Observe that $S_{m+1} = S_m - (\zeta(m+1) - 1)$.

7.139. (a) We have

$$\int_1^\infty \left(e - 1 - \frac{1}{1!} - \cdots - \frac{1}{\lfloor x \rfloor !} \right) dx = \sum_{n=1}^\infty \int_n^{n+1} \left(e - 1 - \frac{1}{1!} - \cdots - \frac{1}{\lfloor x \rfloor !} \right) dx$$

$$= \sum_{n=1}^\infty \left(e - 1 - \frac{1}{1!} - \cdots - \frac{1}{n!} \right)$$

$$\overset{(*)}{=} \sum_{n=1}^\infty \frac{n}{(n+1)!}$$

$$= 1.$$

We used at step (*) Abel's summation formula with $a_n = 1$ and $b_n = e - 1 - \frac{1}{1!} - \cdots - \frac{1}{n!}$.

The other parts of the problem can be solved similarly. The calculation of the series that are equal to the integrals which appear in parts (b)–(f) can be found in [59].

Remark. Problem 7.139 refers to the calculation of improper integrals of the form $\int_1^\infty f(x)\,dx$, which verify the equality

$$\int_1^\infty f(x)\,dx = \sum_{n=1}^\infty f(n).$$

(continued)

The preceding equality can be viewed as *sophomore's dream* for integrals of the form $\int_1^\infty f(x)dx$ (see also Problem 3.52).

7.143.

(a) Let k be the degree of f. We integrate by parts and we have that

$$\int_0^\infty e^{-x} f(x)dx = f(0) + \int_0^\infty e^{-x} f'(x)dx$$

$$= f(0) + f'(0) + \int_0^\infty e^{-x} f''(x)dx$$

$$= \cdots$$

$$= f(0) + f'(0) + \cdots + f^{(k)}(0).$$

(b) First we observe that if $n \geq 0$ is an integer, then $\int_0^\infty e^{-x} \frac{x^n}{n!}dx = 1$. Let $S_n = \sum_{i=0}^n f^{(i)}(0)$. A calculation shows that

$$S_n = \int_0^\infty e^{-x} T_n(x)dx,$$

where $T_n(x) = f(0) + \frac{f'(0)}{1!}x + \frac{f''(0)}{2!}x^2 + \cdots + \frac{f^{(n)}(0)}{n!}x^n$ is the nth degree Taylor polynomial of f at 0.

Now we prove that the integral $\int_0^\infty e^{-x} f(x)dx$ converges. We need to show that $\lim_{A \to \infty} \int_0^A e^{-x} f(x)dx$ exists and is finite. We apply Bolzano–Cauchy criterion for proving that the limit of a function exists and is finite. Let $\epsilon > 0$. Since $\sum_{n=0}^\infty f^{(n)}(0)$ converges absolutely, we get that there exists $n_0 \in \mathbb{N}$ such that $\sum_{n=n_0}^\infty |f^{(n)}(0)| < \frac{\epsilon}{2}$, for all $n \geq n_0$. On the other hand, since $\int_0^\infty e^{-x} T_{n_0}(x)dx$ converges, its value being calculated by formula in part (a) of the problem, we get that $\lim_{A \to \infty} \int_0^A e^{-x} T_{n_0}(x)dx$ exists and it follows that, for $\epsilon > 0$ there exists $\delta > 0$ such that for all $A', A'' > \delta$ one has

$$\left| \int_{A'}^{A''} e^{-x} T_{n_0}(x)dx \right| < \frac{\epsilon}{2}.$$

We have, for $A', A'' > \delta$, that

$$\left| \int_{A'}^{A''} e^{-x} f(x) dx \right| \leq \int_{A'}^{A''} e^{-x} |f(x) - T_{n_0}(x)| dx + \left| \int_{A'}^{A''} e^{-x} T_{n_0}(x) dx \right|$$

$$< \int_{A'}^{A''} e^{-x} \left| \sum_{k=n_0+1}^{\infty} \frac{f^{(k)}(0)}{k!} x^k \right| dx + \frac{\epsilon}{2}$$

$$\leq \sum_{k=n_0+1}^{\infty} |f^{(k)}(0)| \int_{A'}^{A''} e^{-x} \frac{x^k}{k!} dx + \frac{\epsilon}{2}$$

$$\leq \sum_{k=n_0+1}^{\infty} |f^{(k)}(0)| \int_{0}^{\infty} e^{-x} \frac{x^k}{k!} dx + \frac{\epsilon}{2}$$

$$\leq \sum_{k=n_0+1}^{\infty} |f^{(k)}(0)| + \frac{\epsilon}{2}$$

$$\leq \epsilon,$$

which proves that $\lim\limits_{A \to \infty} \int_0^A e^{-x} f(x) dx$ exists and $\lim\limits_{A \to \infty} \int_0^A e^{-x} f(x) dx =$ $\int_0^{\infty} e^{-x} f(x) dx$.

Now we show that $\sum\limits_{n=0}^{\infty} f^{(n)}(0) = \int_0^{\infty} e^{-x} f(x) dx$. We calculate

$$\left| \int_0^{\infty} e^{-x} f(x) dx - S_n \right| = \left| \int_0^{\infty} e^{-x} (f(x) - T_n(x)) dx \right|$$

$$\leq \int_0^{\infty} e^{-x} |f(x) - T_n(x)| dx$$

$$\leq \int_0^{\infty} e^{-x} \sum_{k=n+1}^{\infty} \frac{f^{(k)}(0)}{k!} x^k dx$$

$$\leq \sum_{k=n+1}^{\infty} |f^{(k)}(0)|$$

$$\leq \epsilon,$$

which implies that $\lim\limits_{n \to \infty} S_n = \int_0^{\infty} e^{-x} f(x) dx$ and the problem is solved.

Remark.

(a) An example of a function which verifies the statement of the problem is $f(x) = e^{-\alpha x}$, $\alpha \in (-1, 1)$, and a function which has a power series expansion at 0, with radius of convergence $R = \infty$, for which the problem fails to hold is $f(x) = \sin x$, the series $\sum_{n=0}^{\infty} f^{(n)}(0)$ being divergent. It is interesting to note that if f does not have a power series expansion at 0, or the Maclaurin series has radius of convergence $R = 0$, then the problem fails to hold. The function $f(x) = e^{-1/x^2}$, when $x \neq 0$ and $f(0) = 0$, has $f^{(n)}(0) = 0$ and $\sum_{n=0}^{\infty} f^{(n)}(0) = 0 \neq \int_0^{\infty} e^{-x-\frac{1}{x^2}} dx$.

(b) *A generalization.* Similarly, one can prove that if f is a function which has a Maclaurin series expansion with radius of convergence $R = \infty$ such that $\sum_{n=0}^{\infty} f^{(n)}(x)$ converges absolutely, then $\int_0^{\infty} e^{-t} f(x+t) dt$ converges and

$$\sum_{n=0}^{\infty} f^{(n)}(x) = \int_0^{\infty} e^{-t} f(x+t) dt.$$

7.144.

(a) Use a similar technique to that in the solution of Problem 7.143, with f replaced by $\frac{1}{2}(f(x) + f(-x))$ (the even part of f).
(b) Use a similar technique to that in the solution of Problem 7.143, with f replaced by $\frac{1}{2}(f(x) - f(-x))$ (the odd part of f).

References

1. Andreescu, T.: Problem U457, Mathematical Reflections 5, 2018, 3
2. Andreoli, M.: Problem 819, Coll. Math. J. **37** (1), 2006, 60
3. Baranenkov, G., Chostak, R., Démidovitch, B., et al., Démidovitch, B (éd.): Recueil d'exercices et de problèmes d'analyse mathématique, Cinquième édition, Éditions Mir, Moscou, 1974
4. Bataille, M.: Solution to problem 3965, Crux Mathematicorum 41 (7), 2015, 311–313
5. Becker, M.: Problem 874, Coll. Math. J. 39 (2), 2008, 154
6. Berndt, B.C.: Ramanujan's Notebooks, Part 1, Springer-Verlag, New York, 1985
7. Biler, P., Witkowski, A.: Problems in Mathematical Analysis, Marcel Dekker, Inc., New York and Basel, 1990
8. Borwein, D., Borwein, J.M.: On an intriguing integral and some series related to $\zeta(4)$, Proc. Amer. Math. Soc. 123 (4), 1995, 1191–1198
9. Borwein, J.M., Borwein, P.B.: Pi and the AGM, John Wiley & Sons, New York, 1987
10. Bromwich, T.J.I'A.: An Introduction to the Theory of Infinite Series, Third Edition, American Mathematical Society, Providence, Rhode Island, 1991
11. Castellanos, D.: The ubiquitous π, Math. Mag. Part I, 61 (2), 1988, 67–98; Part II, 61 (3), 1988, 148–163
12. Chen, H.: A new ratio test for positive monotone series, Coll. Math. J. 44 (2), 2013, 139–141
13. Comtet, L.: Advanced Combinatorics. The Art of Finite and Infinite Expansions, D. Reidel Publishing Company, Dordrecht, 1974
14. De Doelder, P.J.: On some series containing $\psi(x) - \psi(y)$ and $(\psi(x) - \psi(y))^2$ for certain values of x and y, J. Comput. Appl. Math. 37, 1991, 125–141
15. de Souza, P. N., Silva, J.-N.: Berkeley Problems in Mathematics, Third Edition, Springer, New York, 2004
16. Dutta, R.: Solution to problem H-790, Fibonacci Q. 56 (2), 2018, 190–191
17. Edgar, T.: Proof without words: Sums of reciprocals of binomial coefficients, Math. Mag. 89 (3), 2016, 212–213
18. Eggleton, R., Kustov, V.: The product and the quotient rules revisited, Coll. Math. J. 42 (4), 2011, 323–326
19. Eydelzon, A.: Problem 12186, Amer. Math. Monthly 127 (5), 2020, 462
20. Fanego, B.S.: Problem 355, Soluciones, Gac. R. Soc. Mat. Esp. 22 (3), 2019, 539–540 http://gaceta.rsme.es/abrir.php?id=1543
21. Finch, S.R.: Mathematical Constants. Encyclopedia of Mathematics and Its Applications 94, Cambridge University Press, New York, 2003
22. Flajolet, P., Salvy, B.: Euler sums and contour integral representations, Exp. Math. 7 (1), 1998, 15–35
23. Freitas, P.: Integrals of polylogarithmic functions, recurrence relations, and associated Euler sums, Math. Comp. 74, 2005, 1425–1440

© The Author(s), under exclusive license to Springer Nature Switzerland AG 2021
A. Sîntămărian, O. Furdui, *Sharpening Mathematical Analysis Skills*, Problem Books in Mathematics, https://doi.org/10.1007/978-3-030-77139-3

24. Furdui, O.: From Lalescu's sequence to a Gamma function limit, Austral. Math. Soc. Gaz. 35 (5), 2008, 339–344
25. Furdui, O.: Series involving products of two harmonic numbers, Math. Mag. 84 (5), 2011, 371–377
26. Furdui, O.: Limits, Series, and Fractional Part Integrals. Problems in Mathematical Analysis, Springer, New York, 2013
27. Furdui, O.: The evaluation of a quadratic and a cubic series with trigamma function, Applied Math. E Notes, 15, 2015, 187–196
28. Furdui, O.: Two surprising series with harmonic numbers and the tail of $\zeta(2)$, Gazeta Matematică, Seria A, 33 (112) (1-2), 2015, 1–8 https://ssmr.ro/gazeta/gma/2015/gma1-2-2015-continut.pdf
29. Furdui, O.: A note on problem 7 of day 2 of IMC 2015, Gazeta Matematică, Seria A, 33 (112) (1-2), 2015, 41–44 https://ssmr.ro/gazeta/gma/2015/gma1-2-2015-continut.pdf
30. Furdui, O.: Harmonic series with polygamma functions, J. Classical Anal. 8 (2), 2016, 123–130
31. Furdui, O.: Problem 11810, Amer. Math. Monthly 122 (1), 2015, 75
32. Furdui, O.: Problem 11941, Amer. Math. Monthly 123 (9), 2016, 942
33. Furdui, O.: Problem 893, Coll. Math. J. 40 (1), 2009, 55–56
34. Furdui, O.: Solution to problem 874, Coll. Math. J. 40 (2), 2009, 137–138
35. Furdui, O.: Problem 1042, Coll. Math. J. 46 (1), 2015, 61
36. Furdui, O.: Problem 1117, Coll. Math. J. 49 (1), 2018, 60
37. Furdui, O.: Problem 4010, Crux Mathematicorum 41 (1), 2015, 28
38. Furdui, O.: Problem 158, EMS Newsletter March 2016, 62
39. Furdui, O.: Problem H-790, Fibonacci Q. 54 (2), 2016, 186
40. Furdui, O.: Problem 341, Gac. R. Soc. Mat. Esp. 21 (1), 2018, 110
41. Furdui, O.: Problem 346, Gac. R. Soc. Mat. Esp. 21 (2), 2018, 331
42. Furdui, O.: Problem 433, Gazeta Matematică, Seria A, 33 (112) (1-2), 2015, 45 https://ssmr.ro/gazeta/gma/2015/gma1-2-2015-continut.pdf
43. Furdui, O.: Problem 436, Gazeta Matematică, Seria A, 33 (112) (1-2), 2015, 46 https://ssmr.ro/gazeta/gma/2015/gma1-2-2015-continut.pdf
44. Furdui, O.: Problem 482, Gazeta Matematică, Seria A, 36 (115) (3-4), 2018, 44 https://ssmr.ro/gazeta/gma/2018/gma3-4-2018-continut.pdf
45. Furdui, O.: Problem Q1014, Math. Mag. 84 (4), 2011, 297
46. Furdui, O.: Problem Q1040, Math. Mag. 87 (2), 2014, 152
47. Furdui, O.: Problem 53, Problems, Mathproblems, 2 (4), 2012, 119 http://www.mathproblems-ks.org/?wpfb_dl=16
48. Furdui, O.: Problem 58, Problems, Mathproblems, 3 (1), 2013, 118 http://www.mathproblems-ks.org/?wpfb_dl=9
49. Furdui, O.: Problem 5306, Sch. Sci. Math. 114 (4), 2014, 2
50. Furdui, O.: Problem 5336, Sch. Sci. Math. 115 (1), 2015, 2
51. Furdui, O.: Problem 5354, Sch. Sci. Math. 115 (4), 2015, 2
52. Furdui, O.: Problem 5384, Sch. Sci. Math. 116 (1), 2016, 2
53. Furdui, O.: Problem 5426, Sch. Sci. Math. 116 (8), 2016, 3
54. Furdui, O.: Problem 5456, Sch. Sci. Math. 117 (5), 2017, 2
55. Furdui, O., Ivan, M., Sîntămărian, A.: A note on a UTCN SEEMOUS selection test problem, Gazeta Matematică, Seria A, 34 (113) (3-4), 2016, 34–39 https://ssmr.ro/gazeta/gma/2016/gma3-4-2016-continut.pdf
56. Furdui, O., Ivan, D.M., Sîntămărian, A.: Problem 11982, Amer. Math. Monthly 124 (5), 2017, 465
57. Furdui, O., Ivan, D.M., Sîntămărian, A.: Problem 1097, Coll. Math. J. 48 (2), 2017, 138
58. Furdui, O., Sîntămărian, A.: Quadratic series involving the tail of $\zeta(k)$, Integral Transforms Spec. Funct. 26 (1) 2015, 1–8

59. Furdui, O., Sîntămărian, A.: Exotic series with fractional part function, Gazeta Matematică, Seria A, 35 (114) (3-4), 2017, 1–12 https://ssmr.ro/gazeta/gma/2017/gma3-4-2017-continut. pdf
60. Furdui, O., Sîntămărian, A.: A new proof of the quadratic series of Au–Yeung, Gazeta Matematică, Seria A, 37 (116), (1-2), 2019, 1–6 https://ssmr.ro/gazeta/gma/2019/gma1-2-2019-continut.pdf
61. Furdui, O., Sîntămărian, A.: Pearls of quadratic series, Gazeta Matematică, Seria A, 38 (117), (3-4), 2020, 1–12 https://ssmr.ro/gazeta/gma/2020/gma3-4-2020-continut.pdf
62. Furdui, O., Sîntămărian, A.: Problem 12012, Amer. Math. Monthly 124 (10), 2017, 971
63. Furdui, O., Sîntămărian, A.: Problem 12045, Amer. Math. Monthly 125 (5), 2018, 467
64. Furdui, O., Sîntămărian, A.: Problem 12060, Amer. Math. Monthly 125 (7), 2018, 661
65. Furdui, O., Sîntămărian, A.: Problem 12102, Amer. Math. Monthly 126 (3), 2019, 281
66. Furdui, O., Sîntămărian, A.: Problem 12207, Amer. Math. Monthly 127 (8), 2020, 753
67. Furdui, O., Sîntămărian, A.: Problem 12215, Amer. Math. Monthly 127 (9), 2020, 853
68. Furdui, O., Sîntămărian, A.: Problem 12241, Amer. Math. Monthly 128 (3), 2021, 276
69. Furdui, O., Sîntămărian, A.: Problem 1154, Coll. Math. J. 50 (3), 2019, 224
70. Furdui, O., Sîntămărian, A.: Problem 1163, Coll. Math. J. 50 (5), 2019, 379
71. Furdui, O., Sîntămărian, A.: Problem 3965, Crux Mathematicorum 40 (7), 2014, 300
72. Furdui, O., Sîntămărian, A.: Problem 4368, Crux Mathematicorum 44 (7), 2018, 303
73. Furdui, O., Sîntămărian, A.: Problem 308, Gac. R. Soc. Mat. Esp. 19 (3), 2016, 590
74. Furdui, O., Sîntămărian, A.: Problem 355, Gac. R. Soc. Mat. Esp. 21 (3), 2018, 544
75. Furdui, O., Sîntămărian, A.: Problem 387, Gac. R. Soc. Mat. Esp. 23 (1), 2020, 136
76. Furdui, O., Sîntămărian, A.: Problem 403, Gac. R. Soc. Mat. Esp. 23 (3), 2020, 536
77. Furdui, O., Sîntămărian, A.: Problem 474, Gazeta Matematică, Seria A, 36 (115) (1-2), 2018, 48 https://ssmr.ro/gazeta/gma/2018/gma1-2-2018-continut.pdf
78. Furdui, O., Sîntămărian, A.: Problem 486, Gazeta Matematică, Seria A, 36 (115) (3-4), 2018, 45 https://ssmr.ro/gazeta/gma/2018/gma3-4-2018-continut.pdf
79. Furdui, O., Sîntămărian, A.: Problem 493, Gazeta Matematică, Seria A, 37 (116) (1-2), 2019, 35 https://ssmr.ro/gazeta/gma/2019/gma1-2-2019-continut.pdf
80. Furdui, O., Sîntămărian, A.: Solution to problem 474, Gazeta Matematică, Seria A, 37 (116) (1-2), 2019, 44–45 https://ssmr.ro/gazeta/gma/2019/gma1-2-2019-continut.pdf
81. Furdui, O., Sîntămărian, A.: Problem 2068, Math. Mag. 91 (2), 2019, 151
82. Furdui, O., Sîntămărian, A.: Problem 825, The Pentagon 77 (2), 2018, 29
83. Furdui, O., Sîntămărian, A.: Problem 5504, Sch. Sci. Math. 118 (5), 2018, 2
84. Furdui, O., Sîntămărian, A.: Problem 5510, Sch. Sci. Math. 118 (6), 2018, 2
85. Furdui, O., Sîntămărian, A.: Problem 5516, Sch. Sci. Math. 118 (7), 2018, 2
86. Furdui, O., Sîntămărian, A.: Problem 5546, Sch. Sci. Math. 119 (4), 2019, 2
87. Furdui, O., Sîntămărian, A.: Problem 5558, Sch. Sci. Math. 119 (6), 2019, 1–2
88. Furdui, O., Sîntămărian, A.: Problem 5576, Sch. Sci. Math. 120 (1), 2020, 2
89. Furdui, O., Sîntămărian, A.: Problem 5588, Sch. Sci. Math. 120 (3), 2020, 2
90. Furdui, O., Sîntămărian, A.: Problem 5618, Sch. Sci. Math. 120 (8), 2020, 2
91. Furdui, O., Sîntămărian, A.: Problem 5642, Sch. Sci. Math. 121 (4), 2021, 2
92. Furdui, O., Vălean, C.: Evaluation of series involving the product of the tail of $\zeta(k)$ and $\zeta(k+1)$, Mediterr. J. Math. 13 2016, 517–526
93. Garaschuk, K. (ed.): Problem CC95, Crux Mathematicorum 39 (9), 2013, 393
94. Garnir, H.G.: Fonctions de variables réelles, Tomme I, Librairie Universitaire, Lauvain & Gauthiers-Villars, Paris, 1963
95. Gavrea, I., Ivan, M.: On an extension of Pólya–Szegö formula, Mediterr. J. Math. 13 (5), 2016, 3409–3416
96. Gelbaum, B.R., Olmsted, J.M.H.: Theorems and Counterexamples in Mathematics, Springer-Verlag, New York, 1990
97. Grigorieva, E.: Methods of Solving Sequences and Series Problems, Birkhäuser/Springer, Cham, 2016

98. Holland, F.: $\lim_{m\to\infty} \sum_{k=0}^{m}(k/m)^m = e/(e-1)$, Math. Mag. 83 (1), 2010, 51–54
99. Ioachimescu, A.G.: Problem 16, Gazeta Matematică 1 (2), 1895, 39
100. Jeffrey, A., Dai, H-H.: Handbook of Mathematical Formulas and Integrals, 4th ed., Elsevier Academic Press, Amsterdam, 2008
101. Kotronis, A., Kouba, O.: Problem 53, Solutions, Mathproblems, 3 (1), 2013, 126–128 http://www.mathproblems-ks.org/?wpfb_dl=9
102. Krijan, I.: Problem 6, IMC, Day 2, Blagoevgrad, Bulgaria, 30 July 2015
103. Lalescu, T.: Problem 579, Gazeta Matematică 6 (6), 1900-1901, 148
104. Levin, V.I.: About a problem of S. Ramanujan (in Russian), Uspekhi Mat. Nauk. 5 (3), 1950, 161–166
105. Levy, M.: Problem 5406, Solution 3, Solutions, Sch. Sci. Math. 116 (8), 2016, 17–18
106. Loya, P.: Amazing and Aesthetic Aspects of Analysis, Springer, New York, 2017
107. Mabry, R.: Problem 893, Problems and Solutions, Coll. Math. J., 41 (1), 2010, 67–69
108. Martínez Fernández, A.R.: Solution to problem 387, Gac. R. Soc. Mat. Esp. 24 (1), 2021, 136–137
109. Mező, I.: Nonlinear Euler sums, Pac. J. Math., 272 (1), 2014, 201–226
110. Mincu, G., Pop, V., Rus, M.: Traian Lalescu national mathematics contest for university students, 2018 edition, Gazeta Matematică, Seria A, 36 (115) (3-4), 2018, 24–37 https://ssmr.ro/gazeta/gma/2018/gma3-4-2018-continut.pdf
111. Mortici, C.: A power series approach to some inequalities, Amer. Math. Monthly 119 (2), 2012, 147–151
112. Nahin, P.J.: Inside Interesting Integrals, Springer, New York, 2015
113. Papacu, N.: Problem 27079, Gazeta Matematică, Seria B, 120 (5), 2015, 269
114. Pârşan, L.: Problema 17263, Gazeta Matematică, Seria B, 83 (6), 1978, 250
115. Perfetti, P.: Problem 3815, Crux Mathematicorum 39 (2), 2013, 92
116. Perfetti, P.: Solution to problem 58, Mathproblems, 3 (2), 2013, 142–143 http://www.mathproblems-ks.org/?wpfb_dl=10
117. Plaza, Á.: Proof without words: Arctangent of two and the golden ratio, Math. Mag. 90 (3), 2017, 179
118. van der Poorten, A.J.: Some wonderful formulae... Footnotes to Apéry's proof of the irrationality of $\zeta(3)$, Séminaire Delange-Pisot-Poitou. Théorie des nombres, tome 20 (2), 1978-1979, 1–7
119. Pop, V., Furdui, O.: Square Matrices of Order Two. Theory, Applications, and Problems, Springer, Cham, 2017
120. Popa, D.: Problem C: 1789, Gazeta Matematică, Seria B, 101 (2), 1996, 124
121. Pólya, G., Szegö, G.: Problems and Theorems in Analysis I. Series. Integral Calculus. Theory of Functions, Springer-Verlag, Berlin, 1998
122. Rădulescu, T.-L.T., Rădulescu, V.D., Andreescu, T.: Problems in Real Analysis: Advanced Calculus on the Real Axis, New York, 2009
123. Ricardo, H.: Solution to problem 1154, Coll. Math. J. 51 (3), 2020, 231
124. Rizzoli, I.: O teoremă Stolz–Cesaro, Gazeta Matematică, Seria B, 95 (10-11-12), 1990, 281–284
125. Sandham, H.F.: Advanced Problem 4305, Amer. Math. Monthly 55 (7), 1948, 431
126. Senum, G.I.: Problem E 3352, Amer. Math. Monthly 96 (9), 1989, 838
127. Singh, J.: A noninductive proof of de Moivre's formula, Amer. Math. Monthly 125 (1), 2018, 80
128. Sîntămărian, A.: Probleme selectate cu şiruri de numere reale (in Romanian), Editura U. T. Press, Cluj-Napoca, 2008
129. Sîntămărian, A.: A Generalization of Euler's Constant, Editura Mediamira, Cluj-Napoca, 2008
130. Sîntămărian, A.: Some convergent sequences and series, Int. J. Pure Appl. Math. 57 (6), 2009, 885–902
131. Sîntămărian, A.: Some applications of a limit problem, Gazeta Matematică, Seria B, 115 (11), 2010, 517–520

132. Sîntămărian, A.: Solution to problem CC95, Crux Mathematicorum 40 (9), 2014, 372–373
133. Sîntămărian, A.: Problem 11528, Amer. Math. Monthly 117 (8), 2010, 742
134. Sîntămărian, A.: Problem 1083, Coll. Math. J. 47 (4), 2016, 301
135. Sîntămărian, A.: Problem 3760, Crux Mathematicorum 38 (6), 2012, 243, 244
136. Sîntămărian, A.: Problem 266, Gac. R. Soc. Mat. Esp. 18 (1), 2015, 121
137. Sîntămărian, A.: Problem C: 3226, Gazeta Matematică, Seria B, 112 (10), 2007, 554, 556
138. Sîntămărian, A.: Problem Q1030, Math. Mag. 86 (2), 2013, 148
139. Sîntămărian, A., Furdui, O.: Problem Q1055, Math. Mag. 88 (5), 2015, 378
140. Sondow, J.: Double integrals for Euler's constant and $\ln \frac{4}{\pi}$ and an analog of Hadjicostas's formula, Amer. Math. Monthly 112 (1), 2005, 61–65
141. Stadler, A.: Solution to problem 5306, Sch. Sci. Math. 114 (7), 2014, 11-13
142. Stoica, G.: Problem 1991, Math. Mag. 89 (2), 2016, 147
143. Trif, T.: Probleme de calcul diferenţial şi integral în $\mathbb{R}^n$ (in Romanian), Casa Cărţii de Ştiinţă, Cluj-Napoca, 2003
144. Ulrich, A.: Solution to problem 346, Gac. R. Soc. Mat. Esp. 22 (2), 2019, 318–319 http://gaceta.rsme.es/abrir.php?id=1517
145. Vălean, C.I.: Problem 5406, Sch. Sci. Math. 116 (5), 2016
146. Vălean, C.I., Furdui, O.: Reviving the quadratic series of Au–Yeung, J. Class. Anal. 6 (2) 2015, 113-118
147. Vălean, C.I., Levy, M.: Euler sum involving tail, ResearchGate, 2019
148. Weisstein, E. W.: Apéry's constant http://mathworld.wolfram.com/AperysConstant.html
149. Weisstein, E. W.: Riemann Zeta Function zeta(2) http://mathworld.wolfram.com/RiemannZetaFunctionZeta2.html
150. Whittaker, E.T., Watson, G.N.: A Course of Modern Analysis, Fourth Edition, University Press, Cambridge, 1927
151. Woeginger, G.: Problem 3, IMC, Day 2, Blagoevgrad, Bulgaria, 31 July 2011
152. Xu, C.: A unique area property of the quadratic function, Math. Mag. 87 (1), 2014, 52–56

Index

© The Author(s), under exclusive license to Springer Nature Switzerland AG 2021   535
A. Sîntămărian, O. Furdui, *Sharpening Mathematical Analysis Skills*, Problem Books
in Mathematics, https://doi.org/10.1007/978-3-030-77139-3

Printed in the United States
by Baker & Taylor Publisher Services